Infosys Science Foundation Series

Infosys Science Foundation Series in Mathematical Sciences

The **Infosys Science Foundation Series in Mathematical Sciences**, a Scopus-indexed book series, focuses on high-quality content in the domain of pure and applied mathematics, biomathematics, financial mathematics, operations research and theoretical computer science. With this series, Springer hopes to provide readers with monographs, textbooks, edited volume, handbooks and professional books of the highest academic quality on current topics in relevant disciplines. Literature in this series will appeal to a wide audience of researchers, students, educators and professionals across the world.

Mohamed Jleli · Bessem Samet · Calogero Vetro
Editors

Advances in Nonlinear Evolution Equations

Springer

Editors
Mohamed Jleli
Department of Mathematics
King Saud University
Riyadh, Saudi Arabia

Bessem Samet
Department of Mathematics
King Saud University
Riyadh, Saudi Arabia

Calogero Vetro
Department of Mathematics and Computer Science
University of Palermo
Palermo, Italy

ISSN 2363-6149 ISSN 2363-6157 (electronic)
Infosys Science Foundation Series
ISSN 2364-4036 ISSN 2364-4044 (electronic)
Infosys Science Foundation Series in Mathematical Sciences
ISBN 978-981-97-9811-7 ISBN 978-981-97-9812-4 (eBook)
https://doi.org/10.1007/978-981-97-9812-4

Mathematics Subject Classification: 35B44, 35B33, 35L05, 35J05, 35K05, 58J35

This Springer imprint is published by the registered company Springer Nature Singapore Pte Ltd.
The registered company address is: 152 Beach Road, #21-01/04 Gateway East, Singapore 189721, Singapore

Dedicated to Professor Pasquale Vetro on the occasion of his 80th birthday

Preface

The theory of nonlinear evolution equations represents a crucial branch of nonlinear analysis, hence it has many applications in mathematical physics. Clearly, it allows us to deal with settings exhibiting time evolution, for example that involving wave equations of Schrödinger type. Those equations, as well as their stationary counterparts and related ones, arise naturally in the analysis of complex systems to understand and control the behavior of their dynamical processes. Hence, nonlinear evolution equations are widely considered in science and technology but are also useful in qualitative investigations connected with the theories of mappings and operators, where a variety of topological and variational techniques has been successfully involved. This book focuses on some special features of the theory of nonlinear evolution equations and its applications, as well as on underlying the role of some milestone results in the wide topic of partial differential equations. The book is addressed to the broad audience of advanced students and researchers in mathematics interested in working with the techniques of nonlinear analysis and in exploring various issues related to evolution processes. This book indeed collects known results, complete references and new contributions associated with the qualitative study of nonlinear differential problems. Apart from Preface, this book is organized into seven chapters dealing with a variety of topics of nonlinear analysis in which the theory of evolution equations is central. For readers' convenience, each chapter is self-contained and contributed by different authors who are all experts in the methods and tools associated with the theory of nonlinear evolution equations. Each chapter depicts the appropriate mathematical background with references for the specific topic considered in it.

In Chap. 1 entitled "Initial Trace of Positive Solutions of Some Diffusion Equations with Absorption", the author provides a wide and detailed study about the existence of an initial trace of positive solutions to certain classes of diffusion equations. Representation issues and regularity properties of solutions are also addressed. Finally, the author gives a vision on some directions for further research, as well as on several open problems to deal with. Chapter 2 entitled "Graph Gradient Flows: From Discrete to Continuum" provides a framework to study a continuum limit of a gradient flow on a graph where the number of vertices increases in an appropriate way. As examples, the authors establish the convergence of a discrete total variation flow and a discrete Allen-Cahn flow on discretized tori to their respective continuum limits. "An Overview on Some Nonlinear Evolution Problems" is Chap. 3 of our book. It contains a tribute to the memory of Prof. Vasilii V. Zhikov. Hence, the chapter presents results concerning the Kirchhoff wave initial boundary value problems in the so-called degenerate case and involving the fractional Laplacian, nonlinear damping and source terms. Next, Chap. 4 entitled "A Little Tribute to Zeev Nehari" discusses a result, the so-called Nehari method, which has influenced a wide variety of studies on differential equations, in the last decades. The author also provides some generalizations and highlights certain current results involving the Nehari method. Chapter 5 entitled "Fractional Constant Exponent or Variable-Order and Variable Exponent Problems" collects relatively complex and representative nonlinear fractional p-Laplace type problems, in the setting of constant exponents and involving Kirchhoff terms. Particular attention is devoted to special continuous and compact embedding results. The subsequent Chap. 6 entitled "Nonlinear Periodic Systems with Variable Exponents" studies a second order periodic system driven by a nonhomogeneous differential operator with variable growth and a reaction that involves competing concave-convex nonlinearities. Using variational tools and critical groups, the authors show that the system has at least two distinct nontrivial solutions. The last Chap. 7 "A Quasi-Linear Parabolic Inequality with Convection in Riemannian Manifolds" deals with inhomogeneous quasi-linear parabolic inequalities involving a convection term, and posed in a complete noncompact Riemannian manifold. Under suitable hypotheses on the Ricci curvature of the manifold, the authors establish the nonexistence of weak solutions.

The editors would like to express their gratitude to all the contributing authors. Special thanks are also addressed to Mr. Shamim Ahmad, Senior Editor, Sciences Books in Springer India, for taking interest in publishing this book.

Riyadh, Saudi Arabia — Mohamed Jleli
Riyadh, Saudi Arabia — Bessem Samet
Palermo, Italy — Calogero Vetro
February 2025

Contents

Contributors

Giovany Figueiredo Universidade de Brasília—UnB, Brasília, Brazil

Yves van Gennip Delft Institute of Applied Mathematics, Technische Universiteit Delft, Delft, The Netherlands

Yoshikazu Giga Graduate School of Mathematical Sciences, The University of Tokyo, Meguro-ku, Tokyo, Japan

Mohamed Jleli Department of Mathematics, College of Science, King Saud University, Riyadh, Saudi Arabia

Zhenhai Liu Guangxi Colleges and Universities Key Laboratory of Complex System Optimization and Big Data Processing, Yulin Normal University, Yulin, People's Republic of China

Jun Okamoto Faculty of Medicine Bldg. B, Institute for the Advanced Study of Human Biology, Kyoto University, Sakyo-ku, Kyoto, Japan

Nikolaos S. Papageorgiou Guangxi Key Laboratory of Universities Optimization Control and Engineering Calculation, Guangxi Minzu University, Nanning, Guangxi, People's Republic of China;
Department of Mathematics, National Technical University, Athens, Greece

Patrizia Pucci Dipartimento di Matematica e Informatica—Università degli Studi di Perugia, Perugia, Italy

Bessem Samet Department of Mathematics, College of Science, King Saud University, Riyadh, Saudi Arabia

Calogero Vetro Department of Mathematics and Computer Science, University of Palermo, Palermo, Italy

Laurent Véron Institut Denis Poisson CNRS, UMR 7013, Université de Tours, Tours, France

Jiabin Zuo School of Mathematics and Information Science, Guangzhou University, Guangzhou, China

Chapter 1
Initial Trace of Positive Solutions of Some Diffusion Equations with Absorption

Laurent Véron

Abstract We prove the existence of an initial trace of a positive solution of diffusion equations with a nonlinear reaction term in the class of outer regular Borel measures. We study the reverse problem of characterising a solution from its initial trace.

1.1 Introduction

If we consider a nonnegative $C^{2,1}$ function $(x,t) \mapsto u(x,t)$ satisfying a diffusion equation

$$\partial_t u - A(u, \nabla u, D^2 u) + b(u, \nabla u) = 0 \tag{1.1}$$

in $Q_T := \mathbb{R}^N \times (0, T)$, a natural question is to understand what the data are on which the function depends, besides the structural assumptions on the functions A and B. One important approach is to associate with this function a general notion of initial value that we call the *initial trace problem*. The initial trace problem is two-fold:

1- Is it possible to define *in a suitable way* the limit value of $u(.,t)$ when $t \to 0$? This limit, whenever it exists, is called the *initial trace of* u, noted $tr\,(u)$. In most cases it is an outer regular nonnegative Borel measure in $\mathbb{R}^N$.
2- Given an outer regular nonnegative Borel measure ν in $\mathbb{R}^N$, is it possible to find a nonnegative solution u of (1.1) in Q_T such that $tr\,(u) = \nu$? This step requires a sharp study of the possible initial traces.
3- Is the correspondence between the set of initial traces ν and the set of positive solutions of (1.1) in Q_T one-to-one? More simply, is a positive solution defined in a unique way by its initial trace.

In this formulation, A is a real valued Caratheodory function defined in $\mathbb{R} \times \mathbb{R}^N \times M^N(\mathbb{R})$ and B a real valued Caratheodory function defined in $\mathbb{R}^N \times \mathbb{R}$.

L. Véron (✉)
Institut Denis Poisson CNRS, UMR 7013, Université de Tours, Tours, France
e-mail: veronl@univ-tours.fr

M. Jleli et al. (eds.), *Advances in Nonlinear Evolution Equations*, Infosys Science Foundation Series, https://doi.org/10.1007/978-981-97-9812-4_1

In this full generality, the problem is hard to analyse except for the mere diffusion equation

$$\partial_t u - A(u, \nabla u, D^2 u) = 0, \tag{1.2}$$

where the two cases of the p-Laplace diffusion equation (with $A(u, \nabla u, D^2 u) = div(|\nabla u|^{p-2}\nabla u)$) and the porous-media equation (with $A(u, \nabla u, D^2 u) = \Delta u^m$) are fairly well understood. In these cases, the initial trace is a nonnegative Radon measure with some growth at infinity. When the equation contains a reaction term, the situation is completely changed, even in the mere case where $A(u, \nabla u, D^2 u) = \Delta u$:

$$\partial_t u - \Delta u + b(u, \nabla u) = 0. \tag{1.3}$$

The sign of the reaction term plays an important role. Surprisingly the question of identifying the initial trace of a solution of (1.3) is much easier if $b(u, \nabla u)$ is nonpositive, e.g. $b(u, \nabla u) = -u^q$. In that case the function u is super-caloric and it always admits an initial trace in the class of nonnegative Radon measures in $\mathbb{R}^N$. The second question of reconstructing the solution from its initial trace is more involved, and the associated question of uniqueness is even deeper. In this paper, we will concentrate on the case where the perturbation term is a nonlinear power term under the form

$$b(u, \nabla u) = \epsilon u^q \tag{1.4}$$

where $\epsilon = \pm 1$. The first easy to prove results dealing with the case $\epsilon = -1$ and $q > 0$, or $\epsilon = 1$ and $0 < q \leq 1$ are the following.

Theorem 1.1 *Let $\epsilon = -1$ and $q > 0$ or $\epsilon = 1$ and $0 < q \leq 1$. If u is a nonnegative solution of (1.3) in Q_T, there exists a nonnegative Radon measure μ in $\mathbb{R}^N$ such that*

$$\lim_{t \to 0} \int_{\mathbb{R}^N} u(x,t)\zeta(x)dx = \int_{\mathbb{R}^N} \zeta d\mu(x) \quad \textit{for all } \zeta \in C_c^\infty(\mathbb{R}^N). \tag{1.5}$$

The problems arising from the study of the model case

$$\partial_t u - \Delta u + |u|^{q-1}u = 0 \text{ in } Q_T \tag{1.6}$$

is now fairly well understood after the initial work of Marcus and Véron [30] who put into light that the initial trace has to be understood in the sense of Borel measures and the exhaustive study of the supercritical case by Marcus and Véron [32] and Gkikas and Véron [22, 23]. Note that this study followed the very complete analysis of the boundary trace of positive solutions of

$$-\Delta u + u^q = 0 \text{ in } \Omega, \tag{1.7}$$

which was carried on by Marcus and Véron [32, 33] and concluded by Marcus in the remarkable paper [28] to which the construction of Gkikas and Véron [23] that we will develop thoroughly in the sequel is much indebted.

Concerning (1.6), Marcus and Véron pointed out the key role of the critical exponent $q_c := 1 + \frac{2}{N}$ and showed that the analysis is very different according to the position of q with respect to q_c Their starting result concerning this equation is the following.

Theorem 1.2 *Let $q > 1$ and u be a positive solution of (1.5) in Q_T. Then there exist a closed set $\mathcal{S} \subset \mathbb{R}^N$ and a nonnegative Radon measure μ in $\mathcal{R} := \mathbb{R}^N \setminus \mathcal{S}$ such that*

(i) For any $\zeta \in C_c^\infty(\mathcal{R})$ there holds

$$\lim_{t \to 0} \int_{\mathbb{R}^N} u(x,t)\zeta(x)dx = \int_{\mathbb{R}^N} \zeta d\mu(x). \tag{1.8}$$

(ii) For any $y \in \mathcal{S}$ and any $\epsilon > 0$, there holds

$$\lim_{t \to 0} \int_{B_\epsilon(y)} u(x,t)dx = \infty. \tag{1.9}$$

The set $\mathcal{S} := \mathcal{S}ing(u)$ (resp. $\mu := \mu(u)$) is called the singular (resp. regular) part of the initial trace of u. Conversely we have an existence and uniqueness result in the subcritical case.

Theorem 1.3 *Let $1 < q < q_c$. Then for any couple $(\mathcal{S}, \mu)$ where $\mathcal{S}$ is a closed subset of $\mathbb{R}^N$ and μ a nonnegative Radon measure in $\mathcal{R} := \mathbb{R}^N \setminus \mathcal{S}$, there exists a unique positive solution u of (1.5) in $Q_\infty := \mathbb{R}^N \times (0, \infty)$ with initial trace $(\mathcal{S}, \mu)$.*

When $q \geq q_c$ not every measure is admissible for being the measure part of the initial trace of a positive solution of (1.5), neither every closed set can be the singular part. To answer this question, it is necessary to introduce the Riesz $(\frac{2}{q}, q')$-capacity of a Borel set $E \subset \mathbb{R}^N$.

$$\dot{R}_{\frac{2}{q},q'}(E) = \inf\left\{\int_{\mathbb{R}^N \times \mathbb{R}^N} \frac{|\zeta(x) - \zeta(y)|^{q'}}{|x-y|^{N - \frac{2}{q-1}}} dxdy : \zeta \in C_c^\infty(\mathbb{R}^N), 0 \leq \zeta \leq 1, \zeta \geq \mathbf{1}_E\right\}. \tag{1.10}$$

This capacity plays a fundamental role since Baras and Pierre proved in [6, 7] two fundamental results which assert the following:

A Borel set $S \subset \mathbb{R}^N$ is a removable singularity for any solution u of (1.5) in $Q_\infty := \mathbb{R}^N \times (0, \infty)$ which is continuous in $\overline{Q}_\infty \setminus (S \times \{0\})$ if and only if

$$\dot{R}_{\frac{2}{q},q'}(S) = 0. \tag{1.11}$$

Similarly there exists a solution u of (1.6) in Q_∞ such that $u(.,0) = \mu$, where μ is a bounded measure in $\mathbb{R}^N$ if and only if

$$\text{If } S \subset \mathbb{R}^N \text{ is a Borel set, } \dot{R}_{\frac{2}{q},q'}(S) = 0 \Longrightarrow |\mu(S)| = 0. \tag{1.12}$$

For the sake of completeness, we sketch a proof in Theorem 1.11. Note that the boundedness assumption on μ is not necessary.

If $\mathcal{R} \subset \mathbb{R}^N$ is open, we denote by $\mathfrak{M}_q(\mathcal{R})$ the space of Radon measures μ in $\mathcal{R}$ satisfying

$$\text{If } S \subset \mathcal{R} \text{ is a Borel set, } \dot{R}_{\frac{2}{q},q'}(S) = 0 \Longrightarrow |\mu(S)| = 0. \tag{1.13}$$

The positive cone of this space is denoted by $\mathfrak{M}_q^+(\mathcal{R})$.

If $\mathcal{S} \subset \mathbb{R}^N$ is closed and μ is a positive Radon measure in $\mathcal{S}^c$ we define

$$\partial_\mu \mathcal{S} := \left\{y \in \mathcal{S} : \mu(B_\epsilon(y) \cap \mathcal{S}^c) = \infty, \ \forall \epsilon > 0\right\}, \tag{1.14}$$

which is the set of blowing points of the measure, and

$$\mathcal{S}^* := \left\{y \in \mathcal{S} : \dot{R}_{\frac{2}{q},q'}(B_\epsilon(y) \cap \mathcal{S}) > 0, \ \forall \epsilon > 0\right\}, \tag{1.15}$$

which is the set of intrinsically non-removable points of $\mathcal{S}$.

Theorem 1.4 *Let $q \geq q_c$. A couple $(\mathcal{S}, \mu)$ where $\mathcal{S}$ is a closed subset of $\mathbb{R}^N$ and $\mu \in \mathfrak{M}_q^+(\mathcal{R})$ where $\mathcal{R} := \mathbb{R}^N \setminus \mathcal{S}$ is the initial trace of a positive solution u of (1.5) in $Q_\infty := \mathbb{R}^N \times (0, \infty)$ if and only if $\mathcal{S} = \partial_\mu \mathcal{S} \cup \mathcal{S}^*$.*

A striking aspect of the supercritical case, i.e. $q \geq q_c$, is that there exist infinitely many solutions when $\mathcal{S}$ is not empty and the solution constructed in Theorem 1.2 is actually the maximal solution with any initial trace $(\mathcal{S}, 0)$. This has resulted in a finer definition of the initial trace called the *precise trace*. This non uniqueness has been observed by Le Gall (see Appendix of [17]) in the probabilistic approach of the study of the fine boundary trace of equations of the type (1.16) below by using the superdiffusion technique [19]. The main idea of this extension is to replace the Euclidean topology which served as a basic tool in the definition of the trace process by the thin $\mathfrak{T}_q$-topology associated with the $(\frac{2}{q}, q')$-capacity. Note that this process was developed by Marcus and Véron [33] in a similar way for analysing the boundary trace of positive solutions of

$$-\Delta u + |u|^{q-1}u = 0 \quad \text{in } \Omega \subset \mathbb{R}^N. \tag{1.16}$$

When $q \geq q_c$ it is proved in [23] that any nonnegative solution u of (1.5) in Q_T admits a precise singular initial set $\mathcal{S}_q(u)$ which is the set of $\xi \in \mathbb{R}^N$ such that for any thin-neighbourhood U (for the $\mathfrak{T}_q$-topology) of ξ there holds

$$\int_0^T \int_{\mathbb{R}^N} u^q \left(\mathbb{H}[\mathbf{1}_U]\right)^{2q'} dxdt = \infty \tag{1.17}$$

where $\mathbb{H}[\mathbf{1}_U]$ is the heat potential in Q_∞ of the characteristic function of U. The set $\mathcal{R}_q(u) := \mathcal{S}_q^c(u)$ is the fine regular set of the initial trace. It is the set of the $\xi \in \mathbb{R}^N$ such that there exists a thin-neighbourhood U of ξ with the property that

$$\int_0^T \int_{\mathbb{R}^N} u^q \, (\mathbb{H}[\mathbf{1}_U])^{2q'} \, dxdt < \infty. \tag{1.18}$$

Essentially the precise regular set of the initial trace carries a nonnegative Radon measure $\mu_{\mathcal{R}_q(u)}$, absolutely continuous with respect to the Riesz capacity $\dot{R}_{\frac{2}{q},q'}$, and such that for any bounded test function η belonging to the Besov space $B^{\frac{2}{q},q'}(\mathbb{R}^N)$ with "support" in $\mathcal{R}_q(u)$ (more precisely $\mathfrak{T}_q$-support in a sense which will be defined in the text), there holds

$$\lim_{t\to 0} \int_{\mathbb{R}^N} u(x,t)\eta_+^{2q'} dx = \int_{\mathbb{R}^N} \eta_+^{2q'} d\mu_{\mathcal{R}_q(u)}. \tag{1.19}$$

This allows to define a solution $v_{\mathcal{R}_q(u)}$ of (1.5) corresponding to this measure $\mu_{\mathcal{R}_q(u)}$. It is called the *regular component of* u. For defining the *singular component of* u, we first denote by $U_{\mathcal{S}_q(u)}$ the maximal solution of (1.5) with an initial trace vanishing in $\mathcal{S}_q^c(u)$. The singular component is $[u]_{\mathcal{S}_q(u)}$ which is the maximal solution of (1.5) bounded from above by u and with initial trace vanishing in $\mathcal{S}_q^c(u)$. The couple $(\mu_{\mathcal{R}_q(u)}, \mathcal{S}_q(u))$ is called *the precise initial trace*.

The main results in the supercritical case ($q \geq q_c$) are summarised by the following statement.

Theorem 1.5 *1- If u is a nonnegative solution of (1.5) in Q_T, then the function $v_{\mathcal{R}_q(u)} \oplus [u]_{\mathcal{S}_q(u)}$, which is the largest solution dominated by the supersolution $v_{\mathcal{R}_q(u)} + [u]_{\mathcal{S}_q(u)}$, admits for precise initial trace $(\mu_{\mathcal{R}_q(u)}, \mathcal{S}_q(u))$.*
2- The solution $v_{\mathcal{R}_q(u)} \oplus [u]_{\mathcal{S}_q(u)}$ is σ-moderate in the sense that it is the increasing limit of solutions u_{μ_n} with initial data μ_n which are nonnegative bounded measures belonging to $B^{\frac{2}{q},q'}(\mathbb{R}^N)$. It is the unique σ-moderate solution with such a trace.
3- Any positive solution u of (1.5) is σ-moderate.

As a consequence, there exists a one-to-one correspondence between the set of nonnegative solutions u of (1.5) and the set of couples $(\mu_{\mathcal{R}_q(u)}, \mathcal{S}_q(u))$.

These notes are based upon the articles [30, 33, 34] *written jointly with M. Marcus and* [22, 23] *written with K. Gkikas, in both cases in a very fruitful collaboration.*

1.2 Equation with Sublinear or Reaction Forcing Term

In order to point out the differences between the absorption case which is the main object of this survey and the sublinear or forcing cases which are indeed much easier to treat, we present some standard results in these later cases.

1.2.1 The Heat Equation

We first consider the basic approach of the trace problem for the heat equation. Let $u \in C^{2,1}(Q_T)$ be a positive solution of

$$\partial_t u - \Delta u = 0 \quad \text{in } Q_T. \tag{1.20}$$

If $G \subset \mathbb{R}^N$ is any bounded domain, we denote by λ_G the first eigenvalue of $-\Delta$ in $H_0^1(G)$ and by ϕ_G the corresponding first positive eigenfunction normalised by $\max \phi_G = 1$. Then

$$\frac{d}{dt}\int_G u(x,t)\phi_G^2(x)dx + 2\lambda_G \int_G u(x,t)\phi_G^2(x)dx = 2\int_G u(x,t)|\nabla \phi_G|^2(x)dx.$$

Therefore the function

$$t \mapsto e^{2\lambda_G t}\int_G u(x,t)\phi_G^2(x)dx$$

is nondecreasing. It admits a finite nonnegative limit $M_u(G)$ when $t \to 0$ and

$$e^{2\lambda_G \tau}\int_G u(x,\tau)\phi_G^2(x)dx - M_u(G) = \iint_{Q_\tau^G} u(x,t)|\nabla \phi_G|^2(x)dxdt < \infty,$$

where $Q_T^G = G \times (0,T)$. This implies in particular that $u \in L^1(Q_\tau^G)$ for any $\tau < T$. Then, if $\zeta \in C_c^\infty(\mathbb{R}^N)$ there exists $\ell(\zeta)$ with the property that

$$\ell(\zeta) = \lim_{t\to 0}\int_{\mathbb{R}^N} u(x,t)\zeta(x) = \int_{\mathbb{R}^N} u(x,\tau)\zeta(x) - \int_0^\tau \int_{\mathbb{R}^N} u(x,s)\Delta\zeta(x)dxds.$$

The mapping $\zeta \mapsto \ell(\zeta)$ is a positive linear functional; hence it extends as a unique Radon measure in $\mathbb{R}^N$ that we denote by μ. The following characterisation of the measures μ is proved in [4, 5].

Let u be a nonnegative solution of (1.20) in Q_T and μ be the initial trace of u, then

$$\int_{\mathbb{R}^N} e^{-a|x|^2} d\mu(x) < \infty \quad \text{for all } a < \frac{1}{4T}. \tag{1.21}$$

Conversely, if μ is a nonnegative Radon measure in $\mathbb{R}^N$ satisfying (1.21) the function u defined in Q_T by

$$u(t,x) = \frac{1}{(4\pi t)^{\frac{N}{2}}}\int_{\mathbb{R}^N} e^{-\frac{|x-y|^2}{4t}} d\mu(y) < \infty \tag{1.22}$$

is the unique positive solution of (1.20) with initial trace μ.

Definition 1.1 If μ is a Radon measure in $\mathbb{R}^N$, we denote by $\mathbb{H}[\mu]$ the heat potential of μ, defined by

$$\mathbb{H}[\mu](x,t) = \frac{1}{(4\pi t)^{\frac{N}{2}}} \int_{\mathbb{R}^N} e^{-\frac{|x-y|^2}{4t}} d\mu(y) = \int_{\mathbb{R}^N} H(x,y,t) d\mu(y), \tag{1.23}$$

provided this formula has a meaning, e.g. if μ is bounded. The function $H(x,y,t) := \frac{1}{(4\pi t)^{\frac{N}{2}}} e^{-\frac{|x-y|^2}{4t}}$ is called the heat kernel in Q_∞.

This result is the extension to higher dimension of Widder representation theorem proved in 1-D in [39].

1.2.2 *Equations with Sublinear or Forcing Reaction: Proof of Theorem 1.1*

1.2.2.1 The Sublinear Case

We assume that $u \in C^{2,1}(Q_T)$ is a positive solution of (1.3) where b is continuous in $\mathbb{R} \times \mathbb{R}^N$ and satisfies

$$0 \le b(r,\xi) \le cr + d \quad \text{for all } r \ge 0 \text{ and } \xi \in \mathbb{R}^N, \tag{1.24}$$

for some $c > 0$ and $d \ge 0$. As in the case of heat equation we call ϕ_G the first normalised eigenfunction of a bounded smooth domain $G \subset \mathbb{R}^N$, with associated eigenfunction λ_G. Then

$$\frac{d}{dt} \int_G u(x,t)\phi_G^2(x)dx + (2\lambda_G + c) \int_G u(x,t)\phi_G^2(x)dx + d \int_G \phi_G^2(x)dx \ge 0.$$

This implies that the function

$$t \mapsto e^{(2\lambda_G + c)t} \left(\int_G u(x,t)\phi_G^2(x)dx + \frac{d}{2\lambda_G + c} \int_G \phi_G^2(x)dx \right)$$

is increasing. Therefore it admits a finite limit (positive) when $t \to 0$ and thus there exists

$$M_u(G) := \lim_{t \to 0} \int_G u(x,t)\phi_G^2(x)dx. \tag{1.25}$$

Since for any $0 < \tau < T$

$$\int_0^\tau \int_G b(u, \nabla u)(x,t)\phi_G^2(x)dxdt$$
$$\leq c \int_0^\tau \int_G u(x,t)\phi_G^2(x)dxdt + d \int_0^\tau \int_G \phi_G^2(x)dxdt < \infty,$$

and G is arbitrary, it follows that $u + b(u, \nabla u) \in L^1(Q_G^\tau)$ where $Q_G^\tau := G \times (0, \tau)$. As in the case of heat equation we obtain that there exists a Radon measure μ in $\mathbb{R}^N$ such that for any $\zeta \in C_c^\infty(\mathbb{R}^N)$ there holds

$$\lim_{t\to 0} \int_{\mathbb{R}^N} u(x,t)\zeta(x)dx = \int_{\mathbb{R}^N} \zeta d\mu(x), \tag{1.26}$$

and also for any $t \in (0, T)$,

$$\begin{aligned} \int_{\mathbb{R}^N} u(x,t)\zeta(x) - \int_0^t \int_{\mathbb{R}^N} u(x,s)\Delta\zeta(x)dxds \\ + \int_0^t \int_{\mathbb{R}^N} b(u, \nabla u)(x,t)\zeta(x)dxds = \int_{\mathbb{R}^N} \zeta d\mu(x). \end{aligned} \tag{1.27}$$

This implies that u admits μ as an initial trace in the sense defined by (1.26), and it is a solution of

$$\begin{aligned} \frac{\partial u}{\partial t} - \Delta u + b(u, \nabla u) = 0 \quad \text{in } Q_T \\ u(.,0) = \mu \quad \text{in } \mathbb{R}^N. \end{aligned} \tag{1.28}$$

Finally, this problem can be solved for any nonnegative bounded measure μ, and more generally for any nonnegative Radon measure μ such that the heat potential $\mathbb{H}[\mu]$ can be defined.

1.2.2.2 The Forcing Reaction Case

We assume that b satisfies

$$b(r, \xi) \leq 0 \quad \text{for all } r \geq 0 \text{ and } \xi \in \mathbb{R}^N. \tag{1.29}$$

As in the previous case the function

$$t \mapsto e^{2\lambda_G t} \int_G u(x,t)\phi_G^2(x)dx$$

is increasing, thus (1.25) holds. From (1.3) we have for any $0 < t < \tau < T$,

$$\int_G u(x,\tau)\phi_G^2(x)dx + 2\lambda_G \int_t^\tau \int_G u(x,s)\phi_G^2(x)dxds - \int_G u(x,t)\phi_G^2(x)dx$$
$$= 2\int_t^\tau \int_G u(x,s)|\nabla\phi_G|^2(x)dxds + \int_t^\tau \int_G b(u,\nabla u)(x,s)\phi_G^2(x)dxds.$$

Because (1.3) holds $u\phi_G^2 \in L^1(Q_\tau^G)$, hence, by the monotone convergence theorem both $b(u,\nabla u)\phi_G^2$ and $u|\nabla\phi_G|^2$ are integrable in Q_τ^G. If $\zeta \in C_c^\infty(\mathbb{R}^N)$ we obtain that (1.27), (1.5) and (1.28) hold for some nonnegative Radon measure μ.

1.3 The Rough Trace

This section is devoted to the construction and the study of the properties of the *rough initial trace* of positive solution of

$$\partial_t u - \Delta u + u^q = 0 \quad \text{in } Q_\infty, \tag{1.30}$$

when $q > 1$. The qualifier of *rough* will be justified later on in connection with a surprising non-uniqueness result.

Proof of Theorem 1.2. Let u be a nonnegative solution of 1.30 in Q_∞ and $y \in \mathbb{R}^N$, then the following alternative holds:
(i) either there exists $\alpha > 0$ such that

$$\iint_{Q_1^{B_\alpha(y)}} u^q(x,t)dxdt < \infty, \tag{1.31}$$

(ii) or for any $\alpha > 0$

$$\iint_{Q_1^{B_\alpha(y)}} u^q(x,t)dxdt = \infty. \tag{1.32}$$

If (1.31) holds, then $u \in L^1(Q_T^B)$ for any open ball $B \subset \overline{B} \subset B_\alpha(y)$, and for any $\zeta \in C_c^\infty(B)$ there holds

$$\frac{d}{dt}\left(\int_B u(x,t)\zeta dx + \int_t^1 \int_B \left(u\Delta\zeta - u^q\zeta\right)dxd\tau\right) = 0. \tag{1.33}$$

Consequently there exists $\ell(\zeta)$ defined by

$$\ell(\zeta) := \lim_{t\to 0} \int_B u(x,t)\zeta dx = \int_B u(x,1)\zeta dx + \int_0^1 \int_B \left(u\Delta\zeta - u^q\zeta\right)dxd\tau. \tag{1.34}$$

The mapping $\zeta \mapsto \ell(\zeta)$ is a positive linear functional on $C_c^\infty(B)$; hence it is a Radon measure μ_B in B.

If (1.49) holds, let B as above and ϕ_B be the first eigenfunction of $-\Delta$ in $H_0^1(B)$ with maximal value 1 and corresponding eigenfunction λ_B. Then

$$\begin{aligned}\frac{d}{dt}\int_B u(x,t)\phi_B^{2q'}dx + 2q'\lambda_B\int_B u(x,t)\phi_B^{2q'}dx \\ - 2q'(2q'-1)\int_B u(x,t)\phi_B^{2q'-2}|\nabla\phi_B|^2\,dx + \int_B u^q(x,t)\phi_B^{2q'}dx = 0,\end{aligned}$$

where we have set $B = B_\alpha(y)$ and $q' = \frac{q}{q-1}$. Since

$$\int_B u(x,t)\phi_B^{2q'-2}|\nabla\phi_B|^2\,dx \le \frac{\delta^q}{q}\int_B u^q(x,t)\phi_B^{2q'}dx + \frac{1}{\delta^{q'}q'}\int_B |\nabla\phi_B|^{2q'}\,dx,$$

for suitable $\delta > 0$ and $c > 0$, we have that

$$\frac{d}{dt}\left(e^{2q'\lambda_B t}\int_B u(x,t)\phi_B^{2q'}dx\right) + \frac{e^{2q'\lambda_B t}}{2}\int_B u^q(x,t)\phi_B^{2q'}dx \le ce^{2q'\lambda_B t}\int_B |\nabla\phi_B|^{2q'}dx.$$

Then

$$\begin{aligned}e^{2q'\lambda_B}\int_B u(x,1)\phi_B^{2q'}dx + \frac{1}{2}\int_t^1 e^{2q'\lambda_B\tau}\int_B u^q(x,\tau)\phi_B^{2q'}dxd\tau \\ \le e^{2q'\lambda_B t}\int_B u(x,t)\phi_B^{2q'}dx + c\int_t^1 e^{2q'\lambda_B\tau}\int_B |\nabla\phi_B|^{2q'}\,dxd\tau.\end{aligned} \tag{1.35}$$

Therefore

$$\lim_{t\to 0}\int_B u(x,t)\phi_B^{2q'}dx = \infty. \tag{1.36}$$

The set of points y such that (i) holds is clearly open and its union is called *the regular set* $\mathcal{R}$. By a partition of unity, there exists a unique nonnegative Radon measure μ on $\mathcal{R}$ such that for any $\zeta \in C_c^\infty(\mathcal{R})$ there holds

$$\lim_{t\to 0}\int_{\mathbb{R}^N} u(x,t)\zeta dx = \int_{\mathbb{R}^N}\zeta d\mu(x). \tag{1.37}$$

For any $y \in \mathcal{S}$ and any $\alpha > 0$ we have (1.36); therefore we define a Borel measure ν such that for any Borel set $E \subset \mathbb{R}^N$

$$\nu(E) = \begin{cases}\displaystyle\int_E d\mu(x) & \text{if } E \subset \mathcal{R} \\ \infty & \text{if } E \cap \mathcal{S} \ne \emptyset,\end{cases} \tag{1.38}$$

and it is outer regular.

1.3.1 The a Priori Estimate

The function ϕ_∞ defined on $(0,\infty)$ by

$$\phi_\infty(t) = \left(\frac{1}{t(q-1)}\right)^{\frac{1}{q-1}} \tag{1.39}$$

is the maximal solution of the differential equation $u' + u^q = 0$ on $(0,\infty)$.

For any $R>0$, let w_R be unique solution of

$$\begin{aligned} -\Delta w + w^q &= 0 \quad \text{in } B_R \\ \lim_{|x|\to R} w(x) &= \infty. \end{aligned} \tag{1.40}$$

Existence follows from the Keller-Osserman upper construction and uniqueness from the scaling identity

$$w_R(x) = R^{-\frac{2}{q-1}} w_1(x/R). \tag{1.41}$$

For any $y\in\mathbb{R}^N$, $R>0$ and $\epsilon>0$ the function

$$u_{\epsilon,R,y}(x,t = \phi_\infty(t-\epsilon) + w_R(x-y) \tag{1.42}$$

is a supersolution of (1.30) in $B_R(y)\times(\epsilon,\infty)$. Hence it dominates u therein. Letting $\epsilon\to 0$ and $R\to\infty$ yields

$$u(x,t) \le \phi_\infty(t) \quad \text{for all } (x,t)\in\mathbb{R}^N\times(0,\infty). \tag{1.43}$$

This a priori estimate admits a localised version.

Proposition 1.1 *Let $q>1$ and $R>0$.*
1- There exists a unique nonnegative solution $u := u_{\infty,R}$ of (1.30) in Q_∞ such that

$$\lim_{t\to 0} u(x,t) = \infty \quad \textit{uniformly in } B_R, \tag{1.44}$$

and

$$\lim_{t\to 0} u(x,t) = 0 \quad \textit{locally uniformly in } \overline{B}^c_R. \tag{1.45}$$

Furthermore

$$\lim_{t\to 0} t^{\frac{1}{q-1}} u_{\infty,R}(x,t) = \left(\frac{1}{q-1}\right)^{\frac{1}{q-1}} \quad \textit{locally uniformly in } B_R, \tag{1.46}$$

and for any $\alpha > \frac{3-q}{q-1}$ *there exists* $C_\alpha > 0$ *such that*

$$u_{\infty,R}(x,t) \le C_\alpha t^{-\frac{1}{q-1}} \left(\frac{|x|-R}{\sqrt{t}}\right)^\alpha e^{-\frac{(|x|-R)^2}{4t}} \quad \text{for all } (x,t) \in Q_\infty \text{ s.t. } |x|-R \ge \sqrt{t}. \tag{1.47}$$

2- There exists a unique nonnegative solution $u := u_{\infty,R^c}$ *of (1.30) in* Q_∞ *such that*

$$\lim_{t\to 0} u(x,t) = 0 \quad \text{uniformly in } B_R, \tag{1.48}$$

and

$$\lim_{t\to 0} u(x,t)dx = \infty \quad \text{locally uniformly in } \overline{B}_R^c. \tag{1.49}$$

Furthermore

$$\lim_{t\to 0} t^{\frac{1}{q-1}} u_{\infty,R^c}(x,t) = \left(\frac{1}{q-1}\right)^{\frac{1}{q-1}} \quad \text{uniformly in } B^c_{R+\epsilon} \tag{1.50}$$

for any $\epsilon > 0$. *Next, for any* $\theta \in (0,1)$ *and* $\alpha < \frac{3-q}{q-1}$, *there exists* $C_{\alpha,\theta} > 0$ *such that*

$$u_{\infty,R^c}(x,t) \le C_{\alpha,\theta} t^{-\frac{1}{q-1}} \left(\frac{\theta R-|x|}{\sqrt{t}}\right)^\alpha e^{-\frac{(\theta R-|x|)^2}{4t}} \quad \text{for all } (x,t) \in Q_\infty \text{ s.t. } |x| \le \theta R - \sqrt{t}. \tag{1.51}$$

Proof *Step 1-1- There exists a unique* $C^\infty((0,\infty))$ *function* W *with positive value satisfying*

$$\begin{aligned} \partial_t W - \partial_{xx} W + W^q &= 0 \quad \text{in } (0,\infty) \\ \lim_{x\to 0} W(x,t) &= \infty \text{ for all } t > 0 \\ \lim_{t\to 0} W(x,t) &= 0 \text{ for all } x > 0. \end{aligned} \tag{1.52}$$

This function is self-similar and endows the form

$$W(x,t) = t^{-\frac{1}{q-1}} \widetilde{W}\left(\frac{x}{\sqrt{t}}\right), \tag{1.53}$$

where $\widetilde{W}$ is the unique positive solution of

$$\begin{aligned} \widetilde{W}'' + \tfrac{\eta}{2}\widetilde{W}' + \tfrac{1}{q-1}\widetilde{W} - \widetilde{W}^q &= 0 \quad \text{in } (0,\infty) \\ \lim_{\eta\to 0} \widetilde{W}(\eta) &= \infty \\ \lim_{\eta\to\infty} \eta^{\frac{2}{q-1}} \widetilde{W}(\eta) &= 0. \end{aligned} \tag{1.54}$$

The construction is as follows. Let $k > 1$ and $\zeta = \zeta_k$ be the solution of

$$\begin{aligned} \partial_t \zeta - \partial_{xx}\zeta + \zeta^q &= 0 \qquad \text{in } \mathbb{R} \\ \zeta(.,0) &= \mathbf{1}_{[-k,0]} \text{ on } \mathbb{R}. \end{aligned} \tag{1.55}$$

The correspondence $k \mapsto \zeta_k$ is increasing. Since ζ_k is bounded from above by 1, $\{\zeta_k\}$ converges to ζ_∞ which is the unique solution of

$$\begin{aligned} \partial_t \zeta - \partial_{xx}\zeta + \zeta^q &= 0 \qquad \text{in } \mathbb{R}\times(0,\infty) \\ \zeta(.,0) &= \mathbf{1}_{(-\infty,0]} \text{ on } \mathbb{R}. \end{aligned} \tag{1.56}$$

For $\ell > 0$ we denote by T_ℓ the scaling transformation which leaves (1.30) equivariant,

$$T_\ell[\phi](x,t)) = \ell^{\frac{2}{q-1}}\phi\left(\ell x, \ell^2 t\right). \tag{1.57}$$

Then $T_\ell[\zeta_\infty] := \zeta_{\infty,\ell}$ is the solution of (1.30) in Q_∞ with initial data $\ell^{\frac{2}{q-1}}\mathbf{1}_{(-\infty,0]}$. Again $\ell \mapsto \zeta_{\infty,\ell}$ is increasing. Since the function

$$x \mapsto \left(\frac{2(q+1)}{(q-1)^2}\right)^{\frac{1}{q-1}} x^{-\frac{2}{q-1}} := C_q x^{-\frac{2}{q-1}} \quad \text{for all } x > 0$$

is a solution of

$$\begin{aligned} \partial_t v - \partial_{xx} v + v^q &= 0 \qquad \text{in } (0,\infty)\times(0,\infty) \\ v(0,t) &= \infty \qquad \text{in } (0,\infty), \end{aligned} \tag{1.58}$$

we have

$$\zeta_\infty(x,t) \le C_q x^{-\frac{2}{q-1}} \qquad \text{in } (0,\infty)\times(0,\infty),$$

which implies that, for all $\ell > 1$,

$$\zeta_\infty(x,t) \le \zeta_{\infty,\ell}(x,t) \le \phi_\infty(t)\mathbf{1}_{(-\infty,0]}(x) + \min\left\{C_q|x|^{-\frac{2}{p-1}}, \phi_\infty(t)\right\}\mathbf{1}_{(0,\infty)}(x). \tag{1.59}$$

Thus $\zeta_{\infty,\ell}$ converges to some function W when $\ell \to \infty$, and W satisfies (1.30). Because there holds for any $\phi \in C_c^\infty((0,\infty))$,

$$\int_0^\infty \int_{\mathbb{R}} \left(-\zeta_{\infty,\ell}\left(\partial_t\phi + \Delta\phi\right) + \zeta_{\infty,\ell}^q\phi\right) dxdt = 0,$$

the function W satisfies the same upper bound (1.59) as $\zeta_{\infty,\ell}$ and it is a solution of the equation

$$\begin{aligned} \partial_t \zeta - \partial_{xx}\zeta + \zeta^q &= 0 \quad \text{in } \mathbb{R} \\ \zeta(.,0) &= 0 \quad \text{on } (0,\infty) \\ \lim_{t\to 0} \zeta(x,t) &= \infty \text{ for all } x \le 0. \end{aligned} \tag{1.60}$$

Finally, for any $k > 0$ we have $T_k \circ T_\ell = T_{k\ell}$, thus $T_k[\zeta_{\infty,\ell}] = \zeta_{\infty,k\ell}$, which implies that $T_k[W] = W$ for all $k > 0$, and therefore W is self-similar. This implies that $W(x,t) = t^{-\frac{1}{q-1}}\widetilde{W}\left(\frac{x}{\sqrt{t}}\right)$ and W satisfies

$$\begin{aligned}\widetilde{W}'' + \frac{\eta}{2}\widetilde{W}' + \frac{1}{q-1}\widetilde{W} - \widetilde{W}^q = 0 \quad \text{in } (0,\infty)\\ \lim_{\eta\to\infty}\eta^{\frac{2}{q-1}}\widetilde{W}(\eta) = 0\\ \lim_{\eta\to 0}\widetilde{W}(\eta) = \infty.\end{aligned} \tag{1.61}$$

The behaviour of W can be obtained by matching asymptotic expansion, if we consider the function $\eta \mapsto \widetilde{W}_\alpha := \eta^\alpha e^{-\frac{\eta^2}{4}}$ which is a supersolution (resp. subsolution) when $\eta \to \infty$ if $\alpha > \frac{3-q}{q-1}$ (resp. $\alpha < \frac{3-q}{q-1}$). Thus for any $\alpha > \frac{3-q}{q-1}$ there exists $C_\alpha > 0$ such that

$$\widetilde{W}(\eta) \leq C_\alpha \eta^\alpha e^{-\frac{\eta^2}{4}} \quad \text{for all } \eta \text{ in } [1,\infty). \tag{1.62}$$

Inequality (1.51) follows from this estimate.

Step 1-2- We claim that there exists a unique positive function $u_{\infty,R}$ which satisfies (1.30)

$$\lim_{t\to 0} u_{\infty,R}(x,t) = \infty \qquad \text{locally uniformly in } B_R, \tag{1.63}$$

and

$$\lim_{t\to 0} u_{\infty,R}(x,t) = 0 \qquad \text{uniformly in } B^c_{R+\epsilon} \tag{1.64}$$

for any $\epsilon > 0$.

Since the equation and the initial conditions are invariant under the transformation T_ℓ, we can assume that $R = 1$. If $\mathbf{e}$ is a unit vector we denote by $v_{\mathbf{e}}$ the function defined by

$$v_{\mathbf{e}}(x,t) = \begin{cases} W(\langle x-\mathbf{e},\mathbf{e}\rangle, t) & \text{if } (\langle x-\mathbf{e},\mathbf{e}\rangle > 0, t > 0\\ \infty & \text{if } (\langle x-\mathbf{e},\mathbf{e}\rangle \leq 0, t > 0,\end{cases} \tag{1.65}$$

and by $H^+_{\mathbf{e}}$ (resp. $H^+_{\mathbf{e}}$) the half space $\{x : \langle x-\mathbf{e},\mathbf{e}\rangle > 0\}$ (resp. $\{x : \langle x-\mathbf{e},\mathbf{e}\rangle \leq 0\}$). Then $v_{\mathbf{e}}$ satisfies (1.30) in Q_∞ with initial data $v_{\mathbf{e}}(x,0) = 0$ if $x \in H^+_{\mathbf{e}}$ and $v_{\mathbf{e}}(x,0) = \infty$ if $x \in H^-_{\mathbf{e}}$. Therefore

$$v_1 = \inf\{v_{\mathbf{e}} : |\mathbf{e}| = 1\} \tag{1.66}$$

is a supersolution of (1.30) in Q_∞ which satisfies $v_1(x,0) = 0$ if $|x| > 1$ and $v_1(x,0) = \infty$ if $|x| \leq 1$. For $k > 0$ let u_k be the solution of (1.30) in Q_∞ with initial data $k\mathbf{I}_{B_1}$. Then $u_k \leq v_1$. Since $k \mapsto u_k$ is increasing, there exists a nonnegative function u which is a solution of (1.30) in Q_∞ such that

$$\lim_{t\to 0} u(x,t) = 0 \quad \text{if } |x| > 1 \text{ and } \lim_{t\to 0} u(x,t) = \infty \quad \text{if } |x| \leq 1. \tag{1.67}$$

By construction u is a minimal solution, and by (1.62) inequality (1.47) holds. Let $\widetilde{u}$ be another nonnegative function solution of (1.30) in Q_∞ satisfying (1.67). For $\ell < 1$ and $R > \ell^{-1}$, there exists $\epsilon_{\ell,R}$ such that

$$u(x,t) \leq w_R(x) \quad \text{for all } (x,t) \in Q_\infty \text{ s.t. } \ell^{-1} \leq |x| < R \text{ and } 0 < t \leq \epsilon_{\ell,R},$$

where w_R is defined in (1.40). Therefore the supersolution $(x,t) \mapsto T_\ell[\widetilde{u}](x, t - \epsilon_{\ell,R}) + w_R(x)$ defined in $B_R \times (\epsilon_{\ell,R}, \infty)$ is larger than u on $\partial B_R \times (\epsilon_{\ell,R}, \infty)$ and for $t = \epsilon_{\ell,R}$. Hence

$$u(x,t) \leq T_\ell[\widetilde{u}](x, t - \epsilon) + w_R(x) \quad \text{for all} \quad (x,t) \in B_R \times (\epsilon_{\ell,R}, \infty).$$

When $R \to \infty$, $\epsilon_{\ell,R} \to 0$ and $w_R(x) \to 0$. This implies

$$u(x,t) \leq T_\ell[\widetilde{u}](x,t) \quad \text{for all} \quad (x,t) \in Q_\infty.$$

Letting $\ell \to 1$ yields $u \leq \widetilde{u}$. Similarly $\widetilde{u} \leq u$.
Step 1-3- The function $u_{\infty,R}$ satisfies

$$\lim_{t\to 0} t^{\frac{1}{q-1}} u_{\infty,R}(x,t) = \left(\frac{1}{q-1}\right)^{\frac{1}{q-1}} \quad \text{locally uniformly in } B_R. \tag{1.68}$$

In order to prove this claim, for any $R' < R$ we construct a function $\psi \in C_c^2(\mathbb{R}^N)$ such that $0 \leq \psi \leq 1$, $\psi = 1$ in $B_{R'}$, $\psi = 0$ in $\overline{B}_R^c$ and $-\Delta\psi \leq C\psi$ for some $C = CR, R') > 0$. For any $\delta > 0$, the function

$$(x,t) \mapsto X(x,t) := (1-\delta)\psi(x)\phi_\infty(t)$$

satisfies

$$\partial_t X - \Delta X + X^q \leq (1-\delta)\phi_\infty\psi\left(\phi_\infty\left((1-\delta)^{q-1} - 1\right) - \frac{\Delta\psi}{\psi}\right).$$

Then there exists $\epsilon > 0$ such that the above expression is negative for $0 < t \leq \epsilon$. Therefore $u_{\infty,R} \geq X$ in $\mathbb{R}^N \times (0, \epsilon]$. This implies

$$\liminf_{t\to 0} t^{\frac{1}{q-1}} \inf\left\{u_{\infty,R}(x,t) : x \in B_{R'}\right\} \geq (1-\delta)\left(\frac{1}{q-1}\right)^{\frac{1}{q-1}}. \tag{1.69}$$

Since δ is arbitrary, we obtained the claim from (1.43).

Step 2-1- We claim that there exists a unique positive function u_{∞,R^c} which satisfies (1.30),

$$\lim_{t\to 0} u_{\infty,R^c}(x,t) = \infty \qquad \text{uniformly in } B_R^c, \tag{1.70}$$

and

$$\lim_{t\to 0} u_{\infty,R^c}(x,t) = 0 \qquad \text{locally uniformly in } B_R. \tag{1.71}$$

The proof uses the previous constructions. For any $k > 0$ we denote by v_k the solution of

$$\begin{aligned} \partial_t v - \Delta v + v^q &= 0 \qquad \text{in } Q_\infty \\ v(.,0) &= k\mathbf{1}_{B_R^c} \text{ in } \mathbb{R}^N. \end{aligned} \tag{1.72}$$

The sequence $\{v_k\}$ is increasing and it satisfies

$$v_k(x,t) \leq \phi_\infty(t)\mathbf{1}_{\overline{B}_R^c}(x) + \inf\{\phi_\infty(t), w_R(x)\}\mathbf{1}_{B_R}(x). \tag{1.73}$$

Then it converges to a positive solution of (1.30) in Q_∞ that we denote by u_{∞,R^c}. Therefore u_{∞,R^c} satisfies

$$\iint_{Q_\infty} \left((-\partial_t\zeta - \Delta\zeta)u_{\infty,R^c} + \zeta u^p_{\infty,R^c}\right) dxdt = 0 \tag{1.74}$$

for all $\zeta \in C_c^\infty(B_R)$. Using (1.70) and (1.74) it implies that u_{∞,R^c} vanishes on B_R. Uniqueness of such a solution is obtained by the same scaling and shifting argument as in Steps 1 and 2.

Step 2-2- Improved estimates. As a subsolution of (1.30) in Q_∞ we take

$$v_2(x,t) = \sup\{v_{-\mathbf{e}}(-x,t) : \mathbf{e} \in \partial B_1\}. \tag{1.75}$$

With the same notations as in Step 1-2, v_2 is a subsolution, and it satisfies $v_2(t,x) \to 0$ when $t \to 0$ if $x \in B_1$ and $v_2(t,x) \to \infty$ when $t \to 0$ and $x \in B_1^c$. This implies that (1.50) holds. The construction of the supersolution is more subtle: for $0 < \theta < 1$ there exists an integer n_θ such that

$$B_\theta \subset \bigcap_{1\leq j\leq n_\theta} H_{j,\theta} \subset B_1,$$

where

$$H_{j,\theta} = \left\{x \in \mathbb{R}^N : \langle x - \theta\mathbf{e}_j, \mathbf{e}_j\rangle < 0\right\} \quad \text{with } \mathbf{e}_j \in \partial B_1.$$

Hence the function

$$v_{2,\theta}(x,t) = \sum_{1\leq j\leq n_\theta} W(\langle -x + \theta\mathbf{e}_j, \mathbf{e}_j\rangle, t) \tag{1.76}$$

is a supersolution of (1.30) in Q_∞ which dominates u_{∞,R^c}. If $x \in B_\theta$, we have that $\operatorname{dist}(x, H^c_{j,\theta}) \leq \theta - |x|$, therefore

$$u(x,t) \leq n_\theta t^{-\frac{1}{q-1}} \widetilde{W}(\frac{\theta - |x|}{\sqrt{t}}), \tag{1.77}$$

which implies, thanks to (1.47),

$$u(x,t) \leq n_\theta C_\alpha t^{-\frac{1}{q-1}} \left(\frac{\theta - |x|}{\sqrt{t}}\right)^\alpha e^{-\frac{(\theta-|x|)^2}{4t}} \quad \text{for } (x,t) \in Q_\infty \text{ s.t. } |x| \leq \theta - \sqrt{t}. \tag{1.78}$$

From this inequality estimate (1.51) follows by rescaling. □

1.3.2 The Subcritical Case

For a given $q > 1$ it not always possible to find a solution of (1.30) belonging to $C(\overline{Q}_\infty \setminus \{0\})$ vanishing on $\mathbb{R}^N \times \{0\} \setminus \{0\}$. Indeed Brezis and Friedman [15] proved the following results.

Theorem 1.6 *Let*

$$q_c = 1 + \tfrac{2}{N}. \tag{1.79}$$

If $q \geq q_c$ any solution u of (1.30) belonging to $C(\overline{Q}_\infty \setminus \{0\})$ and vanishing on $\mathbb{R}^N \times \{0\} \setminus \{0\}$ is identically 0.
If $1 < q < q_c$, for any $c \in \mathbb{R}$ there exists a unique solution $u := u_{c\delta_0}$ of

$$\begin{aligned} \partial_t u - \Delta u + |u|^{q-1}u = 0 \quad &\text{in } Q_\infty \\ u(.,0) = c\delta_0 \ &\text{in } \mathcal{D}'(\mathbb{R}^N) \end{aligned} \tag{1.80}$$

where δ_0 is the Dirac mass at 0. Furthermore if $\{\rho_n\}$ is a sequence of positive integrable functions which converges weakly to $c\delta_0$ in the sense of distributions in $\mathbb{R}^N$, then the sequence of functions $\{u_{\rho_n}\}$ which satisfy

$$\begin{aligned} \partial_t u - \Delta u + |u|^{q-1}u = 0 \quad &\text{in } Q_\infty \\ u(.,0) = \rho_n \ &\text{in } \mathcal{D}'(\mathbb{R}^N) \end{aligned} \tag{1.81}$$

converges to $u_{c\delta_0}$ locally uniformly in Q_∞.

An important consequence of the previous result is the existence of *very singular solutions* which was first discovered by Brezis, Peletier and Terman in [16]. We give below an alternative proof of its construction and uniqueness.

Theorem 1.7 *Suppose $1 < q < q_c$. Then there exists a unique positive C^∞ function f defined on $[0,\infty)$ such that*

$$\begin{aligned} & f'' + \left(\frac{N-1}{\eta} + \frac{\eta}{2}\right) f' + \frac{1}{q-1} f - f^p = 0 \quad on\ (0,\infty) \\ & f'(0) = 0 \quad and \quad \lim_{\eta\to\infty} \eta^{\frac{2}{q-1}} f(\eta) = 0. \end{aligned} \tag{1.82}$$

Furthermore

$$f(\eta) = Ae^{-\frac{\eta^2}{4}} \eta^{\frac{2}{q-1}-N} \left(1 + \frac{2q}{q-1}\left(\frac{2}{q-1} - N\right)\eta^{-2} + o(\eta^{-2})\right) \quad as\ \eta \to \infty. \tag{1.83}$$

Proof For any $\epsilon > 0$, u_c is bounded from above by the solution $u_{\infty,\epsilon}$ of (1.30) with initial data $u_{\infty,\epsilon}(x,0) = \infty \times \mathbf{1}_{B_\epsilon}(x)$ which is defined in the proof of Proposition 1.1. When $c \to \infty$, u_c increases and converges to some solution u_∞ which is a positive solution of (1.30) and is bounded from above by $u_{\infty,\epsilon}$. Because of uniqueness there holds $T_\ell[u_c] = u_{c\ell^{\frac{2}{q-1}-N}}$ where T_ℓ is defined in (1.57). Therefore

$$T_\ell[u_\infty] = u_\infty \quad \text{for any } \ell > 0.$$

Hence u_∞ is self-similar and radial because of uniqueness as u_c is, thus it takes the form

$$u_\infty(x,t) = t^{-\frac{2}{q-1}} f\left(\frac{|x|}{\sqrt{t}}\right), \tag{1.84}$$

and f satisfies the ODE (1.83). Because $u_\infty(x,t) \to 0$ for $x \neq 0$ when $t \to 0$, it implies that $\lim_{\eta\to\infty} \eta^{\frac{2}{q-1}} f(\eta) = 0$. The function f is a positive radial and bounded solution of

$$-\Delta_\eta f - \frac{1}{2}\eta.\nabla f - \frac{1}{q-1} f + f^p = 0 \quad \text{in } \mathbb{R}^N \setminus \{0\}.$$

Hence the singularity at $\eta = 0$ is removable. Thus f is C^∞ in $\mathbb{R}^N$ and $f'(0) = 0$. Similarly $T_\ell[u_{\infty,\epsilon}] = u_{\infty,\ell^{-1}\epsilon}$. Therefore $u_{\infty,\epsilon}$ decreases and converges when $\epsilon \to 0$ to the function $u_{\infty,0}$ which is a positive self-similar solution of (1.30), say

$$u_{\infty,0}(x,t) = t^{-\frac{2}{q-1}} \tilde{f}\left(\frac{|x|}{\sqrt{t}}\right), \tag{1.85}$$

and $\tilde{f}$ is a positive solution of (1.80). Since $u_\infty \leq u_{\infty,0}$, one has $f \leq \tilde{f}$. Actually, f (resp. $\tilde{f}$) is the minimal (resp. maximal) solution of (1.82). Estimate (1.83) is obtained by the classical method of matching asymptotic expansion.
For uniqueness, it follows from the fact that $0 < f(0) \leq \tilde{f}(0)$ combined to the expansion (1.83) that there exists $A > 1$ such that

$$f(\eta) \leq \tilde{f}(\eta) \leq Af(\eta) \quad \text{for all } \eta \geq 0. \tag{1.86}$$

Actually, only the truncated expansion

$$Ae^{-\frac{\eta^2}{4}}\eta^{\frac{2}{q-1}-N}(1+o(1)) \quad \text{as } \eta \to \infty,$$

which is easily obtained as in Proposition 1.1-step 2 is needed. If $f \neq \widetilde{f}$, then $f < \widetilde{f}$ by the maximum principle. We set

$$W = f - \frac{1}{2A}(\widetilde{f} - f).$$

By convexity

$$\left(1+\frac{1}{2A}\right) f^p \leq \left(\left(1+\frac{1}{2A}\right) f - \frac{1}{2A}\widetilde{f}\right)^p + \frac{1}{2A}\widetilde{f}^p.$$

Hence W satisfies

$$-\Delta_\eta W - \frac{1}{2}\eta.\nabla W - \frac{1}{q-1}W + W^p \geq 0 \quad \text{in } \mathbb{R}^N.$$

Since $\widetilde{W} = \left(\frac{1}{2}+\frac{1}{2A}\right) f$ is smaller than W and satisfies

$$-\Delta_\eta W - \frac{1}{2}\eta.\nabla W - \frac{1}{q-1}W + W^p \leq 0 \quad \text{in } \mathbb{R}^N,$$

there exists a positive and radial function f^* satisfying

$$-\Delta_\eta f^* - \frac{1}{2}\eta.\nabla f^* - \frac{1}{q-1}f^* + f^{*p} = 0 \quad \text{in } \mathbb{R}^N,$$

and such that $0 < f^* < f$, which contradicts the minimality of f. □.

The following result is fundamental in the study of the singular points of the initial trace of a solution u of (1.30) in the subcritical case.

Lemma 1.1 *Suppose $1 < q < q_c$, u is a positive solution of (1.30) and $y \in \mathbb{R}^N$ is such that*

$$\limsup_{t\to 0} \int_{B_\epsilon(y)} u(x,t)dx = \infty \quad \textit{for all } \epsilon > 0. \tag{1.87}$$

Then

$$u(x,t) \geq t^{-\frac{2}{q-1}} f\left(\frac{|x-y|}{\sqrt{t}}\right). \tag{1.88}$$

Proof For any $\epsilon > 0$, there exists a sequence $\{t_n\}$ decreasing to 0 such that

$$\int_{B_\epsilon(y)} u(x,t_n)dx = M(\epsilon,n) \to \infty \quad \text{as } n \to \infty.$$

Let $c > 0$, then for $n \geq n_0 = n_0(\epsilon, c)$, $M(\epsilon, n) > c$, hence there exist ϵ_n and $k_n > 0$, both depending on c such that

$$\int_{B_{\epsilon_n}(y)} \min\{u(x, t_n), k_n\}dx = c.$$

Let u_n be the solution of (1.30) with initial data $u_n(x, 0) = \min\{u(x, t_n), k_n\}\mathbf{1}_{B_{\epsilon_n}(y)}$. By the maximum principle

$$u(x, t + t_n) \geq u_n(x, t) \quad \text{for } (x, t) \in Q_\infty.$$

By Theorem 1.6 u_n converges to $u_{c\delta_y}$ when $n \to \infty$. Hence

$$u(x, t) \geq u_{c\delta_y}(x, t) \quad \text{for } (x, t) \in Q_\infty.$$

Since $c > 0$ is arbitrary, the claim follows from the fact that $\lim_{c \to \infty} u_{c\delta_y}(x, t) \to u_\infty(x - y, t)$ by Theorem 1.7. □

Proposition 1.2 *Suppose* $1 < q < q_c$. *Then for any closed set* $\mathcal{S} \subset \mathbb{R}^N$ *there exists a unique positive solution of (1.30) with initial trace* $(\mathcal{S}, 0)$.

Proof *Step 1- Construction of the minimal solution* $\underline{u}_{\mathcal{S},0}$. Let $\{a_n\} \subset \mathcal{S}$ be a sequence of points dense in $\mathcal{S}$ and $\mu_n = n \sum_{j=1}^{n} \delta_{a_j}$. Then the sequence u_{μ_n} of solutions of (1.30) is increasing. By Lemma 1.1

$$u_{\mu_n} \geq \sup\{u_{n\delta_{a_j}} : 1 \leq j \leq n\}. \tag{1.89}$$

Furthermore, by (1.33), for any $y \in \mathcal{S}^c$ and $R = \text{dist}\,(y, \mathcal{S})$ there holds

$$u_{\mu_n}(x, t) \leq Ct^{-\frac{1}{q-1}} \left(\frac{|x - y|}{\sqrt{t}}\right)^\alpha e^{-\frac{(R-[x-y])^2}{4t}} \quad \text{for all } (x, t) \in B_R(y) \times (0, \infty), \tag{1.90}$$

where $C = C(\alpha, q) > 0$ and $\alpha > \frac{3-p}{p-1}$, and classically, $u_{\mu_n}(x, t) \leq \phi_\infty(t)$. Therefore the sequence $\{u_{\mu_n}\}$ increases and converges to some function denoted by $\underline{u}_{\mathcal{S},0}$ which is a positive solution of (1.30) and satisfies the same estimate from above (1.90) as u_{μ_n}. By (1.88) and (1.89) there holds

$$\underline{u}_{\mathcal{S},0}(x, t) \geq t^{-\frac{2}{q-1}} f\left(\frac{|x - a_n|}{\sqrt{t}}\right) \quad \text{for all } (x, t) \in B_R(y) \times (0, \infty) \text{ and } n \in \mathbb{N}. \tag{1.91}$$

Because $\{a_n\}$ is dense in $\mathcal{S}$, this last inequality implies that for any $y \in \mathcal{S}$ and $\epsilon > 0$,

$$\int_{B_\epsilon(y)} \underline{u}_{\mathcal{S},0}(x, t)dx \to \infty \quad \text{as } t \to 0. \tag{1.92}$$

Step 2- We claim that the function $\underline{u}_{\mathcal{S},0}$ *is the minimal solution with initial trace* $(\mathcal{S}, 0)$. Let u be such a solution. For $n \in \mathbb{N}^*$ we consider a double sequence of real numbers $\{\epsilon_{n,\ell}\}$ such that

$$\epsilon_{n,\ell} \leq \min\{|a_j - a_i| : 1 \leq i, j \leq n, \, i \neq j\} \quad \text{for all } \ell \in \mathbb{N}^*.$$

Since the set $\{a_j\}_{j\in\mathbb{N}}$ is dense in $\mathcal{S}$, for any ℓ there holds

$$\lim_{n\to\infty} \epsilon_{n,\ell} = 0.$$

We assume also

$$\lim_{\ell\to\infty} \epsilon_{n,\ell} = 0 \quad \text{for all } n \in \mathbb{N}^*.$$

For any $n \in \mathbb{N}^*$, $\ell \in \mathbb{N}^*$ and $j = 1, ..., n$, there holds

$$\int_{B_{\epsilon_{n,\ell}}(a_j)} u(x,t)dx \to \infty \quad \text{as } t \to 0.$$

Then for fixed $n \in \mathbb{N}^*$ there exists $t_{n,\ell} > 0$ such that

$$\int_{B_{\epsilon_{n,\ell}}(a_j)} u(x,t_{n,\ell})dx \geq 2n \quad \text{for all } j = 1, ..., n.$$

Since $\epsilon_{n,\ell} \to 0$ when $\ell \to \infty$, it follows that $t_{n,\ell} \to 0$ under the same condition. Consequently there exist positive numbers $m_{j,n,\ell}$ for $j = 1, ..., n$ such that

$$\int_{B_{\epsilon_{n,\ell}}(a_j)} \min\{u(x,t_{n,\ell}), m_{j,n,\ell}\}dx = n.$$

We set

$$\rho_{n,\ell}(x) = \sum_{j=1}^{n} \min\{u(x,t_{n,\ell}), m_{j,n,\ell}\}\mathbf{1}_{B_{\epsilon_{n,\ell}}(a_j)}(x). \tag{1.93}$$

Then

$$\lim_{\ell\to\infty} \rho_{n,\ell} = \mu_n := n \sum_{j=1}^{n} \delta_{a_j} \quad \text{in the sense of distributions in } \mathbb{R}^N. \tag{1.94}$$

Since $u(x,t_{n,\ell}) \geq \rho_{n,\ell}(x)$ we have that $u(x,t+t_{n,\ell}) \geq u_{n,\ell}(x,t)$ in Q_∞ where $u_{n,\ell}$ is the solution of (1.30) with initial data $\rho_{n,\ell}$. By the stability result in [15] $u_{n,\ell}$ converges to u_{μ_n} defined in Step 1. Hence $u \geq u_{\mu_n}$. Letting $n \to \infty$ implies $u \geq \underline{u}_{\mathcal{S},0}$.

Step 3- Construction of the maximal solution $\overline{u}_{\mathcal{S},0}$. For $\epsilon > 0$ we set

$$\mathcal{S}_\epsilon = \{x \in \mathbb{R}^N : \operatorname{dist}(x, \mathcal{S}) \le \epsilon\}.$$

For $R > 0$ we also define $\mathcal{S}_{\epsilon,R} = \mathcal{S}_\epsilon \cap \overline{B}_R$. Let $u = u_{\epsilon,R,n}$ be the solution of (1.30) with initial data $n\mathbf{1}_{\mathcal{S}_{\epsilon,R}}$. The mapping $(R, n) \mapsto u_{\epsilon,R,n}$ is increasing and bounded from above by ϕ_∞, hence there exists

$$u_\epsilon := \lim_{\substack{n\to\infty \\ R\to\infty}} u_{\epsilon,R,n}.$$

The mapping $R \mapsto u_{\epsilon,R}$ is increasing therefore there exists a limit u_ϵ when $R \to \infty$ which satisfies

$$\lim_{t\to 0} t^{\frac{1}{q-1}} u_\epsilon(x,t) = \left(\frac{1}{q-1}\right)^{\frac{1}{q-1}}, \tag{1.95}$$

uniformly on any ball B_θ interior to $\mathcal{S}_\epsilon$ and

$$u_\epsilon(x,t) \le Ct^{-\frac{1}{q-1}} \left(\frac{|x-y|}{\sqrt{t}}\right)^\alpha e^{-\frac{(R-|x-y|)^2}{4t}} \quad \text{for all } (x,t) \in B_R(y) \times (0,\infty), \tag{1.96}$$

for all $y \in \mathcal{S}_\epsilon^c$ where $R = \operatorname{dist}(y, \mathcal{S}_\epsilon)$. This implies that the initial trace of u_ϵ is $(\mathcal{S}_\epsilon, 0)$. It is a consequence of the construction of u_ϵ as the limit of $u_{\epsilon,R,n}$ when $(n, R) \to (\infty, \infty)$ that the mapping $\epsilon \mapsto u_\epsilon$ is decreasing with limit $\overline{u}_{\mathcal{S}}$. Furthermore $\overline{u}_{\mathcal{S}} \ge \underline{u}_{\mathcal{S},0}$. Using (1.96) applied with $y \in \mathcal{S}^c$ and $R = \operatorname{dist}(y, \mathcal{S})$ we deduce that $\overline{u}_{\mathcal{S}}$ has initial trace $(\mathcal{S}, 0)$, and from now it is denoted by $\overline{u}_{\mathcal{S},0}$.

Step 4- We claim that the function $\overline{u}_{\mathcal{S},0}$ is the maximal solution with initial trace $(\mathcal{S}, 0)$. Assume u is any positive solution of (1.30) with initial trace $(\mathcal{S}, 0)$ and for $R > 0$ let w_R be the solution of (1.40). For $\epsilon > 0$ the function $u_\epsilon + w_R$ is a supersolution of the equation in $B_R \times (0,\infty)$, thus for any $\delta > 0$ the function $(1+\delta)(u_\epsilon + w_R)$ is also a supersolution of the equation in $B_R \times (0,\infty)$. Since $u(x,t) \to 0$ when $t \to 0$ uniformly in $B_R \setminus \mathcal{S}_\epsilon$ we obtain that $u \le (1+\delta)(u_\epsilon + w_R)$ in $B_R \times (0,\infty)$. Letting successively $\delta \to 0$, $R \to \infty$, here we use (1.41) and $\epsilon \to 0$, we infer that $u \le \overline{u}_{\mathcal{S},0}$.

Step 5- We claim that there exists $K > 1$ such that $\overline{u}_{\mathcal{S},0} \le K\underline{u}_{\mathcal{S},0}$. If $y \in \mathcal{S}$ there holds by (1.43) and (1.88) that

$$f(0)t^{-\frac{1}{q-1}} \le \underline{u}_{\mathcal{S},0}(y,t) \le \overline{u}_{\mathcal{S},0}(y,t) \le \left(\frac{1}{q-1}\right)^{\frac{1}{q-1}} t^{-\frac{1}{q-1}}. \tag{1.97}$$

Thus the claim follows with $K = \left(\frac{1}{q-1}\right)^{\frac{1}{q-1}} (f(0))^{-1}$.
If $y \in \mathcal{S}^c$ let $z \in \mathcal{S}$ such that $|z-y| = \operatorname{dist}(y, \mathcal{S}) := d_y$. Then by (1.88) and (1.49),

$$f\left(\frac{d_y}{\sqrt{t}}\right) t^{-\frac{1}{q-1}} \le \underline{u}_{\mathcal{S},0}(y,t) \le \overline{u}_{\mathcal{S},0}(y,t) \le Ce^{-\frac{d_y^2}{4t}} t^{-\frac{1}{q-1}} \left(\frac{d_y}{\sqrt{t}}\right)^\alpha, \tag{1.98}$$

where $\alpha > \frac{3-q}{q-1}$. For $\sigma > 0$ we put

$$\mathcal{P}_\sigma = \left\{(y,t) : \frac{d_y}{\sqrt{t}} \leq \sigma\right\}.$$

If $(y,t) \in \mathcal{P}_\sigma$ there exists $K_\sigma > 0$ such that

$$\overline{u}_{S,0}(y,t) \leq K_\sigma \underline{u}_{S,0}(y,t). \tag{1.99}$$

Next we prove that for any $c > 1$ there exists σ_c such that for any $\sigma \geq \sigma_c$ there holds

$$\overline{u}_{S,0}(y,t) \leq K_\sigma \underline{u}_{S,0}(y,Ct) \quad \text{for all } (x,t) \in Q_\infty \setminus \mathcal{P}_\sigma. \tag{1.100}$$

It follows from expansion (1.83) that we have

$$e^{-\frac{d_y^2}{4t}} t^{-\frac{1}{q-1}} \left(\frac{d_y}{\sqrt{t}}\right)^\alpha \leq f\left(\frac{d_y}{\sqrt{ct}}\right)(ct)^{-\frac{1}{q-1}}, \tag{1.101}$$

which implies (1.100).
Next, for $\tau > 0$, let $u_{1,\tau}$ and $u_{2,\tau}$ be the solutions of (1.30) with respective initial data

$$\begin{aligned} u_{1,\tau} &= K_\sigma \mathbf{1}_{\mathcal{P}_\sigma}(x,\tau)\underline{u}_{S,0}(x,\tau) \\ u_{2,\tau} &= (1 - \mathbf{1}_{\mathcal{P}_\sigma}(x,\tau))\underline{u}_{S,0}(x,C\tau). \end{aligned} \tag{1.102}$$

It is known and easy to prove that the solutions of (1.30) are uniquely determined by their initial data ([14]). The function $u_{1,\tau} + u_{2,\tau}$ is a supersolution and

$$\begin{aligned} (u_{1,\tau} + u_{2,\tau})(x,0) &= K_\sigma \mathbf{1}_{\mathcal{P}_\sigma}(x,\tau)\underline{u}_{S,0}(x,\tau) + (1 - \mathbf{1}_{\mathcal{P}_\sigma}(x,\tau))\underline{u}_{S,0}(x,C\tau) \\ &\geq \overline{u}_{S,0}(x,\tau). \end{aligned}$$

Since $K_\sigma \underline{u}_{S,0}(x,\tau) \geq u_{1,\tau}(x,0)$ it follows that

$$K_\sigma \underline{u}_{S,0}(x,\tau+t) \geq u_{1,\tau}(x,t) \quad \text{for all } (x,t) \in Q_\infty.$$

Similarly $\underline{u}_{S,0}(x,C\tau) \geq u_{2,\tau}(x,0)$, therefore

$$\underline{u}_{S,0}(x,C\tau+t) \geq u_{2,\tau}(x,t) \quad \text{for all } (x,t) \in Q_\infty.$$

Combining these two inequalities we have that

$$\overline{u}_{S,0}(x,t+\tau) \leq u_{2,\tau}(x,t) + u_{1,\tau}(x,t) \leq K_\sigma \underline{u}_{S,0}(x,\tau+t) + \underline{u}_{S,0}(x,C\tau+t). \tag{1.103}$$

Letting $\tau \to 0$ yields

$$\overline{u}_{\mathcal{S},0}(x,t) \leq (1+K_\sigma)\underline{u}_{\mathcal{S},0}(x,t) \quad \text{for all } (x,t) \in Q_\infty. \tag{1.104}$$

Next we set $K = 1 + K_\sigma$ and

$$W = \underline{u}_{\mathcal{S},0} - \frac{1}{2K}\left(\overline{u}_{\mathcal{S},0} - \underline{u}_{\mathcal{S},0}\right).$$

If $\overline{u}_{\mathcal{S},0} \neq \underline{u}_{\mathcal{S},0}$, then $\overline{u}_{\mathcal{S},0} > \underline{u}_{\mathcal{S},0}$ and W is a supersolution of (1.30) by the same convexity argument used in the proof of Theorem 1.7. Note also that $\left(\frac{1}{2} + \frac{1}{2K}\right)\underline{u}_{\mathcal{S},0}$ is a subsolution of (1.30) smaller than W. Hence there exists a solution $\widetilde{u}$ of (1.30) satisfying

$$\left(\frac{1}{2} + \frac{1}{2K}\right)\underline{u}_{\mathcal{S},0} \leq \widetilde{u} \leq W. \tag{1.105}$$

This implies that the initial trace of $\widetilde{u}$ is also $(\mathcal{S}, 0)$. Since $W < \underline{u}_{\mathcal{S},0}$, we have a contradiction with the minimality of $\underline{u}_{\mathcal{S},0}$. □

The next result shows that the initial trace provides a one-to-one correspondence between the set of nonnegative solutions of (1.30) and the set of couples $(\mathcal{S}, \mu)$ where $\mathcal{S}$ is a closed subset of $\mathbb{R}^N$ and μ a nonnegative Radon measure on $\mathcal{R} := \mathbb{R}^N \setminus \mathcal{S}$.

Theorem 1.8 *Suppose $1 < q < q_c$. Then for any closed set $\mathcal{S} \subset \mathbb{R}^N$ and any positive Radon measure μ on $\mathcal{R} := \mathbb{R}^N \setminus \mathcal{S}$, there exists a unique positive solution of (1.30) with initial trace $(\mathcal{S}, \mu)$.*

Proof *Step 1- Construction of the minimal solution.* The principle is standard. We set $\mathcal{S}_\epsilon = \{x \in \mathbb{R}^N : \text{dist}\,(x, \mathcal{S}) \leq \epsilon\}$. For $R > 0$ we define

$$\mu_{\epsilon,R} = \mathbf{1}_{\mathcal{S}_\epsilon^c \cap B_R}\mu.$$

and denote by $u_{\mu_n+\mu_{\epsilon,R}}$ the solution of (1.30) with initial data $\mu_n + \mu_{\epsilon,R}$ where μ_n has been defined in the proof of Proposition 1.2-Step 1. Clearly $(\epsilon, R, n) \mapsto u_{\mu_n+\mu_{\epsilon,R}}$ is increasing in n and R and decreasing with respect to ϵ and we have

$$\max\{u_{\mu_n}, u_{\mu_{\epsilon,R}}\} \leq u_{\mu_n+\mu_{\epsilon,R}} \leq u_{\mu_n} + u_{\mu_{\epsilon,R}}. \tag{1.106}$$

If we set

$$\underline{u}_{\mathcal{S},\mu} = \lim_{\epsilon\to 0, n\to\infty, R\to\infty} u_{\mu_n+\mu_{\epsilon,R}}, \tag{1.107}$$

then

$$\max\{u_{(\mathcal{S},0)}, u_\mu\} \leq \underline{u}_{\mathcal{S},\mu} \leq u_{(\mathcal{S},0)} + u_\mu. \tag{1.108}$$

Note that we have used Brezis' uniqueness result to obtain that u_μ is uniquely determined by μ. Inequality (1.108) implies that the initial trace of $\underline{u}_{\mathcal{S},\mu}$ is $(\mathcal{S}, \mu)$.

Let u be any positive solution of (1.30) with initial trace $(\mathcal{S}, \mu)$. If $k > R$ we denote by w_k the solution of (1.40) in B_k. Then $u + w_k$ is a supersolution of (1.30) in $Q_\infty^{B_k}$. Because $B_k = (B_k \cap \mathcal{S}_\epsilon) \cup (B_k \cap \mathcal{S}_\epsilon^c)$, there holds

$$\lim_{t \to 0} \int_{B_k \cap S_\epsilon} (u_{\mu_n + \mu_{\epsilon,R}} - u)_+ dx = \lim_{t \to 0} \int_{B_k \cap S_\epsilon} (u_{\mu_n} - u)_+ dx = 0,$$

and

$$\lim_{t \to 0} \int_{B_k \cap S_\epsilon^c} (u_{\mu_n + \mu_{\epsilon,R}} - u)_+ dx = \lim_{t \to 0} \int_{B_k \cap S_\epsilon^c} (u_{\mu_{\epsilon,R}} - u)_+ dx = 0.$$

Therefore the subsolution $(u_{\mu_n + \mu_{\epsilon,R}} - u - w_k)_+$ has zero initial data and it vanishes on $\partial_\ell(Q_\infty^{B_k}) := \partial B_k \times (0, \infty)$; thus it is identically zero. Therefore

$$u_{\mu_n + \mu_{\epsilon,R}} \leq u - w_k \quad \text{in } Q_\infty^{B_k}. \tag{1.109}$$

Letting $k \to \infty$, $\epsilon \to 0$ and $n \to \infty$ yields

$$\underline{u}_{S,\mu} \leq u \quad \text{in } Q_\infty. \tag{1.110}$$

Step 2- Construction of the maximal solution. For $n, \epsilon > 0$ we set $\mu_\epsilon = \mathbf{1}_{S_\epsilon^c}\mu$ and $\mu_\epsilon^n = \mathbf{1}_{S_\epsilon^c}\mu + n\mathbf{1}_{S_\epsilon}dx$. Let $u_{n,\epsilon} = u_{\emptyset,\mu_\epsilon^n}$. When $n \to \infty$, the sequence $\{u_{n,\epsilon}\}$ increases and converges to a solution with initial trace $(S_\epsilon, \mu_\epsilon)$ denoted by $u_{S_\epsilon,\mu_\epsilon}$.
Let $k, \theta > 0$, then $u_{\theta,\epsilon,k} := (1 + \theta)u_{S_\epsilon,\mu_\epsilon} + w_k$ is a supersolution of (1.30) in Q_∞. By (1.63), the function $(u - u_{\theta,\epsilon,k})_+$ satisfies

$$\lim_{t \to 0} \int_{S_\epsilon \cap B_k} (u - u_{\theta,\epsilon,k})_+(x, t) dx = 0.$$

In the cylinder $Q_\infty^{B_k}$ the function $u_{\theta,\epsilon,k}$ is a supersolution of (1.30) with initial data $(1 + \theta)\mu$ and infinite boundary data. Hence it dominates u therein. Consequently

$$\lim_{t \to 0} \int_{S_\epsilon^c \cap B_k} (u - u_{\theta,\epsilon,k})_+(x, t) dx = 0,$$

which yields

$$\lim_{t \to 0} \int_{B_k} (u - u_{\theta,\epsilon,k})_+(x, t) dx = 0.$$

Because $u_{\theta,\epsilon,k}$ has infinite value on $\partial_\ell Q_\infty^{B_k}$ we deduce that $u_{\theta,\epsilon,k} \geq u$ in $Q_\infty^{B_k}$. Letting successively $k \to \infty$, $\delta \to 0$ and $\epsilon \to 0$ we obtain that

$$u_{S_\epsilon,\mu_\epsilon} \geq u \quad \text{in } Q_\infty.$$

When $\epsilon \to 0$, $u_{S_\epsilon,\mu_\epsilon}$ is decreasing and it converges to a solution $\overline{u}_{S,\mu}$ of (1.30) in Q_∞ with initial trace (S, μ) and is larger than any positive solution u with the same initial trace.

Step 3- End of the proof. With the notations of Steps 1 and 2, we set

$$Z_{\epsilon,\mu_\epsilon} = u_{\mathcal{S}_\epsilon,\mu_\epsilon} - u_{\mu_n+\mu_\epsilon} \quad \textit{and} \quad Z_{\epsilon,0} = u_{\mathcal{S}_\epsilon,0} - u_{\mu_n}.$$

Then

$$\partial_t(Z_{\epsilon,\mu_\epsilon} - Z_{\epsilon,0}) - \Delta(Z_{\epsilon,\mu_\epsilon} - Z_{\epsilon,0}) + u^q_{\mathcal{S}_\epsilon,\mu_\epsilon} - u^q_{\mu_n+\mu_\epsilon} - (u^q_{\mathcal{S}_\epsilon,0} - u^q_{\mu_n}) = 0.$$

Now

$$u^q_{\mathcal{S}_\epsilon,\mu_\epsilon} - u^q_{\mu_n+\mu_\epsilon} = \frac{u^q_{\mathcal{S}_\epsilon,\mu_\epsilon} - u^q_{\mu_n+\mu_\epsilon}}{u_{\mathcal{S}_\epsilon,\mu_\epsilon} - u_{\mu_n+\mu_\epsilon}} Z_{\epsilon,\mu_\epsilon},$$

and

$$u^q_{\mathcal{S}_\epsilon,0} - u^q_{\mu_n} = \frac{u^q_{\mathcal{S}_\epsilon,0} - u^q_{\mu_n}}{u_{\mathcal{S}_\epsilon,0} - u_{\mu_n}} Z_{\epsilon,0}.$$

Since

$$u_{\mathcal{S}_\epsilon,\mu_\epsilon} \geq \max\left\{u_{\mu_n+\mu_\epsilon}, u_{\mathcal{S}_\epsilon,0}\right\} \quad \textit{and} \quad u_{\mu_n} \leq \min\left\{u_{\mathcal{S}_\epsilon,0}, u_{\mu_n+\mu_\epsilon}\right\},$$

the convexity of the function $r \mapsto r^q$ on $\mathbb{R}_+$ implies that

$$\frac{u^q_{\mathcal{S}_\epsilon,\mu_\epsilon} - u^q_{\mu_n+\mu_\epsilon}}{u_{\mathcal{S}_\epsilon,\mu_\epsilon} - u_{\mu_n+\mu_\epsilon}} \geq \frac{u^q_{\mathcal{S}_\epsilon,0} - u^q_{\mu_n}}{u_{\mathcal{S}_\epsilon,0} - u_{\mu_n}} \geq 0.$$

Therefore

$$\partial_t(Z_{\epsilon,\mu_\epsilon} - Z_{\epsilon,0}) - \Delta(Z_{\epsilon,\mu_\epsilon} - Z_{\epsilon,0}) + \frac{u^q_{\mathcal{S}_\epsilon,0} - u^q_{\mu_n}}{u_{\mathcal{S}_\epsilon,0} - u_{\mu_n}}(Z_{\epsilon,\mu_\epsilon} - Z_{\epsilon,0}) \leq 0.$$

Since

$$\lim_{t\to\infty} \int_{\mathbb{R}^N} (Z_{\epsilon,\mu_\epsilon} - Z_{\epsilon,0})_+(x,t) = 0,$$

it follows by the maximum principle that $Z_{\epsilon,\mu_\epsilon} \leq Z_{\epsilon,0}$. Letting $n \to \infty$ and $\epsilon \to 0$ implies

$$\overline{u}_{\mathcal{S},\mu} - \underline{u}_{\mathcal{S},\mu} \leq \overline{u}_{\mathcal{S},0} - \underline{u}_{\mathcal{S},0}.$$

Uniqueness follows by Proposition 1.2. □

Extensions and comments. The initial trace of positive solutions of (1.30) in the cylinder Q^Ω_∞ can be defined similarly. If u is such a solution, it admits an initial trace in Ω which consists in a closed subset $\mathcal{S} \subset \Omega$ and a Radon measure μ defined in $\Omega \setminus \mathcal{S}$. Furthermore the value of u on the parabolic boundary $\partial_\ell Q^\Omega_\infty$ has to be taken into account in order to prove results of existence and uniqueness. This theory is

developed in [30] in the following framework:
(i) $\Omega \subset \mathbb{R}^N$ is a smooth domain.
(ii) $u\lfloor_{\partial_\ell Q^\Omega_\infty} = f \in L^1(\partial_\ell Q^\Omega_\infty)$.
(iii) μ is a positive Radon measure in Ω which is bounded in a neighbourhood of $\partial\Omega$.
(iv) $1 < q < q_c$.

Under these conditions and the subcriticality assumption, the initial trace provides a one-to-one correspondence between the sets of positive solutions u of

$$\begin{aligned} \partial_t u - \Delta u + u^q &= 0 \text{ in } Q^\Omega_\infty \\ u &= f \text{ on } \partial_\ell Q^\Omega_\infty, \end{aligned} \tag{1.111}$$

and the set of couples $(\mathcal{S}, \mu)$ where $\mathcal{S}$ is a closed subset of Ω and μ a nonnegative Radon measure μ in $\mathcal{R} := \Omega \setminus \mathcal{S}$ which is bounded in a neighbourhood of $\partial\Omega$.

1.3.3 The Supercritical Case

The next lemma shows that a very singular solution of (1.30) cannot exist if $q \geq q_c$.

Lemma 1.2 *Let $q \geq q_c$, then problem (1.82) admits no positive solution.*

Proof Let f be such a solution. Since $f(\eta) = o(\eta^{-\frac{2}{q-1}})$ as $\eta \to \infty$, by matching asymptotic expansion we obtain that for any $\alpha > \frac{2}{q-1} - N$ there exists $c_\alpha > 0$ such that

$$f(\eta) \leq c_\alpha \eta^\alpha e^{-\frac{\eta^2}{4}} \quad \text{for all } \eta \geq 1. \tag{1.112}$$

Then it follows from the equation that

$$f'(\eta) \leq c'_\alpha \eta^{\alpha+1} e^{-\frac{\eta^2}{4}} \quad \text{for all } \eta \geq 1. \tag{1.113}$$

Set $\phi_1(\eta) = e^{-\frac{\eta^2}{4}}$, then

$$\left(\eta^{N-1} e^{\frac{\eta^2}{4}} \phi_1'(\eta)\right)' = \frac{N}{2} \eta^{N-1} e^{\frac{\eta^2}{4}} \phi_1(\eta).$$

We write (1.82) under the form

$$\left(\eta^{N-1} e^{\frac{\eta^2}{4}} \phi_1'(\eta)\right)' + \eta^{N-1} e^{\frac{\eta^2}{4}} \left(\frac{1}{q-1} f - f^q\right) = 0.$$

Multiplying by ϕ_1 and integrating on $(0, \infty)$, which is justified by (1.112) and (1.113), we infer that

$$\int_0^\infty \left(\left(\frac{1}{q-1} - \frac{N}{2}\right) f - f^q\right) \eta^{N-1} e^{\frac{\eta^2}{4}} \phi_1 d\eta = 0.$$

This leads to a contradiction because $\frac{1}{q-1} - \frac{N}{2} \leq 0$. □

The following result proved by Brezis and Friedman [15] points out the role of the exponent q_c in the study of singularities of solutions of (1.30).

Theorem 1.9 *Let $q \geq q_c$, $\Omega \subset \mathbb{R}^N$ be a domain containing 0 and $u \in C(\overline{Q_T^\Omega} \setminus \{(0,0)\})$ be a solution of (1.30) in Q_T^Ω vanishing at $t = 0$ except at $x = 0$. Then u can be extended as a continuous function in $C(\overline{Q_T^\Omega})$.*

Proof We can assume that $\overline{B}_R \subset \Omega$ and we first assume that u_+ vanishes on $\partial B_R \times (0, T)$. Then for any $\epsilon > 0$ the function u_+ is bounded from above by the function $u_{\epsilon,\infty}$ which satisfies (1.30) in Q_∞ and has initial trace $(\overline{B}_\epsilon, 0)$. By scaling

$$T_\ell[u_{\epsilon,\infty}] = u_{\ell^{-1}\epsilon,\infty} \quad \text{for all } \ell > 0,$$

and since $\epsilon \mapsto u_{\epsilon,\infty}$ is increasing, there exists $u_{0,\infty} = \lim_{\epsilon \to 0} u_{\epsilon,\infty}$. Furthermore $u_{0,\infty}$ is self-similar and $u_+ \leq u_{0,\infty}$. By Lemma 1.2 $u_{0,\infty} = 0$, thus $u_+ = 0$.
In the general case we denote by ϕ the boundary value of u on $\partial_\ell Q_T^{B_R}$, and by ψ the solution of

$$\begin{aligned} \partial_t \psi - \Delta \psi &= 0 \quad \text{in } Q_T^{B_R} \\ \psi &= \phi_+ \text{ on } \partial_\ell Q_T^{B_R} \\ \psi(.,0) &= 0 \quad \text{in } B_R. \end{aligned}$$

Then $(u - \psi)_+$ is a subsolution of (1.30) in $Q_T^{B_R}$. By the previous argument, $(u - \psi)_+ = 0$. Hence u_+ is bounded from above. Similarly u_- is bounded; this implies that u remains bounded in $Q_T^{B_R}$. Standard regularity results imply that u vanishes on $B_R \times \{0\}$ and the claim follows. □

When $q \geq q_c$ there exists no solution of (1.30) in Q_∞ with a Dirac measure as an initial data. This phenomenon is general, and the next result proved in [7] shows that if $\mu \in \mathfrak{M}(\mathbb{R}^N)$ the problem with measure initial data

$$\begin{aligned} \partial_t u - \Delta u + |u|^{q-1} u &= 0 \qquad \text{in } Q_\infty \\ u(.,0) &= \mu \qquad \text{in } \mathbb{R}^N \end{aligned} \tag{1.114}$$

can be solved only if the measure μ is not too concentrated.

Definition 1.2 Let $q > 1$ and $\mu \in \mathfrak{M}(\mathbb{R}^N)$. A function $u \in L^q_{loc}(\overline{Q}_\infty) \cap C(Q_\infty)$ is a weak solution of (1.114) if for all $\zeta \in C^2_c(\overline{Q}_\infty)$ there holds

$$\int_0^\infty \int_{\mathbb{R}^N} \left(-u\,(\partial_t \zeta + \Delta \zeta) + |u|^{q-1} u \zeta\right) dx dt = \int_{\mathbb{R}^N} \zeta(x,0) d\mu(x). \tag{1.115}$$

A measure μ for which (1.114) is solvable is called q-admissible.

For the sake of completeness we introduce the Bessel capacity $cap_{\frac{2}{q},q'}$ (relative to $\mathbb{R}^N$) defined by

$$cap_{\frac{2}{q},q'}(E) = \inf\left\{\int_{\mathbb{R}^N\times\mathbb{R}^N} \frac{|\zeta(x)-\zeta(y)|^{q'}}{|x-y|^{N-\frac{2}{q-1}}}dxdy + \|\zeta\|^{q'}_{L^{q'}(\mathbb{R}^N)} \right. \\ \left. : \zeta\in C_c^\infty(\mathbb{R}^N), 0\le\zeta\le 1, \zeta\ge \mathbf{1}_E\right\}. \tag{1.116}$$

The properties of Bessel capacities $cap_{s,p}$ which are associated with the Besov space $B^{s,p}$ and the Bessel kernel G_s are fully developed in [1]. It is noticeable that when $q>q_c$ the zero-capacity Borel sets of the Riesz and the Bessel capacities coincide since there holds [1, 5.6.1]

$$\dot{R}_{\frac{2}{q},q'}(E) \le cap_{\frac{2}{q},q'}(E) \le A\left(\dot{R}_{\frac{2}{q},q'}(E) + \left(\dot{R}_{\frac{2}{q},q'}(E)\right)^{\frac{N}{N-\frac{2}{q-1}}}\right). \tag{1.117}$$

Theorem 1.10 *A measure $\mu\in\mathfrak{M}(\mathbb{R}^N)$ is q-admissible if and only if*

$$cap_{\frac{2}{q},q'}(F)=0 \Longrightarrow |\mu|(F)=0, \tag{1.118}$$

for all Borel set $F\subset\mathbb{R}^N$.

Before proving this result we give an equivalence of norms estimate which will be used in the sequel.

Lemma 1.3 *Assume $q\ge q_c$. Then for any $T>0$ there exists $c=c(n,q,T)>0$ such that for any bounded measure $\mu\in B^{-\frac{2}{q},q}(\mathbb{R}^N)$ there holds*

$$c^{-1}\|\mu\|_{B^{-\frac{2}{q},q}(\mathbb{R}^N)} \le \|\mathbb{H}[\mu]\|_{L^q(Q_T)} \le c\,\|\mu\|_{B^{-\frac{2}{q},q}(\mathbb{R}^N)}. \tag{1.119}$$

Furthermore, if $q>q_c$, there holds

$$c^{-1}\|\mu\|_{B^{-\frac{2}{q},q}(\mathbb{R}^N)} \le \|\mathbb{H}[\mu]\|_{L^q(Q_\infty)} \le c\left(\|\mu\|_{B^{-\frac{2}{q},q}(\mathbb{R}^N)} + \|\mu\|_{\mathfrak{M}(\mathbb{R}^N)}\right). \tag{1.120}$$

Proof If $\mu\in B^{-2/q,q}(\mathbb{R}^N)$, there exists a unique $\omega\in B^{2-2/q,q}(\mathbb{R}^N)$ such that $\mu=(I-\Delta)\omega$, and $\|\mu\|_{B^{-2/q,q}}\approx\|\omega\|_{B^{2-2/q,q}}$. Applying standard interpolation methods to the analytic semigroup $e^{-t(I-\Delta)}=e^{-t}e^{t\Delta}$ (see, for example, [8, 38]) we obtain

$$\left(\int\int_{Q_\infty}\left|t^{1/q}(I-\Delta)\mathbb{H}[\omega]\right|^q dx\frac{e^{-qt}dt}{t}\right)^{1/q} = \left(\int\int_{Q_\infty}\left|t^{1/q}\mathbb{H}[\mu]\right|^q dx\frac{e^{-qt}dt}{t}\right)^{1/q} \\ \approx \|\omega\|_{B^{2-2/q,q}} \\ \approx \|\mu\|_{B^{-2/q,q}}. \tag{1.121}$$

Clearly

$$e^{-qT}\int\int_{Q_T}\left|t^{1/q}\mathbb{H}[\mu]\right|^q dx\frac{dt}{t}\leq\int\int_{Q_\infty}\left|t^{1/q}\mathbb{H}[\mu]\right|^q dx\frac{e^{-qt}dt}{t},$$

and

$$\begin{aligned}\int\int_{Q_\infty}\left|t^{1/q}\mathbb{H}[\mu]\right|^q dx\frac{e^{-qt}dt}{t}&=\sum_{n=0}^{\infty}\int\int_{Q_{T+n+1}\setminus Q_{T+n}}\left|t^{1/q}\mathbb{H}[\mu]\right|^q dx\frac{e^{-qt}dt}{t}\\&=\sum_{n=0}^{\infty}\int\int_{Q_T}\left|\mathbb{H}[\mu](s+n)\right|^q e^{-q(s+n)}ds\\&\leq\left(\sum_{n=0}^{\infty}e^{-qn}\right)\int\int_{Q_T}\left|t^{1/q}\mathbb{H}[\mu]\right|^q\frac{dt}{t},\end{aligned}$$

and (1.119) follows. Notice that $\||\mathbb{H}[\mu](.,t)|\|^q_{L^q}\leq ct^{-N(q-1)/2}\|\mu\|^q_{\mathfrak{M}}$; therefore $\mathbb{H}[\mu]$ belongs to $L^q(Q_\infty)$ if $q>q_c$ (but not if $q=q_c$). If $q>q_c$ (equivalently $N(q-1)/2>1$),

$$\begin{aligned}\int\int_{Q_\infty}\left|t^{1/q}\mathbb{H}[\mu]\right|^q dx\frac{dt}{t}&=\sum_{n=0}^{\infty}\int\int_{Q_{T+n+1}\setminus Q_{T+n}}\left|t^{1/q}\mathbb{H}[\mu]\right|^q dx\frac{dt}{t}\\&=\int\int_{Q_T}\left|t^{1/q}\mathbb{H}[\mu]\right|^q dx\frac{dt}{t}+\int\int_{Q_T}\sum_{n=1}^{\infty}\left|\mathbb{H}[\mu](s+n)\right|^q dxds\\&\leq\int\int_{Q_T}\left|t^{1/q}\mathbb{H}[\mu]\right|^q dx\frac{dt}{t}+C\left(\sum_{n=1}^{\infty}n^{-N(q-1)/2}\right)\|\mu\|^q_{\mathfrak{M}}.\end{aligned}$$

Thus we obtain (1.120). □

Proof of Theorem 1.10. We present here an abridged proof slightly different from the original one due to Baras and Pierre [7]. We first notice that if $1<q<q_c$ any function in $B^{\frac{2}{q},q'}(\mathbb{R}^N)$ coincides with a continuous function. Hence only the empty set has zero $cap_{\frac{2}{q},q'}$-capacity. Therefore any measure in $\mathbb{R}^N$ is q-admissible. From now on we assume that $q\geq q_c$. Let F be a compact set with zero $cap_{\frac{2}{q},q'}$-capacity and $\{\zeta_n\}$ a sequence as in the previous theorem. We take $\widetilde{\phi}_n=\widetilde{\phi}\mathbb{H}[\zeta_n]$ for test functions, where now $\widetilde{\phi}\in C_c^\infty(\mathbb{R}^N\times[0,\infty))$ is nonnegative, takes value in $[0,1]$ and is equal to 1 in a neighbourhood of F. Then (1.133) is replaced by

$$\int_0^\infty\int_{\mathbb{R}^N}\left(u^q\widetilde{\phi}_n-u\left(\partial_t\widetilde{\phi}_n+\Delta\widetilde{\phi}_n\right)\right)dxdt=\int_{\mathbb{R}^N}(\widetilde{\phi}\mathbb{H}[\zeta_n])(.,0)d\mu(x)\geq\mu(F).\tag{1.122}$$

Since $\zeta_n \to 0$ and $0 \leq \zeta_n \leq 1$, $\widetilde{\phi}_n \to 0$ a.e. and $\partial_t \widetilde{\phi}_n + \Delta \widetilde{\phi}_n \to 0$ in $L^{q'}_{loc}(\mathbb{R}^N \times [0, \infty)))$. Thus the left-hand side of (1.122) converges to 0, which implies $\mu(F) = 0$.

Conversely, if μ is a nonnegative measure which vanishes on Borel sets with zero $cap_{\frac{2}{q},q'}$-capacity, it can be proved by the Hahn-Banach theorem (see [21]) that there exists an increasing sequence $\{\mu_n\}$ of nonnegative bounded measures belonging to $B^{-\frac{2}{q},q}(\mathbb{R}^N)$ which converges to μ. We first prove that a nonnegative bounded measure μ belonging to $B^{-\frac{2}{q},q}(\mathbb{R}^N)$ is q-admissible. By the previous lemma, $\mathbb{H}[\mu]$ belongs to $L^q_{loc}(\mathbb{R}^N \times [0, \infty))$. Next, for $k > 0$, we set $g_k(r) = \text{sign}(u) \min\{|u|^p, k^p\}$ and we denote by u_k the solution of

$$\begin{aligned} \partial_t u - \Delta u + g_k(u) = 0 \quad &\text{in } Q_\infty \\ u(., 0 = \mu \text{ in } \mathbb{R}^N. \end{aligned} \tag{1.123}$$

For $0 < k < \ell$ one has $0 < u_\ell < u_k < \mathbb{H}[\mu]$. We denote by u the limit of the u_k. Since for any $\zeta \in C_c^\infty(\mathbb{R}^N \times [0, \infty))$ there holds

$$\int_0^\infty \int_{\mathbb{R}^N} (g_k(u_k)\zeta - (\partial_t \zeta + \Delta \zeta)\, u_k)\, dxdt = \int_{\mathbb{R}^N} \zeta(., 0) d\mu, \tag{1.124}$$

and $g_k(u_k) \leq (\mathbb{H}[\mu])^p \in L^1_{loc}(\mathbb{R}^N \times [0, \infty))$, we deduce by the Lebesgue dominated convergence theorem that

$$\int_0^\infty \int_{\mathbb{R}^N} \left(u^p \zeta - (\partial_t \zeta + \Delta \zeta)\, u\right) dxdt = \int_{\mathbb{R}^N} \zeta(., 0) d\mu. \tag{1.125}$$

This prove that u is a nonnegative solution of (1.114) and μ is q-admissible. Finally if μ is a nonnegative measure satisfying (1.118), there exists an increasing sequence of q-admissible measures $\{\mu_n\}$ converging to μ. For each n, let u_n be the solution of (1.114) with initial data μ_n. Then the sequence $\{u_n\}$ is nondecreasing. For any nonnegative $\zeta \in C_c^2(\overline{Q}_\infty)$ there holds

$$\int_0^\infty \int_{\mathbb{R}^N} \left(-u_n (\partial_t \zeta + \Delta \zeta) + u_n^q \zeta\right) dxdt = \int_{\mathbb{R}^N} \zeta(x, 0) d\mu_n(x). \tag{1.126}$$

Let u be the limit of the increasing sequence $\{u_n\}$. By the Beppo-Levi convergence theorem one has (1.115). This implies in particular that $u \in L^q_{loc}(\overline{Q}_\infty)$. If ζ is no longer nonnegative then

$$\lim_{n \to \infty} \int_0^\infty \int_{\mathbb{R}^N} u_n^q \zeta dxdt,$$

by the dominated convergence theorem. Since

$$\lim_{n \to \infty} \int_0^\infty \int_{\mathbb{R}^N} u_n (\partial_t \zeta + \Delta \zeta)\, dxdt = \int_0^\infty \int_{\mathbb{R}^N} u\, (\partial_t \zeta + \Delta \zeta)\, dxdt,$$

it follows that u is a weak solution of (1.114) and therefore μ is q-admissible. For a general measure μ satisfying (1.118), we write the Jordan decomposition $\mu = \mu_+ - \mu_-$ and the proof follows. □

Baras and Pierre proved in [7] a general removability result which involves the Bessel capacities of a set.

Theorem 1.11 *Let $q > 1$ and $F \subset \mathbb{R}^N$ a closed set. A function $u \in C(\overline{Q}_\infty \setminus F)$ which is a solution of (1.30) in Q_∞ can be extended continuously to a function in $C(\overline{Q}_\infty)$ if and only if*

$$cap_{\frac{2}{q},q'}(F) = 0 \quad where \ q' = \tfrac{q}{q-1}. \tag{1.127}$$

Proof We give an abridged proof in order to point out the duality method introduced in [7]. We recall that the heat potential of a measure ω is

$$\mathbb{H}[\omega](x,t) = \frac{1}{(4\pi t)^{\frac{N}{2}}} \int_{\mathbb{R}^N} e^{-\frac{|x-y|^2}{4t}} d\omega(y) = \int_{\mathbb{R}^N} H(x,y,t) d\omega(y). \tag{1.128}$$

Without loss of generality, we can assume that F is a compact subset of B_R. Since $cap_{\frac{2}{q},q'}(F) = 0$, there exists a sequence $\{\zeta_n\} \subset C_c^\infty(\mathbb{R}^N)$ such that $\zeta_n = 1$ on F, $0 < \zeta_n \le 1$ and

$$\|\zeta_n\|_{B^{\frac{2}{q},q'}} \to 0 \quad \text{as } n \to \infty.$$

We can assume that the support of ζ_n is included into B_{R+1}. Let $\theta \in C_c^\infty(\mathbb{R}^N)$ such that $0 \le \theta \le 1$, $\theta = 1$ in B_{R+1} and $\theta = 0$ in B_{R+2}^c. We set $\eta_n := \theta\mathbb{H}[1-\zeta_n]$ and take η_n^α for test function where $\alpha > 0$.

By a straightforward computation based on Hölder's inequality we get

$$\begin{aligned}\int_0^1 \int_{\mathbb{R}^N} u^q \eta_n^\alpha dx dt \le c_\alpha \int_0^1 \int_{\mathbb{R}^N} \Big(\eta_n^{\alpha-q'} \left(|\partial_t \eta_n|^{q'} + |\Delta \eta_n|^{q'}\right) + |\nabla\theta|^{q'} |\nabla \eta_n|^{q'} \\ + \eta_n^{\alpha-2q'} |\nabla \eta_n|^{2q'}\Big) dx dt + \int_{\mathbb{R}^N} (u\eta_n^\alpha)(.,1) dx.\end{aligned} \tag{1.129}$$

We fix $\alpha = 2q'$. Replacing η_n by its value, we obtain

$$\eta_n^{\alpha-q'} |\partial_t \eta_n|^{q'} \le \theta^{q'} |\partial_t \mathbb{H}[\zeta_n]|^{q'} \le |\partial_t \mathbb{H}[\zeta_n]|^{q'}.$$

At this point we use the interpolation results associated with the analytic semigroup in $L^{q'}(\mathbb{R}^N)$ generated by $-\Delta$; see, for example, [38, Section 1.14.5]. We get

$$\int_0^1 \int_{\mathbb{R}^N} \eta_n^{\alpha-q'} |\partial_t \mathbb{H}[\zeta_n]|^{q'} dt \le c \, \|\eta_n\|^{q'}_{B^{\frac{2}{q},q'}}. \tag{1.130}$$

Similarly

$$\int_0^1 \int_{\mathbb{R}^N} \eta_n^{\alpha-q'} |\Delta \eta_n|^{q'} dt \leq c \, \|\eta_n\|^{q'}_{B^{\frac{2}{q},q'}} . \tag{1.131}$$

For the last term, we use Triebel's result combined with Gagliardo-Nirenberg inequality

$$\int_0^1 \int_{\mathbb{R}^N} \eta_n^{\alpha-2q'} |\nabla \eta_n|^{2q'} dt \leq c \, \|\eta_n\|^{q'}_{L^\infty} \|\eta_n\|^{q'}_{B^{\frac{2}{q},2q'}} \leq c \, \|\eta_n\|^{q'}_{B^{\frac{2}{q},q'}} . \tag{1.132}$$

Letting $n \to \infty$ and using the fact that $\eta_n \to 0$ in $B^{\frac{2}{q},q'}$, we infer that $u \in L^q(\mathbb{R}^N \times (0,1))$. In order to prove that u is a solution, we take $\phi_n = \phi \mathbb{H}[1-\zeta_n]$ for test function where $\phi \in C_c^\infty(\mathbb{R}^N \times [0,\infty))$. Then

$$\int_{\mathbb{R}^N} (u\phi_n)(.,0)dx = \int_0^\infty \int_{\mathbb{R}^N} \left(u^q \phi_n - u\left(\partial_t \phi_n + \Delta \phi_n\right)\right) dxdt. \tag{1.133}$$

By computation,

$$\begin{aligned}\partial_t \phi_n + \Delta \phi_n = \mathbb{H}[1-\zeta_n]\partial_t \phi - \phi \partial_t \mathbb{H}[\zeta_n] \\ + \mathbb{H}[1-\zeta_n]\Delta \phi - \phi \Delta \mathbb{H}[\zeta_n] - 2\nabla \phi . \nabla \mathbb{H}[\zeta_n].\end{aligned}$$

When $n \to \infty$, we have that

$$\mathbb{H}[1-\zeta_n]\left(\partial_t \phi + \Delta \phi\right) \to \partial_t \phi + \Delta \phi$$

in $L^\infty_{loc}(\mathbb{R}^N \times [0,\infty))$, and

$$\phi \partial_t \mathbb{H}[\zeta_n] + \phi \Delta \mathbb{H}[\zeta_n] + 2\nabla \phi . \nabla \mathbb{H}[\zeta_n] \to 0$$

in $L^p(\mathbb{R}^N \times [0,\infty))$. Hence, we infer that

$$\int_{\mathbb{R}^N} (u\phi)(.,0)dx = \int_0^\infty \int_{\mathbb{R}^N} \left(u^q \phi - u\left(\partial_t \phi + \Delta \phi\right)\right) dxdt \tag{1.134}$$

from (1.133).

The converse is a consequence of the fact that any compact set with positive $cap_{\frac{2}{q},q'}$-capacity is the support of a nonnegative measure (*the capacitary measure*) μ belonging to the space $B^{-\frac{2}{q},q}(\mathbb{R}^N)$; see [1, Theorem 2.2.7]. By Theorem 1.10 any nonnegative bounded measure belonging to $B^{-\frac{2}{q},q}(\mathbb{R}^N)$ is q-admissible. The result follows. □

In order to prove some analogue of Theorem 1.8 in the case $q \geq q_c$, there are conditions both on the measure μ which has to satisfy a non-concentration condition such as (1.118) and on the singular set $\mathcal{S}$ which cannot be locally removable.

Furthermore the singular set $\mathcal{S}$ can locally be created because the measure μ may be locally unbounded.

Definition 1.3 Assume $q > 1$. Let $\mathcal{S} \subset \mathbb{R}^N$ be a closed set and μ a nonnegative Radon measure on $\mathcal{R} := \mathbb{R}^N \setminus \mathcal{S}$ satisfying (1.118) for all Borel sets $F \subset \mathcal{S}$. We denote

$$\partial_\mu \mathcal{S} = \left\{x \in \mathcal{S} : \mu(B_\epsilon(x) \cap \mathcal{S}^c) = \infty \text{ for all } \epsilon > 0\right\}, \tag{1.135}$$

and

$$\mathcal{S}^* = \left\{x \in \mathcal{S} : cap_{\frac{2}{q},q'}(B_\epsilon(x) \cap \mathcal{S}) > 0 \text{ for all } \epsilon > 0\right\}. \tag{1.136}$$

The next result is proved in [30].

Theorem 1.12 *Let $q \geq q_c$. There exists a maximal positive solution u of (1.30) in Q_∞ with initial trace $(\mathcal{S}, \mu)$ if and only if μ satisfies (1.118) or all Borel set $F \subset \mathcal{S}$ and*

$$\mathcal{S} = \partial_\mu \mathcal{S} \cup \mathcal{S}^*. \tag{1.137}$$

Proof *Step 1: Construction of $\overline{u}_\mu$.* Let $\{K_n\}$ be an increasing sequence of compact subsets of $\mathcal{R} := \mathbb{R}^N \setminus \mathcal{S}$ such that $\cup_n K_n = \mathcal{R}$, $\mu_n = \mathbf{1}_{K_n}\mu$. Since $\mu_n \leq \mu$, it follows from Theorem 1.10 that there exists a unique solution u_n to (1.114) with initial data $\mu = \mu_n$. The sequence $\{u_n\}$ is increasing and it converges to some nonnegative solution $\overline{u}_\mu$ of (1.30) in Q_∞. By Proposition 1.1-2, $\overline{u}_\mu(.,t)$ converges to 0 when $t \to 0$ locally uniformly in the interior of $\mathcal{S}$. For any $y \in \mathcal{R}$ and $R > 0$ such that $\overline{B}_R(y) \subset \mathcal{R}$, $\overline{u}_\mu$ is bounded from above in $Q_\infty^{B_R(y)}$ by $w_R(y - .) + u_{\mathbf{1}_{B_R(y)}\mu}$ where $u_{\mathbf{1}_{B_R(y)}\mu}$ is the solution of (1.30) in $Q_\infty^{B_R(y)}$ with initial data $\mathbf{1}_{B_R(y)}\mu$ and vanishing on $\partial_\ell Q_\infty^{B_R(y)}$, and w_R is defined in (1.40). Since $w_R(y - .) + u_{\mathbf{1}_{B_R(y)}\mu}$ is bounded in $L^q(Q_T^{B_{R'}(y)})$ for any $T > 0$ and $R' < R$ it follows that $\overline{u}_\mu$ satisfies (1.115) for any $\zeta \in C_c^\infty(Q_\infty^{B_R(y)})$.x
Step 2: Characterisation of $\partial_\mu \mathcal{S}$. For any $x \in \partial_\mu \mathcal{S}$, $\epsilon > 0$ and $n \in \mathbb{N}$, $\overline{u}_\mu$ is bounded from below by the solution $u_{n,\epsilon}$ of (1.30) in Q_∞ with initial data $\mathbf{1}_{K_n \cap B_\epsilon(y)}\mu$. Furthermore for any nonnegative $\zeta \in C_c^\infty(B_\epsilon(y))$,

$$\begin{aligned}\liminf_{t\to 0} \int_{B_\epsilon(y)} u(x,t)\zeta(x)dx &\geq \lim_{t\to 0} \int_{B_\epsilon(y)} u_{n,\epsilon}(x,t)\zeta(x)dx \\ &= \int_{K_n \cap B_\epsilon(y)} \zeta(x)d\mu(x).\end{aligned} \tag{1.138}$$

We can take ζ such that $\zeta = 1$ on $B_{\epsilon'}(y)$ for some $0 < \epsilon' < \epsilon$. When $n \to \infty$, we have

$$\liminf_{n\to\infty} \int_{K_n \cap B_\epsilon(y)} \zeta(x)d\mu(x) = \int_{B_\epsilon(y)\cap \mathcal{S}^c} \zeta(x)d\mu(x) = \infty.$$

Hence y belongs to the singular set of the initial trace of $\overline{u}_\mu$ that we denote by $Sing(\overline{u}_\mu)$. Therefore

$$\partial_\mu \mathcal{S} \subset Sing(\overline{u}_\mu).$$

Conversely, if $y \notin \partial_\mu \mathcal{S}$ there exists $\delta > 0$ such that $\mu(B_\delta(y) \cap S^c) = m_{\delta,y} < \infty$. Then for any nonnegative $\zeta \in C_c^\infty(B_\delta(y))$, one has

$$\lim_{t\to 0} \int_{\mathbb{R}^N} \overline{u}_\mu(x,t)\zeta(x)dx = \int_{\mathbb{R}^N} \zeta(x)d\mu(x) < \infty.$$

This implies that $y \notin \mathcal{S}ing(\overline{u}_\mu)$. Thus

$$\partial_\mu \mathcal{S} = \mathcal{S}ing(\overline{u}_\mu). \tag{1.139}$$

Step 3: Construction of $\overline{u}_\mathcal{S}$. By thickening $\mathcal{S}$ into $\mathcal{S}_\epsilon = \{x \in \mathbb{R}^N : \text{dist}\,(x, \mathcal{S}) \le \epsilon\}$, we construct an increasing sequence of solutions $\{u_{\mathcal{S}_\epsilon}\}$ with initial trace $(\mathcal{S}_\epsilon, 0)$. When $\epsilon \downarrow 0$, $\{u_{\mathcal{S}_\epsilon}\}$ decreases and converges to some nonnegative solution $\overline{u}_\mathcal{S}$ of (1.30) in Q_∞. Let $y \in \mathcal{S}^*$. Then for any $\epsilon > 0$ the set $B_\epsilon(y) \cap \mathcal{R}$ has positive $cap_{\frac{2}{q},q'}$-capacity. Hence there exists a positive measure $\mu_{\epsilon,y}$ in the dual space $B^{-\frac{2}{q},q}(\mathbb{R}^N)$ with support in $B_\epsilon(y) \cap \mathcal{R}$. For $n \in \mathbb{N}_*$ we denote by $u_{n\mu_{\epsilon,y}}$ the solution of (1.30) in $Q_\infty^{B_\epsilon(y)}$ with initial data $n\mu_{\epsilon,y}$ and which vanishes on $\partial_\ell Q_\infty^{B_\epsilon(y)}$. Then $\overline{u}_\mathcal{S} \ge u_{n\mu_{\epsilon,y}}$ in $Q_\infty^{B_\epsilon(y)}$. Hence

$$\liminf_{t\to 0} \int_{B_\epsilon(y)} \overline{u}_\mathcal{S}(x,t)dx \ge n \int_{B_\epsilon(y)} d\mu_{\epsilon,y}.$$

Since n is arbitrary this implies that

$$\lim_{t\to 0} \int_{B_\epsilon(y)} \overline{u}_\mathcal{S}(x,t)dx = \infty. \tag{1.140}$$

Hence y belongs to the singular set of the initial trace of $\overline{u}_\mathcal{S}$ that we denote by $\mathcal{S}ing(\overline{u}_\mathcal{S})$. Thus

$$\mathcal{S}^* \subset \mathcal{S}ing(\overline{u}_\mathcal{S}).$$

Conversely, if $y \in \mathcal{S} \setminus \mathcal{S}^*$, there exists $\delta > 0$ such that $cap_{\frac{2}{q},q'}(B_\delta(y) \cap \mathcal{S}) = 0$. For $0 < \epsilon < \delta' < \delta$ we denote by $u_{1,\epsilon}$ (resp. $u_{2,\epsilon}$) the solution of (1.30) in Q_∞ with initial trace $\mathcal{S}_\epsilon \cap \overline{B}_{\delta'}(y)$ (resp $\mathcal{S}_\epsilon \cap B_{\delta'}^c(y)$). Then

$$u_{\mathcal{S}_\epsilon} \le u_{1,\epsilon} + u_{2,\epsilon}.$$

When $\epsilon \to 0$, $u_{1,\epsilon} \downarrow u_{1,0}$ (resp. $u_{2,\epsilon} \downarrow u_{2,0}$) where $u_{1,0}$ is a solution of (1.30) in Q_∞ with $\mathcal{S}ing(u_{1,0}) \subset \mathcal{S} \cap \overline{B}_{\delta'}(y)$ (resp $\mathcal{S}ing(u_{2,0}) \subset \mathcal{S} \cap B_{\delta'}^c(y)$) and with no regular part. By Theorem 1.9 $u_{1,0} = 0$. Since

$$\overline{u}_\mathcal{S} \le u_{1,0} + u_{2,0} = u_{2,0},$$

we have for any $0 < \delta'' < \delta'$,

$$\lim_{t\to 0}\int_{B_{\delta''}(y)} \overline{u}_{\mathcal{S}}(x,t)dx \le \lim_{t\to 0}\int_{B_{\delta''}(y)} u_{2,0}(x,t)dx = 0,$$

hence $y \notin \mathcal{S}ing(\overline{u}_{\mathcal{S}})$. We conclude that

$$\mathcal{S}^* = \mathcal{S}ing(\overline{u}_{\mathcal{S}}). \tag{1.141}$$

Step 4: Construction of a solution u with initial trace $(\mathcal{S},\mu)$. Since $\max\{\overline{u}_{\mathcal{S}},\overline{u}_\mu\}$ is a subsolution of (1.30) and $\overline{u}_{\mathcal{S}}+\overline{u}_\mu$ a supersolution, and since $\max\{\overline{u}_{\mathcal{S}},\overline{u}_\mu\} \le \overline{u}_{\mathcal{S}}+\overline{u}_\mu$, there exists a solution u such that

$$\max\{\overline{u}_{\mathcal{S}},\overline{u}_\mu\} \le u \le \overline{u}_{\mathcal{S}}+\overline{u}_\mu. \tag{1.142}$$

Then

$$\mathcal{S}ing(u) = \mathcal{S}ing(\overline{u}_{\mathcal{S}}) \cup \mathcal{S}ing(\overline{u}_\mu) = \mathcal{S}^* \cup \partial_\mu\mathcal{S}. \tag{1.143}$$

Therefore $\mathcal{S}ing(u) = \mathcal{S}$ if and only if (1.137) holds.
As in the proof of Theorem 1.8-Step 2, the function constructed above is the maximal solution of (1.30) with initial trace $(\mathcal{S},\mu)$. □

We end this section with a non-uniqueness result which asserts that in the supercritical case there could exist many positive solutions of (1.30) with the same initial trace with a non-empty singular set. This was proved first by Le Gall [27] in the framework of the *Brownian Snake*, with $q = 2$ and $N \ge 3$.

Theorem 1.13 *If $q \ge q_c$, there exist infinitely many solutions with initial trace* $(\mathbb{R}^N, 0)$.

Proof Let $\{a_n\}$ be a dense sequence in $\mathbb{R}^N$, $\{\epsilon_n\}$ a sequence of positive numbers such that the series $\sum_n \epsilon_n$ is convergent and $\{u_n\}$ the sequence of maximal solutions of (1.30) in Q_∞ with initial trace $(\overline{B}_{\epsilon_n}(a_n), 0)$. We have

$$u_n(x,t) = \widetilde{u}_n(|x-a_n|,t).$$

The function $\widetilde{u}_n$ is radial and radially decreasing for fixed t. Furthermore $t \mapsto \widetilde{u}_n(z,t)$ is decreasing if $z \in B_{\epsilon_n}(a_n)$. We set

$$\eta_n = \sup\{\widetilde{u}_n(x,t) : (x,t) \in \mathbb{R}^N \times [1,\infty)\} = \widetilde{u}_n(0,1).$$

Since $q \ge q_c$, $\widetilde{u}_n \to 0$ uniformly on $\mathbb{R}^N \times [\epsilon,\infty)$ when $n \to \infty$, for any $\epsilon > 0$. For any $E > 0$ we can choose the ϵ_n such that the following inequality holds:

$$\sum_{n\ge 0} \eta_n \le E.$$

Since $|u_n(x,t)|^{q-1} \le \frac{1}{t(q-1)}$, it follows by the parabolic Harnack inequality that the series $\sum_{n\ge 0} u_n$ converges normally on any compact subset of $\mathbb{R}^N \times (0,\infty)$, and we denote by $\overline{U}$ its sum. Since $(a+b)^q \geq a^q + b^q$ for any $a, b \geq 0$, $\overline{U}$ is a supersolution of (1.30). We set

$$\underline{U} = \sup\{u_n : n \in \mathbb{N}\}.$$

Then $\underline{U}$ is a subsolution of (1.30) and it is smaller than $\overline{U}$. Therefore there exists a positive solution U of (1.30) in Q_∞ such that

$$\underline{U} \leq U \leq \overline{U}. \tag{1.144}$$

For any $y \in \mathbb{R}^N$ and $\epsilon > 0$ there exist infinitely many a_n such that $B_{\epsilon_n}(a_n) \subset B_\epsilon(y)$, for such an n

$$\lim_{t \to 0} \int_{B_{\epsilon_n}(a_n)} u_n(x,t)dx = \infty.$$

Hence $y \in \mathcal{S}ing(u)$. Since the sequence $\{\epsilon_n\}$ can be chosen such that $\sum_{n \geq 0} \eta_n \leq E$, we obtain

$$0 < U(0,1) \leq E. \tag{1.145}$$

This ends the proof. □

This result can be improved in the following way ([30, Proposition 4.14]).

Theorem 1.14 *If $q > q_c$, for any $\epsilon > 0$ there exists a positive solution u of (1.30) in Q_∞ with initial trace $(\mathbb{R}^N, 0)$ and a Borel set $F \subset \mathbb{R}^N$ such that*

$$\int_{\mathbb{R}^N} \mathbf{1}_E dx < \epsilon,$$

and

$$\lim_{t \to 0} u(x,t) = 0 \quad \textit{for all } x \in \mathbb{R}^N \setminus E.$$

Starting from this result, it appeared clear that the definition of the initial trace performed by an averaging of the function $u(.,t)$ in an Euclidean neighbourhood of a point y is not suitable to distinguish between the different solutions of (1.30). The idea of using the fine topology associated with the $cap_{\frac{2}{q},q'}$-topology is due to S. Kuznetsov. It was first used in [20] in the framework of the study of the boundary trace of positive solutions of

$$-\Delta u + u^q = 0 \tag{1.146}$$

in a domain $\Omega \subset \mathbb{R}^N$. In [32, 33], a sharper definition, suitable for all the supercritical exponents in the semilinear elliptic problem (1.16), was introduced and developed. It is this method adapted to the parabolic framework in [23] that we present in the next section. It will apply to all the exponents $q \geq q_c$.

1.4 The Capacitary Representation

1.4.1 σ-Moderate Solutions

We denote by $\mathfrak{M}_+^b(\mathbb{R}^N)$ the set of nonnegative bounded Radon measures in $\mathbb{R}^N$. If μ is a q-admissible measure, we denote by u_μ the solution of (1.114).

Definition 1.4 A positive solution u of (1.30) in Q_∞ is called σ-moderate if there exists an increasing sequence $\{\mu_n\} \subset B^{-\frac{2}{q},q'}(\mathbb{R}^N) \cap \mathfrak{M}_+^b(\mathbb{R}^N)$ such that the corresponding solution $u := u_{\mu_n}$ of (1.114) converges to u locally uniformly in Q_∞.

If F is a closed subset of $\mathbb{R}^N$ we set

$$\underline{u}_F = \max\{u_\mu : \mu \in B^{-\frac{2}{q},q'}(\mathbb{R}^N) \cap \mathfrak{M}_+^b(\mathbb{R}^N), \mu(F^c) = 0\}. \tag{1.147}$$

1.4.1.1 Besov and Bessel Capacitary Potentials

The main goal of this section is to prove that $\underline{u}_F$ coincides with the maximal solution $\overline{u}_F$ of (1.30) in Q_∞ with initial trace $(F, 0)$. We introduce several tools linked to Bessel and Besov capacities relative to a domain $\Omega \subset \mathbb{R}^N$. If K is a compact subset of the domain $\Omega \subset \mathbb{R}^N$, we set

$$\mathcal{T}_\Omega(K) = \{\eta \in C_c^\infty(\Omega), 0 \leq \eta \leq 1, \ \eta = 1 \text{ on } K\}. \tag{1.148}$$

Definition 1.5 Let $s \in (0, 1)$, $p > 1$ such that $sp \leq N$ and $\Omega \subset \mathbb{R}^N$ be a domain. The Besov capacity $\dot{R}^\Omega_{s,p}$ of a compact set $K \subset \Omega$ is

$$\dot{R}^\Omega_{s,p}(K) = \inf\left\{\|\phi\|^p_{\dot{B}_{s,p}} : \phi \in \mathcal{T}_\Omega(K)\right\}, \tag{1.149}$$

where $\dot{B}_{s,p}$ is the Aronszajn-Slobodeckij norm defined by

$$\|\phi\|_{\dot{B}_{s,p}} = \left(\int_\Omega \int_\Omega \frac{|\phi(x) - \phi(y)|^p}{|x - y|^{N+sp}} dxdy\right)^{\frac{1}{p}}. \tag{1.150}$$

If $\Omega = \mathbb{R}^N$ the Besov capacity coincides with the Riesz capacity defined in (1.10). The Bessel capacity $cap^\Omega_{s,p}$ relative to Ω is defined by

$$cap^\Omega_{s,p}(K) = \inf\left\{\|\phi\|^p_{B_{s,p}} : \phi \in \mathcal{T}_\Omega(K)\right\}, \tag{1.151}$$

and $cap^{\mathbb{R}^N}_{s,p} = cap_{s,p}$.

In the sequel we will see that the capacity $\dot{R}^\Omega_{\frac{2}{q},q'}(K)$ is more suitable for the computations in our problem than the Bessel capacity $cap_{\frac{2}{q},q'}$.

Definition 1.6 Let $q > 1$. If $F \subset \mathbb{R}^N$ is a closed set we denote for $n \in \mathbb{N}$ and $(x,t) \in Q_\infty$

$$F_n := F_n(x,t) = \{y \in F : d_n \leq |x-y| \leq d_{n+1}\} \quad \text{where } d_n = \sqrt{nt},$$

$$\Gamma_n = B_{d_{n+1}} \setminus B_{d_n} = \left\{y \in \mathbb{R}^N : d_n \leq |y| \leq d_{n+1}\right\}.$$

The Bessel-capacitary potential of F is the function W_F defined by

$$W_F(x,t) = \frac{1}{t^{\frac{N}{2}}} \sum_{n=0}^{\infty} d_{n+1}^{N-\frac{2}{q-1}} e^{-\frac{n}{4}} cap_{\frac{2}{q},q'}\left(\frac{F_n}{d_{n+1}}\right) \quad \text{for all } (x,t) \in Q_\infty. \tag{1.152}$$

Similarly, the Besov capacitary potential of F is $\widetilde{W}_F$ defined by

$$\widetilde{W}_F(x,t) = \frac{1}{t^{\frac{N}{2}}} \sum_{n=0}^{\infty} d_{n+1}^{N-\frac{2}{q-1}} e^{-\frac{n}{4}} \dot{R}^{\Gamma_n}_{\frac{2}{q},q'}\left(\frac{F_n}{d_{n+1}}\right) \quad \text{for all } (x,t) \in Q_\infty. \tag{1.153}$$

The Besov capacitary potential of F is invariant by the scaling T_ℓ in the sense that for any $\ell > 0$,

$$\ell^{\frac{1}{q-1}} \widetilde{W}_F(\sqrt{\ell}x, \ell t) := T_\ell[\widetilde{W}_F](x,t) = \widetilde{W}_{\frac{F}{\sqrt{\ell}}}(x,t) \quad \text{for all } (x,t) \in Q_\infty. \tag{1.154}$$

The Besov capacity is linked to the Bessel capacity through the following directional Poincaré inequality [30].

Lemma 1.4 *Let $b > a > 0$ and Ω be a domain in $\mathbb{R}^N$ such that $\Omega \subset H_{a,b} := \{x = (x_1, x') : a < x_1 < b\}$. If $s \in (0,1)$ and $p > 1$ verify $sp \leq N$, there exists $\lambda = \lambda(N, s, p, \frac{b}{a}) > 0$ such that*

$$\int\int_{\Omega\times\Omega} \frac{|\eta(x)-\eta(y)|^p}{|x-y|^{N+sp}} dxdy \geq \lambda(b-a)^{-sp} \int_\Omega |\eta(x)|^p dx \quad \textit{for all } \eta \in C_c^\infty(\Omega). \tag{1.155}$$

It is noticeable that the above domain Ω is not necessarily bounded, in which case the standard Poincaré inequality is easy to prove, but it is only contained in a strip of finite thickness. The Aronszajn-Slobodeckij norm in $C_c^\infty(\Omega)$ is smaller than the standard $B^{s,p}$-norm associated with the Bessel potential $G_s := \mathcal{F}[((1+|\xi|^2)^{-\frac{S}{2}}]$ (see [1]) and defined by

$$\|\phi\|_{B^{s,p}} = \|\phi\|_{\dot{B}_{s,p}} + \|\phi\|_{L^p} \quad \text{for all } \phi \in C_c^\infty(\Omega)$$

However, thanks to Lemma 1.4, there holds the following.

Lemma 1.5 *Let $b > a > 0$ and Ω be a domain in $\mathbb{R}^N$ such that $\Omega \subset \Gamma_{a,b} = B_b \setminus \overline{B}_a$. If $s \in (0,1)$ and $p > 1$ verify $sp \leq N$, there exists $\lambda = \lambda(N, s, p, \frac{b}{a}) > 0$ such that*

$$\|\phi\|_{\dot{B}^{s,p}} \leq \|\phi\|_{B^{s,p}} \leq \left(1 + C(b-a)^s\right) \|\phi\|_{\dot{B}^{s,p}} \quad \text{for all } \eta \in C_c^\infty(\Omega). \tag{1.156}$$

The following properties of Bessel capacities $cap^{\Omega}_{s,p}$ relative to Ω and Besov capacities relative to $\mathbb{R}^N$ are classical and easy to establish.

Lemma 1.6 *For any $\tau > 0$ and any Borel set $K \subset \Omega$ there holds*

$$\dot{R}^{\Omega}_{s,p}(K) = \tau^{N-sp} \dot{R}^{\tau^{-1}\Omega}_{s,p}(\tau^{-1}K). \tag{1.157}$$

If $b > a$ and $\Omega \subset B_b \setminus \overline{B}_a$ there exists $c = c(b-a, \frac{b}{a}, N, s, p) > 0$ such that

$$\frac{1}{c} cap^{\Omega}_{s,p}(K) \leq \dot{R}^{\Omega}_{s,p}(K) \leq c\, cap^{\Omega}_{s,p}(K). \tag{1.158}$$

Finally if $K \subset \Omega' \subset \overline{\Omega}' \subset \Omega$, there exists $c = c(\text{dist}\,(\Omega', \Omega^c), N, s, p) > 0$ such that

$$\frac{1}{c} cap^{\Omega}_{s,p}(K) \leq cap^{\Omega}_{s,p}(K) \leq c\, cap_{s,p}(K). \tag{1.159}$$

1.4.1.2 Heat Potential and Besov Space

If Ω is a bounded domain in $\mathbb{R}^N$, $p > 1$ and $s \in (0, 1)$, we extend any $\eta \in C_c^\infty(\Omega)$ by zero in Ω^c and set

$$\|\eta\|_{\widetilde{B}^{s,p}} = \left(\int\int_{Q_\infty} |t^{1-\frac{s}{2}} \partial_t \mathbb{H}[\eta]|^p dx \frac{dt}{t} \right)^{\frac{1}{p}}. \tag{1.160}$$

It is proved in [8] that the following equivalence of norms holds for the Besov space $B^{s,p}(\Omega)$,

$$C^{-1} \|\eta\|_{B^{s,p}} := C^{-1} \left(\|\eta\|_{L^p} + \|\eta\|_{\dot{B}^{s,p}} \right) \leq \|\eta\|_{L^p} + \|\eta\|_{\widetilde{B}^{s,p}} \leq C \left(\|\eta\|_{L^p} + \|\eta\|_{\dot{B}^{s,p}} \right) \tag{1.161}$$

for all $\eta \in C_c^\infty(\Omega)$ for some $C = C(s, p, N) > 0$. Actually it is easy to see by scaling that the two norms $\|.\|_{\dot{B}^{s,p}}$ and $\|.\|_{\widetilde{B}^{s,p}}$ are universally equivalent in the sense that there exists $C = C(s, p, N) > 0$ such that for any domain Ω and any $\eta \in C_c^\infty(\Omega)$,

$$C^{-1} \|\eta\|_{\dot{B}^{s,p}} \leq \|\eta\|_{\widetilde{B}^{s,p}} \leq C \|\eta\|_{\dot{B}^{s,p}}. \tag{1.162}$$

If K is a compact subset of Ω and $\eta \in \mathcal{T}_\Omega(K)$ we set

$$R[\eta] = |\partial_t \mathbb{H}[\eta]| + |\nabla \mathbb{H}[\eta]|^2. \tag{1.163}$$

Lemma 1.7 *There exists $C = C(N, q) > 0$ such that for every $\eta \in \mathcal{T}_\Omega(K)$ there holds*

$$\|\eta\|^{q'}_{\widetilde{B}^{\frac{2}{q},q'}} \leq \|R[\eta]\|^{q'}_{L^{q'}} := \int\int_{Q_\infty} (R[\eta])^{q'} dxdt \leq C\, \|\eta\|^{q'}_{\widetilde{B}^{\frac{2}{q},q'}} . \tag{1.164}$$

Proof Using the Gagliardo-Nirenberg inequality in $\mathbb{R}^N$, an elementary elliptic estimate and the fact that $0 \leq \mathbb{H}[\eta] \leq 1$, we see that

$$\begin{aligned}\int_{\mathbb{R}^N} |\nabla \mathbb{H}[\eta](.,t)|^{2q'} dx &\leq C \left\| D^2\mathbb{H}[\eta](.,t)\right\|^{q'}_{L^{q'}} \|\mathbb{H}[\eta](.,t)\|^{q'}_{L^\infty} \\ &\leq C\, \|\Delta\mathbb{H}[\eta](.,t)\|^{q'}_{L^{q'}} ,\end{aligned}$$

for all $t > 0$. Since $\partial_t \mathbb{H}[\eta] = \Delta\mathbb{H}[\eta]$, (1.164) follows. □

1.4.2 Estimate from Above

The main result that we prove in this section is the following upper estimate

Theorem 1.15 *Let $q \geq q_c$. There exists a positive constant $c = c(N, q)$ such that for any closed subset $F \subset \mathbb{R}^N$, any nonnegative function $u \in C^{2,1}(Q_\infty) \cap C(\overline{Q}_\infty \setminus F \times \{0\})$ verifying*

$$\begin{aligned}&\partial_t u - \Delta u + u^p = 0 \;\; in \;\; Q_\infty \\ &\lim_{t\to 0} u(x,t) = 0 \; locally\ uniformly\ in \;\; F^c\end{aligned} \tag{1.165}$$

satisfies

$$u(x,t) \leq CW_F(x,t) \;\; for\ all \;\; (x,t) \in Q_\infty, \tag{1.166}$$

where W_F is the capacitary potential defined in (1.152).

We will first consider the case where $F = K$ is a compact set.

1.4.2.1 Global L^q Estimates

Let $K \subset B_r \subset \overline{B}_r \subset B_{r+\rho}$ where $r, \rho > 0$ is a compact set. We set

$$\mathcal{T}_{r,\rho}(K) = \{\eta \in C^\infty_c(B_{r+\rho}),\; 0 \leq \eta \leq 1,\; \eta = 1 \text{ on } K\}.$$

If $\eta \in \mathcal{T}_{r,\rho}(K)$ we set

$$\eta^* = 1 - \eta \quad \text{and} \quad \zeta = (\mathbb{H}[\eta^*])^{2q'}.$$

Lemma 1.8 *If u is a positive function satisfying (1.165), there exists $C = C(N, q) > 0$ such that for every $T > 0$ and every compact set $K \subset B_r$,*

$$\iint_{Q_T} u^q \zeta dx\,dt + \int_{\mathbb{R}^N} (u\zeta)(x, T)dx \leq C\|R[\eta]\|_{L^{q'}}^{q'} \quad \forall \eta \in \mathcal{T}_{r,\rho}(K). \tag{1.167}$$

Proof By assumption η^* vanishes in an open neighbourhood $\mathcal{N}_1$ of K, for any open set $\mathcal{N}_2$ such that $K \subset \mathcal{N}_2 \subset \overline{\mathcal{N}}_2 \subset \mathcal{N}_1$ there exists $C_{\mathcal{N}_2}, c_{\mathcal{N}_2} > 0$ such that

$$\mathbb{H}[\eta^*](x, t) \leq C_{\mathcal{N}_2} e^{-\frac{c_{\mathcal{N}_2}}{t}} \qquad \text{for all } (x, t) \in Q_\infty^{\mathcal{N}_2}.$$

By Proposition 1.1-2, this implies

$$\lim_{t\to 0} \int_{\mathbb{R}^N} (u\zeta)(x, t)dt = 0.$$

Taking ζ as a test function, we obtain

$$\iint_{Q_T} u^q \zeta dxdt + \int_{\mathbb{R}^N} (u\zeta)(x, T)dx = \iint_{Q_T} (\partial_t \zeta + \Delta\zeta)\, udxdt. \tag{1.168}$$

Since

$$\partial_t \zeta + \Delta\zeta = 2q' H[\eta^*]^{2q'-1} (\partial_t H[\eta] + \Delta\mathbb{H}[\eta]) + 2q'(2q' - 1)\mathbb{H}[\eta*]^{2q'-2} \left|\nabla\mathbb{H}[\eta^1|\right|^2,$$

we deduce

$$\left|\iint_{Q_T} (\partial_t \zeta + \Delta\zeta)\, udxdt\right| \leq c(q) \left(\iint_{Q_T} u^q \zeta dxdt\right)^{\frac{1}{q}} \left(\iint_{Q_T} R[\eta]^{q'} dxdt\right)^{\frac{1}{q'}},$$

where $R[\eta]$ is defined in (1.163). The proof follows from Lemma 1.7. □

Proposition 1.3 *Under the assumptions of Lemma 1.8, let $r, \rho > 0$ and $T \geq (r + \rho)^2$. Set*

$$\mathcal{E}_{r+\rho} = \{(x, t) \in Q_\infty : |x|^2 + t \leq (r + \rho)^2\},$$

and $Q_{r+\rho,T} = Q_T \setminus \mathcal{E}_{r+\rho}$. Then there exists $C = C(N, q, T) > 0$ such that

$$\iint_{Q_{r+\rho,T}} u^q dxdt + \int_{\mathbb{R}^N} u(x, T)dx \leq C \|R[\eta]\|_{L^{q'}}^{q'} \quad \text{for all } \eta \in \mathcal{T}_{r,\rho}. \tag{1.169}$$

Proof In view of the previous lemma we have to show that under the above assumptions on T and η, there exists some $C = C(N, q, T) > 0$ such that

$$\zeta = \mathbb{H}[\eta^*]^{2q'} \geq C.$$

Since by assumption $K \subset B_r$, $\eta^* = 1$ outside $B_{r+\rho}$ and $0 \leq \eta^* \leq 1$, we have

$$\begin{aligned}\mathbb{H}[\eta^*](x,t) \geq \mathbb{H}[1 - \mathbf{1}_{B_{r+\rho}}](x,t) &= \frac{1}{(4\pi t)^{\frac{N}{2}}} \int_{|y|>r+\rho} e^{\frac{|x-y|^2}{4t}} dy \\ &= 1 - \frac{1}{(4\pi t)^{\frac{N}{2}}} \int_{|y|\leq r+\rho} e^{\frac{|x-y|^2}{4t}} dy.\end{aligned}$$

If $(x,t) \in Q_{r+\rho,T}$, we write $x = (r+\rho)\xi$, $y = (r+\rho)\upsilon$ and $t = (r+\rho)^2\tau$. Then $(\xi,\tau) \in Q_{1,\frac{T}{(r+\rho)^2}}$ and

$$\frac{1}{(4\pi t)^{\frac{N}{2}}} \int_{|y|\leq r+\rho} e^{-\frac{|x-y|^2}{4t}} dy = \frac{1}{(4\pi \tau)^{\frac{N}{2}}} \int_{|\upsilon|\leq 1} e^{-\frac{|\xi-\upsilon|^2}{4\tau}} d\upsilon.$$

It is therefore easy to verify that

$$\max \left\{ \frac{1}{(4\pi \tau)^{\frac{N}{2}}} \int_{|\upsilon|\leq 1} e^{-\frac{|\xi-\upsilon|^2}{4\tau}} d\upsilon : (\xi,\tau) \in Q_{1,\frac{T}{(r+\rho)^2}} \right\} = \ell, \tag{1.170}$$

and $\ell = \ell(N, \frac{T}{(r+\rho)^2}) \in (0,1)$. Actually ℓ is independent of $\frac{T}{(r+\rho)^2}$ Ê if this quantity is larger than 1. Putting $C = (1-\ell)^{-1}$ we deduce (1.169). □

1.4.2.2 Pointwise Upper Estimates

In this section the assumptions of Lemma 1.8 are fulfilled.

Lemma 1.9 *There exists a constant $C = C(N,q) > 0$ such that, for any $\eta \in \mathcal{T}_{r,\rho}(K)$*

$$u(x,(r+\rho)^2) \leq C \frac{\|R[\eta]\|^{q'}_{L^{q'}}}{(r(r+\rho))^{\frac{N}{2}}} \quad \textit{for all } x \in \mathbb{R}^N. \tag{1.171}$$

Proof Integrating the equation

$$\int_s^T \int_{\mathbb{R}^N} u^q dx d\tau + \int_{\mathbb{R}} u(x,T)dx = \int_{\mathbb{R}} u(x,s) \quad \text{for all } T\ s > 0, \tag{1.172}$$

and by Proposition 1.3 we have that

$$\int_{\mathbb{R}^N} u(x,s)dx \leq C \int\int_{Q_T} (R[\eta])^{q'} dxdt \quad \text{for all } T > s > (r+\rho)^2. \tag{1.173}$$

Since

$$u(x, s+\tau) \leq \mathbb{H}[u(s,.)](x,\tau) \leq \frac{1}{(4\pi\tau)^{\frac{N}{2}}}\int_{\mathbb{R}^N} u(y,s)dy,$$

we obtain (1.171) from (1.172) and (1.173) with $s = (r+\rho)^2$ and $\tau = (r+2\rho)^2 - (r+\rho)^2 \approx r(r+\rho)$ if $\rho = o(r)$. □

In the next result we show an integral estimate of u on the lateral boundary of $Q_t^{B_r}$.

Lemma 1.10 *L4 Let $\gamma \geq r+2\rho$ and $c>0$, and either $N=1,2$ and $0\leq t\leq c\gamma^2$, or $N\geq 3$ and $t>0$. Then, for any $\eta \in \mathcal{T}_{r,\rho}(K)$, there holds*

$$\int_0^t \int_{\partial B_\gamma} u(x,\tau)dSd\tau \leq C\,\|R[\eta]\|_{L^{q'}}^{q'}, \tag{1.174}$$

where $C>0$ depends on N, q and c if $N=1,2$ or N and q if $N\geq 3$.

Proof Assume first that $N=1,2$ and set $G^\gamma := B_\gamma^c \times (-\infty,0)$ and $\partial_\ell G^\gamma := \partial B_\gamma^c \times (-\infty, 0)$. Let h_γ be the function

$$h_\gamma(x) = 1 - \frac{\gamma}{|x|},$$

and ψ_γ the solution of

$$\begin{aligned} \partial_t\psi_\gamma + \Delta\psi_\gamma &= 0 \quad \text{in } G^\gamma\\ \psi_\gamma &= 0 \quad \text{in } \partial_\ell G^\gamma\\ \psi_\gamma(.,0) &= h_\gamma \ \text{ in } B_\gamma^c. \end{aligned} \tag{1.175}$$

Then the function $\widetilde{\psi}(x,\tau) = \psi_\gamma(\gamma x, \gamma^2\tau)$ satisfies

$$\begin{aligned} \partial_t\widetilde{\psi} + \Delta\widetilde{\psi} &= 0 \quad \text{in } G^1\\ \widetilde{\psi} &= 0 \quad \text{in } \partial_\ell G^1\\ \widetilde{\psi}(.,0) &= h_1 \ \text{ in } B_1^c. \end{aligned} \tag{1.176}$$

By the maximum principle $\widetilde{\psi} \leq 1$ and by Hopf lemma

$$-\frac{\partial\widetilde{\psi}}{\partial\mathbf{n}}\lfloor_{\partial B_1\times[-c,0]} \geq \theta > 0, \tag{1.177}$$

where $\theta = \theta(N,c)$. Thus

$$-\frac{\partial\psi_\gamma}{\partial\mathbf{n}}\lfloor_{\partial B_\gamma\times[-\gamma^2,0]} \geq \frac{\theta}{\gamma}. \tag{1.178}$$

Multiplying the equation by $\psi_\gamma(x,\tau-t) = \psi_\gamma^*(x,\tau)$ and integrating on $B_\gamma^c \times (0,t)$ implies

$$\begin{aligned}\int_0^t\int_{B_\gamma^c} u^q\psi_r^* dx d\tau + \int_{B_\gamma^c}(uh_\gamma)(x,t)dx - \int_0^t\int_{\partial B_\gamma}\frac{\partial u}{\partial \mathbf{n}}\psi_\gamma^* dS d\tau \\ = -\int_0^t\int_{\partial B_\gamma}\frac{\partial \psi_\gamma^*}{\partial \mathbf{n}} u d\sigma d\tau.\end{aligned} \tag{1.179}$$

Since $0 \le \psi_\gamma^* \le 1$, we derive (1.174) from (1.178) and Proposition 1.3 since $B_\gamma^c \times (0,t) \subset \mathcal{E}_\gamma^c$, first by taking $t = T = \gamma^2 \ge (r+2\rho)^2$, and then for any $t \le \gamma^2$t if $N \ge 3$, we proceed as above except that we introduce a new function

$$h_\gamma(x) = 1 - \left(\frac{\gamma}{|x|}\right)^{N-2}.$$

This function is harmonic; thus the solution ψ_γ of (1.175) coincides with h_γ, and $\theta = N-2$ is independent of the length of the time interval. This ends the proof. □

The following estimates concerning solution of the heat equation are easy to obtain from the Gaussian integral representation and left to the reader.

Lemma 1.11 *L5I- Let M, $a > 0$ and $\eta \in L^\infty(\mathbb{R}^N)$ such that*

$$0 \le \eta(x) \le Me^{-a|x|^2} \qquad a.e.\ in\ \mathbb{R}^N. \tag{1.180}$$

Then, for any $t > 0$,

$$0 \le \mathbb{H}[\eta](x,t) \le \frac{M}{(4at+1)^{\frac{N}{2}}}e^{-\frac{a|x|^2}{4at+1}} \qquad for\ all\ x \in \mathbb{R}^N. \tag{1.181}$$

II- Let M, a, $b > 0$ and $\eta \in L^\infty(\mathbb{R}^N)$ such that

$$0 \le \eta(x) \le Me^{-a(|x|-b)_+^2} \qquad a.e.\ in\ \mathbb{R}^N. \tag{1.182}$$

Then, for any $t > 0$,

$$0 \le \mathbb{H}[\eta](x,t) \le \frac{Me^{-\frac{a(|x|-b)_+^2}{4at+1}}}{(4at+1)^{\frac{N}{2}}} \qquad for\ all\ x \in \mathbb{R}^N. \tag{1.183}$$

Lemma 1.12 *There exists a constant $C = C(N,q) > 0$ such that, for any $\eta \in \mathcal{T}_{r,\rho}(K)$, there holds*

$$\begin{aligned}&u(x,(r+2\rho)^2) \le \\ &C\max\left\{\frac{r+\rho}{(|x|-r-2\rho)^{N+1}}, \frac{|x|-r-2\rho}{(r+\rho)^{N+1}}\right\} e^{-\frac{(|x|-(r+2\rho))^2}{4(r+2\rho)^2}} \|R[\eta]\|_{L^{q'}}^{q'},\end{aligned} \tag{1.184}$$

for any $x \in \mathbb{R}^N \setminus B_{r+3\rho}$.

Proof The heat kernel in $B_1^c \times (0, \infty)$ with Dirichlet data on $\partial B_1^c \times (0, \infty)$ satisfies

$$H^{B_1^c}(x', x'; t', s') \leq C(t'-s')^{-\frac{N}{2}-1}(|x|-1)e^{-\frac{|x'-y'|^2}{4(t'-s')}} \quad \text{for } t' > s'. \tag{1.185}$$

If we denote $x = (r+2\rho)x'$ and $t = (r+2\rho)^2 t'$ for $(x, t) \in B_{r+2\rho}^c \times (0, T)$, then

$$u(x, t) \leq (|x| - r - 2\rho) \int_0^t \int_{\partial B_{r+\rho}} \frac{e^{-\frac{|x-y|^2}{4(t-s)}}}{(t-s)^{\frac{N}{2}+1}} dS(y)ds. \tag{1.186}$$

The right-hand side term in (1.186) is bounded from above by

$$\max\left\{\frac{C(|x|-r-2\rho)}{(t-s)^{1+\frac{N}{2}}} e^{-\frac{(|x|-r-2\rho)^2}{4(t-s)}} : s \in (0, t)\right\} \int_0^t \int_{\partial B_{r+2\rho}} u(y, s)d\sigma(y)ds. \tag{1.187}$$

We fix $t = (r+2\rho)^2$ and $|x| \geq r + 3\rho$. Since

$$\begin{aligned}
&\max\left\{\frac{e^{-\frac{(|x|-r-2\rho)^2}{4s}}}{s^{1+\frac{N}{2}}} : s \in (0, (r+2\rho)^2)\right\} \\
&\quad = (|x|-r-2\rho)^{-2-N} \max\left\{\frac{e^{-\frac{1}{4\sigma}}}{\sigma^{\frac{N}{2}+1}} : 0 < \sigma < \left(\frac{r+2\rho}{(|x|-r-2\rho)}\right)^2\right\},
\end{aligned}$$

a direct technical computation shows that

$$\max\left\{\frac{e^{-\frac{(|x|-r-2\rho)^2}{4s}}}{s^{1+\frac{N}{2}}} : s \in (0, (r+2\rho)^2)\right\} \leq C(N)\rho^{-2-N} e^{\left(\frac{|x|-r-2\rho}{2r+4\rho}\right)^2}. \tag{1.188}$$

Combining this estimate with (1.186), (1.187) and Lemma 1.20, one gets (1.184). □

Remark Since there exists $C > 0$ such that

$$(|x|-r-2\rho)e^{\left(\frac{|x|-r-2\rho}{2r+4\rho}\right)^2} \leq C\frac{(r+\rho)^2}{\rho} e^{-\left(\frac{|x|-r-3\rho}{2r+4\rho}\right)^2} \quad \text{for all } x \in B_{r+3\rho}^c, \tag{1.189}$$

the following variant of (1.184) holds for all $x \in B_{r+3\rho}^c$,

$$\begin{aligned}
&u(x, (r+2\rho)^2) \leq \\
&C \max\left\{\frac{(r+\rho)^3}{\rho(|x|-r-2\rho)^{N+1}}, \frac{1}{\rho(r+\rho)^{N+1}}\right\} e^{-\left(\frac{|x|-r-3\rho}{2r+4\rho}\right)^2} \|R[\eta]\|_{L^{q'}}^{q'}.
\end{aligned} \tag{1.190}$$

Next, we give a sharp pointwise upper bound of $u(x,t)$ when t is bounded from below.

Lemma 1.13 *There exists a constant $C = C(N,q) > 0$ such that for any $\eta \in \mathcal{T}_{r+\rho}(K)$ the following estimate holds:*

$$u(x,t) \leq \frac{C\widetilde{M}e^{-\frac{(|x|-r-3\rho)_+^2}{4t}}}{t^{\frac{N}{2}}} \|R[\eta]\|_{L^{q'}}^{q'} \quad \textit{for all } (x,t) \in \mathbb{R}^N \times [(r+\rho)^2, \infty), \tag{1.191}$$

where

$$\widetilde{M} = \widetilde{M}(x,r,\rho) = \begin{cases} \left(1+\frac{r}{\rho}\right)^{\frac{N}{2}} & \textit{if } |x| < r+3\rho \\ \frac{(r+\rho)^{N+3}}{\rho(|x|-r-2\rho)^{N+2}} & \textit{if } r+3\rho \leq |x| \leq c_N^*(r+2\rho) \\ 1+\frac{r}{\rho} & \textit{if } |x| \geq c_N^*(r+2\rho), \end{cases} \tag{1.192}$$

with $c_N^* = 1+\sqrt{4+2N}$.

Proof By the maximum principle

$$u(x,t) \leq \mathbb{H}[u(.,(r+2\rho)^2](x,t-(r+2\rho)^2) \quad \text{for any } t \geq (r+2\rho)^2.$$

By Lemma 1.9 and (1.190),

$$u(x,(r+2\rho)^2) \leq C\widetilde{M}e^{-\frac{(|x|-r-3\rho)^2}{4(r+2\rho)^2}} \|R[\eta]\|_{L^{q'}}^{q'},$$

where

$$M^\dagger = M^\dagger(x,r,\rho) = \begin{cases} (r(r+\rho)^{-\frac{N}{2}} & \text{if } |x| < r+3\rho \\ \frac{(r+\rho)^{N+3}}{\rho(|x|-r-2\rho)^{N+2}} & \text{if } r+3\rho \leq |x| \leq c_N^*(r+2\rho) \\ 1+\frac{r}{\rho} & \text{if } |x| \geq c_N^*(r+2\rho). \end{cases}$$

Applying Lemma 1.27 with $a = (2r+4\rho)^{-2}$, $b = r+3\rho$ and t replaced by $t - (r+2r)^2$ implies

$$u(x,t) \leq C\frac{M^\dagger(r+2\rho)^N}{t^{\frac{N}{2}}}e^{-\frac{(|x|-r-3\rho)^2}{4t}} \|R[\eta]\|_{L^{q'}}^{q'} \tag{1.193}$$

for $|x| \geq r+3\rho$ and $t \geq (r+2r)^2$, which implies (1.191). □

Finally we obtain an upper bound of $u(x,t)$ when t is not bounded from below.

Lemma 1.14 *There exists a constant* $C = C(N, q) > 0$ *such that for any* $\eta \in \mathcal{T}_{r+\rho}(K)$ *the following estimate holds when* $0 < t \leq (r + 2\rho)^2$,

$$u(x,t) \leq C(r+\rho)\max\left\{\frac{1}{(|x|-r-2\rho)^{N+1}}, \frac{1}{\rho t^{\frac{N}{2}}}\right\} e^{-\frac{(|x|-r-3\rho)^2}{4t}} \|R[\eta]\|^{q'}_{L^{q'}}, \tag{1.194}$$

for any $(x,t) \in \left(\mathbb{R}^N \setminus B_{r+3\rho}\right) \times (0, (r+2\rho)^2]$.

Proof From Lemma 1.20 we deduce by a simple modification of (1.184) that for any $|x| \geq r + 2\rho$, there holds

$$u(x,t) \leq C(|x|-r-2\rho)(r+2\rho)\max\left\{\frac{e^{-\frac{(|x|-r-2\rho)^2}{4s}}}{s^{1+\frac{N}{2}}} : 0 < s \leq t\right\} \|R[\eta]\|^{q'}_{L^{q'}}. \tag{1.195}$$

Next,

$$\max\left\{\frac{e^{-\frac{(|x|-r-2\rho)^2}{4s}}}{s^{1+\frac{N}{2}}} : 0 < s \leq t\right\}$$
$$= \begin{cases} (2N+4)^{1+\frac{N}{2}}(|x|-r-2\rho)^{-N-2}e^{-\frac{N+2}{2}} & \text{if } 0 < |x| \leq r+2\rho+\sqrt{2t(N+2)} \\ \dfrac{e^{-\frac{(|x|-r-2\rho)^2}{4t}}}{t^{1+\frac{N}{2}}} & \text{if } |x| > r+2\rho+\sqrt{2t(N+2)}. \end{cases}$$

When $x \in B^c_{r+3\rho}$, we have that

$$(|x|-r-2\rho)e^{-\frac{(|x|-r-2\rho)^2}{4t}} \leq e^{-\frac{(|x|-r-3\rho)^2}{4t}} \begin{cases} \rho e^{-\frac{\rho^2}{4t}} & \text{if } 2t < \rho^2 \\ \dfrac{2t}{\rho}e^{-1+\frac{\rho^2}{4t}} & \text{if } \rho^2 \leq 2t \leq 2(r+2\rho)^2. \end{cases}$$

However, since

$$\frac{\rho}{t}e^{-\frac{\rho^2}{4t}} \leq \frac{4}{\rho},$$

we derive

$$(|x|-r-2\rho)e^{-\frac{(|x|-r-2\rho)^2}{4t}} \leq \frac{Ct}{\rho}e^{-\frac{(|x|-r-3\rho)^2}{4t}},$$

and (1.194) follows. □

1.4.2.3 The Upper Wiener Test Estimate

Definition 1.7 We denote by δ_2 and δ_∞ the two parabolic distances

$$\begin{array}{ll} (i) & \delta_2[(x,t),(y,s)] = \sqrt{(x-y)^2 + |t-s|} \\ (ii) & \delta_\infty[(x,t),(y,s)] = \max\{|x-y|, \sqrt{|t-s|}\}. \end{array} \tag{1.196}$$

If $K \subset \mathbb{R}^N$ and $i = 2$ or ∞,

$$\delta_i[(x,t),K] = \inf\{\delta_i[(x,t),(y,0)] : y \in K\} = \begin{cases} \sqrt{\text{dist}^2(x,K) + |t|} & \text{if } i = 2 \\ \max\{\text{dist}(x,K), \sqrt{t}\} & \text{if } i = \infty. \end{cases}$$

For $\beta > 0$ and $i = 2$ or ∞, we denote by $\mathcal{B}^i_\beta$ the parabolic ball with centre $m = (x,t)$ and radius β in the metric δ_i.

If $K \subset \mathbb{R}^N$ is any compact we denote by $\overline{u}_K$ the maximal solution of (1.30) with initial trace $(K, 0)$.

If $m = (x,t) \in Q_T$ we set $d_K = \text{dist}(x,K)$, $D_K = \max\{|x-y| : y \in K\}$ and $\lambda = \sqrt{d_K^2 + t} = \delta_2(m,K)$. We define the slicing of K by setting $d_n = d_n(K,t) := \sqrt{nt}\ n \in \mathbb{N}_*$, $d_n^{\pm} = \left(\sqrt{nt} \pm \frac{\sqrt{t}}{\sqrt{n}}\right)_+$ and

$$T_n^* = \overline{B}_{d_{n+1}^+}(x) \setminus B_{d_n^-}(x)\,,\ T_n = \overline{B}_{d_{n+1}}(x) \setminus B_{d_n}(x) \quad \text{for all } \ n \in \mathbb{N},$$

thus $T_0^* = \overline{B}_{2\sqrt{t}}(x)$, $T_0 = \overline{B}_{\sqrt{t}}(x)$, and set

$$K_n := K_n(x,t) = K \cap T_n \quad \text{and} \quad Q_n := Q_n(x,t) = K \cap T_n.$$

The main result of this section is the following upper estimate.

Theorem 1.16 *Assume $q \geq q_c$, then there exists a positive constant $C = C(N,q,T)$ such that*

$$\overline{u}_K \leq \frac{C}{t^{\frac{N}{2}}} \sum_{n=0}^{a_j} d_{n+1}^{N-\frac{2}{q-1}} cap_{\frac{2}{q},q'}\left(\frac{K_n}{d_{n+1}}\right) \quad \textit{for all } (x,t) \in Q_T, \tag{1.197}$$

where a_j is the largest integer such that $K_j \neq \emptyset$.

We can assume that $x = 0$. Furthermore, in considering the scaling transformation T_ℓ with $\ell > 0$ we can assume $t = 1$. Thus the new compact singular set of the initial trace becomes $\frac{1}{\sqrt{\ell}}K$ that we still denote by K. For $n \in \mathbb{N}_*$ set $\delta_n = d_{n+1} - d_n$, then $\frac{1}{2\sqrt{n+1}} \leq \delta_n \leq \frac{1}{2\sqrt{n}}$. By convention $\delta_0 = 1$. It is possible to exhibit a collection Θ_n of points $a_{n,j}$ with centre on the sphere $\Sigma_n = \{y \in \mathbb{R}^N : |y| = (d_{n+1} + d_n)/2\}$, such that

$$T_n \subset \bigcup_{a_{n,j} \in \Theta_n} B_{\delta_n}(a_{n,j}), \quad \left|a_{n,j} - a_{n,k}\right| \geq \delta_n \ \text{ and } \ \#\Theta_n \leq Cn^{N-1},$$

for some constant $C = C(N)$. If $K_{n,j} = K_n \cap B_{\delta_n}(a_{n,j})$, there holds

$$K = \bigcup_{0 \leq n \leq a_K} \bigcup_{a_{n,j} \in \Theta_n} K_{n,j}.$$

The first intermediate step is based on the *quasi-additivity* property of capacities developed in [2].

Lemma 1.15 *Let $q \geq q_c$. There exists a constant $C = C(N,q)$ such that*

$$\sum_{a_{n,j}\in\Theta_n} \dot{R}^{B_{2\delta_n}(a_{n,j})}_{2/q,q'}(K_{n,j}) \leq C d_{n+1}^{N-\frac{2}{q-1}} cap_{\frac{2}{q},q'}\left(\frac{K_n}{d_{n+1}}\right) \qquad \forall n \in \mathbb{N}_*. \tag{1.198}$$

Proof The following result is proved in [2, Th 3]: if the spheres $B_{\rho_j^\theta}(b_j)$, $\theta = 1 - \frac{2}{N(q-1)}$, are disjoint in $\mathbb{R}^N$ and G is a Borel (more generally an analytic) subset of $\bigcup_j B_{\rho_j}(b_j)$ where the ρ_j are positive numbers smaller than some $\rho^* > 0$, there holds

$$cap_{\frac{2}{q},q'}(G) \leq \sum_j cap_{\frac{2}{q},q'}(G \cap B_{\rho_j}(b_j)) \leq A cap_{\frac{2}{q},q'}(G), \tag{1.199}$$

for some A depending on N, q and ρ^*. This property is called *quasi-additivity*. We define for $n \in \mathbb{N}_*$,

$$\widetilde{T}_n = d_{n+1} T_n, \quad \widetilde{K}_n = d_{n+1} K_n \quad \text{and} \quad \widetilde{Q}_n = d_{n+1} Q_n.$$

Since $K_{n,j} \subset B_{\delta_n}(a_{n,j})$, it follows that

$$\widetilde{K}_{n,j} := d_{n+1} K_{n,j} \subset B_{d_{n+1}\delta_n}(\widetilde{a}_{n,j}).$$

Note that by Lemma 1.6

$$\begin{aligned} \dot{R}^{B_{2\delta_n}(a_{n,j})}_{2/q,q'}(K_{n,j}) &= d_{n+1}^{\frac{2}{q-1}-N} \dot{R}^{B_{2\delta_n d_{n+1}}(d_{n+1}a_{n,j})}_{2/q,q'}(\widetilde{K}_{n,j}) \\ &\approx d_{n+1}^{\frac{2}{q-1}-N} cap^{B_{2\delta_n d_{n+1}}(d_{n+1}a_{n,j})}_{2/q,q'}(\widetilde{K}_{n,j}) \\ &\approx d_{n+1}^{\frac{2}{q-1}-N} cap_{\frac{2}{q},q'}(\widetilde{K}_{n,j}), \end{aligned} \tag{1.200}$$

where $\widetilde{K}_{n,j} = d_{n+1} K_{n,j}$. For a fixed $n > 0$ and each repartition Λ of points $\widetilde{a}_{n,j} = d_{n+1}\, a_{n,j}$ such that the balls $B_{2^\theta}(\widetilde{a}_{n,j})$ are disjoint, the quasi-additivity property holds: if we set

$$K_{n,\Lambda} = \bigcup_{a_{n,j}\in\Lambda} K_{n,j}\,, \quad \widetilde{K}_{n,\Lambda} = d_{n+1}\, K_{n,\Lambda} = \bigcup_{a_{n,j}\in\Lambda} \widetilde{K}_{n,j} \quad \text{and} \quad \widetilde{K}_n = d_{n+1}\, K_n,$$

then

$$\sum_{a_{n,j}\in\Lambda} cap_{\frac{2}{q},q'}(\widetilde{K}_{n,j}) \approx cap_{\frac{2}{q},q'}(\widetilde{K}_{n,\Lambda}). \tag{1.201}$$

The maximal cardinal of any such repartition Λ is of the order of Cn^{N-1} for some positive constant $C = C(N)$; therefore the number of repartitions needed for a full covering of the set $\widetilde{T}_n$ is of finite order depending only on the dimension. Because $\widetilde{K}_n$ is the union of the $\widetilde{K}_{n,\Lambda}$,

$$\sum_{a_{n,j}\in\Theta_n} cap_{\frac{2}{q},q'}(\widetilde{K}_{n,j}) = \sum_{\Lambda}\sum_{a_{n,j}\in\Lambda} cap_{\frac{2}{q},q'}(\widetilde{K}_{n,j}) \approx cap_{\frac{2}{q},q'}(\widetilde{K}_n). \qquad (1.202)$$

By Lemma 1.6,

$$\begin{aligned} cap_{\frac{2}{q},q'}(\widetilde{K}_n) \leq cap_{2/q,q'}^{B_{2d_{n+1}}}(\widetilde{K}_n) &\approx d_{n+1}^{N-\frac{1}{q-1}} cap_{2/q,q'}^{B_2}\left(\frac{K_n}{d_{n+1}}\right) \\ &\approx d_{n+1}^{N-\frac{1}{q-1}} cap_{\frac{2}{q},q'}\left(\frac{K_n}{d_{n+1}}\right), \end{aligned}$$

we obtain (1.198) by combining this last inequality with (1.200) and (1.202). □

Proof of Theorem 1.16. *Step 1.* We first notice that

$$\overline{u}_K \leq \sum_{0\leq n\leq a_K}\sum_{a_{n,j}\in\Theta_n} \overline{u}_{K_{n,j}}. \qquad (1.203)$$

Actually, since $K = \bigcup_n \bigcup_{a_{n,j}} K_{n,j}$, there holds $\overline{K_{\epsilon'}} \subset \bigcup_n \bigcup_{a_{n,j}} K_{n,j\,\epsilon}$ for any $0 < \epsilon' < \epsilon$. Because a finite sum of positive solutions of (1.30) is a supersolution,

$$\overline{u}_{K_{\epsilon'}} \leq \sum_{0\leq n\leq a_K}\sum_{a_{n,j}\in\Theta_n} \overline{u}_{K_{n,j\,\epsilon}}. \qquad (1.204)$$

Letting successively ϵ' and ϵ go to 0 implies (1.203).

Step 2. Let $n \in \mathbb{N}$. Since $K_{n,j} \subset B_{\delta_n}(a_{n,j})$ and $|x - a_{n,j}| = (d_n + d_{n+1})/2$, we can apply the previous lemmas with $r = \delta_n$ and $\rho = r$. For $n \geq n_N$, there holds $t = 1 \geq (r + 2\rho)^2 = 9/(n+1)$ and $|x - a_{n,j}| = (\sqrt{n+1} - \sqrt{n})/2 \geq (2 + C_N)(3/\sqrt{n+1})$ (notice that $n_N \geq 8$). Thus

$$u_{K_{n,j}}(0,1) \leq Ce^{(\sqrt{n}-3/\sqrt{n+1})^2/4} R_{2/q,q'}^{B_{2\delta_n}(a_{n,j})}(K_{n,j}) \leq Ce^{3/2}e^{-\frac{n}{4}} R_{2/q,q'}^{B_{2\delta_n}(a_{n,j})}(K_{n,j}). \qquad (1.205)$$

Using Lemma 1.15 we obtain, with $d_n = d_n(1) = \sqrt{n+1}$,

$$\sum_{n=n_N}^{a_K}\sum_{a_{n,j}\in\Theta_n} u_{K_{n,j}}(0,1) \leq C\sum_{n=n_N}^{a_K} d_{n+1}^{N-\frac{2}{q-1}} e^{-\frac{n}{4}} cap_{\frac{2}{q},q'}\left(\frac{K_n}{d_{n+1}}\right). \qquad (1.206)$$

Finally, we apply Lemma 1.9 if $1 \leq n < n_N$ and get

$$
\begin{aligned}
\sum_{1}^{n_N-1}\sum_{a_{n,j}\in\Theta_n} u_{K_{n,j}}(0,1) &\leq C\sum_{1}^{n_N-1} cap_{\frac{2}{q},q'}\left(\frac{K_n}{d_{n+1}}\right)\\
&\leq C'\sum_{1}^{n_N-1} d_{n+1}^{N-\frac{2}{q-1}} e^{-\frac{n}{4}} cap_{\frac{2}{q},q'}\left(\frac{K_n}{d_{n+1}}\right).
\end{aligned}
\tag{1.207}
$$

For $n=0$, we proceed similarly, in splitting K_1 in a finite number of sets $K_{1,i}$, depending only on the dimension, such that $\operatorname{diam} K_{1,i} < 1/3$. Combining (1.206) and (1.207), we derive

$$
\overline{u}_K(0,1) \leq C\sum_{n=0}^{a_K} d_{n+1}^{N-\frac{2}{q-1}} e^{-\frac{n}{4}} cap_{\frac{2}{q},q'}\left(\frac{K_n}{d_{n+1}}\right). \tag{1.208}
$$

In order to derive the same result for any $t>0$, we notice that

$$
\overline{u}_K(y,t) = t^{-\frac{1}{q-1}}\overline{u}_{K/\sqrt{t}}(y/\sqrt{t},1).
$$

Going back to the definition of $d_n = d_n(K,t) = \sqrt{nt} = d_n(K\sqrt{t},1)$, we derive from (1.208) and the fact that $a_{K,t} = a_{K\sqrt{t},1}$

$$
\overline{u}_K(0,t) \leq Ct^{-\frac{N}{2}}\sum_{n=0}^{a_K} d_{n+1}^{N-\frac{2}{q-1}} e^{-\frac{n}{4}} cap_{\frac{2}{q},q'}\left(\frac{K_n}{d_{n+1}}\right), \tag{1.209}
$$

with $d_n = d_n(t) = \sqrt{t(n+1)}$. This is (1.197) with $x=0$, and a space translation leads to the final result. □

Proof of Theorem 1.15. Let $m>0$ and $F_m = F\cap\overline{B}_m$. We denote by $U_{B_m^c}$ the maximal solution of (1.30) in Q_∞ the initial trace of which vanishes on B_m. It is straightforward by scaling to verify that such a solution is actually the unique positive solution of (1.30) which satisfies

$$
\lim_{t\to 0} u(x,t) = \infty
$$

uniformly on $B_{m'}^c$, for any $m'>m$. Furthermore

$$
\lim_{m\to\infty} U_{B_m^c}(y,t) = \lim_{m\to\infty} m^{-\frac{2}{q-1}} U_{B_1^c}(y/m, t/m^2) = 0,
$$

uniformly on any compact subset of $\overline{Q}_\infty$. Since $\overline{u}_{F_m} + U_{B_m^c}$ is a supersolution, it is larger that $\overline{u}_F$ and therefore $\overline{u}_{F_m}\uparrow\overline{u}_F$. Because $W_{F_m}(x,t)\leq W_F(x,t)$ and $\overline{u}_{F_m}\leq C_1W_{F_m}(x,t)$, the result follows. □

Remark It is clear that Theorem 1.15 still holds if u is a positive subsolution of (1.30) satisfying the initial trace condition (1.165).

The Bessel capacitary potential admits an integral form. The next result is a variant of Theorem 1.15.

Theorem 1.17 *Assume $q \geq q_c$. Then there exists a positive constant $C_1^* = C_1^*(N, q, T)$ such that for any closed subset $F \subset \mathbb{R}^N$ there holds for all $(x,t) \in Q_T$,*

$$\overline{u}_F(x,t) \leq \frac{C_1^*}{t^{1+\frac{N}{2}}} \int_{\sqrt{t}}^{\sqrt{t(a_t+2}} e^{-\frac{s^2}{4t}} s^{N-\frac{2}{q-1}} cap_{\frac{2}{q},q'}\left(\frac{1}{s} B \cap B_1(x)\right) sds, \tag{1.210}$$

where $a_t = \min\left\{n \in \mathbb{N} : F \subset B_{\sqrt{(n+1)t}}(x)\right\}$.

Proof We use the inequality

$$cap_{\frac{2}{q},q'}\left(\frac{F_n}{d_{n+1}}\right) \leq cap_{\frac{2}{q},q'}\left(\frac{F}{d_{n+1}} \cap B_1\right),$$

and we set

$$\phi(s) = cap_{\frac{2}{q},q'}\left(\frac{F}{s} \cap B_1\right) \quad \text{for all } s > 0. \tag{1.211}$$

Step 1. By [1, 32], there exists $c = c(N, q) > 0$ such that

$$\frac{1}{c}\phi(\alpha s) \leq \phi(s) \leq c\phi(\beta s) \quad \text{for all } s > 0 \text{ and } \frac{1}{2} \leq \alpha \leq 1 \leq \beta \leq 2. \tag{1.212}$$

Actually, if $\beta \in [1, 2]$,

$$\phi(\beta s) = cap_{\frac{2}{q},q'}\left(\frac{1}{\beta}\left(\frac{1}{s}F \cap B_\beta(x)\right)\right) \approx cap_{\frac{2}{q},q'}\left(\frac{1}{s}F \cap B_\beta(x)\right) \geq \frac{1}{c}\phi(s),$$

and if $\alpha \in [\frac{1}{2}, 1]$,

$$\phi(\alpha s) = cap_{\frac{2}{q},q'}\left(\frac{1}{\alpha}\left(\frac{1}{s}F \cap B_\alpha(x)\right)\right) \approx cap_{\frac{2}{q},q'}\left(\frac{1}{s}F \cap B_\alpha(x)\right) \leq c\phi(s).$$

Step 2. By (1.212)

$$cap_{\frac{2}{q},q'}\left(\frac{F}{d_{n+1}} \cap B_1(x)\right) \leq c cap_{\frac{2}{q},q'}\left(\frac{F}{s} \cap B_1(x)\right) \quad \text{for all } s \in [d_{n+1}, d_{n+2}],$$

and $n \leq a_t$. Then

$$\begin{aligned} c\int_{d_{n+1}}^{d_{n+2}} & s^{N-\frac{2}{q-1}} e^{-\frac{s^2}{4t}} cap_{\frac{2}{q},q'}\left(\frac{F}{s} \cap B_1(x)\right) sds \\ & \geq cap_{\frac{2}{q},q'}\left(\frac{F}{d_{n+1}} \cap B_1(x)\right) \int_{d_{n+1}}^{d_{n+2}} s^{N-\frac{2}{q-1}} e^{-\frac{s^2}{4t}} sds. \end{aligned}$$

Because $N - \frac{2}{q-1} \geq 0$ as $q \geq q_c$, we obtain

$$\int_{d_{n+1}}^{d_{n+2}} s^{N-\frac{2}{q-1}} e^{-\frac{s^2}{4t}} s ds \geq e^{-\frac{n+2}{4}} d_{n+1}^{N-\frac{2}{q-1}} (d_{n+2} - d_{n+1}) \\ \geq \frac{t d_{n+1}^{N-\frac{2}{q-1}} e^{-\frac{n}{2}}}{4e^2}, \tag{1.213}$$

which implies (1.210). □

1.4.3 Estimate from Below

If μ is a bounded nonnegative q-admissible measure, we recall that u_μ is the solution of (1.114). The maximal σ-moderate solution of (1.30) with an initial trace vanishing outside a closed set $F \subset \mathbb{R}^N$ is denoted by $\underline{u}_F$ and defined by

$$\underline{u}_F = \sup\{u_\mu : \mu \in \mathfrak{M}_+^b, \text{ and } q\text{-admissible s.t. } \mu(F^c) = 0\}. \tag{1.214}$$

The main result of this section is as follows.

Theorem 1.18 *Let $q \geq q_c$ and $T > 0$. Then there exists a positive constant $C = C(N, q, T)$ such that for any closed set $F \subset \mathbb{R}^N$,*

$$\underline{u}_F(x, t) \geq C W_F(x, t) \quad \text{for all } (x, t) \in Q_\infty. \tag{1.215}$$

We first assume that F is compact and we denote it by K. If μ is q- admissible and nonnegative, then $u_\mu \leq \mathbb{H}[\mu]$. Since

$$u_\mu = \mathbb{H}[\mu] - \mathbb{G}[u^q],$$

where $\mathbb{G}$ is the Green parabolic heat potential, defined by

$$\mathbb{G}[f](x, t) = \int_0^t \mathbb{H}[f(., s)](x, t - s) ds \\ = \int_0^t \int_{\mathbb{R}^N} \frac{1}{(4\pi(t-s))^{\frac{N}{2}}} e^{-\frac{|x-y|^2}{4(t-s)}} f(y, s) dy ds \tag{1.216}$$

for all $(x, t) \in Q_\infty$, there holds

$$
\begin{aligned}
u_\mu(x,t) &\geq \mathbb{H}[\mu](x,t) - \mathbb{G}[(\mathbb{H}[\mu])^q](x,t) \\
&\geq \left(\frac{1}{4\pi t}\right)^{\frac{N}{2}} \int_{\mathbb{R}^N} e^{-\frac{|x-y|^2}{4t}} d\mu(y) \\
&- \int_0^t \int_{\mathbb{R}^N} \frac{1}{(4\pi(t-s))^{\frac{N}{2}}} e^{-\frac{|x-y|^2}{4(t-s)}} \left(\frac{1}{(4\pi s)^{\frac{N}{2}}} \int_{\mathbb{R}^N} e^{-\frac{|y-z|^2}{4s}} d\mu(z)\right)^q dyds
\end{aligned} \tag{1.217}
$$

for all $(x,t) \in Q_\infty$. The main idea of the proof is as follows: for any $(x,t) \in Q_T$ we construct a q - admissible bounded measure $\mu = \mu_{x,t}$ such that

$$
\mathbb{H}[\mu_{x,t}](x,t) \geq C W_K(x,t), \tag{1.218}
$$

and

$$
\mathbb{G}[(\mathbb{H}[\mu_{x,t}])^q] \leq C\mathbb{H}[\mu_{x,t}] \quad \text{in } Q_\infty, \tag{1.219}
$$

with constants C depending only on N, q and T. From this first estimate we replace $\mu_{x,t}$ by $\epsilon\mu_{x,t}$ with $\epsilon = (2C)^{-\frac{1}{q-1}}$ in order to obtain

$$
u_{\epsilon\mu_{x,t}}(x,t) \geq \frac{1}{2}\mathbb{H}[\epsilon\mu_{x,t}](x,t) \geq \frac{C}{2} W_K(x,t).
$$

If such an estimate holds for any $(x,t) \in Q_\infty$, it will follow by the definition of $\underline{u}_K$ that

$$
\underline{u}_K \geq \frac{C}{2} W_K. \tag{1.220}
$$

1.4.3.1 Estimate from Below of the Solution of the Heat Equation

The slicing of $\mathbb{R}^N$ used in the previous section is the intersection with $\mathbb{R}^N \times \{0\}$ of an extended slicing of Q_T that we construct as follows: if K is a compact subset of $\mathbb{R}^N$, $m = (x,t)$, we define d_K, λ, d_n as in Definition 1.7 and a_t as in Theorem 1.17. Let $\alpha \in (0,1)$ to be fixed later on. For $n \in \mathbb{Z}$ we set

$$
\mathcal{T}_n = \begin{cases} \mathcal{B}^2_{\sqrt{t(n+1)}}(m) \setminus \mathcal{B}^2_{\sqrt{tn}}(m) & \text{if } n \geq 1 \\ \mathcal{B}^2_{\alpha^{-n}\sqrt{t}}(m) \setminus \mathcal{B}^2_{\alpha^{1-n}\sqrt{t}}(m) & \text{if } n \leq 0, \end{cases}
$$

and

$$
\mathcal{T}^*_n = \mathcal{T}_n \cap Q_t \quad \text{if } n \in \mathbb{Z}.
$$

For any $n \in \mathbb{N}_*$ and $m = (x,t) \in Q_T$, we recall that

$$
Q_n = K \cap \mathcal{B}^2_{\sqrt{t(n+1)}}(m) = K \cap B_{d_{n+1}}(x),
$$

and

$$K_n = K \cap \mathcal{T}_{n+1} = K \cap \left(B_{d_{n+1}}(x) \setminus B_{d_n}(x)\right).$$

Let $\nu_n \in \mathfrak{M}_+^b(\mathbb{R}^N) \cap B^{-\frac{2}{q},q'}(\mathbb{R}^N)$ be the capacitary measure of the set $d_{n+1}^{-1}K_n$ (see [1, Section 2.2]). Then ν_n vanishes outside $d_{n+1}^{-1}K_n$ and satisfies

$$\nu_n(d_{n+1}^{-1}K_n) = cap_{\frac{2}{q},q'}(d_{n+1}^{-1}K_n) \quad \text{and} \quad \|\nu_n\|_{B^{-\frac{2}{q},q'}(\mathbb{R}^N)} = \left(cap_{\frac{2}{q},q'}(d_{n+1}^{-1}K_n)\right)^{\frac{1}{q}}. \tag{1.221}$$

Let μ_n be defined on any Borel set $A \subset K_n$ by

$$\mu_n(A) = d_{n+1}^{N-\frac{2}{q-1}} \nu_n(d_{n+1}^{-1}A). \tag{1.222}$$

We set

$$\mu_{t,K} = \sum_{n=0}^{a_t} \mu_n,$$

and

$$\mathbb{H}[\mu_{t,K}] = \sum_{n=0}^{a_t} \mathbb{H}[\mu_n].$$

Proposition 1.4 *Let $q \geq q_c$, then there holds*

$$\mu_{t,K} \geq \frac{1}{(4\pi t)^{\frac{N}{2}}} \sum_{n=0}^{a_t} e^{-\frac{n+1}{4}} d_{n+1}^{N-\frac{2}{q-1}} cap_{\frac{2}{q},q'}\left(\frac{d_{n+1}}{K_n}\right) \tag{1.223}$$

for all $(x,t) \in Q_T$.

Proof We have

$$\mathbb{H}[\mu_n](x,t) = \frac{1}{(4\pi t)^{\frac{N}{2}}} \int_{K_n} e^{-\frac{|x-y|^2}{4t}} d\mu_n(y).$$

Furthermore

$$\begin{aligned}\int_{K_n} e^{-\frac{|x-y|^2}{4t}} d\mu_n(y) &\leq \left(\max\{e^{-\frac{|x-y|^2}{4t}} : y \in K_n\}\right) \mu_n(K_n) \\ &\leq \left(\max\{e^{-\frac{|x-y|^2}{4t}} : y \in K_n\}\right) d_{n+1}^{N-\frac{2}{q-1}} cap_{\frac{2}{q},q'}(d_{n+1}^{-1}K_n),\end{aligned}$$

by (1.221) and (1.222). Thus

$$\int_{K_n} e^{-\frac{|x-y|^2}{4t}} d\mu_n(y) \leq e^{-\frac{n+1}{4}} d_{n+1}^{N-\frac{2}{q-1}} cap_{\frac{2}{q},q'}(d_{n+1}^{-1}K_n),$$

by the definition of K_n and d_n, and (1.223) follows by the definition of $\mu_{t,K}$. □

1.4.3.2 Estimate from Above of the Nonlinear Term

We write (1.217) under the form

$$\begin{aligned} u_\mu(x,t) &\geq \sum_{n\in\mathbb{Z}} \mathbb{H}[\mu_n](x,t) - \int_0^t \int_{\mathbb{R}^N} H(x,y,t-s) \left(\sum_{n\in A_K} \mathbb{H}[\mu_n](y,s) \right)^q dyds \\ &= I_1 - I_2. \end{aligned} \tag{1.224}$$

We recall that $\mu_n = 0$ if $n \notin A_K = \mathbb{N} \cap [1, a_t]$. Then

$$\begin{aligned} I_2 &= \frac{1}{(4\pi)^{\frac{N}{2}}} \int_0^t \int_{\mathbb{R}^N} (t-s)^{\frac{N}{2}} e^{-\frac{|x-y|^2}{4(t-s)}} \left(\sum_{n\in A_K} \mathbb{H}[\mu_n](y,s) \right)^q dyds \\ &\leq \frac{2^{q-1}}{(4\pi)^{\frac{N}{2}}} (J_\ell + J'_\ell), \end{aligned} \tag{1.225}$$

where

$$J_\ell = \sum_{p\in\mathbb{Z}} \iint_{\mathcal{T}_p^*} (t-s)^{\frac{N}{2}} e^{-\frac{|x-y|^2}{4(t-s)}} \left(\sum_{n\leq p+\ell} \mathbb{H}[\mu_n](y,s) \right)^q dyds,$$

and

$$J'_\ell = \sum_{p\in\mathbb{Z}} \iint_{\mathcal{T}_p^*} (t-s)^{\frac{N}{2}} e^{-\frac{|x-y|^2}{4(t-s)}} \left(\sum_{n> p+\ell} \mathbb{H}[\mu_n](y,s) \right)^q dyds.$$

In these expressions $\ell \in \mathbb{N}$ will be fixed later on.

Lemma 1.16 *Let* $0 < a < b$ *and* $t > 0$, *then*

$$\max\left\{ \sigma^{-\frac{N}{2}} e^{-\frac{\rho^2}{4\sigma}} : 0 \leq \sigma \leq t,\ at \leq \rho^2 + \sigma \leq bt \right\} = \begin{cases} t^{-\frac{N}{2}} e^{-\frac{a}{4}} & \text{if } \frac{a}{2N} > 1 \\ \left(\frac{2N}{at}\right)^{\frac{N}{2}} e^{-\frac{N}{2}} & \text{if } \frac{a}{2N} \leq 1. \end{cases} \tag{1.226}$$

Proof Set

$$\mathcal{J}(\rho,\sigma) = \sigma^{-\frac{N}{2}} e^{-\frac{\rho^2}{4\sigma}},$$

and

$$\mathcal{K}_{a,b,t} = \{(\rho,\sigma) \in [0,\infty) \times (0t] : at \leq \rho^2 + \sigma \leq bt\}.$$

We notice that, for fixed σ, the maximum of $\mathcal{J}(.,\sigma)$ is achieved when ρ is minimal. If $\sigma \in [at, bt]$, the minimal value of ρ is zero, while if $\sigma \in (0, at)$, the minimal value is $\sqrt{at - s}$.

- Assume first $a \geq 1$, then $\mathcal{J}(at-\sigma,\sigma) = e^{\frac{1}{4}}\sigma^{-\frac{N}{4}}e^{-\frac{at}{4\sigma}}$. Thus if $1 \leq \frac{a}{2N}$, the minimal value of $\mathcal{J}(\sqrt{at-\sigma},\sigma)$ is $e^{\frac{1-2N}{4}}\left(\frac{2N}{at}\right)^{\frac{N}{2}}$, while if $\frac{a}{2N} < 1$, this minimum is $e^{\frac{1}{4}}t^{-\frac{N}{2}}e^{-\frac{a}{4}}$.

- Assume now $a < 1$, then

$$\begin{aligned}\max\left\{\mathcal{J}(\rho,\sigma) : (\rho,\sigma) \in \mathcal{K}_{a,b,t}\right\} &= \max\left\{\max_{\sigma\in(at,t}\mathcal{J}(0,\sigma),\ \max_{\sigma\in(0,at}\mathcal{J}(\sqrt{at-\sigma},\sigma)\right\}\\ &= \max\left\{(at)^{-\frac{N}{2}}, e^{-\frac{1-2N}{4}}\left(\frac{2N}{at}\right)^{\frac{N}{2}}\right\}\\ &= e^{-\frac{1-2N}{4}}\left(\frac{2N}{at}\right)^{\frac{N}{2}}.\end{aligned}$$

From these two estimates, (1.226) follows. □

Remark The following variant of Lemma 1.16 will be useful in the sequel: *For any* $\theta > \frac{1}{2N}$, *there holds*

$$\max\left\{\mathcal{J}(\rho,\sigma) : (\rho,\sigma) \in \mathcal{K}_{a,b,t}\right\} \leq e^{\frac{1}{4}}\left(\frac{2N\theta}{t}\right)^{\frac{N}{2}} e^{\frac{a}{4}} \quad \text{if } \theta\, a \geq 1. \tag{1.227}$$

Lemma 1.17 *There exists a positive constant* $C = C(N, q, \ell, T)$ *such that*

$$J_\ell \leq Ct^{-\frac{N}{2}}\sum_{n=1}^{a_t} d_{n+1}^{N-\frac{2}{q-1}} e^{-\frac{1+(n-\ell)_+}{4}} cap_{\frac{2}{q},q'}\left(\frac{K_n}{d_{n+1}}\right). \tag{1.228}$$

Proof The set of the indices p for the summation in J_ℓ is reduced to $\mathbb{Z}\cap[-\ell+2,\infty)$, thus there holds $J_\ell = J_{1,\ell} + J_{2,\ell}$ where

$$J_{1,\ell} = \sum_{p=2-\ell}^{0}\iint_{\mathcal{T}_p^*}(t-s)^{-\frac{N}{2}}e^{-\frac{|x-y|^2}{4(t-s)}}\left(\sum_{n<p+\ell}\mathbb{H}[\mu_n](y,s)\right)^q dyds$$

and

$$J_{2,\ell} = \sum_{p=1}^{\infty}\iint_{\mathcal{T}_p^*}(t-s)^{-\frac{N}{2}}e^{-\frac{|x-y|^2}{4(t-s)}}\left(\sum_{n<p+\ell}\mathbb{H}[\mu_n](y,s)\right)^q dyds.$$

If $p = 2 - p, ..., 0$,

$$(y,s) \in \mathcal{T}_p^* \Longrightarrow t\alpha^{2-2p} \leq |x-y|^2 + t - s \leq t\alpha^{-2p},$$

while if $p \geq 1$,

$$(y, s) \in \mathcal{T}_p^* \Longrightarrow pt \leq |x-y|^2 + t - s \leq (p+1)t.$$

By Lemma 1.16 and (1.227), there exists a positive constant $C = C(N, \ell, \alpha)$ such that

$$\max\left\{(t-s)^{-\frac{N}{2}} e^{-\frac{|x-y|^2}{4(t-s)}} : (y, s) \in \mathcal{T}_p^*\right\} \leq Ct^{-\frac{N}{2}} e^{-\frac{\alpha^{2-p}}{4}}, \tag{1.229}$$

whenever $p = 2-\ell, \ldots, 0$, and

$$\max\left\{(t-s)^{-\frac{N}{2}} e^{-\frac{|x-y|^2}{4(t-s)}} : (y, s) \in \mathcal{T}_p^*\right\} \leq Ct^{-\frac{N}{2}} e^{-\frac{p}{4}}, \tag{1.230}$$

when $p \geq 1$.
When $p = 2-\ell, \ldots, 0$

$$\left(\sum_{n=1}^{p-\ell-1} \mathbb{H}[\mu_n](y, s)\right)^q \leq C \sum_{n=1}^{p-\ell-1} \left(\mathbb{H}[\mu_n]\right)^q (y, s), \tag{1.231}$$

where $C = C(q, \ell) > 0$, therefore

$$\begin{aligned} J_{1,\ell} &\leq Ct^{-\frac{N}{2}} \sum_{p=2-\ell}^{0} e^{-\frac{\alpha^{2-p}}{4}} \sum_{n=1}^{p-\ell-1} \|\mathbb{H}[\mu_n]\|^q_{L^q(Q_T)} \\ &\leq Ct^{-\frac{N}{2}} \sum_{n=1}^{\ell-1} \|\mathbb{H}[\mu_n]\|^q_{L^q(Q_T)} \sum_{p=n-\ell+1}^{0} e^{-\frac{\alpha^{2-p}}{4}} \\ &\leq Ct^{-\frac{N}{2}} e^{-\frac{\alpha^{2\ell-2}}{4}} \sum_{n=1}^{\ell-1} \|\mathbb{H}[\mu_n]\|^q_{L^q(Q_T)} . \end{aligned} \tag{1.232}$$

When the set of indices p is not upper bounded, we introduce some extra parameter to be made precise later on. Then

$$\left(\sum_{n=1}^{p-\ell-1} \mathbb{H}[\mu_n](y, s)\right)^q \leq \left(\sum_{n=1}^{p-\ell-1} e^{\frac{q'\delta n}{4}}\right)^{\frac{q}{q'}} \sum_{n=1}^{p-\ell-1} e^{-\frac{q\delta n}{4}} \left(\mathbb{H}[\mu_n]\right)^q (y, s). \tag{1.233}$$

Remembering that $\mu_n = 0$ if $n \geq a_t$, we obtain that there exists $C > 0$ depending also on δ such that

$$
\begin{aligned}
J_{2,\ell} &\leq Ct^{-\frac{N}{2}}\sum_{p=1}^{\infty}e^{\frac{\delta(p+\ell-1)q-p}{4}}\sum_{n=1}^{p-\ell-1}e^{-\frac{q\delta n}{4}}\|\mathbb{H}[\mu_n]\|^q_{L^q(Q_T)}\\
&\leq Ct^{-\frac{N}{2}}\sum_{n=1}^{\infty}e^{-\frac{q\delta n}{4}}\|\mathbb{H}[\mu_n]\|^q_{L^q(Q_T)}\sum_{p=n-\ell+1\vee 1}^{\infty}e^{\frac{\delta(p+\ell-1)q-p}{4}}\\
&\leq Ct^{-\frac{N}{2}}\sum_{n=1}^{\infty}e^{-\frac{1+(n-\ell)_+}{4}}\|\mathbb{H}[\mu_n]\|^q_{L^q(Q_T)}\,.
\end{aligned}
\tag{1.234}
$$

We chose δ such that $\delta\ell q < 1$. Combining (1.232) and (1.234) and using Lemma 1.3 and (1.221) and (1.222) we obtain (1.228). □

The set of indices p such that the term μ_n is not zero in the summation J'_ℓ is $\mathbb{Z}\cap(-\infty, a_t-\ell]$. We write

$$
J'_\ell = J'_{1,\ell} + J'_{2,\ell}
$$

with

$$
J'_{1,\ell} = \sum_{-\infty}^{p=0}\int\!\!\int_{\mathcal{T}^*_p}(t-s)^{-\frac{N}{2}}e^{-\frac{|x-y|^2}{4(t-s)}}\left(\sum_{n=1\vee p+\ell}^{\infty}\mathbb{H}[\mu_n](y,s)\right)^q
$$

and

$$
J'_{2,\ell} = \sum_{p=1}^{a_t-\ell}\int\!\!\int_{\mathcal{T}^*_p}(t-s)^{-\frac{N}{2}}e^{-\frac{|x-y|^2}{4(t-s)}}\left(\sum_{n=p+\ell}^{\infty}\mathbb{H}[\mu_n](y,s)\right)^q.
$$

Lemma 1.18 *There exists a positive constant $C = C(N, q, \ell)$ such that*

$$
J'_{1,\ell} \leq Ct^{1-\frac{N}{q}2}\sum_{n=0}^{a_t}e^{-\frac{(1+\beta_0)(n-h)_+}{4}}d_{n+1}^{Nq-2q'}cap^q_{2q,q'}\left(\frac{K_n}{d_{n+1}}\right),
\tag{1.235}
$$

where $\beta_0 = \frac{q-1}{4}$ *and* $h = \frac{2q(q+1)}{(q-1)^2}$.

Proof Since

$$
(y,s)\in\mathcal{T}^*_p \quad\text{and}\quad (z,0)\in K_n \Longrightarrow |y-z|\geq(\sqrt{n}-\alpha^{-p})\sqrt{t},
\tag{1.236}
$$

there holds by Lemma 1.16,

$$
\mathbb{H}[\mu_n](y,s)\leq\frac{1}{(4\pi s)^{\frac{N}{2}}}e^{-\frac{(\sqrt{n}-\alpha^{-p})^2 t}{4s}}\mu_n(K_n)\leq Ct^{-\frac{N}{2}}e^{-\frac{(\sqrt{n}-\alpha^{-p})^2}{4}}\mu_n(K_n).
$$

Let $\{\epsilon_n\}$ be a sequence of positive numbers such that

$$
A_\epsilon := \sum_{n=0}^{\infty}\epsilon_n < \infty,
$$

then

$$
\begin{aligned}
J'_{1,\ell} &\leq CA_\epsilon^{\frac{q'}{q}} t^{-\frac{Nq}{2}} \sum_{p=-\infty}^{0} \iint_{\mathcal{T}_p^*} (t-s)^{-\frac{N}{2}} e^{-\frac{|x-y|^2}{4(t-s)}} \sum_{n=1\vee p+\ell}^{\infty} \epsilon_n^{-q} e^{-\frac{q(\sqrt{n}-\alpha^- p)^2}{4}} (\mu_n(K_n))^q \, dyds \\
&\leq CA_\epsilon^{\frac{q'}{q}} t^{-\frac{Nq}{2}} \sum_{n=1}^{\infty} \epsilon_n^{-q} (\mu_n(K_n))^q \sum_{p=-\infty}^{0\wedge n-\ell} e^{-\frac{q(\sqrt{n}-\alpha^- p)^2}{4}} \iint_{\mathcal{T}_p^*} (t-s)^{-\frac{N}{2}} e^{-\frac{|x-y|^2}{4(t-s)}} dyds \\
&\leq CA_\epsilon^{\frac{q'}{q}} t^{-\frac{Nq}{2}} \sum_{n=1}^{\infty} \epsilon_n^{-q} (\mu_n(K_n))^q e^{-\frac{q(\sqrt{n}-1)^2}{4}} \iint_{\bigcup_{p\leq 0}\mathcal{T}_p^*} (t-s)^{-\frac{N}{2}} e^{-\frac{|x-y|^2}{4(t-s)}} dyds \\
&\leq CA_\epsilon^{\frac{q'}{q}} t^{1-\frac{Nq}{2}} \sum_{n=1}^{\infty} \epsilon_n^{-q} e^{-\frac{q(\sqrt{n}-1)^2}{4}} (\mu_n(K_n))^q .
\end{aligned}
\tag{1.237}
$$

Set $h = \frac{2q(q+1)}{(q-1)^2}$ and $Q = \frac{q+1}{2}$, then $q(\sqrt{n}-1)^2 \geq Q(n-h)_+$ for $n \geq 1$. Then

$$
\epsilon_n = e^{\frac{(q-1)(n-h)_+}{16q}} \Longrightarrow \epsilon_n^{-q} e^{-\frac{q(\sqrt{n}-1)^2}{4}} \leq e^{-\frac{(q+3)(n-h)_+}{16}} .
$$

Therefore

$$
J'_{1,\ell} \leq Ct^{1-\frac{N}{2}} \sum_{n=1}^{\infty} e^{-\frac{(1+\beta_0)(n-h)_+}{4}} (\mu_n(K_n))^q .
$$

This implies (1.235) from the properties of μ_n. □

The estimate of the term $J'_{2,\ell}$ is more involved. In order to help the reader to follow the idea, we first give a proof in dimension 1.

Lemma 1.19 *Assume $N = 1$, $q \geq 3$ and ℓ is an integer larger than 1. Then there exists a positive constant $C = C(q, \ell)$ such that*

$$
J'_{2,\ell} \leq Ct^{-\frac{1}{2}} \sum_{n=\ell}^{a_t} e^{-\frac{n}{4}} d_{n+1}^{\frac{q-3}{q-1}} cap_{\frac{2}{q},q'} \left(\frac{K_n}{d_{n+1}}\right). \tag{1.238}
$$

Proof If $p \geq 1$, $n \geq p = \ell$ and if $(y,s) \in \mathcal{T}_p^*$ and $z \in K_n$ there holds $|x-y| \geq \sqrt{t}\sqrt{p}$ and $|y-z| \geq \sqrt{t}\left(\sqrt{n}-\sqrt{p+1}\right)$. Therefore

$$
J'_{2,\ell} \leq C\sqrt{t} \sum_{p=1}^{a_t-\ell} \frac{1}{\sqrt{p}} \int_0^t e^{-\frac{pt}{4(t-s)}} \left(\sum_{n=p+\ell}^{a_t} \frac{1}{\sqrt{s}} e^{-\frac{(\sqrt{n}-\sqrt{p+1})^2 t}{4s}} \mu_n(K_n) \right)^q ds.
$$

Let $\epsilon \in (0, q)$ be some parameter to be made more precise later on, then

$$\left(\sum_{n=p+\ell}^{a_t}\frac{1}{\sqrt{s}}e^{-\frac{(\sqrt{n}-\sqrt{p+1})^2t}{4s}}\mu_n(K_n)\right)^q$$
$$\leq\left(\sum_{n=p+\ell}^{a_t}e^{-\frac{\epsilon q'(\sqrt{n}-\sqrt{p+1})^2t}{4s}}\right)^{\frac{q'}{q}}\sum_{n=p+\ell}^{a_t}s^{-\frac{q}{2}}e^{-(q-\epsilon)\frac{(\sqrt{n}-\sqrt{p+1})^2t}{4s}}(\mu_n(K_n))^q\,.$$

By comparison between series and integrals we have

$$\begin{aligned}\sum_{n=p+\ell}^{a_t}e^{-\frac{\epsilon q'(\sqrt{n}-\sqrt{p+1})^2t}{4s}}&\leq\int_{p+\ell}^{\infty}e^{-\frac{\epsilon q'(\sqrt{x}-\sqrt{p+1})^2t}{4s}}dx\\&\leq 2\int_{\sqrt{p+\ell}-\sqrt{p+1}}^{\infty}e^{-\frac{\epsilon q'x^2t}{4s}}(x+\sqrt{p+1})dx\\&\leq\frac{4s}{\epsilon q't}e^{-\epsilon q'\frac{(\sqrt{p+\ell}-\sqrt{p+1})^2t}{4s}}+2\sqrt{p+1}\int_{\sqrt{p+\ell}-\sqrt{p+1}}^{\infty}e^{-\frac{\epsilon q'x^2t}{4s}}dx\\&\leq C\sqrt{\frac{(p+1)s}{t}}e^{-\epsilon q'\frac{(\sqrt{p+\ell}-\sqrt{p+1})^2t}{2s}}\\&\leq C\sqrt{\frac{(p+1)s}{t}}.\end{aligned}$$

Set $q_\epsilon=q-\epsilon$, then

$$J'_{2,\ell}\leq C\epsilon^{-\frac{q'}{q}}t^{1-\frac{q}{2}}\sum_{n=\ell+1}^{\infty}(\mu_n(K_n))^q\sum_{p=1}^{n-\ell}\int_0^t\sqrt{s(t-s)}e^{-\frac{-pt}{4(t-s)}}e^{-q_\epsilon\frac{(\sqrt{n}-\sqrt{p+1})^2t}{4s}}ds,$$

where $C=C(\epsilon,q)>0$. Since

$$\int_0^t\sqrt{s(t-s)}e^{-\frac{-pt}{4(t-s)}}e^{-q_\epsilon\frac{(\sqrt{n}-\sqrt{p+1})^2t}{4s}}ds=\int_0^1\sqrt{s(1-s)}e^{-\frac{-p}{4(t-s)}}e^{-q_\epsilon\frac{(\sqrt{n}-\sqrt{p+1})^2}{4s}}ds,$$

we can apply Lemma 1.21 with $a=\frac{1}{2}$, $b=\frac{1}{2}$, $A=\sqrt{p}$, $B=\sqrt{q_\epsilon}(\sqrt{n}-\sqrt{p+1})$. For such a choice,

$$B\geq\sqrt{q_\epsilon}(\sqrt{p+\ell}-\sqrt{p+1})\geq\frac{(\ell-1)\sqrt{q_\epsilon}}{\sqrt{p}}\Longrightarrow\kappa=(\ell-1)\sqrt{q_\epsilon},$$

and

$$\sqrt{\frac{A}{A+B}}\sqrt{\frac{B}{A+B}}\leq\frac{\sqrt[4]{p}\sqrt{\sqrt{n}-\sqrt{p}}}{\sqrt{n}}.$$

Therefore

$$\int_0^t \frac{e^{-\frac{pt}{4(t-s)}} e{-q}\frac{(\sqrt{n}-\sqrt{p+1})^2 t}{4t}}{\sqrt{s^q(t-s)}} ds \leq C \frac{\sqrt[4]{p}\sqrt{\sqrt{n}-\sqrt{p}}}{\sqrt{n}} e^{-\frac{(\sqrt{p}+\sqrt{q\epsilon}(\sqrt{n}-\sqrt{p+1}))^2}{4}}.$$

This implies

$$J'_{2,\ell} \leq C t^{1-\frac{q}{2}} \sum_{n=\ell+1}^{a_t} \frac{(\mu_n(K_n))^q}{\sqrt{n}} \sum_{p=1}^{n-\ell} p^{\frac{2q-3}{4}} \sqrt{\sqrt{n}-\sqrt{p}}\, e^{-\frac{(\sqrt{p}+\sqrt{q\epsilon}(\sqrt{n}-\sqrt{p+1}))^2}{4}}, \tag{1.239}$$

where $C = C(\epsilon, q, \ell) > 0$. Then, by Lemma 1.22,

$$J'_{2,\ell} \leq C t^{1-\frac{q}{2}} \sum_{n=\ell+1}^{a_t} n^{\frac{q-3}{4}} e^{-\frac{n}{4}} (\mu_n(K_n))^q. \tag{1.240}$$

Replacing $\mu_n(K_n)$ by its value $d_{n+1}^{\frac{q-3}{q-1}} cap_{\frac{2}{q}q'}\left(\frac{K_n}{d_{n+1}}\right)$, the expression when $N = 1$, and since $diam\left(\frac{K_n}{d_{n+1}}\right) \leq \frac{1}{n}$, we obtain

$$(\mu_n(K_n))^q \leq C\left(\frac{t}{n}\right)^{\frac{q-3}{2}} \mu_n(K_n) = C\left(\frac{t}{n}\right)^{\frac{q-3}{2}} d_{n+1}^{\frac{q-3}{q-1}} cap_{\frac{2}{q}q'}\left(\frac{K_n}{d_{n+1}}\right), \tag{1.241}$$

and the proof follows. □

Next we give the proof for $N \geq 2$. For this task we will use again the quasi-additivity property.

Lemma 1.20 *Assume $N \geq 2$ and ℓ is an integer larger than* 1. *There exists a positive constant $C_1 = C_1(N, q, \ell)$ such that*

$$J'_{2,\ell} \leq C_1 t^{-\frac{N}{2}} \sum_{n=\ell}^{a_t} e^{-\frac{n}{4}} d_{n+1}^{N-\frac{2}{q-1}} cap_{\frac{2}{q},q'}\left(\frac{K_n}{d_{n+1}}\right). \tag{1.242}$$

Proof As in the proof Theorem 1.16 there exists a finite number J depending only on the dimension N of separated sub-partitions $\{\Theta^h_{t,n}\}_{h=1}^J$ of the rescaled sets $\widetilde{T}_n = \sqrt{\frac{n+1}{t}} T_n$ by the N-dimensional balls $B_2(\widetilde{a}_{n,j})$ where $\widetilde{a}_{n,j} = \sqrt{\frac{n+1}{t}} a_{n,j}$, $|a_{n,j}| = \frac{1}{2}(d_n + d_{n+1})$ and $|a_{n,j} - a_{n,k}| \geq \sqrt{\frac{4t}{n+1}}$. Furthermore $\#\Theta^h_{t,n} \leq C n^{N-1}$. We denote $K_{n,j} = K_n \cap B_{\sqrt{\frac{t}{n+1}}}(a_{n,j})$. We can write $\mu_n = \sum_{h=1}^J \mu_n^h$ and accordingly $J'_{2,\ell} \sum_{h=1}^J \mu_n^h J'^h_{2,\ell}$ where $\mu_n^h = \sum_{j\in\Theta^h_{t,n}} \mu_{n,j}$ and the $\mu_{n,j}$ are the capacitary measures of $K_{n,j}$ relative to $B_{n,j} := B_{\frac{6t}{5}}(a_{n,j}$, which means

$$\nu_{n,j}(K_{n,j}) = cap^{B_{n,j}}_{\frac{2}{q},q'}(K_{n,j} \quad \text{and} \quad \left\|\nu_{n,j}\right\|_{B^{-\frac{2}{q},q}(B_{n,j})} = \left(cap^{B_{n,j}}_{\frac{2}{q},q'}(K_{n,j}\right)^{\frac{1}{q}}. \quad (1.243)$$

Thus

$$J'_{2,\ell} = \sum_{p=1}^{a_t-\ell} \int\int_{\mathcal{T}^*_p} (-s)^{-\frac{N}{2}} e^{-\frac{|x-y|^2}{4(t-s)}} \left(\sum_{n=p+\ell}^{\infty} \sum_{h=1}^{J} \sum_{j\in Gth^h_{t,n}} \mathbb{H}[\mu_{n,j}(y,s)] \right)^q dyds. \quad (1.244)$$

We denote

$$J'^h_{2,\ell} = \sum_{p=1}^{a_t-\ell} \int\int_{\mathcal{T}^*_p} (-s)^{-\frac{N}{2}} e^{-\frac{|x-y|^2}{4(t-s)}} \left(\sum_{n=p+\ell}^{\infty} \sum_{j\in Gth^h_{t,n}} \mathbb{H}[\mu_{n,j}(y,s)] \right)^q dyds.$$

Since J depends only on N and q,

$$J'_{2,\ell} \le C \sum_{h=1}^{J} J'^h_{2,\ell}.$$

If n and p are such that $n \ge \ell + 1$, we set

$$\lambda_{n,j,y} = \inf\left\{|y-z| : z \in B_{\sqrt{t}n+1}(a_{n,j}\right\} = |y - a_{n,j}| - \sqrt{t}n + 1.$$

Then

$$\begin{aligned}\sum_{n=p+\ell}^{a_t} \int_{K_n} e^{-\frac{|y-z|^2}{4t}} d\mu^h_n(z) &= \sum_{n=p+\ell}^{a_t} \sum_{j\in Gth^h_{t,n}} \mathbb{H}[\mu_{n,j}(z)] \\ &\le \left(\sum_{n=p+\ell}^{a_t} \sum_{j\in Gth^h_{t,n}} e^{-\epsilon q' \frac{\lambda^2_{n,j,y}}{4s}} \right)^{\frac{1}{q'}} \\ &\quad \times \left(\sum_{n=p+\ell}^{a_t} \sum_{j\in Gth^h_{t,n}} e^{-q\lambda^2_{n,j,y}\frac{1-\epsilon}{4s}} \left(\mu_{n,j}(K_{n,j})\right)^q \right)^{\frac{1}{q}}\end{aligned}$$

where $\epsilon > 0$ will be made precise later on.

Step 1. We claim that

$$\sum_{n=p+\ell}^{a_t} \sum_{j\in Gth^h_{t,n}} e^{-\epsilon q' \frac{\lambda^2_{n,j,y}}{4s}} \le C\sqrt{\frac{ps}{t}}, \quad (1.245)$$

where $C = C(\epsilon, q, N) > 0$. If $y \in T_p$, let $z_y \in T_n$ such that $|y - z| = \operatorname{dist}(y, T_n)$ hence

$$\sqrt{t}(\sqrt{n} - \sqrt{p+1}) \leq |y - z| \leq \sqrt{t}(\sqrt{n} - \sqrt{p+1}).$$

Let $Y = \frac{\sqrt{t(p+1)}}{|y|}y$, $\mathbf{e} = \frac{Y}{|Y|}$ and, for integers $k \in [-n, n]$, $b_k = \frac{k\sqrt{t}}{\sqrt{n}}\mathbf{e}$. We denote by $H_{n,k}$ the domain in $\mathbb{R}^N$ limited by the hyperplanes orthogonal to $\mathbf{e}$ going through the points $\frac{(k+1)\sqrt{t}}{\sqrt{n}}\mathbf{e}$ and $\frac{(k-1)\sqrt{t}}{\sqrt{n}}\mathbf{e}$, and by $G_{n,k}$ the spherical shell obtained by intersecting the spherical shell T_n with $H_{n,k}$. The number of points $a_{n,j}$ belonging to $G_{n,k}$ is smaller than $C(n+1-|k|)^{N-2}$ where $C = C(N) > 0$. Let $\Lambda_{n,k}$ be the set of indices $j \in \Theta_{t,n}$ such that $a_{n,j} \in G_{n,k}$. Note that in $a_{n,j} \in G_{n,k}$, it is a consequence of Pythagoras' theorem that $\lambda^2_{n,j,y}$ is larger than $t(n+p+1-2k\frac{p+1}{n})$. Therefore

$$\sum_{n=p+\ell}^{a_t} \sum_{j\in\Theta_{t,n}} e^{-\epsilon q' \frac{\lambda^2_{n,j,y}}{4s}} \leq C \sum_{n=p+\ell}^{a_t} \sum_{k=-n}^{n} (n+1-|k|)^{N-2} e^{\frac{\epsilon q'(n+p+1-2k\sqrt{p+1})t}{4s\sqrt{n}}}. \tag{1.246}$$

Case $N = 2$. Summing a geometric series and using the inequality $\frac{e^u}{e^u - 1} \leq 1 + \frac{1}{u}$ on $(0, \infty)$, we obtain

$$\sum_{k=-n}^{n} e^{\frac{\epsilon q' k\sqrt{p+1}t}{2s\sqrt{n}}} \leq e^{\frac{\epsilon q'\sqrt{p+1}t}{2s}} \frac{e^{\frac{\epsilon q'\sqrt{p+1}t}{2s\sqrt{n}}}}{e^{\frac{\epsilon q'\sqrt{p+1}t}{2s\sqrt{n}}} - 1} \leq e^{\frac{\epsilon q'\sqrt{p+1}t}{2s}} \left(1 + \frac{2s\sqrt{n}}{\epsilon q' t\sqrt{p+1}}\right). \tag{1.247}$$

Therefore using comparison between series and integrals,

$$\begin{aligned}
\sum_{n=p+\ell}^{a_t} \sum_{j\in Gth^h_{t,n}} e^{-\epsilon q' \frac{\lambda^2_{n,j,y}}{4s}} &\leq C \sum_{n=p+\ell}^{a_t} \left(1 + \frac{s\sqrt{n}}{t\sqrt{p}}\right) e^{-\epsilon q' \frac{(\sqrt{n}-\sqrt{p+1})^2 t}{4s}} \\
&\leq C \int_{p+1}^{\infty} e^{-\epsilon q' \frac{(\sqrt{x}-\sqrt{p+1})^2 t}{4s}} dx \\
&\quad + \frac{Cs}{t\sqrt{n}} \int_{p+1}^{\infty} \sqrt{x} e^{-\epsilon q' \frac{(\sqrt{x}-\sqrt{p+1})^2 t}{4s}} dx.
\end{aligned} \tag{1.248}$$

Next

$$\begin{aligned}
\int_{p+1}^{\infty} e^{-\epsilon q' \frac{(\sqrt{x}-\sqrt{p+1})^2 t}{4s}} dx &= 2\int_{\sqrt{p+1}}^{\infty} e^{-\epsilon q' \frac{(y-\sqrt{p+1})^2 t}{4s}} y dy \\
&= 2\int_0^{\infty} e^{-\epsilon q' \frac{y^2 t}{4s}} y dy + 2\sqrt{p+1} \int_0^{\infty} e^{-\epsilon q' \frac{y^2 t}{4s}} dy \\
&= \frac{2s}{t} \int_0^{\infty} e^{-\epsilon q' \frac{z^2}{4}} z dz + \sqrt{\frac{(p+1)s}{t}} \int_0^{\infty} e^{-\epsilon q' \frac{z^2}{4}} dz,
\end{aligned} \tag{1.249}$$

and

$$
\begin{aligned}
\int_{p+1}^{\infty} e^{-\epsilon q' \frac{(\sqrt{x}-\sqrt{p+1})^2 t}{4s}} \sqrt{x}dx &= 2\int_{\sqrt{p+1}}^{\infty} e^{-\epsilon q' \frac{(y-\sqrt{p+1})^2 t}{4s}} y^2 dy \\
&= 2\int_0^{\infty} e^{-\epsilon q' \frac{y^2 t}{4s}} (y+\sqrt{p+1})^2 dy \\
&\leq 4\int_0^{\infty} e^{-\epsilon q' \frac{y^2 t}{4s}} y^2 dy + 4(p+1)\int_0^{\infty} e^{-\epsilon q' \frac{y^2 t}{4s}} dy \\
&\leq 4\left(\frac{s}{t}\right)^{\frac{3}{2}} \int_0^{\infty} e^{-\epsilon q' \frac{z^2}{4}} z^2 dz + 4(p+1)\sqrt{\frac{s}{t}} \int_0^{\infty} e^{-\epsilon q' \frac{z^2}{4}} dz.
\end{aligned}
\tag{1.250}
$$

Combined with (1.248), these inequalities imply

$$
\sum_{n=p+\ell}^{a_t} \sum_{j\in\Theta_{t,n}} e^{\frac{\epsilon q' \lambda^2_{n,j,y}}{4}} \leq C\sqrt{\frac{ps}{t}}. \tag{1.251}
$$

Case $N > 2$. The value of the right-hand side of (1.246) is clearly an increasing function of N, hence it is sufficient to prove (1.246) when $N = 2 + 2d$ with $d \in \mathbb{N}_*$. There holds

$$
\sum_{k=-n}^{n} (n+1-|k|)^d e^{\frac{\epsilon q' k t \sqrt{p+1}}{2s\sqrt{n}}} \leq 2\sum_{k=0}^{n} (n+1-k)^d e^{\frac{\epsilon q' k t \sqrt{p+1}}{2s\sqrt{n}}}. \tag{1.252}
$$

We set

$$
\alpha = \epsilon q' \frac{t\sqrt{p+1}}{2s\sqrt{n}} \quad \text{and } I_d = \sum_{k=0}^{n} (n+1-k)^d e^{k\alpha}.
$$

Since

$$
e^{k\alpha} = \frac{e^{(k+1)\alpha} - e^{k\alpha}}{e^{\alpha} - 1},
$$

we use Abel's transform and obtain

$$
\begin{aligned}
I_d &= \frac{1}{e^{\alpha}-1}\left(e^{(n+1)\alpha} - (n+1)^d + \sum_{k=1}^{n} \left((n+2-k)^d - (n+1-k)d\right) e^{k\alpha}\right) \\
&\leq \frac{1}{e^{\alpha}-1}\left((1-d)e^{(n+1)\alpha} - (n+1)^d + de^{\alpha} \sum_{k=1}^{n} (n+1-k)^{\prime} e^{k\alpha}\right).
\end{aligned}
$$

Therefore the following induction relation holds

$$
I_d \leq \frac{de^{\alpha}}{e^{\alpha}-1} I_{d-1}. \tag{1.253}
$$

We use again the fact that

$$\frac{de^{\alpha}}{e^{\alpha}-1} \le C\left(1+\frac{s\sqrt{n}}{t\sqrt{p}}\right)$$

as in (1.247), and

$$I_d \le C\left(1+C\left(\frac{s\sqrt{n}}{t\sqrt{p}}\right)^{d+1}\right)I_0.$$

Therefore (1.248) is replaced by

$$\begin{aligned}\sum_{n=p+\ell}^{a_t}\sum_{j\in Gth^h_{t,n}} e^{-\epsilon q'\frac{\lambda^2_{n,j,y}}{4s}} &\le C\sum_{n=p+\ell}^{a_t}\left(1+\frac{s\sqrt{n}}{t\sqrt{p}}\right)^{d+1} e^{-\epsilon q'\frac{(\sqrt{n}-\sqrt{p+1})^2 t}{4s}}\\ &\le C\int_{p+1}^{\infty} e^{-\epsilon q'\frac{(\sqrt{x}-\sqrt{p+1})^2 t}{4s}}dx+\left(\frac{Cs}{t\sqrt{p}}\right)^{d+1}\int_{p+1}^{\infty} x^{\frac{d+1}{2}}e^{-\epsilon q'\frac{(\sqrt{x}-\sqrt{p+1})^2 t}{4s}}dx.\end{aligned} \tag{1.254}$$

Using the estimate of the first integral of the right-hand side of (1.254) that we have obtained in (1.249), we can concentrate on the second integral,

$$\begin{aligned}\int_{p+1}^{\infty} x^{\frac{d+1}{2}}e^{-\epsilon q'\frac{(\sqrt{x}-\sqrt{p+1})^2 t}{4s}}dx &= \int_0^{\infty}(y+\sqrt{p+1})^{d+2}e^{-\epsilon q'\frac{y^2 t}{4s}}dy\\ &\le C\int_0^{\infty} y^{d+2}e^{-\epsilon q'\frac{y^2 t}{4s}}dy+Cp^{1+\frac{d}{2}}\int_0^{\infty}e^{-\epsilon q'\frac{y^2 t}{4s}}dy\\ &\le C\left(\frac{s}{t}\right)^{2+\frac{d}{2}}\int_0^{\infty}e^{-\epsilon q'\frac{z^2}{4}}dz\\ &\qquad +C\left(\frac{s}{t}\right)^{\frac{3}{2}}p^{1+\frac{d}{2}}\int_0^{\infty}e^{-\epsilon q'\frac{z^2}{4}}dz.\end{aligned} \tag{1.255}$$

We obtain (1.245) from (1.250), (1.254) and (1.255).

Step 2. Since $\mathcal{T}^*_p \subset \Gamma_p \times [0,t]$, where we recall it $\Gamma_p = B_{d+1}(x)\setminus B_{d_n}(x)$, the fact that $(y,s)\in\mathcal{T}^*_p$ implies $|x-y|^2 \ge (p-1)t$. Therefore $J^{th}_{2,\ell}$ satisfies

$$\begin{aligned}J^{th}_{2,\ell} &\le Ct^{\frac{1-q}{2}}\sum_{p=1}^{\infty}p^{\frac{q-1}{2}}\int_0^t\int_{\Gamma_p}(t-s)^{-\frac{N}{2}}s^{\frac{-q(N-1)+1}{2}}e^{-4\frac{|x-y|^2}{4(t-s}}\\ &\qquad\times\sum_{n=p+\ell}^{a_t}\sum_{j\in\Theta^h_{t,n}}e^{-\frac{q(1-\epsilon)\lambda^2_{n,j,y}}{4s}}\left(\mu_{n,j}(K_{n,j})\right)^q dyds\\ &\le Ct^{\frac{1-q}{2}}\sum_{n=p+\ell}^{a_t}\sum_{j\in\Theta^h_{t,n}}\left(\mu_{n,j}(K_{n,j})\right)^q\\ &\qquad\times\sum_{p=1}^{n-\ell}\int_0^t\int_{\Gamma_p}(t-s)^{-\frac{N}{2}}s^{\frac{-q(N-1)+1}{2}}e^{-4\frac{|x-y|^2}{4(t-s}}e^{-\frac{q(1-\epsilon)\lambda^2_{n,j,y}}{4s}}dyds,\end{aligned} \tag{1.256}$$

where $C = C(N, q, \epsilon) > 0$. Next we set $q_\epsilon = (1 - \epsilon)q$. If we write

$$
\begin{aligned}
|y - a_{n,j}|^2 &= |x - y|^2 + |x - a_{n,j}|^2 - 2\langle y - x, a_{n,j} - x\rangle \\
&\geq pt + |x - a_{n,j}|^2 - 2\langle y - x, a_{n,j} - x\rangle,
\end{aligned}
$$

we deduce

$$
\int_{\Gamma_p} e^{-\frac{q_\epsilon |y-a_{n,j}|^2}{4s}} dy = e^{-\frac{|x-a_{n,j}|^2}{4s}} \int_{\sqrt{tp}}^{\sqrt{t(p+1}} e^{-\frac{q_\epsilon r^2}{4s}} \int_{\partial B_r(x)} e^{\frac{\langle y-x, a_{n,j}-x\rangle}{2s}} dS_r(y)dr.
$$

Since the value of the spherical integral is invariant by rotations in $\mathbb{R}^N$, we can assume that $a_{n,j} - x = (0, 0, 0, ..., |a_{n,j} - x|$. We then use the spherical coordinates in $\mathbb{R}^N$ with centre x and the representation of $S^{N-1} = \{(\sin\phi.\sigma, \phi) : \sigma \in S^{N-2}, \phi \in [0, \pi]\}$. With this representation $\langle y - x, a_{n,j} - x\rangle = |y - x||a_{n,j} - x| \cos\phi$. This yields

$$
\int_{\partial B_r(x)} e^{\frac{\langle y-x, a_{n,j}-x\rangle}{2s}} dS_r(y) = r^{N-1}|S^{N-2} \int_0^\pi e^{q_\epsilon \frac{|a_{n,j}-x| r \cos\phi}{2s}} \sin^{N-2}\phi d\phi.
$$

By Lemma 1.23

$$
\begin{aligned}
\int_{\partial B_r(x)} e^{-q_\epsilon \frac{\langle y-x, a_{n,j}-x\rangle}{2s}} dS_r(y) &\leq C \frac{r^{N-1} e^{q_\epsilon r \frac{r|a_{n,j}-x|}{2s}}}{\left(1 + \frac{r|a_{n,j}-x|}{s}\right)^{\frac{N-1}{2}}} \\
&\leq C s^{\frac{N-1}{2}} \left(\frac{r}{|a_{n,j} - x|}\right)^{\frac{N-1}{2}}.
\end{aligned}
\tag{1.257}
$$

Therefore

$$
\int_{\Gamma_p} e^{-\frac{q_\epsilon |y-a_{n,j}|^2}{4s}} dy \leq C t^{\frac{N-1}{4}} p^{\frac{N-3}{4}} \frac{s^{\frac{N-1}{2}} e^{-q_\epsilon \frac{(|a_{n,j}-x| - \sqrt{t(p+1)})^2}{4s}}}{|a_{n,j} - x|^{\frac{N-1}{2}}}.
\tag{1.258}
$$

Since $|a_{n,j} - x| \geq \sqrt{tn}$ we obtain

$$
\begin{aligned}
&\int_0^t \int_{\Gamma_p} (t - s)^{-\frac{N}{2}} s^{-\frac{q(N-1)+1}{2}} e^{-\frac{|x-y|^2}{4(t-s)}} e^{-q_\epsilon \frac{\lambda^2_{n,j,y}}{4s}} dyds \\
&\quad \leq C \frac{\sqrt{t}\, p^{\frac{N-3}{4}}}{n^{\frac{N-1}{4}}} \int_0^t (t - s)^{-\frac{N}{2}} s^{-\frac{q(N-1)+1}{2}} e^{-\frac{p}{4(t-s)}} e^{-q_\epsilon \frac{(\sqrt{tn} - \sqrt{(p+1)t})^2}{4t}} ds \\
&\quad \leq C \frac{t^{\frac{1-q(N-2)}{2}}}{n^{\frac{N-1}{4}}} \int_0^1 (1 - \tau)^{-\frac{N}{2}} \tau^{-\frac{q(N-1)+1}{2}} e^{-\frac{p}{4(1-\tau)}} e^{-q_\epsilon \frac{(\sqrt{n} - \sqrt{p+1})^2}{4\tau}} d\tau.
\end{aligned}
\tag{1.259}
$$

We apply Lemma 1.21 with $A=\sqrt{p}$, $B=\sqrt{q_\epsilon}(\sqrt{n}-\sqrt{p+1})$, $b=\frac{(q-1)(N-1)+1}{2}$, $a=\frac{N}{2}$ and $\kappa=\frac{\sqrt{q_\epsilon}(\ell-1)}{8}$, as in the case $N=1$. For these specific values

$$\begin{aligned}A^{1-a}B^{1-b}(A+B)^{a+b-2} &= p^{\frac{2-N}{4}}\left(\sqrt{q_\epsilon}\left(\sqrt{n}-\sqrt{p+1}\right)\right)^{\frac{1-(q-1)(N-1)}{2}} \\ &\quad\times\left(\sqrt{p}+\sqrt{q_\epsilon}\left(\sqrt{n}-\sqrt{p+1}\right)\right)^{\frac{(q-1)(N-1)+N-3}{2}} \\ &\le C\left(\frac{n}{p}\right)^{\frac{N}{4}-\frac{1}{2}}\left(\frac{\sqrt{n}-\sqrt{p}}{\sqrt{n}}\right)^{\frac{1-(q-1)(N-1)}{2}},\end{aligned}$$

where $C=C(N,q,\kappa)>0$. Hence

$$\begin{aligned}&\int_0^t\int_{\Gamma_p}(t-s)^{-\frac{N}{2}}s^{-\frac{N}{2}}e^{-\frac{q_\epsilon|x-y|^2}{4s}}dyds \\ &\quad\le C\frac{t^{\frac{1-q(N-1)}{2}}p^{\frac{N-3}{4}}}{n^{\frac{N-1}{4}}}\left(\frac{n}{p}\right)^{\frac{N-2}{4}}\left(\frac{\sqrt{n}-\sqrt{p}}{\sqrt{n}}\right)^{\frac{1-(q-1)(N-1)}{2}}e^{-\frac{(\sqrt{p}+\sqrt{q_\epsilon}(\sqrt{n}-\sqrt{p+1}))^2}{4}} \\ &\quad\le Ct^{\frac{1-q(N-1)}{2}}p^{-\frac{1}{4}}n^{\frac{(q-1)(N-1)-2}{4}}\left(\sqrt{n}-\sqrt{p}\right)^{\frac{1-(q-1(N-1))}{2}}e^{-\frac{(\sqrt{p}+\sqrt{q_\epsilon}(\sqrt{n}-\sqrt{p+1}))^2}{4}}.\end{aligned}\tag{1.260}$$

Then we deduce from (1.256), (1.260)

$$\begin{aligned}J'^{h}_{2,\ell} &\le Ct^{1-\frac{Nq}{2}}\sum_{n=\ell+1}^{a_l}\sum_{j\in\Theta^h_{l,n}}n^{\frac{(q-1)(N-1)-2}{4}}\left(\mu_{n,j}(K_{n,j})\right)^q \\ &\quad\times\sum_{p=1}^{n-\ell}p^{\frac{2q-3}{4}}\left(\sqrt{n}-\sqrt{p}\right)^{\frac{1-(q-1)(N-1)}{2}}e^{-\frac{(\sqrt{p}+\sqrt{q_\epsilon}(\sqrt{n}-\sqrt{p+1}))^2}{4}}.\end{aligned}\tag{1.261}$$

By Lemma 1.22 with $\alpha=\frac{2q-3}{4}$, $\beta=\frac{1-(q-1(N-1)}{2}$, $\delta=\frac{1}{4}$ and $\gamma=q_\epsilon$, we obtain

$$\sum_{p=1}^{n-\ell}p^{\frac{2q-3}{4}}\left(\sqrt{n}-\sqrt{p}\right)^{\frac{1-(q-1)(N-1)}{2}}e^{-\frac{(\sqrt{p}+\sqrt{q_\epsilon}(\sqrt{n}-\sqrt{p+1}))^2}{4}}\le Cn^{\frac{N(q-1)+q-3}{4}}e^{-\frac{n}{4}},\tag{1.262}$$

thus

$$J'^{h}_{2,\ell}\le Ct^{1-\frac{Nq}{2}}\sum_{n=\ell+1}^{a_l}n^{\frac{N(q-1)}{2}-1}e^{-\frac{n}{4}}\sum_{j\in\Theta^h_{l,n}}\left(\mu_{n,j}(K_{n,j})\right)^q.\tag{1.263}$$

Because

$$\mu_{n,j}(K_{n,j})=cap^{B_{n,j}}_{\frac{2}{q},q'}(K_{n,j}),$$

we use the rescaling procedure of Lemma 1.15 except that the scaling factor is $\sqrt{(N+1)t}$ instead of $\sqrt{N+1}$, so that the sets $\widetilde{T}_n$, $\widetilde{K}_n$, $\widetilde{K}_{n,j}$ and $\widetilde{Q}_n$ remain unchanged. Using the quasi-additivity and the fact that $J'_{2,\ell}=\sum_{h=1}^{J} J'^{h}_{2,\ell}$, we deduce

$$J'_{2,\ell}\leq Ct^{-\frac{N}{2}}\sum_{n=\ell+1} a_t d_{n+1}^{N-\frac{2}{q-1}} cap_{\frac{2}{q},q'}\left(\frac{K_n}{d_{n+1}}\right), \tag{1.264}$$

which implies (1.242). □

The proof of Theorem 1.18 follows from the previous estimates on J_1 and J_2.

In the same way as for Theorem 1.15, the estimate in Theorem 1.18 admits an integral form. Fortunately it yields the same form as for Theorem 1.17.

Theorem 1.19 *Assume $q\geq q_c$. Then there exists a positive constant $C_2^*=C_2^*(N,q,T$ such for any closed subset $F\subset\mathbb{R}^N$ there holds for all $(x,t)\in Q_T$,*

$$\underline{u}_F(x,t)\geq\frac{C_2^*}{t^{1+\frac{N}{2}}}\int_{\sqrt{t}}^{\sqrt{t(a_t+2}} e^{-\frac{s^2}{4t}} s^{N-\frac{2}{q-1}} cap_{\frac{2}{q},q'}\left(\frac{1}{s}B\cap B_1(x)\right) sds, \tag{1.265}$$

where a_t is the smallest integer j such that $F\subset B_{\sqrt{jt}}(x)$.

Proof We distinguish accordingly $q=q_c$, or $q>q_c$, and for simplicity we denote $B_r=B_r(x)$ for the various values of r.
Case 1: $q=q_c\iff N-\frac{2}{q-1}=0$. Because $F_n=F\cap(B_{d_{n+1}}\setminus B_{d_n})$ there holds

$$cap_{\frac{2}{q},q'}\left(\frac{F_n}{d_{n+1}}\right)\geq cap_{\frac{2}{q},q'}\left(\frac{F}{d_{n+1}}\cap B_1\right)-cap_{\frac{2}{q},q'}\left(\frac{F\cap B_{d_n}}{d_{n+1}}\right).$$

Furthermore, since $d_{n+1}\geq d_n$,

$$cap_{\frac{2}{q},q'}\left(\frac{F\cap B_{d_n}}{d_{n+1}}\right)=cap_{\frac{2}{q},q'}\left(\frac{d_n}{d_{n+1}}\frac{F\cap B_{d_n}}{d_n}\right)\leq cap_{\frac{2}{q},q'}\left(\frac{F}{d_n}\cap B_1\right),$$

thus

$$cap_{\frac{2}{q},q'}\left(\frac{F_n}{d_{n+1}}\right)\geq cap_{\frac{2}{q},q'}\left(\frac{F}{d_{n+1}}\cap B_1\right)-cap_{\frac{2}{q},q'}\left(\frac{F}{d_n}\cap B_1\right),$$

it follows

$$\begin{aligned}
&\sum_{n=1}^{a_t} e^{-\frac{n}{4}} cap_{\frac{2}{q},q'}\left(\frac{F_n}{d_{n+1}}\right) \\
&\quad \geq \sum_{n=1}^{a_t} e^{-\frac{n}{4}} cap_{\frac{2}{q},q'}\left(\frac{F}{d_{n+1}} \cap B_1\right) - \sum_{n=1}^{a_t} e^{-\frac{n}{4}} cap_{\frac{2}{q},q'}\left(\frac{F}{d_n} \cap B_1\right) \\
&\quad \geq \sum_{n=1}^{a_t} e^{-\frac{n}{4}} cap_{\frac{2}{q},q'}\left(\frac{F}{d_{n+1}} \cap B_1\right) - e^{-\frac{1}{4}} \sum_{n=0}^{a_t-1} e^{-\frac{n}{4}} cap_{\frac{2}{q},q'}\left(\frac{F}{d_{n+1}} \cap B_1\right) \\
&\quad \geq (1-e^{-\frac{1}{4}}) \sum_{n=1}^{a_t-1} e^{-\frac{n}{4}} cap_{\frac{2}{q},q'}\left(\frac{F}{d_{n+1}} \cap B_1\right) - e^{-\frac{1}{4}} cap_{\frac{2}{q},q'}\left(\frac{F}{\sqrt{t}} \cap B_1\right).
\end{aligned}$$

Since, by (1.212),

$$cap_{\frac{2}{q},q'}\left(\frac{F}{s'} \cap B_1\right) \geq cap_{\frac{2}{q},q'}\left(\frac{F}{d_{n+1}} \cap B_1\right) \geq cap_{\frac{2}{q},q'}\left(\frac{F}{s} \cap B_1\right),$$

for any $s' \in [d_{n+1}, d_{n+2}]$ and $s \in [d_n, d_{n+1}]$, there holds

$$\begin{aligned}
te^{-\frac{n}{4}} cap_{\frac{2}{q},q'}\left(\frac{F}{d_{n+1}} \cap B_1\right) &\geq cap_{\frac{2}{q},q'}\left(\frac{F}{d_{n+1}} \cap B_1\right) \int_{d_n}^{d_{n+1}} e^{-s^2/4t} s\, ds \\
&\geq \int_{d_n}^{d_{n+1}} e^{-s^2/4t} cap_{\frac{2}{q},q'}\left(\frac{F}{s} \cap B_1\right) s\, ds.
\end{aligned}$$

This implies

$$W_F(x,t) \geq (1-e^{-\frac{1}{4}}) t^{-(1+\frac{N}{2})} \int_0^{\sqrt{t a_t}} e^{-s^2/4t} cap_{\frac{2}{q},q'}\left(\frac{F}{s} \cap B_1\right) s\, ds.$$

Case 2: $q > q_c \iff N - \frac{2}{q-1} > 0$. In that case it follows from Lemma 1.6 that

$$cap_{\frac{2}{q},q'}\left(\frac{F_n}{d_{n+1}}\right) \approx d_{n+1}^{\frac{2}{q-1}-N} cap_{\frac{2}{q},q'}(F_n).$$

Thus

$$W_F(x,t) \approx t^{-1-\frac{N}{2}} \sum_{n=0}^{a_t} e^{-\frac{n}{4}} cap_{\frac{2}{q},q'}(F_n).$$

Since

$$cap_{\frac{2}{q},q'}(F_n) \geq cap_{\frac{2}{q},q'}\left(F \cap B_{d_{n+1}}\right) - cap_{\frac{2}{q},q'}\left(F \cap B_{d_n}\right),$$

we obtain, using again Abel's transform,

$$t^{-\frac{N}{2}}\sum_{n=0}^{a_t} e^{-\frac{n}{4}} cap_{\frac{2}{q},q'}(F_n) \geq (1-e^{-\frac{1}{4}})t^{-\frac{N}{2}}\sum_{n=0}^{a_t-1} e^{-\frac{n}{4}} cap_{\frac{2}{q},q'}\left(F\cap B_{d_{n+1}}\right)$$
$$\geq (1-e^{-\frac{1}{4}})t^{-(1+\frac{N}{2})}\int_0^{\sqrt{ta_t}} e^{-\frac{s^2}{4t}} cap_{\frac{2}{q},q'}(F\cap B_s)\, s\, ds.$$

Because $cap_{\frac{2}{q},q'}(F\cap B_s) \approx s^{N-\frac{2}{q-1}} cap_{\frac{2}{q},q'}\left(s^{-1}F\cap B_1\right)$, (1.265) follows. □

1.4.4 Applications

The main result of this section is the following.

Theorem 1.20 *Let $N \geq 1$, $q > 1$ and F be a closed subset of $\mathbb{R}^N$. Then $\overline{u}_F = \underline{u}_F$.*

Proof When $1 < q < q_c$ this is proved in Proposition 1.2. The principle of the proof uses convexity and the integral forms of Theorems 1.15 and 1.18. The technique is an adaptation that we recall for the sake of completeness of the proof in the subcritical case. By Theorems 1.17 and 1.19 there exists a positive constant C, depending on N, q and T such that

$$\underline{u}_F \leq \overline{u}_F \leq C\underline{u}_F \quad \text{in } Q_T. \tag{1.266}$$

Let us assume that $\overline{u}_F \neq \underline{u}_F$. By the strong maximum principle $\overline{u}_F > \underline{u}_F$. By convexity $\tilde{u} = \overline{u}_F - \frac{1}{2C}(\overline{u}_F - \underline{u}_F)$ is a supersolution, which is smaller than $\underline{u}_F$. If we set $\theta := \left(\frac{1}{2}+\frac{1}{2C}\right)$, then $0 < \theta < 1$ and $\theta\underline{u}_F$ is a subsolution smaller than $\underline{u}_F$. There exists a solution u^* of (1.30) which satisfies

$$\theta\underline{u}_F \leq u^* \leq \tilde{u} < \underline{u}_F \quad \text{in } Q_T.$$

Hence u^* is a solution of (1.174). If μ is an admissible measure vanishing outside F, then $u_{\theta\mu}$ is the smallest solution above the subsolution θu_μ. Thus $u_{\theta\mu} \leq u^* < \underline{u}_F$. Since μ is arbitrary, we deduce $\underline{u}_F \leq u^* < \underline{u}_F$, which is a contradiction. □

Another consequence of the uniqueness result is the following equivalence of the discrete and integral capacitary potentials.

Proposition 1.5 *Assume $q \geq q_c$. Then there exist two positive constants $C_1^\dagger$, $C_2^\dagger$, depending only on N, q and T such that*

$$C_2^\dagger t^{-(1+\frac{N}{2})}\int_0^{\sqrt{ta_t}} s^{N-\frac{2}{q-1}} e^{-\frac{s^2}{4t}} cap_{\frac{2}{q},q'}\left(\frac{F}{s}\cap B_1(x)\right) s\,ds \leq W_F(x,t)$$
$$\leq C_1^\dagger t^{-(1+\frac{N}{2})}\int_{\sqrt{t}}^{\sqrt{t(a_t+2)}} s^{N-\frac{2}{q-1}} e^{-\frac{s^2}{4t}} cap_{\frac{2}{q},q'}\left(\frac{F}{s}\cap B_1(x)\right) s\,ds \tag{1.267}$$

for any $(x,t) \in Q_T$.

Definition 1.8 If F is a closed subset of $\mathbb{R}^N$, we define the $(\frac{2}{q}, q')$-integral parabolic capacitary potential $\mathcal{W}_F$ by

$$\mathcal{W}_F(x,t) = t^{-1-\frac{N}{2}} \int_0^{D_F(x)} s^{N-\frac{2}{q-1}} e^{-\frac{s^2}{4t}} cap_{\frac{2}{q},q'}\left(\frac{F}{s} \cap B_1(x)\right) s\, ds \quad \forall (x,t) \in Q_\infty, \tag{1.268}$$

where $D_F(x) = \max\{|x-y| : y \in F\}$.

By an easy computation we obtain that

$$\begin{aligned} 0 \leq \mathcal{W}_F(x,t) - t^{-(1+\frac{N}{2})} \int_0^{\sqrt{ta_t}} s^{N-\frac{2}{q-1}} e^{-\frac{s^2}{4t}} cap_{\frac{2}{q},q'}\left(\frac{F}{s} \cap B_1(x)\right) s\, ds \\ \leq C \frac{t^{(q-3)/2(q-1)}}{D_F(x)} e^{-\frac{D_F^2(x)}{4t}}, \end{aligned} \tag{1.269}$$

and

$$\begin{aligned} 0 \leq t^{-(1+\frac{N}{2})} \int_0^{\sqrt{t(a_t+2)}} s^{N-\frac{2}{q-1}} e^{-\frac{s^2}{4t}} cap_{\frac{2}{q},q'}\left(\frac{F}{s} \cap B_1(x)\right) s\, ds - \mathcal{W}_F(x,t) \\ \leq C \frac{t^{(q-3)/2(q-1)}}{D_F(x)} e^{-\frac{D_F^2(x)}{4t}}, \end{aligned} \tag{1.270}$$

for some $C = C(N,q) > 0$. Furthermore

$$\mathcal{W}_F(x,t) = t^{-\frac{1}{q-1}} \int_0^{D_F(x)/\sqrt{t}} s^{N-\frac{2}{q-1}} e^{-\frac{s^2}{4t}} cap_{\frac{2}{q},q'}\left(\frac{F}{s\sqrt{t}} \cap B_1(x)\right) s\, ds. \tag{1.271}$$

The following result gives a sufficient condition in order that $\overline{u}_F$ has a strong blow-up (i.e. of the maximal order $t^{-1/(q-1)}$) at a point x.

Proposition 1.6 *Assume $q \geq q_c$ and F is a closed subset of $\mathbb{R}^N$. If there exists $\gamma \in [0,\infty)$ such that*

$$\lim_{\tau \to 0} cap_{\frac{2}{q},q'}\left(\frac{F}{\tau} \cap B_1(x)\right) = \gamma, \tag{1.272}$$

then

$$\lim_{t \to 0} t^{\frac{1}{q-1}} \overline{u}_F(x,t) = C\gamma, \tag{1.273}$$

for some $C = C(N,q) > 0$.

Proof Clearly, condition (1.272) implies

$$\lim_{t \to 0} cap_{\frac{2}{q},q'}\left(\frac{F}{\sqrt{t}} \cap B_1(x)\right) = \gamma$$

for any $s > 0$. Then (1.273) follows by Lebesgue's theorem. Notice also that the set of γ is bounded from above by a constant depending on N and q. □

In the next result we give a condition in order that the solution remains bounded at a point x. The proof is similar to the previous one.

Proposition 1.7 *Assume $q \geq q_c$ and F is a closed subset of $\mathbb{R}^N$. If*

$$\limsup_{\tau \to 0} \tau^{-\frac{2}{q-1}} cap_{\frac{2}{q},q'}\left(\frac{F}{\tau} \cap B_1(x)\right) < \infty, \tag{1.274}$$

then $\overline{u}_F(x,t)$ remains bounded when $t \to 0$.

Remark If we assume that f is a convex function on $\mathbb{R}^+$ satisfying

$$c_2 r^q \leq f(r) \leq c_1 r^q \qquad \forall r \geq 0, \tag{1.275}$$

for some $0 < c_2 \leq c_1$ we can construct in the same way as for (1.30) the solutions $\underline{u}_F$ and $\overline{u}_F$ for equation

$$\partial_t u - \Delta u + f(u) = 0 \qquad \text{in } Q_T. \tag{1.276}$$

The bilateral estimate (1.266) is still valid (up to change of the C_i). Since only convexity of f is used in the proof of Theorem 1.20, there still holds $\underline{u}_F = \overline{u}_F$. Similar extensions of Proposition 1.6 and Proposition 1.7 are also clear.

1.4.5 Appendix

We present here some highly technical computations which are not of particular interest for the trace theory but are useful in the proof of the results.

1.4.5.1 Generalised Beta Integrals

Lemma 1.21 *Let a and b be two real numbers, $a > 0$ and $\kappa > 0$. Then there exists a constant $C = C(a,b,\kappa) > 0$ such that for any $A > 0$, $B > \kappa/A$ there holds*

$$\int_0^1 (1-x)^{-a} x^{-b} e^{-A^2/4(1-x)} e^{-B^2/4x} dx \leq C e^{-(A+B)^2/4} A^{1-a} B^{1-b} (A+B)^{a+b-2}. \tag{1.277}$$

Proof We first notice that

$$\max\left\{e^{-A^2/4(1-x)} e^{-B^2/4x} : 0 \leq x \leq 1\right\} = e^{-(A+B)^2/4}, \tag{1.278}$$

and it is achieved for $x_0 = B/(A+B)$. Set

$$\phi(x) = (1-x)^{-a}x^{-b}e^{-A^2/4(1-x)}e^{-B^2/4x},$$

then

$$\int_0^1 \phi(x)dx = \int_0^{x_0} \phi(x)dx + \int_{x_0}^1 \phi(x)dx = I_{a,b} + J_{a,b}.$$

Put

$$u = \frac{A^2}{4(1-x)} + \frac{B^2}{4x}, \tag{1.279}$$

then

$$4ux^2 - (4u + B^2 - A^2)x + B^2 = 0. \tag{1.280}$$

If $0 < x < x_0$ this equation admits the solution

$$x = x(u) = \frac{1}{8u}\left(4u + B^2 - A^2 - \sqrt{16u^2 - 8u(A^2+B^2) + (A^2-B^2)^2}\right)$$

$$\begin{aligned}&\int_0^{x_0} (1-x)^{-a}x^{-b}e^{-A^2/4(1-x)-B^2/4x}dx\\ &= -\int_{(A+B)^2/4}^{\infty} (1-x(u))^{-a}x(u)^{-b}e^{-u}x'(u)du.\end{aligned}$$

Putting $x' = x'(u)$ and differentiating (1.280),

$$4x^2 + 8uxx' - (4u + B^2 - A^2)x' - 4x = 0 \Longrightarrow -x' = \frac{4x(1-x)}{4u + B^2 - A^2 - 8ux}.$$

Thus

$$\int_0^{x_0} \phi(x)dx = 4\int_{(A+B)^2/4}^{\infty} \frac{(1-x(u))^{-a+1}x(u)^{-b+1}e^{-u}du}{4u + B^2 - A^2 - 8ux(u)}. \tag{1.281}$$

Using the explicit value of the root $x(u)$, we finally get

$$\int_0^{x_0} \phi(x)dx = 4\int_{(A+B)^2/4}^{\infty} \frac{(1-x(u))^{-a+1}x(u)^{-b+1}e^{-u}du}{\sqrt{16u^2 - 8u(A^2+B^2) + (A^2-B^2)^2}}, \tag{1.282}$$

and the factorisation below holds

$$16u^2 - 8u(A^2+B^2) + (A^2-B^2)^2 = 16(u - (A+B)^2/4)(u - (A-B)^2/4).$$

We set $u = \upsilon + (A+B)^2/4$ and obtain

$$x(u)=\frac{v+(AB+B^2)/2-\sqrt{v(v+AB)}}{2\left(v+(A+B)^2/4\right)},$$

and

$$1-x(u)=\frac{v+(A^2+AB)/2+\sqrt{v(v+AB)}}{2\left(v+(A+B)^2/4\right)}.$$

We introduce the relation $\approx$ linking two positive quantities depending on A and B. It means that the two-sided inequalities up to multiplicative constants independent of A and B. Therefore

$$\int_0^{x_0}\phi(x)dx=2^{a-b-4}e^{-(A+B)^2/4}\int_0^{\infty}\widetilde{\phi}(v)dv \quad \text{where}$$
$$\widetilde{\phi}(v)=\frac{\left(v+(AB+B^2)/2-\sqrt{v(v+AB)}\right)^{1-b}\left(v+(A^2+AB)/2+\sqrt{v(v+AB)}\right)^{1-a}}{e^v\left(v+(A+B)^2/4\right)^{2-a-b}\sqrt{v(v+AB)}}. \tag{1.283}$$

Case 1: $a\geq 1$, $b\geq 1$. First

$$\frac{\left(v+(A+B)^2/4\right)^{a+b-2}}{\sqrt{v(v+AB)}}\leq\frac{\left(v+(A+B)^2/4\right)^{a+b-2}}{\sqrt{v(v+\kappa)}}\approx\frac{\left(v+(A+B)^2\right)^{a+b-2}}{\sqrt{v(v+\kappa)}} \tag{1.284}$$

since $a+b-2\geq 0$ and $AB\geq\kappa$. Next

$$\left(v+(A^2+AB)/2+\sqrt{v(v+AB)}\right)^{1-a}\approx(v+A(A+B))^{1-a}. \tag{1.285}$$

Furthermore

$$\begin{aligned}v+(AB+B^2)/2-\sqrt{v(v+AB)}&=B^2\frac{v+(A+B)^2/4}{v+B(A+B)/2+\sqrt{v(v+AB)}}\\&\approx B^2\frac{v+(A+B)^2}{v+B(A+B)}.\end{aligned} \tag{1.286}$$

Then

$$\left(v+(AB+B^2)/2-\sqrt{v(v+AB)}\right)^{1-b}\approx B^{2-2b}\left(\frac{v+B(A+B)}{v+(A+B)^2}\right)^{b-1}. \tag{1.287}$$

It follows

$$\begin{aligned}\widetilde{\phi}(v)&\leq CB^{2-2b}\left(\frac{v+(A+B)^2}{v+A(A+B)}\right)^{a-1}\frac{(v+B(A+B))^{b-1}}{\sqrt{v(v+\kappa)}}\\&\leq CB^{2-2b}\left(\frac{v+(A+B)^2}{v+A(A+B)}\right)^{a-1}\frac{v^{b-1}+(B^2+AB)^{b-1}}{\sqrt{v(v+\kappa)}},\end{aligned} \tag{1.288}$$

where C depends on a, b and κ. The function $v \mapsto (v + (A + B)^2)/(v + A(A + B))$ is decreasing on $(0, \infty)$. If we set

$$C_1 = \int_0^\infty \frac{v^{b-1} e^{-v} dv}{\sqrt{v(v+\kappa)}} \quad \text{and} \quad C_2 = \int_0^\infty \frac{e^{-v} dv}{\sqrt{v(v+\kappa)}},$$

then

$$C_1 \leq K(B^2 + AB)^{b-1} C_2$$

with $K = C_1 \kappa^{1-b}/C_2$. Therefore

$$\int_0^{x_0} \phi(x)dx \leq Ce^{-(A+B)^2/4} B^{1-b} A^{1-a} (A + B)^{a+b-2}. \tag{1.289}$$

The estimate of $J_{a,b}$ is obtained by exchanging (A, a) with (B, b) and replacing x by $1 - x$. *Mutatis mutandis*, this leads directly to the same expression as in (1.289) and finally

$$\int_0^1 \phi(x)dx \leq Ce^{-(A+B)^2/4} A^{1-a} B^{1-b} (A + B)^{a+b-2}. \tag{1.290}$$

Case 2: $a \geq 1$, $b < 1$. Estimates (1.283), (1.284), (1.285), (1.286) and (1.287) are valid. Because $v \mapsto (v + B(A + B))^{b-1}$ is decreasing, (1.288) has to be replaced by

$$\widetilde{\phi}(v) \leq CB^{2-2b} \left(\frac{v + (A + B)^2}{v + A(A + B)} \right)^{a-1} \frac{(AB + B^2)^{b-1}}{\sqrt{v(v+\kappa)}}. \tag{1.291}$$

This implies (1.289) directly. The estimate of $J_{a,b}$ is performed by the change of variable $x \mapsto 1 - x$. If $x_1 = 1 - x_0$, there holds

$$J_{a,b} = \int_0^{x_1} x^{-a}(1 - x)^{-b} e^{-A^2/4x} e^{-B^2/4(1-x)} dx = \int_0^{x_1} \Psi(x)dx.$$

Then

$$\int_0^{x_1} \Psi(x)dx = 2^{b-a-4} e^{-(A+B)^2/4} \int_0^{x_1} \widetilde{\Psi}(v)dv \quad \text{where}$$
$$\widetilde{\Psi}(v) = \frac{\left(v + (AB + A^2)/2 - \sqrt{v(v + AB)}\right)^{1-a} \left(v + (B^2 + AB)/2 + \sqrt{v(v + AB)}\right)^{1-b}}{e^v \left(v + (A + B)^2/4\right)^{2-a-b} \sqrt{v(v + AB)}}. \tag{1.292}$$

Equivalence (1.284) is unchanged; (1.285) is replaced by

$$\left(v + (B^2 + AB)/2 + \sqrt{v(v + AB)}\right)^{1-b} \approx (v + B(A + B))^{1-b}, \tag{1.293}$$

(1.286) by

$$v + (AB + A^2)/2 - \sqrt{v(v + AB)} \approx A^2 \frac{v + (A + B)^2}{v + A(A + B)}, \tag{1.294}$$

and (1.287) by

$$\left(v + (AB + A^2)/2 - \sqrt{v(v + AB)}\right)^{1-a} \approx A^{2-2a} \left(\frac{v + A(A + B)}{v + (A + B)^2}\right)^{a-1}. \tag{1.295}$$

Because $a > 1$, (1.288) turns into

$$\begin{aligned}\widetilde{\Psi}(v) &\leq CA^{2-2b}(v + (A + B)^2)^{b-1} \frac{(v + A^2 + AB)^{a-1}(v + B^2 + AB)^{1-b}}{\sqrt{v(v + \kappa)}} \\ &\leq Ce^{-(A+B)^2/4} A^{2-2b}(A + B)^{2b-2} \\ &\times \frac{v^{a-b} + (A^2 + AB)^{a-1}v^{1-b} + (B^2 + AB)^{1-b}v^{a-1} + A^{a-1}B^{1-b}(A + B)^{a-b}}{\sqrt{v(v + \kappa)}}.\end{aligned} \tag{1.296}$$

Because $AB \geq \kappa$, there exists a positive constant C, depending on κ, such that

$$\begin{aligned}\int_0^\infty & \frac{v^{a-b} + (A^2 + AB)^{a-1}v^{1-b} + (B^2 + AB)^{1-b}v^{a-1}}{\sqrt{v(v + \kappa)}} e^{-v} dv \\ &\leq CA^{a-1}B^{1-b}(A + B)^{a-b} \int_0^\infty \frac{e^{-v}dv}{\sqrt{v(v + \kappa)}}.\end{aligned} \tag{1.297}$$

Combining (1.296) and (1.297) leads to

$$\int_0^{x_1} \Psi(x)dx \leq Ce^{-(A+B)^2/4} A^{1-a}B^{1-b}(A + B)^{a+b-2}. \tag{1.298}$$

This, again, implies that (1.277) holds.

Case 3: $\max\{a, b\} < 1$. Inequalities (1.283)–(1.287) hold, but (1.288) has to be replaced by

$$\begin{aligned}\widetilde{\phi}(v) &\leq CB^{2-2b} \left(\frac{v + (A + B)^2}{v + A(A + B)}\right)^{a-1} \frac{\left(v + B^2 + AB\right)^{b-1}}{\sqrt{v(v + \kappa)}} \\ &\leq CB^{1-b}(A + B)^{2a+b-3} \frac{v^{1-a} + \left(A^2 + AB\right)^{1-a}}{\sqrt{v(v + \kappa)}}.\end{aligned} \tag{1.299}$$

Noticing that

$$\int_0^\infty \frac{v^{1-a}e^{-v}dv}{\sqrt{v(v + \kappa)}} \leq C\left(A^2 + AB\right)^{1-a} \int_0^\infty \frac{e^{-v}dv}{\sqrt{v(v + \kappa)}},$$

it follows that (1.289) holds. Finally (1.290) holds by exchanging (A, a) and (B, b). □

1.4.5.2 Discrete Generalised Beta Series

Lemma 1.22 . *Let α, β, γ, δ be real numbers and ℓ an integer. We assume $\gamma > 1$, $\delta > 0$ and $\ell \geq 2$. Then there exists a positive constant C such that for any integer $n > \ell$*

$$\sum_{p=1}^{n-\ell} p^{\alpha}(\sqrt{n}-\sqrt{p})^{\beta} e^{-\delta(\sqrt{p}+\sqrt{\gamma}(\sqrt{n}-\sqrt{p+1}))^2} \leq C n^{\alpha-\beta/2} e^{-\delta n}. \tag{1.300}$$

Proof The function $x \mapsto (\sqrt{x}+\sqrt{\gamma}(\sqrt{n}-\sqrt{x+1}))^2$ is decreasing on $[(\gamma-1)^{-1}, \infty)$. Furthermore there exists $C > 0$ depending on ℓ, α and β such that $p^{\alpha}(\sqrt{n}-\sqrt{p})^{\beta} \leq C x^{\alpha}(\sqrt{n}-\sqrt{x+1})^{\beta}$ for $x \in [p, p+1]$. If we denote by p_0 the smallest integer larger than $(\gamma-1)^{-1}$, we derive

$$\begin{aligned}
S &= \sum_{p=1}^{n-\ell} p^{\alpha}(\sqrt{n}-\sqrt{p})^{\beta} e^{-(\sqrt{p}+\sqrt{\gamma}(\sqrt{n}-\sqrt{p+1}))^2/4} \\
&= \sum_{p=1}^{p_0-1} + \sum_{p_0}^{n-\ell} p^{\alpha}(\sqrt{n}-\sqrt{p})^{\beta} e^{-\delta(\sqrt{p}+\sqrt{\gamma}(\sqrt{n}-\sqrt{p+1}))^2} \\
&\leq \sum_{p=1}^{p_0-1} p^{\alpha}(\sqrt{n}-\sqrt{p})^{\beta} e^{-\delta(\sqrt{p}+\sqrt{\gamma}(\sqrt{n}-\sqrt{p+1}))^2} \\
&\qquad + C\int_{p_0}^{n+1-\ell} x^{\alpha}(\sqrt{n}-\sqrt{x})^{\beta} e^{-\delta(\sqrt{x}+\sqrt{\gamma}(\sqrt{n}-\sqrt{x+1}))^2} dx
\end{aligned}$$

(notice that $\sqrt{n}-\sqrt{x} \approx \sqrt{n}-\sqrt{x+1}$ for $x \leq n-\ell$). Clearly

$$\sum_{p=1}^{p_0-1} p^{\alpha}(\sqrt{n}-\sqrt{p})^{\beta} e^{-\delta(\sqrt{p}+\sqrt{\gamma}(\sqrt{n}-\sqrt{p+1}))^2} \leq C_0 n^{\alpha}(\sqrt{n}-\sqrt{n-\ell})^{\beta} e^{-\delta n}, \tag{1.301}$$

for some C_0 independent of n. We set $y = y(x) = \sqrt{x+1}-\sqrt{x}/\sqrt{\gamma}$. Obviously

$$y'(x) = \frac{1}{2}\left(\frac{1}{\sqrt{x+1}} - \frac{1}{\sqrt{\gamma}\sqrt{x}}\right) \qquad \forall x \geq p_0.$$

Thus there exists $\epsilon = \epsilon(\delta, \gamma) > 0$ such that $\sqrt{2}\sqrt{x} \geq y(x) \geq \epsilon\sqrt{x}$ and $y'(x) \geq \epsilon/\sqrt{x}$. Moreover

$$\sqrt{x} = \frac{\sqrt{\gamma}\left(y+\sqrt{\gamma y^2+1-\gamma}\right)}{\gamma-1},$$

$$\begin{aligned}\sqrt{n}-\sqrt{x} &= \frac{\sqrt{n}(\gamma-1)-\sqrt{\gamma}y-\sqrt{\gamma}\sqrt{\gamma y^2+1-\gamma}}{\gamma-1}\\ &= \frac{n(\gamma-1)+\gamma-2y\sqrt{\gamma n}-\gamma y^2}{\sqrt{n}(\gamma-1)-\sqrt{\gamma}y+\sqrt{\gamma}\sqrt{\gamma y^2+1-\gamma}}\\ &\approx \frac{n(\gamma-1)+\gamma-2y\sqrt{\gamma n}-\gamma y^2}{\sqrt{n}}\end{aligned}$$

since $y(x)\leq\sqrt{n}$. Furthermore

$$\begin{aligned}n(\gamma-1)+\gamma-2y\sqrt{\gamma n}-\gamma y^2 &= \gamma(\sqrt{n+1}+\sqrt{n}/\sqrt{\gamma}+y)(\sqrt{n+1}-\sqrt{n}/\sqrt{\gamma}-y)\\ &\approx \sqrt{n}(\sqrt{n+1}-\sqrt{n}/\sqrt{\gamma}-y),\end{aligned}$$

because y ranges from $\sqrt{n+2-\ell}-\sqrt{n+1-\ell}\sqrt{\gamma}\approx\sqrt{n}$ and $\sqrt{p_0+1}-\sqrt{p_0}\sqrt{\gamma}$. Thus

$$(\sqrt{n}-\sqrt{x})^\beta\approx\left(\sqrt{n+1}-\sqrt{n}/\sqrt{\gamma}-y\right)^\beta.$$

This implies

$$\begin{aligned}&\int_{p_0}^{n+1-\ell}x^\alpha(\sqrt{n}-\sqrt{x})^\beta e^{-\delta(\sqrt{x}+\gamma(\sqrt{n}-\sqrt{x+1}))^2}dx\\ &\quad\leq C\int_{y(p_0)}^{y(n+1-\ell)}y^{2\alpha+1}\left(\sqrt{n+1}-\sqrt{n}/\sqrt{\gamma}-y\right)^\beta e^{-\gamma\delta(\sqrt{n}-y)^2}dy\\ &\quad\leq Cn^{\alpha+\beta/2+1}\int_{1-y(n+1-\ell)/\sqrt{n}}^{1-y(p_0)/\sqrt{n}}(1-z)^{2\alpha+1}(z+\sqrt{1+1/n}-1-1/\sqrt{\gamma})^\beta e^{-\gamma\delta nz^2}dz.\end{aligned}\tag{1.302}$$

Moreover

$$\begin{aligned}1-\frac{y(p_0)}{\sqrt{n}} &= 1-\frac{1}{\sqrt{n}}\left(\sqrt{p_0+1}-\frac{\sqrt{p_0}}{\sqrt{\gamma}}\right),\\ 1-\frac{y(n-\ell+1)}{\sqrt{n}} &= 1-\frac{\sqrt{n-\ell+2}}{\sqrt{n}}+\frac{\sqrt{n-\ell+1}}{\sqrt{n\gamma}}\\ &= \frac{1}{\sqrt{\gamma}}\left(1+\frac{\sqrt{\gamma}(\ell-2)-\ell+1}{2n}+\frac{\sqrt{\gamma}(\ell-2)^2-(\ell-1)^2}{8n^2}\right)+O(n^{-3}).\end{aligned}\tag{1.303}$$

Let θ be fixed such that $1-\dfrac{y(n-\ell+1)}{\sqrt{n}}<\theta<1-\dfrac{y(p_0)}{\sqrt{n}}$ for any $n>p_0$. Then

$$\int_{\theta}^{1-y(p_0)/\sqrt{n}} (1-z)^{2\alpha+1}(z+\sqrt{1+1/n}-1-1/\sqrt{\gamma})^{\beta} e^{-\gamma\delta n z^2} dz$$
$$\leq C_\theta \int_{\theta}^{1-y(p_0)/\sqrt{n}} (1-z)^{2\alpha+1} e^{-\gamma\delta n z^2} dz$$
$$\leq C_\theta\, e^{-\gamma\delta n\theta^2} \int_{\theta}^{1-y(p_0)/\sqrt{n}} (1-z)^{2\alpha+1} dz$$
$$\leq C\, e^{-\gamma\delta n\theta^2} \max\{1, n^{-\alpha-1/2}\}.$$

Because $\gamma\theta^2 > 1$ we derive

$$\int_{\theta}^{1-y(p_0)/\sqrt{n}} (1-z)^{2\alpha+1}(z+\sqrt{1+1/n}-1-1/\sqrt{\gamma})^{\beta} e^{-\gamma\delta n z^2} dz \leq C n^{-\beta} e^{-\delta n}, \tag{1.304}$$

for some constant $C > 0$. On the other hand

$$\int_{1-y(n+1-\ell)/\sqrt{n}}^{\theta} (1-z)^{2\alpha+1}(z+\sqrt{1+1/n}-1-1/\sqrt{\gamma})^{\beta} e^{-\gamma\delta n z^2} dz$$
$$\leq C'_\theta \int_{1-y(n+1-\ell)/\sqrt{n}}^{\theta} (z+\sqrt{1+1/n}-1-1/\sqrt{\gamma})^{\beta} e^{-\gamma\delta n z^2} dz.$$

The minimum of $z \mapsto (z+\sqrt{1+1/n}-1-1/\sqrt{\gamma})^{\beta}$ is achieved at $1-y(n+1-\ell)$ with value

$$\frac{\sqrt{\gamma}(\ell+1)+1-\ell}{2n\sqrt{\gamma}} + O(n^{-2}),$$

and the maximum of the exponential term is achieved at the same point with value

$$e^{-n\delta+((\ell-2)\sqrt{\gamma}+1-\ell)/2}(1+\mathrm{o}(1)) = C_\gamma e^{-n\delta}(1+\mathrm{o}(1)).$$

We denote

$$z_{\gamma,n} = 1+1/\sqrt{\gamma}-\sqrt{1+1/n} \quad \text{and} \quad I_\beta = \int_{1-y(n+1-\ell)/\sqrt{n}}^{\theta} (z-z_{\gamma,n})^{\beta} e^{-\gamma\delta n z^2} dz.$$

Since $1-y(n+1-\ell) \geq 1/\sqrt{2\gamma}$ for n large enough,

$$\begin{aligned} I_\beta &\le \sqrt{2\gamma}\int_{1-y(n+1-\ell)/\sqrt{n}}^{\theta}(z-z_{\gamma,n})^\beta z e^{-\gamma\delta n z^2}dz \\ &\le \frac{-\sqrt{2\gamma}}{2n\gamma\delta}\left[(z-z_{\gamma,n})^\beta e^{-\gamma\delta n z^2}\right]_{1-y(n+1-\ell)/\sqrt{n}}^{\theta} \\ &\qquad + \frac{\beta\sqrt{2\gamma}}{2n\gamma\delta}\int_{1-y(n+1-\ell)/\sqrt{n}}^{\theta}(z-z_{\gamma,n})^{\beta-1} z e^{-\gamma\delta n z^2}dz. \end{aligned}$$

But $1-y(n+1-\ell)/\sqrt{n}-z_{\gamma,n}=(\ell-1)(1-1/\sqrt{\gamma})/2n$, therefore

$$I_\beta \le C_1 n^{-\beta-1}e^{-\delta n}+\beta C_1' n^{-1}I_{\beta-1}. \tag{1.305}$$

If $\beta \le 0$, we derive

$$I_\beta \le C_1 n^{-\beta-1}e^{-\delta n},$$

which inequality, combined with (1.302) and (1.304), yields (1.300). If $\beta > 0$, we iterate and get

$$I_\beta \le C_1 n^{-\beta-1}e^{-\delta n}+C_1' n^{-1}(C_1 n^{-\beta}e^{-\delta n}+(\beta-1)C_1' n^{-1}I_{\beta-2}).$$

If $\beta - 1 \le 0$ we derive

$$I_\beta \le C_1 n^{-\beta-1}e^{-\delta n}+C_1C_1' n^{-1-\beta}e^{-\delta n}=C_2 n^{-\beta-1}e^{-\delta n},$$

which again yields (1.300). If $\beta - 1 > 0$, we iterate this process until we obtain a positive integer k such that $\beta - k \le 0$, which again leads to

$$I_\beta \le C_k n^{-\beta-1}e^{-\delta n},$$

and finally to (1.300). □

1.4.5.3 Generalised Wallis Integrals

Lemma 1.23 *For any integer $N \ge 2$ there exists a constant $c_N > 0$ such that*

$$\int_0^\pi e^{m\cos\theta}\sin^{N-2}\theta\,d\theta \le c_N\frac{e^m}{(1+m)^{(N-1)/2}} \qquad \forall m>0. \tag{1.306}$$

Proof Put $\mathcal{I}_N(m)=\displaystyle\int_0^\pi e^{m\cos\theta}\sin^{N-2}\theta\,d\theta$. Then $\mathcal{I}_2'(m)=\displaystyle\int_0^\pi e^{m\cos\theta}\cos\theta\,d\theta$ and

$$\begin{aligned}
\mathcal{I}_2''(m) &= \int_0^\pi e^{m\cos\theta}\cos^2\theta\, d\theta = \mathcal{I}_2(m) - \int_0^\pi e^{m\cos\theta}\sin^2\theta\, d\theta \\
&= \mathcal{I}_2(m) - \frac{1}{m}\int_0^\pi e^{m\cos\theta}\cos\theta\, d\theta \\
&= \mathcal{I}_2(m) - \frac{1}{m}\mathcal{I}_2'(m).
\end{aligned}$$

Thus $\mathcal{I}_2$ satisfies a Bessel equation of order 0. Since $\mathcal{I}_2(0) = \pi$ and $\mathcal{I}_2'(0) = 0$, $\pi^{-1}\mathcal{I}_2$ is the modified Bessel function of index 0 (usually denoted by I_0) the asymptotic behaviour of which is well known, thus (1.306) holds. If $N = 3$

$$\mathcal{I}_3(m) = \int_0^\pi e^{m\cos\theta}\sin\theta\, d\theta = \left[\frac{-e^{m\cos\theta}}{m}\right]_0^\pi = \frac{2\sinh m}{m}.$$

For $N > 3$ arbitrary

$$\mathcal{I}_N(m) = \int_0^\pi \frac{-1}{m}\frac{d}{d\theta}(e^{m\cos\theta})\sin^{N-3}\theta\, d\theta = \frac{N-3}{m}\int_0^\pi e^{m\cos\theta}\cos\theta\sin^{N-4}\theta\, d\theta. \tag{1.307}$$

Therefore,

$$\mathcal{I}_4(m) = \frac{1}{m}\int_0^\pi e^{m\cos\theta}\cos\theta\, d\theta = \mathcal{I}_2'(m),$$

and again (1.306) holds since $I_0'(m)$ has the same behaviour as $I_0(m)$ at infinity. For $N \geq 5$

$$\mathcal{I}_N(m) = \frac{3-N}{m^2}\left[e^{m\cos\theta}\cos\theta\sin^{N-5}\theta\right]_0^\pi + \frac{N-3}{m^2}\int_0^\pi e^{m\cos\theta}\frac{d}{d\theta}\left(\cos\theta\sin^{N-5}\theta\right)d\theta.$$

Differentiating $\cos\theta\sin^{N-5}\theta$ and using (1.307), we obtain

$$\mathcal{I}_5(m) = \frac{4\sinh m}{m^2} - \frac{4\sinh m}{m^3},$$

while

$$\mathcal{I}_N(m) = \frac{(N-3)(N-5)}{m^2}\left(\mathcal{I}_{N-4}(m) - \mathcal{I}_{N-2}(m)\right), \tag{1.308}$$

for $N \geq 6$. Since the estimate (1.306) for $\mathcal{I}_2, \mathcal{I}_3, \mathcal{I}_4$ and $\mathcal{I}_5$ has already been obtained, a straightforward induction implies the general result. □

Remark Although it does not has any importance for our use, it must be noticed that $\mathcal{I}_N$ can be expressed either with hyperbolic functions if N is odd, or with Bessel functions if N is even. Further develoments can be found in [40].

1.5 The Precise Trace

In the supercritical case $q \geq q_c$, Theorem 1.13 has pointed out the necessity to introduce a finer definition of the initial trace which could distinguish among solutions of (1.30) which have the same initial trace in the sense defined previously.

1.5.1 *Lattice Structure of the Set of Positive Solutions of (1.30)*

The idea of analysing the algebraic structure of the set of positive solutions of the semilinear elliptic equation (1.16) is due to Dynkin [18]. It was intensively used by Marcus and Véron [33] in the construction of the precise boundary trace for such equations.

Definition 1.9 We denote by $\mathcal{U}_+(Q_T)$ the set of nonnegative solutions of (1.30). All the elements of $\mathcal{U}_+(Q_T)$ belong to $C^{2,1}(Q_T)$.

By a subsolution (resp. supersolution) of (1.30) in Q_T we mean a function $u \in L^q_{loc}$ satisfying

$$\iint_{Q_T} \left(-(\partial_t\zeta + \Delta\zeta)u + |u|^{q-1}u\zeta\right) dxdt \leq 0 \quad (\text{resp. } \geq 0), \tag{1.309}$$

for all $\zeta \in C_c^\infty(Q_T)$, $\zeta \geq 0$.

Lemma 1.24 *Let u be a subsolution of (1.30) in Q_T, then*

$$|u(x,t)| \leq \left(\frac{1}{t(q-1)}\right)^{\frac{1}{q-1}} \quad \textit{for almost all } (x,t) \in Q_T. \tag{1.310}$$

Proof Because of Kato's inequality, the function $|u|$ is a subsolution of (1.30). Hence we can assume that u is nonnegative. Let $\{\rho_{\epsilon_n}\}$ $(\epsilon_n > 0)$ be a sequence of $C_c^\infty(\mathbb{R}^{N+1})$ nonnegative functions with support in B_{ϵ_n} and total mass equal to 1. We assume that $\epsilon_n \to 0$; hence $\rho_{\epsilon_n} \to \delta_0$ is the sense of distributions. Such a sequence is called a sequence of mollifiers. If $\epsilon_n < \epsilon$ the distribution $u_n := u * \rho_{\epsilon_n}$ is well defined and is C^∞ in $\mathbb{R}^N \times (\epsilon, T)$ where, by convexity, it satisfies

$$\partial_t u_n - \Delta u_n + u_n^q \leq 0.$$

As in the proof of (1.39), for any $y \in \mathbb{R}^N$, the function $(x,t) \mapsto \phi_\infty(t-\epsilon) + w_R(x-y)$ where w_R is defined in (1.40) is a supersolution of (1.30) in $\mathbb{R}^N \times (e, T)$ which dominates u_n at $t = \epsilon$ and for $|x-y| \to R$. Hence it is larger than u_n in this domain. Letting $R \to \infty$ and $\epsilon \to 0$ yields

$$u_n(x,t) \leq \phi_\infty(t) \quad \text{for all } (x,t) \in Q_T.$$

When $\epsilon_n \to 0$, u_n converges to u a.e. in Q_T and in $L^q_{loc}(Q_T)$. This implies (1.310). □

Proposition 1.8 *Let $T > 0$ and $u \in L^q_{loc}(Q_T)$ be nonnegative.*
(i) If u is a subsolution of (1.30) there exists a minimal solution v above u, that if U is any solution larger than u, then $u \leq v \leq U$.
(ii) If u is a continuous supersolution of (1.30) there exists a maximal solution w dominated by u, i.e. if U is any solution smaller than u, then $U \leq w \leq u$.
All the above inequalities hold both almost everywhere and in the sense of distributions.

Proof (i) We use again the subsolutions $u_n := u * \rho_{\epsilon_n}$, and for $\epsilon, R > 0$ we denote by $v_n := v_{\epsilon_n,\epsilon,R}$ be the solution of

$$\begin{aligned} \partial_t v_n - \Delta v_n + v_n^q &= 0 && \text{in } B_R \times (\epsilon, T) \\ v_n &= u_n && \text{on } \partial B_R \times (\epsilon, T) \\ v_n(.,s) &= u_n(.,s) && \text{in } B_R. \end{aligned} \tag{1.311}$$

Then $v_n \geq u_n$ by the comparison principle. Furthermore v_n satisfies

$$v_n(x,t) \leq \phi_\infty(t-\epsilon) + w_n(x),$$

where w_R is the large solution in B_R defined in (1.40). Hence it is locally bounded in $B_R \times (\epsilon, T)$ for any $\epsilon > 0$ and $R > 0$. Therefore, up to a subsequence $\{R_j\}$ such that $R_j \to \infty$, the sequence $\{v_{\epsilon_n,\epsilon,R_j}\}$ converges locally in $C^{2,1}(\mathbb{R}^N \times (\epsilon, T))$ to a nonnegative solution $v = v_{\epsilon_n,\epsilon}$ of (1.30) in $\mathbb{R}^N \times (\epsilon, T)$. Furthermore

$$v_{\epsilon_n,\epsilon}(x,t) \geq u_n(x,t) \text{ for all } (x,t) \in \mathbb{R}^N \times (\epsilon, T).$$

Since $v_{\epsilon_n,\epsilon}$ satisfies the uniform parabolic a priori estimates and the associated compactness properties, we infer that, up to a subsequence $v_{\epsilon_n,\epsilon} \to v_\epsilon$ locally in $C^{2,1}(\mathbb{R}^N \times (\epsilon, T))$ when $\epsilon_n \to 0$. As for u_n it converges to u a.e. and in $L^p_{loc}(\mathbb{R}^N \times (\epsilon, T))$ for any $p < \infty$. Furthermore

$$v_\epsilon(x,t) \geq u(x,t) \text{ for all } (x,t) \in \mathbb{R}^N \times (\epsilon, T).$$

By letting $\epsilon \to 0$ using again the local compactness of $\{v_\epsilon\}$ in $C^{2,1}(Q_T)$, we obtain that up to a subsequence, v_ϵ converges locally to a nonnegative solution v of (1.30) in Q_T which dominates u therein. By construction v is smaller than any element of $\mathcal{U}_+(Q_T)$ which dominates u.

(ii) For $\epsilon, R > 0$ we denote by $w := w_{\epsilon,R}$ the solution of

$$\begin{aligned} \partial_t w - \Delta w - w^q &= 0 && \text{in } B_R \times (\epsilon, T) \\ w &= u && \text{on } \partial B_R \times (\epsilon, T) \\ w(., \epsilon) &= u(., \epsilon) && \text{in } B_R. \end{aligned} \tag{1.312}$$

Note that the boundary values of w are well defined since u is continuous. By the comparison principle

$$0 \leq w_{\epsilon,R} \leq u \text{ in } B_R \times (\epsilon, T).$$

Furthermore $w_{\epsilon,R}$ dominates in $B_R \times (\epsilon, T)$ any nonnegative solution U smaller than u. Since u is continuous in Q_T, it is locally bounded therein. As in (i) the set of functions $\{w_{\epsilon,R}\}$ is eventually locally compact in $C^{2,1}(\mathbb{R}^N \times (0, T))$. We conclude as in (i). □

The following result has already been proved but we mention it for the sake of completeness.

Proposition 1.9 *Let u and v be nonnegative, locally bounded functions in Q_T.*
(i) If u and v are subsolutions (resp. supersolutions) then $\max\{u, v\}$ *(resp.* $\min\{u, v\}$*) is a subsolution (resp. a supersolution).*
(ii) If u and v are supersolutions then $u + v$ is a supersolution.
(iii) If u is a subsolution and v is a supersolution then $(u - v)_+$ is a subsolution.

The following notations have been introduced by Dynkin [18].

Notations Let u and v be nonnegative, locally bounded functions in Q_T.
(i) If u is a subsolution, $[u]_\dagger$ denotes the smallest solution dominating u.
(ii) If u is a continuous supersolution, $[u]^\dagger$ denotes the largest solution dominated by u.
(iii) If u and v are subsolutions then $u \vee v := [\max\{u, v\}]_\dagger$.
(iv) If u and v are continuous supersolutions, then $u \wedge v := [\min\{u, v\}]^\dagger$ and $u \oplus v = [u + v]^\dagger$.
(v) If u is a subsolution and v is a supersolution then $u \ominus v := [u - v]_\dagger$.

Proposition 1.10 *The following properties hold:*
(i) $(u \vee v) \vee w = u \vee (v \vee w) = [\max\{u, v, w\}]_\dagger$,
(ii) $(u \wedge v) \wedge w = u \wedge (v \wedge w) = [\min\{u, v, w\}]^\dagger$.

Proposition 1.11 *(i) Let $\{u_k\}$ be a sequence of positive, continuous subsolutions of (1.30). Then $U := \sup u_k$ is a subsolution. The statement remains valid if subsolution is replaced by supersolution and* sup *by* inf.
(ii) Let $\mathcal{T}$ be a family of positive solutions of (1.30). Suppose that, for every u_1 and u_2 belonging to $\mathcal{T}$, there exists $v \in \mathcal{T}$ such that

$$\max\{u_1, u_2\} \leq v \quad (\textit{resp. } \min\{u_1, u_2\} \geq v).$$

Then there exists a monotone sequence $\{u_n\} \subset \mathcal{T}$ such that

$$u_n \uparrow \sup_{\mathcal{T}} u, \qquad (resp.\ u_n \downarrow \inf_{\mathcal{T}} u).$$

Therefore $\sup_{\mathcal{T}} u$ *(resp.* $\inf_{\mathcal{T}} u$ *) is a solution.*

Proof (i) We set

$$v_j = \max\{u_1, u_2, ...u_j\}.$$

By induction on j it is clear that v_j is a subsolution and the sequence $\{v_j\}$ is nondecreasing. Because of the universal upper bound (1.43) v_j converges to some function $\bar{v}$ when $j \to \infty$, and $\bar{v}$ is a subsolution which coincides with U. The proof for the min assertion is similar.

(ii) is already proved in [17] and we recall the construction. For every $x \in Q_T$, we set $\ell(x,t) = \sup\{u(x,t) : u \in \mathcal{T}\}$. Let $A = \{(x_n, t_n)\}$ be a countable dense subset of Q_T. For every n there exists a sequence $\{u_{m,n}\} \subset \mathcal{T}$ such that

$$\sup_m \{u_{m,n}(x_n, t_n)\} = \lim_{m\to\infty} u_{m,n}(x_n, t_n) = \ell(x_n, t_n).$$

We set $u_{m,1} = u_{m_1}$. Since $\mathcal{T}$ is closed with respect to the relation $\vee$, $u_{m_2} := u_{m,1} \vee u_{m,2}$ belongs to $\mathcal{T}$ and the sequence $\{u_{m_2}\}$ is increasing and it satisfies

$$\lim_{m_j\to\infty} u_{m_j}(x_j, t_j) = \ell(x_j, t_j) \quad \text{for } j = 1, 2.$$

By induction we construct an increasing subsequence $\{u_{m_n}\}$ of $\mathcal{T}$ such that

$$\lim_{m_n\to\infty} u_{m_n}(x_n, t_n) = \ell(x_n, t_n) \qquad \text{for all } n \in \mathbb{N}_*.$$

Let us denote by $\mathcal{T}_0$ the countable subset of $\mathcal{T}$ of functions $\{u_{m_n}\}$ and set $v = \sup \mathcal{T}_0$. Then $v(x_n, t_n) = \ell(x_n, t_n)$. Using estimate (1.20) and regularity results for parabolic equations, we infer that the set $\mathcal{T}_0$ is relatively compact in the $C^{2,1}_{loc}(Q_T)$-topology. Hence, there exists a subsequence sequence of $\mathcal{T}_0$ still denoted by $\{u_{m_n,}\}$ which converges in this topology to a function w which is a nonnegative solution of (1.30) and such that $u_{m_n}(x_n, t_n) \to w(x_n, t_n)$ as $m_n \to \infty$. Hence $w(x_n, t_n) = \ell(x_n, t_n)$. We claim now that $w = \sup_{\mathcal{T}} u$. Indeed, if $u \in \mathcal{T}$, $w(x_n, t_n) \geq \ell(x_n, t_n) \geq u(x_n, t_n)$. By continuity, $w(x,t) \geq u(x,t)$ for all $(x,t) \in Q_T$. Thus w is an upper bound of $\mathcal{T}$. It is clearly the least upper bound because any other upper bound $\tilde{u} \in \mathcal{U}_+(Q_T)$ is larger than u_{m_n} on A, hence larger than w on A, and thus larger than w by density and continuity.

The proof concerning the existence of the greatest lower bound is similar if $\mathcal{T}$ is stable under $\wedge$. □

The set $\mathcal{U}_+(Q_T)$ is partially ordered for the relation $\leq$. Since for any $u, v \in \mathcal{U}_+(Q_T)$, $u \wedge v$ and $u \vee v$ belong to $\mathcal{U}_+(Q_T)$, it is a *lattice*. Since, by Proposition 3.4, any non-empty subset $\mathcal{T}$ of $\mathcal{U}_+(Q_T)$ admits both a least upper bound (the supremum) and a greatest lower bound (the infimum), it is a *complete lattice*. In the case of semilinear elliptic equations, a similar result is to be found in [17, Theorem 5.1].

Corollary 1.1 *The set $\mathcal{U}_+(Q_T)$ is a complete lattice stable for $\oplus$ and $\ominus$.*

1.5.2 Fine Topology and Besov Spaces

1.5.2.1 The $\mathfrak{T}_q$-Fine Topology

It is classical in potential theory that there exists a topology which is naturally adapted to the study of subharmonic functions. This topology was initially introduced by Henri Cartan and its definition is expressed in terms of the Newtonian capacity $cap_{1,2}$. In the study of the initial trace, the fine topology is the one associated with the $cap_{\frac{2}{q},q'}$ capacity. In this section we assume $q \geq q_c$ and we recall that $q' = \frac{q}{q-1}$.

Definition 1.10 A set $F \subset \mathbb{R}^N$ is $(\frac{2}{q}, q')$-thin at $a \in \mathbb{R}^N$ if

$$\int_0^1 \left(\frac{cap_{\frac{2}{q},q'}(F \cap B_s(a))}{s^{N-\frac{2}{q-1}}} \right)^{q-1} \frac{ds}{s} < \infty. \tag{1.313}$$

If the above integral is infinite, the set F is $(\frac{2}{q}, q')$-thick at a.
A set F is a $(\frac{2}{q}, q')$- fine neighbourhood of one of its points a if F^c is thin at a.
A set F is $(\frac{2}{q}, q')$-finely open, if F^c is thin at any point $a \in F$. It is $(\frac{2}{q}, q')$-finely closed if its complement F^c is $(\frac{2}{q}, q')$-finely open.

Notations and vocabulary For simplicity we will denote by $\mathfrak{T}_q$ the $cap_{\frac{2}{q},q'}$-fine topology associated with these notions (see [1, Chap. 6] for a detailed study of these notions).

Let $A, B \subset \mathbb{R}^N$
(a) A is $\mathfrak{T}_q$-essentially contained in B, denoted by $A \subset^q B$, if

$$cap_{\frac{2}{q},q'}(A \cap B^c) = 0.$$

(b) The sets A and B are $\mathfrak{T}_q$-equivalent, denoted by $A \sim^q B$ if

$$cap_{\frac{2}{q},q'}(A \Delta B) \quad \text{where } A \Delta B := (A \cap B^c) \cup (B \cap A^c).$$

(c) The closure of a set A in the $\mathfrak{T}_q$-topology is called the $\mathfrak{T}_q$-closure and denoted by $\widetilde{A}$. The $\mathfrak{T}_q$-interior of A is denoted by $A^{\diamondsuit}$.
(d) If $\epsilon > 0$, we denote by A_ϵ the ϵ-neighbourhood of A in the standard Euclidean topology associated with the distance function.

(e) The set of all $\mathfrak{T}_q$-thick points of A is denoted by $b_q(A)$. It is the set of points a of A such that A is $(\frac{2}{q},q')$-thick at a. The set of all $\mathfrak{T}_q$-thin points of A is denoted by $e_q(A)$. The next result is essentially due to Kellog ([1, Corollary 6. 3.17]).

Proposition 1.12 *There holds*

$$A \text{ is } \mathfrak{T}_q\text{-open} \Longleftrightarrow A \subset e_q(A^c), \quad B \text{ is } \mathfrak{T}_q\text{-closed} \Longleftrightarrow b_q(B) \subset B.$$

Therefore

$$\widetilde{A} = A \cup b_q(A) \qquad A^{\Diamond} = A \cap e_q(A^c).$$

Furthermore the capacity $cap_{\frac{2}{q},q'}$ *possesses the Kellog property*

$$cap_{\frac{2}{q},q'}(A \cap e_q(A)) = cap_{\frac{2}{q},q'}(A \setminus b^c_q(A)) = 0. \tag{1.314}$$

Proposition 1.13 *(i) If $Q \subset \mathbb{R}$ is $\mathfrak{T}_q$-open, then $e_q(Q^c)$ is the largest $\mathfrak{T}_q$-open set which is equivalent to Q.*
(ii) If $F \subset \mathbb{R}$ is $\mathfrak{T}_q$-closed, then $b_q(F)$ is the smallest $\mathfrak{T}_q$-closed set which is equivalent to F.

It is often easier to use the related notions of quasi-open or quasi-closed sets although these notions are not equivalent. All details are to be found in [1, Chap. 6].

Definition 1.11 A set $F \subset \mathbb{R}^N$ is $\mathfrak{T}_q$-quasi-open if for any $\epsilon > 0$ there exists an open set $G \subset \mathbb{R}^N$ verifying $cap_{\frac{2}{q},q'}(G) < \epsilon$ such that $F \cap G^c$ is open in the relative topology of G^c.
A set F is $\mathfrak{T}_q$-quasi-closed if F^c is $(\frac{2}{q},q')$-quasi-open.
A property $\mathcal{P}$ holds $\mathfrak{T}_q$-quasi everywhere in an open set $\Omega \subset \mathbb{R}^N$ if it holds in Ω except on a set with zero $cap_{\frac{2}{q},q'}$-capacity. Abridged notation: $\mathfrak{T}_q$-q.e.
A function f defined as $\mathfrak{T}_q$-q.e. in an open set $\Omega \subset \mathbb{R}^N$ is $\mathfrak{T}_q$-quasi-continuous if for every $\epsilon > 0$ there exists an open set $G \subset \Omega$ such that $cap_{\frac{2}{q},q'}(G) = 0$ with the property that $f\lfloor_{G^c}$ is continuous in G^c for the induced topology.

Proposition 1.14 *Any function f in $B^{\frac{2}{q},q'}(\Omega)$ is $\mathfrak{T}_q$-quasi-continuous. Thus every element of $B^{\frac{2}{q},q'}(\Omega)$ admits a $\mathfrak{T}_q$-quasi-continuous representative. Let f_1 and f_2 be two $\mathfrak{T}_q$-quasi-continuous functions which coincide a.e. in Ω, then they coincide $\mathfrak{T}_q$-q.e.*

Remark The notion of $(\frac{2}{q},q')$-quasi-openedness defines a quasi-topology. It is not a topology because an arbitrary union of quasi-open sets may not be quasi-open. However a *countable* union of quasi-open sets is quasi-open.

The next result is proved in [33, Proposition 2.1]. We list below a series of results concerning the $\mathfrak{T}_q$-topology and $\mathfrak{T}_q$-quasi-topology which are used throughout this section. Their proofs can be found in [1, Proposition 6.4.13] for assertion (i), in [1, Proposition 6.4.12] for assertion (ii), in [1, Proposition 6.4.9] for assertion (iii) and in [1, Proposition 6.4.11] for assertion (iv). Assertions (v)–(viii) are classical in the theory of capacities as exposed in the same book.

Proposition 1.15 *Assume $q \geq q_c$.*
(i) Every $\mathfrak{T}_q$-closed set is $\mathfrak{T}_q$-quasi-closed.
(ii) If F is $\mathfrak{T}_q$-quasi-closed set, then $E \sim^q \widetilde{E}$.
(iii) A set F is $\mathfrak{T}_q$-quasi-closed if and only if there exists a sequence of closed sets $\{F_n\}$ such that $cap_{\frac{2}{q},q'}(F \cap F_n^c) \to 0$.
(iv) There exists a positive constant $\widetilde{C}$ such that for every set F,

$$cap_{\frac{2}{q},q'}(\widetilde{F}) \leq \widetilde{C} cap_{\frac{2}{q},q'}(F).$$

(v) If E is $\mathfrak{T}_q$-quasi-closed and $F \sim^q E$, then F is $\mathfrak{T}_q$-quasi-closed.
vi) If $\{E_n\}$ is an increasing sequence of Borel sets of $\mathbb{R}^N$, then

$$cap_{\frac{2}{q},q'}\left(\bigcup_n E_n\right) = \lim_{n\to\infty} cap_{\frac{2}{q},q'}(E_n).$$

(vii) If $\{K_n\}$ is a decreasing sequence of compact sets of $\mathbb{R}^N$, then

$$cap_{\frac{2}{q},q'}\left(\bigcap_n K_n\right) = \lim_{n\to\infty} cap_{\frac{2}{q},q'}(K_n).$$

(viii) For every Borel set $F \subset \mathbb{R}^N$ (and more generally for every Suslin set), there holds the following:

$$\begin{aligned} cap_{\frac{2}{q},q'}(F) &= \inf\left\{cap_{\frac{2}{q},q'}(G),\ F \subset G,\ G \text{ open}\right\} \\ &= \sup\left\{cap_{\frac{2}{q},q'}(K),\ K \subset F,\ K \text{ compact}\right\}. \end{aligned}$$

As a consequence of (iii) there holds the following.

Corollary 1.2 *A set F is $\mathfrak{T}_q$-quasi-closed if and only if there exists a sequence $\{F_n\}$ of $\mathfrak{T}_q$-quasi-closed subsets of F such that $cap_{\frac{2}{q},q'}(F \cap F_n^c) \to 0$ as $n \to \infty$.*

Definition 1.12 Let F be a $\mathfrak{T}_q$-quasi-closed set.
(i) An increasing sequence $\{F_n\}$ of closed subsets of F is called a $\mathfrak{T}_q$-stratification of F if $cap_{\frac{2}{q},q'}(F \cap F_n^c) \to 0$ as $n \to \infty$.
(ii) A $\mathfrak{T}_q$-stratification $\{F_n\}$ is called a proper $\mathfrak{T}_q$-stratification if $cap_{\frac{2}{q},q'}(F \cap F_n^c) \leq 2^{-n-1}$. The sets F_n can be chosen to be compact.
(iii) A $\mathfrak{T}_q$-open set V verifying $cap_{\frac{2}{q},q'}(F \cap V^c) = 0$ is called a $\mathfrak{T}_q$-quasi-neighbourhood of F.

The next separation result is valid in any locally compact Hausdorff space.

Proposition 1.16 *Let X be a locally compact Hausdorff space, and $K \subset X$ be a compact set contained in an open set A. Then there exists an open set G such that*

$$K \subset G \subset \overline{G} \subset A.$$

Although the fine topology is not locally compact (even if it is Hausdorff), it admits some separation results which are the counterpart of Proposition 1.16.

Lemma 1.25 *Let $F \subset \mathbb{R}^N$ be $\mathfrak{T}_q$-closed. Then*
(i) If D is an open set such that $cap_{\frac{2}{q},q'}(F \cap D^c) = 0$, then there exists an open set O such that

$$F \subset^q O \subset \tilde{O} \subset^q D. \tag{1.315}$$

(ii) If D is a $\mathfrak{T}_q$-open set that verifies $F \subset^q D$, there exists a $\mathfrak{T}_q$-open set O such that (1.315) holds the following.

Proof Since $F \cap D \sim^q F$, $F \cap D$ is $\mathfrak{T}_q$-quasi-closed and there exists a proper $\mathfrak{T}_q$-stratification $\{F_n\}$ of $F \cap D$ by compact sets such that $F \sim^q F' := \cup_{n=1}^{\infty} F_n$.
If E' is closed, the result follows by Proposition 1.16. If it is not the case, we can assume that $F_{n+1} \setminus F_n \neq \emptyset$ for all integer n. We apply Proposition 1.16 with $K = F_n$ and $G = F'_n$ is the open set containing F_n such that its closure $\overline{F}'_n$ is contained in D: because

$$cap_{\frac{2}{q},q'}(F_n \setminus F_{n-1}) \leq cap_{\frac{2}{q},q'}((E \cap D) \setminus F_n) \leq 2^{-n-1},$$

there exists an open set D_n containing $F_n \setminus F_{n-1}$ such that $cap_{\frac{2}{q},q'}(D_n) < 2^{-n}$. We have also

$$D_n \cap F_n \subset \widetilde{D_n \cap F_n} \subset \tilde{D}_n \subset D \quad \text{for all } n \in \mathbb{N}.$$

Since $F' = F_1 \cup_{n=2}^{\infty} (F_n \setminus F_{n-1})$ we have that

$$F' = \bigcup_{n=1}^{\infty} D_n \cap F'_n \subset \bigcup_{n=1}^{\infty} \widetilde{D_n \cap F'_n} \subset D.$$

It is therefore sufficient to prove that $\bigcup_{n=1}^{\infty} \widetilde{D_n \cap F'_n}$ is $\mathfrak{T}_q$-closed. Actually, for any $n \in \mathbb{N}$ we have

$$\begin{aligned}
cap_{\frac{2}{q},q'}\left(\bigcup_{n=1}^{\infty} \widetilde{D_n \cap F'_n} \setminus \bigcup_{n=1}^{m} \widetilde{D_n \cap F'_n}\right) &\leq cap_{\frac{2}{q},q'}\left(\bigcup_{n=m+1}^{\infty} \widetilde{D_n \cap F'_n}\right) \\
&\leq \sum_{n=m+1}^{\infty} cap_{\frac{2}{q},q'}\left(\tilde{D}_n\right) \\
&\leq c \sum_{n=m+1}^{\infty} cap_{\frac{2}{q},q'}\left(D_n\right) \\
&\leq c \sum_{n=m+1}^{\infty} 2^{-m} = c2^{-m}.
\end{aligned}$$

Because $\bigcup_{n=1}^{m} \widetilde{D_n \cap F'_n}$ is $\mathfrak{T}_q$-quasi-closed the result follows by Corollary 1.2. □

Lemma 1.26 *I- Let F be a $\mathfrak{T}_q$ closed set and $\{F_n\}$ a proper $\mathfrak{T}_q$-stratification of F. Then there exists a decreasing sequence of open sets $\{Q_j\}$ such that $\cup F_n := F' \subset Q_j$ for every $j \in \mathbb{N}$ and*
(i) $\cap_j Q_j = F'$, $\widetilde{Q}_{j+1} \subset Q_j$,
(ii) $\lim_{j\to\infty} cap_{\frac{2}{q},q'}(Q_j) = cap_{\frac{2}{q},q'}(E)$.

II- If A is a $\mathfrak{T}_q$ open set, there exists a decreasing sequence of open sets $\{A_n\}$ such that

$$A \subset \bigcap_n A_n := A' \,,\; cap_{\frac{2}{q},q'}(A_n \setminus A') \to 0 \text{ as } n \to \infty \,,\; A \sim^q A'.$$

Furthermore there exists an increasing sequence of closed sets $\{E_j\}$ such that $E_j \subset A'$ and
(i) $\cup_j E_j = A'$, $E_j \subset^q E^{\diamond}_{j+1}$,
(ii) $cap_{\frac{2}{q},q'}(E_j) \to cap_{\frac{2}{q},q'}(A')$ when $j \to \infty$.

Proof Let $\{D_j\}$ be a decreasing sequence of open sets containing F such that

$$\lim_{j\to\infty} cap_{\frac{2}{q},q'}(D_j) = cap_{\frac{2}{q},q'}(F') = cap_{\frac{2}{q},q'}(F).$$

Case 1: F is closed. We can assume that $F_n = F$ for all n and we set $K_n = B_n(x) \cap F$ for some $x \in F$. By Proposition 1.16 there exists a decreasing sequence $\{\epsilon_{1,n}\}$ converging to 0 such that

$$F \subset Q_1 := \bigcup_{n=1}^{\infty} K_n^{\frac{\epsilon_{1,n}}{2}} \subset \overline{Q}_1 \subset D_1,$$

where $K_n = B_n(x) \cap F$ and, we recall it, $K_n^{\frac{\epsilon_{1,n}}{2}} := \{y \in \mathbb{R}^N : \text{dist}\,(y, K_n) \leq \frac{\epsilon_{1,n}}{2}\}$. By Proposition 1.16 there exists a decreasing sequence $\{\epsilon_{2,n}\}$ converging to 0, such that $\epsilon_{2,n} \leq \epsilon_{1,n}$ for all n and

$$F \subset Q_2 := \bigcup_{n=1}^{\infty} K_n^{\frac{\epsilon_{1,n}}{4}} \subset \overline{Q}_2 \subset D_2.$$

Note that

$$\overline{Q}_2 \subset \overline{\bigcup_{n=1}^{\infty} K_n^{\frac{\epsilon_{1,n}}{4}}} \subset \bigcup_{n=1}^{\infty} K_n^{\frac{\epsilon_{1,n}}{2}}.$$

Since $\overline{K_n^{\frac{\epsilon_{1,n}}{4}}}$ is closed, we have $Q_2 \subset \overline{Q}_2 \subset Q_1$. By induction we construct a double sequence $\{\epsilon_{j,n}\}$ decreasing in n and converging to 0, nonincreasing in j for any fixed n such that

$$F \subset Q_j := \bigcup_{n=1}^{\infty} K_n^{\frac{\epsilon_{j,n}}{2}} \subset \overline{Q}_j \subset D_j,$$

and

$$Q_{j+1} \subset \overline{Q}_{j+1} \subset Q_j \quad \text{for all } j \geq 1.$$

Noting that $F \subset Q_j \subset F^{2-j}$ we deduce that $F = \cap_j Q_j$. Finally,

$$cap_{\frac{2}{q},q'}(F) \leq \lim_{j\to\infty} cap_{\frac{2}{q},q'}(Q_j) \leq \lim_{j\to\infty} cap_{\frac{2}{q},q'}(D_j) = cap_{\frac{2}{q},q'}(F).$$

This yields the result in that case.

Case 2: F is only $\mathfrak{T}_q$ closed. There exists a proper $\mathfrak{T}_q$ stratification $\{F_n\}$ of F such that $F \sim^q F' := \cup_{n=1}^{\infty} F_n$. We can also assume that $F_{n+1} \cap F_n^c \neq \emptyset$ for all integer n. As in Case 1, for each n we construct the sets Q_j^n relative to F_n that were denoted by Q_j and were related to F. Because $cap_{\frac{2}{q},q'}(\widetilde{F_n \setminus F_{n-1}}) \leq c cap_{\frac{2}{q},q'}(F_n \setminus F_{n-1})$, we can choose an open set D_n^1 such that $cap_{\frac{2}{q},q'}(D_n^1) \leq c2^{-n}$. In view of Lemma 1.25 the set

$$Q_1 := \bigcup_{n=1}^{\infty} (D_n^1 \cap Q_n^1)$$

is open and

$$F' \subset Q_1 \subset \widetilde{Q}_1 \subset D_1.$$

Furthermore the set

$$\bigcup_{n=1}^{\infty} \widetilde{D_n^1 \cap Q_1^n}$$

is $\mathfrak{T}_q$-quasi-open. By Lemma 1.25 there exists an open set D_n^2 such that

$$D_n^2 \subset \widetilde{D}_n^2 \subset D_n^1,$$

and by induction we construct a sequence of open sets D_n^j such that

$$D_n^{j+1} \subset \widetilde{D}_n^{j+1} \subset D_n^j \quad \text{and } cap_{\frac{2}{q},q'}(D_n^j) \leq c2^{-n}.$$

By Lemma 1.25 the set

$$Q_j := \bigcup_n = 1^\infty D_n^j \cap Q_j^n$$

is open and the set

$$\bigcup_{n=1}^{\infty} \widetilde{D_n^j \cap Q_j^n}$$

is $\mathfrak{T}_q$-quasi-closed. For any $n \in \mathbb{N}_*$ we have

$$D_n^j \cap Q_j^n \subset \widetilde{D_n^j \cap Q_j^n} \subset \widetilde{D}_n^j \cap \widetilde{Q}_j^n \subset D_n^{j-1} \cap Q_{j-1}^n.$$

Therefore

$$Q_j \subset \widetilde{Q}_j \subset \bigcup_{n=1}^{\infty} \widetilde{D_n^j \cap Q_j^n} \subset \bigcup_{n=1}^{\infty} D_n^{j-1} \cap Q_{j-1}^n \subset D_j.$$

Since the set $\bigcup_{n=1}^{\infty} \widetilde{D_n^j \cap Q_j^n}$ is $\mathfrak{T}_q$-quasi-closed, we have

$$Q_j \subset \widetilde{Q}_j \subset Q_{j-1}.$$

Finally,

$$F' \subset Q_j \subset F'^{2^{-j}} \Longrightarrow F' = \bigcap_{n=1}^{\infty} Q_j.$$

Because we have

$$cap_{\frac{2}{q},q'}(F) \leq \lim_{j\to\infty} cap_{\frac{2}{q},q'}(Q_j) \leq \lim_{j\to\infty} cap_{\frac{2}{q},q'}(D_j) = cap_{\frac{2}{q},q'}(F),$$

the assertion follows. □

The next results are classical in the framework of the $\mathfrak{T}_q$-topology.

Proposition 1.17 *I- Any family $\mathcal{D}$ of $\mathfrak{T}_q$-open sets contains a countable subfamily $\mathcal{D}'$ whose union differs from the union of the sets of the whole family $\mathcal{D}$ by a set with zero $cap_{\frac{2}{q},q'}$-capacity.*
II- Let F be a bounded $\mathfrak{T}_q$-open set and let $\mathcal{D}$ be a covering of F consisting of $\mathfrak{T}_q$-open sets. Then, for every $\epsilon > 0$ there exists an open subset O_ϵ of F such that $cap_{\frac{2}{q},q'}(O_\epsilon) < \epsilon$ and $F \cap O_\epsilon^c$ is covered by a finite subfamily of $\mathcal{D}$.
III- Let F be a $\mathfrak{T}_q$-open set. Then for any $\xi \in F$ there exists a $\mathfrak{T}_q$-open set Q_ξ such that

$$\xi \in Q_\xi \subset \widetilde{Q}_\xi \subset F.$$

Proof Assertion I is the quasi-Lindelöf property; see [1, 6.5.11]. The second assertion is a consequence of the quasi-Lindelöf property and is proved in [33, Lemma 2.5] and the last assertion is a consequence of the fact that any point in F is a $\mathfrak{T}_q$-thin point of F^c and is proved in [33, Lemma 2.7] by using the definition. □

1.5.2.2 Approximations in Besov Spaces

Lemma 1.27 *Let $U \subset \mathbb{R}^N$ be a $\mathfrak{T}_q$-open set and $z \in U$. Then there exists a function f in $B^{\frac{2}{q},q'}(\mathbb{R}^N)$ with support in U such that $f(z) > 0$. In particular, there exists a bounded $\mathfrak{T}_q$-open set V such that $\overline{V} \subset U$.*

Proof The result is clear if z is an interior point of U with respect to the Euclidean topology. Thus we assume that it is not the case. Since U is $\mathfrak{T}_q$-open, U^c is thin at z. By the assumption we have that $z \in \overline{U^c} \setminus \overset{\circ}{U}$. By [1, p. 174] there exists an open set W such that $z \in \overline{W} \cap W^c$ and W is thin at z. We recall (see [1, Theorems 2.2.7, 2.5.6]) that for a Borel set E with positive $cap_{\frac{2}{q},q'}$-capacity, we define the *Besov nonlinear potential* of the capacitary measure μ_E by

$$F_E := \mathcal{V}^{\mu_E} = G_{\frac{1}{q}} * (G_{\frac{1}{q}} * \mu_E)^{\frac{1}{q-1}},$$

where $G_{\frac{1}{q}}$ is the Bessel kernel in $\mathbb{R}^N$. By [1, Theorem 6.3.9] there holds

$$\mathcal{V}^{\mu_E}(z) < \frac{1}{2},$$

if we take for E the set $B_r(z) \cap W$ for $r > 0$ small enough. By [1, Theorem 6.3.9] we have $\mathcal{V}^{\mu_E} \geq 1$-$\mathfrak{T}_q$-q.e. on $B_r(z) \cap W$, and by [1, Theorem 2.6.7] $\mathcal{V}^{\mu_E} \geq 1$ everywhere on $B_r(z) \cap W$. Therefore

$$\mathcal{V}^{\mu_E}(z) < \frac{1}{2} < 1 \leq \mathcal{V}^{\mu_E}(x) \quad \text{for all } x \in B_r(z) \cap W.$$

This implies that for $r > 0$ small enough there holds

$$\mathcal{V}^{\mu_E}(z) < \frac{1}{2} < 1 \leq \inf\{\mathcal{V}^{\mu_E}(x) : x \in B_r(z) \cap W\}.$$

Now let H be a smooth nondecreasing function defined on $\mathbb{R}$, such that $H(t) = 0$ for $t \leq 0$ and $H(t) = t$ for $t > \frac{1}{4}$. If $\eta \in C_c^\infty(\mathbb{R}^N)$ satisfies $0 \leq \eta \leq 1$, $supp(\eta) \subset B_r(z)$ and $\eta(z) = 1$, then the function

$$f := \eta H \circ (1 - \mathcal{V}^{\mu_E})$$

satisfies the requirements of the Lemma. □

Lemma 1.28 *Let U be a $\mathfrak{T}_q$- open set and $z \in U$. Then there exists a $\mathfrak{T}_q$-open set V, such that $z \in V \subset U$, and a function $\psi \in B^{\frac{2}{q},q}(\mathbb{R}^N)$ such that $0 \leq \psi \leq 1$, $\psi = 1$ q-a.e. on V and $\psi = 0$ in U^c.*

Proof We keep the notations of Lemma 1.27 and assume that z is not interior to U. Let μ be the capacitary measure of $B_r(z) \cap U$ with (up to changing r)

$$\mathcal{V}^{\mu}(z) < \frac{1}{4} \quad \text{and } \mathcal{V}^{\mu}(x) = 1 \quad \text{for all } x \in B_r(z) \cap U^c.$$

By [1, Proposition 6.3.10] $\mathcal{V}^{\mu}$ is quasi-continuous; hence there exists a $\mathfrak{T}_q$-open set W which contains z such that

$$\mathcal{V}^{\mu} \leq \frac{1}{4} \quad \text{q. a.e. on } W.$$

Let $\eta \in C_c^{\infty}(\mathbb{R}^N)$ such that $0 \leq \eta \leq 1$, $supp(\eta) \subset B_r(z)$ and $\eta(x) = 1$ for all $x \in B_r(z)$. We set

$$f(x) = 2\eta(x) H \circ \left(1 - H \circ \left(\frac{1}{2} - \mathcal{V}^{\mu}(x)\right) - \mathcal{V}^{\mu}(x)\right).$$

Then $f \in B^{\frac{2}{q},q'}(\mathbb{R}^N)$, $0 \leq f \leq 1$ on $B_r(z) \cap U^c$, $f = 1$ on $B_{\frac{r}{2}}(z) \cap W$ and $f = 0$ outside of $B_r(z) \cap U$. □

Definition 1.13 If ζ is a function defined in $\mathbb{R}^N$ we denote by $\mathfrak{T}_q$-supp(ζ) the closure in the $\mathfrak{T}_q$-topology of the set $\{x \in \mathbb{R}^N : |\zeta(x)| > 0\}$.

Lemma 1.29 *Assume $q \geq 2$. Let K be a compact set and U a $\mathfrak{T}_q$-open set containing K. Let $\{U_j\}$ be a sequence of $\mathfrak{T}_q$- open subsets of U covering U up to a set of zero Z of zero $cap_{\frac{2}{q},q}$-capacity.*

1- If there exists a nonnegative function $u \in B^{\frac{2}{q},q'}(\mathbb{R}^N) \cap L^{\infty}(\mathbb{R}^N)$ with $\mathfrak{T}_q$-supp(u) included in K, then for any $k \in \mathbb{N}_*$ there exist an integer $m(k)$ and nonnegative functions $u_{k,j} \in B^{\frac{2}{q},q'}(\mathbb{R}^N) \cap L^{\infty}(\mathbb{R}^N)$ with $\mathfrak{T}_q$-supp$(u_{k,j})$ included in U_j such that

$$\sum_{j=1}^{m(k)} u_{k,j} \leq u \quad \text{in } \mathbb{R}^N, \tag{1.316}$$

and

$$\left\| u - \sum_{j=1}^{m(k)} u_{k,j} \right\|_{B^{\frac{2}{q},q}} \to 0 \quad \text{as } k \to \infty. \tag{1.317}$$

2- If u is a signed function, and since $q \geq 2$, $u^{\pm}$ belongs to $B^{\frac{2}{q},q'}(\mathbb{R}^N)$. The existence of the $\{u_{k,j}\}$ is replaced by existence of $\{u_{k,j,\pm}\}$. Estimate (1.316) is replaced by

$$\sum_{j=1}^{m(k)} u_{k,j,\pm} \leq u^{\pm} \quad \text{in } \mathbb{R}^N, \tag{1.318}$$

and estimate (1.317) remains valid with $u_{k,j}$ *replaced by* $u_{k,j,+} - u_{k,j,-}$.

Proof We can assume that U and U_j are bounded. For any j, k there exists an open set $G_{k,j}$ such that $cap_{\frac{2}{q},q'}(G_{k,j}) \leq 2^{-k-j}$ for $j \geq 1$, $K \subset G_{k,0}$, and for $j \geq 1$, the sets $U_j \cup G_{k,j}$ are open. Furthermore the sets

$$\bigcup_{j=0}^{\infty} G_{k,j} \quad \text{and} \quad \bigcup_{j=0}^{\infty} G_k \bigcup_j U_j$$

are open, and clearly $cap_{\frac{2}{q},q'}(G_k) \to 0$ when $k \to \infty$. Since G_k is open, its Besov potential $\mathcal{V}^{\mu_{G_k}} := F^{G_k}$ is larger or equal to 1 on G_k [1, Theorems 2.5.6, 2.6.7] and there holds

$$\|\mathcal{V}^{\mu_{G_k}}\|^{q'}_{B^{\frac{2}{q},q'}} \leq \widetilde{C} cap_{\frac{2}{q},q'}(G_k),$$

for some $C = C(N, q) > 0$. Let H be a smooth nondecreasing defined on $\mathbb{R}_+$ function such that $H(t) = 1$ for $t \geq 1$ and $H(t) = t$ for $0 \leq t \leq \frac{1}{2}$. Then the function $\phi_k = H \circ \mathcal{V}^{\mu_{G_k}}$ belongs to $B^{\frac{2}{q},q'}(\mathbb{R}^N)$, satisfies $0 \leq \phi_k \leq 1$, $\phi_k = 1$ on G_k and there exists $C' = C'(N, q) > 0$ such that

$$\|\phi_k\|^{q'}_{B^{\frac{2}{q},q'}} \leq C' cap_{\frac{2}{q},q'}(G_k).$$

We set $\psi_k = 1 - \phi_k$. Then

$$\|u - \psi_k u\|_{B^{\frac{2}{q},q'}} \to 0 \quad \text{as } k \to \infty. \tag{1.319}$$

For $k \in \mathbb{N}_*$ fixed, there exist open balls $B_{k,j,i}$ such that

$$B_{k,j,i} \subset U_j \bigcup G_k \quad \text{and} \quad \bigcup_j \Big(G_k \bigcup U_j\Big) = G_k \bigcup \left(\bigcup_j U_j\right) = \bigcup_{i,j=1}^{\infty} B_{k,j,i}.$$

Since K is compact there exists $m(k) \in \mathbb{N}_*$ such that

$$K \subset \bigcup_{i,j=1}^{m(k)} B_{k,j,i}.$$

Now we consider functions $w_{k,j,i} \in C_c^{\infty}(\mathbb{R}^N)$ such that

$$B_{k,j,i} = \{x \in \mathbb{R}^N : w_{k,j,i}(x) > 0\},$$

and we set

$$u_{k,j} = u\psi_k \frac{\sum_{i=1}^{m(k)} w_{k,j,i}}{\sum_{j,i=1}^{m(k)} w_{k,j,i}}.$$

Then $u_{k,j} \in L^\infty(\mathbb{R}^N) \cap B^{\frac{2}{q},q'}(\mathbb{R}^N)$ and

$$\mathfrak{T}_q\text{-}supp\, u_{k,j} \subset \left(K \bigcap G_k^c\right) \bigcap \left(\bigcup_i B_{k,j,i}\right) \subset U_j,$$

which ends the proof. □

Remark The construction can be made also in the case $1 < q < 2$, but the proof of (1.319) is still pending.

1.5.3 Regular Sets

1.5.3.1 The Regular Initial Set

In order to define the precise trace we recall that for any Borel set $U \subset \mathbb{R}^N$, $\mathbf{1}_U$ denotes the characteristic function of U and

$$\mathbb{H}[\mathbf{1}_U](x,t) = \frac{1}{(4\pi t)^{\frac{N}{2}}} \int_{\mathbb{R}^N} e^{-\frac{|x-y|^2}{4t}} \mathbf{1}_U(y)dy.$$

If $u \in \mathcal{U}_+(Q_T)$ (i.e. a positive solution of (1.30) in Q_T), the following dichotomy occurs for any $\xi \in \mathbb{R}^N$:

(i) either there exists a bounded $\mathfrak{T}_q$-open neighbourhood $U = U_\xi$ of ξ such the

$$\int_0^T \int_{\mathbb{R}^N} u^q \left(\mathbb{H}[\mathbf{1}_U]\right)^{2q'} dxdt < \infty, \tag{1.320}$$

(ii) or for any $\mathfrak{T}_q$-open neighbourhood U of ξ there holds

$$\int_0^T \int_{\mathbb{R}^N} u^q \left(\mathbb{H}[\mathbf{1}_U]\right)^{2q'} dxdt = \infty. \tag{1.321}$$

Definition 1.14 Let $u \in \mathcal{U}_+(Q_T)$. The set of $\xi \in \mathbb{R}^N$ such that (i) occurs is $\mathfrak{T}_q$-open and denoted by $\mathcal{R}_q(u)$. It is called the q-regular set of u. The set $\mathcal{S}_q(u) := \mathbb{R}^N \setminus \mathcal{R}_q(u)$ is $\mathfrak{T}_q$-closed and called the q-singular set of u.

Proposition 1.18 *Let $\eta \in B^{\frac{2}{q},q'}(\mathbb{R}^N) \cap L^\infty(\mathbb{R}^N)$ with $\mathfrak{T}_q$-supp(η) in a bounded $\mathfrak{T}_q$-open set U, and let $u \in \mathcal{U}_+(Q_T)$ satisfy*

$$M_U = \int_0^T \int_{\mathbb{R}^N} u^q \left(\mathbb{H}[\mathbf{1}_U]\right)^{2q'} dxdt < \infty. \tag{1.322}$$

Then there exists the following limit:

$$\ell(\eta) = \lim_{t\to 0} \int_0^T \int_{\mathbb{R}^N} u^q \left(\mathbb{H}[\eta]_+\right)^{2q'} (x,t)dxdt. \tag{1.323}$$

Furthermore there exists $C = C(M_U, q, N) > 0$ such that

$$|\ell(\eta)| \leq C\left(\|\eta\|^{2q'}_{B^{\frac{2}{q},q'}} + \|\eta\|^{2q'}_{L^\infty}\right). \tag{1.324}$$

Proof Set $h = \mathbb{H}[\eta]$ and $\phi(r) = r_+^{2q'}$. Since $|\eta| \leq \|\eta\|_{L^\infty} \mathbf{1}_U]$, there holds

$$\left|\int_0^T \int_{\mathbb{R}^N} u^q \phi(h)dxdt\right| \leq \|\eta\|^{2q'}_{L^\infty} \int_0^T \int_{\mathbb{R}^N} u^q \left(\mathbb{H}[\mathbf{1}_U]\right)^{2q'} dxdt = \|\eta\|^{2q'}_{L^\infty} M_U < \infty. \tag{1.325}$$

Note that for $0 < s < t < T$,

$$\begin{aligned}\int_s^t \int_{\mathbb{R}^N} &\left(-u\left(\partial_t \phi(h) + \Delta\phi(h)\right) + u^q \phi(h)\right) dxd\tau \\ &= \int_{\mathbb{R}^N} u\phi(h)(.,s)dx - \int_{\mathbb{R}^N} u\phi(h)(.,t)dx.\end{aligned} \tag{1.326}$$

But

$$\partial_t \phi(h) + \Delta\phi(h) = 2q'\phi(h)h_+^{-2}\left(2h_+\partial_t h + (2q'-1)|\nabla h|^2\right).$$

By Hölder's inequality,

$$\begin{aligned}&\left|\int_s^t \int_{\mathbb{R}^N} u\left(\partial_t \phi(h) + \Delta\phi(h)\right) dxd\tau\right| \\ &\leq \left(\int_s^t \int_{\mathbb{R}^N} u^q \phi(h)dxd\tau\right)^{\frac{1}{q}} \left(\int_s^t \int_{\mathbb{R}^N} (\phi(h))^{-\frac{q'}{q}} |\partial_t \phi(h) + \Delta\phi(h)|^{q'} dxd\tau\right)^{\frac{1}{q'}} \\ &\leq 4q' \left(\int_s^t \int_{\mathbb{R}^N} u^q \phi(h)dxd\tau\right)^{\frac{1}{q}} \left(\int_s^t \int_{\mathbb{R}^N} \left(h_+|\partial_t h| + |\nabla h|^2\right)^{q'} dxd\tau\right)^{\frac{1}{q'}}.\end{aligned}$$

Since

$$\int_s^t \int_{\mathbb{R}^N} |\partial_t h|^{q'} dxd\tau \leq \int_0^T \int_{\mathbb{R}^N} |\partial_t h|^{q'} dxd\tau \leq c\,\|\eta\|^{q'}_{B^{\frac{2}{q},q'}}$$

and

$$\int_s^t \int_{\mathbb{R}^N} |\nabla h|^{2q'} dx d\tau \le \int_0^T \int_{\mathbb{R}^N} |\nabla h|^{2q'} dx d\tau$$
$$\le C \|\eta\|_{L^\infty}^{q'} \|\Delta \eta\|_{L^{q'}}^{q'} = C \|\eta\|_{L^\infty}^{q'} \|\partial_t \eta\|_{L^{q'}}^{q'}$$

by Gagliardo-Nirenberg's inequality and the maximum principle, we obtain

$$\left|\int_s^t \int_{\mathbb{R}^N} u\left(\partial_t \phi(h) + \Delta\phi(h)\right) dx d\tau\right| \le C \left(\int_s^t \int_{\mathbb{R}^N} u^q \phi(h) dx d\tau\right)^{\frac{1}{q}} \|\eta\|_{L^\infty} \|\eta\|_{B^{\frac{2}{q},q'}}. \tag{1.327}$$

As a consequence of (1.326) and (1.327), we infer the two following inequalities:

$$\begin{aligned} &\int_s^t \int_{\mathbb{R}^N} u^q \phi(h) dx d\tau + C \left(\int_s^t \int_{\mathbb{R}^N} u^q \phi(h) dx d\tau\right)^{\frac{1}{q}} \|\eta\|_{L^\infty} \|\eta\|_{B^{\frac{2}{q},q'}} \\ &\qquad \ge \left|\int_{\mathbb{R}^N} u\phi(h)(.,s) dx - \int_{\mathbb{R}^N} u\phi(h)(.,t) dx\right|, \end{aligned} \tag{1.328}$$

and

$$\begin{aligned} &\int_s^t \int_{\mathbb{R}^N} u^q \phi(h) dx d\tau - C \left(\int_s^t \int_{\mathbb{R}^N} u^q \phi(h) dx d\tau\right)^{\frac{1}{q}} \|\eta\|_{L^\infty} \|\eta\|_{B^{\frac{2}{q},q'}} \\ &\qquad \le \int_{\mathbb{R}^N} u\phi(h)(.,s) dx - \int_{\mathbb{R}^N} u\phi(h)(.,t) dx. \end{aligned} \tag{1.329}$$

Under the assumption (1.322) the left-hand side of (1.328) tends to zero when $s, t \to 0$, therefore, we deduce from (1.326) that the function

$$t \mapsto \int_{\mathbb{R}^N} u\phi(h)(.,t) dx$$

admits a limit that is denoted by $\ell(\eta)$ when $t \to 0$. Using again (1.326) we get

$$\int_0^T \int_{\mathbb{R}^N} \left(-u\left(\partial_t \phi(h) + \Delta\phi(h)\right) + u^q \phi(h)\right) dx d\tau + \int_{\mathbb{R}^N} u\phi(h)(.,T) dx = \ell(\eta). \tag{1.330}$$

Since

$$\left|\int_{\mathbb{R}^N} u\phi(h)(.,T) dx\right| \le C(T) \|\eta\|_{L^\infty}^{2q'}, \tag{1.331}$$

we infer from (1.327)

$$\ell(\eta) \le C_1 \|\eta\|_{L^\infty}^{2q'} + C_2 \|\eta\|_{L^\infty}^{q'} \|\eta\|_{B^{\frac{2}{q},q'}}^{q'} \le C \left(\|\eta\|_{L^\infty} + \|\eta\|_{B^{\frac{2}{q},q'}}\right)^{2q'}. \tag{1.332}$$

□

This estimate can be improved in order to show that the initial trace holds in the usual sense.

Proposition 1.19 *Let the assumptions of Proposition 1.18 be satisfied, then*

$$\lim_{t\to 0}\int_{\mathbb{R}^N} u(x,t)\eta^{2q'}(x)dx = \ell(\eta). \tag{1.333}$$

Proof Using (1.326) with $t = T$ and replacing $h(x)$ by $h_s(x,t) = \mathbb{H}[\eta](x, t-s)$ we have

$$\begin{aligned}\int_s^T\int_{\mathbb{R}^N} \left(-u\left(\partial_t\phi(h_s) + \Delta\phi(h_s)\right) + u^q\phi(h_s)\right)dxd\tau \\ + \int_{\mathbb{R}^N} u\phi(h_s)(.,T)dx = \int_{\mathbb{R}^N} u\phi(h_s)(.,s)dx.\end{aligned} \tag{1.334}$$

When $s\to 0$, one has by the Lebesgue dominated convergence theorem

$$\int_{\mathbb{R}^N} u\phi(h_s)(.,T)dx \to \int_{\mathbb{R}^N} u\phi(h)(.,T)dx$$

and

$$\int_s^T\int_{\mathbb{R}^N} u^q\phi(h_s)dxd\tau \to \int_s^T\int_{\mathbb{R}^N} u^q\phi(h)dxd\tau.$$

Furthermore

$$\begin{aligned}&\left|\int_0^{T-s}\int_{\mathbb{R}^N} \left(u(x,t+s) - u(x,t)\right)\left(\partial_t\phi(h) + \Delta\phi(h)\right)dxd\tau\right| \\ &\qquad \le C\left(\int_0^{T-s}\int_{\mathbb{R}^N} |u(x,t+s) - u(x,t)|^q\, h_+^{2q'}\right)^{\frac{1}{q}} \|\eta\|_{L^\infty}^{q'}\, \|\eta\|^{q'}_{B^{\frac{2}{q},q'}}.\end{aligned}$$

By Proposition 1.18, the right-hand side of the above inequality tends to 0 when $s\to 0$. Clearly

$$\lim_{s\to 0}\int_{T-s}^T u^q\phi(h)dxd\tau = 0.$$

Combining (1.326) and (1.334) we obtain

$$\lim_{s\to 0}\int_{\mathbb{R}^N} u(x,s)\left(\phi(h)(x,s) - \phi(\eta)\right)dx = 0, \tag{1.335}$$

which ends the proof. □

Combining Proposition 1.18 and Proposition 1.19, one obtains the following.

Corollary 1.3 *Assume $U \subset \mathbb{R}^N$ is a bounded $\mathfrak{T}_q$-open set such that*

$$\lim_{s\to 0}\int_{\mathbb{R}^N} u(x,s)\eta^{2q'}(x)dx = \infty, \tag{1.336}$$

for some nonnegative $\eta \in L^\infty(\mathbb{R}^N) \cap B^{\frac{2}{q},q'}(\mathbb{R}^N)$ with $\mathfrak{T}_q$-supp$(\eta) \subset U$. Then

$$\int_0^T \int_{\mathbb{R}^N} u^q\, (\mathbb{H}[\eta])^{2q'}\, dxdt = \infty. \tag{1.337}$$

The next result shows that the q-singular set of u inherits the main properties of the singular set $\mathcal{S}(u)$ of the rough trace of u

Proposition 1.20 *Let $\xi \in \mathcal{S}_q(u)$. Then for any $\mathfrak{T}_q$-open set G containing ξ, there holds*

$$\lim_{t\to 0}\int_G u(x,t)dx = \infty. \tag{1.338}$$

Proof If $\xi \in \mathcal{S}_q(u)$ and if G is a $\mathfrak{T}_q$-open set containing ξ, then by Lemma 1.28 there exists $\eta \in L^\infty(\mathbb{R}^N) \cap B^{\frac{2}{q},q'}(\mathbb{R}^N)$ such that $0 \le \eta \le 1$, and a $\mathfrak{T}_q$-open set $D \subset G$ such that $\eta = 1$ on D and $\eta = 0$ in G^c. Therefore

$$\lim_{s\to 0}\int_s^T \int_{\mathbb{R}^N} u^q\, (\mathbb{H}[\eta])^{2q'}\, dxdt \ge \lim_{s\to 0}\int_s^T \int_{\mathbb{R}^N} u^q\, (\mathbb{H}[\mathbf{1_D}])^{2q'}\, dxdt = \infty.$$

This implies that the left-hand side of (1.329) tends to ∞ when $s \to 0$. Using again (1.331) we obtain

$$\lim_{s\to 0}\int_{\mathbb{R}^N} u\, (\mathbb{H}[\eta])^{2q'}\, (x,s)dx = \infty,$$

which implies

$$\lim_{s\to 0}\int_{\mathbb{R}^N} u\eta^{2q'}(x,s)dx = \infty.$$

Since $\eta = 1$ on D the result follows. □

1.5.3.2 Moderate Solutions

We recall that a solution u of (1.30) in Q_T is called *moderate* if $u \in L^q(K)$ for any compact set $K \subset \mathbb{R}^N \times [0,T)$. Then there exists a Radon measure μ on $\mathbb{R}^N$ such that

$$\lim_{t\to 0}\int_{\mathbb{R}^N} u(x,t)\zeta(x)dx = \int_{\mathbb{R}^N} \zeta d\mu(x) \quad \text{for all } \zeta \in C_c(\mathbb{R}^N). \tag{1.339}$$

Equivalently, for any $\phi \in C_c^{1,2}(\mathbb{R}^N \times [0,T))$, there holds

$$\int_0^T \int_{\mathbb{R}^N} \left(-u\left(\partial_t \phi + \Delta\phi\right) + |u|^{q-1} u\phi\right) dxdt = \int_{\mathbb{R}^N} \phi(x,0) d\mu(x). \tag{1.340}$$

It is proved in [7] that the measure μ vanishes on Borel subsets of $\mathbb{R}^N$ with $cap_{\frac{2}{q},q'}$-capacity zero.

Lemma 1.30 *Let u be a nonnegative moderate solution of u of (1.30) in Q_T with initial trace $\mu \in \mathfrak{M}_+(\mathbb{R}^N)$. Then for any $\mathfrak{T}_q$-open bounded set O one has*

$$\int_0^T \int_{\mathbb{R}^N} u^q(x,t) \left(\mathbb{H}[\mathbf{1}_O]\right)^{2q'} dxdt < \infty. \tag{1.341}$$

Proof Let $\eta \in C_c^\infty(\mathbb{R}^N)$ be a nonnegative function with value 1 on O. We put $h(x,t) = \mathbb{H}[\eta](x,t)$ and for $0 \le s \le t \le 1$, $h_s(x,t) = \mathbb{H}[\eta](x,t-s)$. We also set $\phi(r) = |r|^{2q'}$. Using again the identities in Proposition 1.18 we have that

$$\int_0^T \int_{\mathbb{R}^N} u^q \phi(h_s) dxd\tau + \int_{\mathbb{R}^N} (u\phi(h_s))(.,T)dx$$
$$\le C\left(\int_{\mathbb{R}^N} u(x,s)\phi(\eta)dx + \|\eta\|_{L^\infty}^{q'} \|\eta\|_{B^{\frac{2}{q},q'}}^{q'}\right).$$

Because for any Borel set E, one has

$$\limsup_{s \to 0} \int_E u(x,s)dx < \infty,$$

we obtain (1.341) by Fatou's lemma the following. □

Definition 1.15 A Radon measure μ in $\mathbb{R}^N$ is regular with respect to the $\mathfrak{T}_q$-topology if for any Borel set E one has

$$\mu(E) = \inf\left\{\mu(D) : D \supset E,\ D\ \mathfrak{T}_q\text{-open}\right\} = \sup\left\{\mu(K) : K \subset E,\ K \text{ compact}\right\}. \tag{1.342}$$

Theorem 1.21 *Let u be a nonnegative solution of (1.30) in Q_T with initial data μ. Then*
(i) The measure μ is a regular measure with respect to the $\mathfrak{T}_q$-topology.
(ii) For any quasi-continuous function $\phi \in L^\infty(\mathbb{R}^N)$ with bounded $\mathfrak{T}_q$-support in $\mathbb{R}^N$, we have

$$\lim_{t \to 0} \int_{\mathbb{R}^N} u(x,t)\phi(x)dx = \int_{\mathbb{R}^N} \phi d\mu(x).$$

Proof (i) We recall that a Radon measure is regular with respect to the standard topology. Moreover, if $E \subset \mathbb{R}^N$ is a Borel set and D is open and contains D, then D is open for the $\mathfrak{T}_q$-topology, hence

$$\mu(E) \leq \inf\left\{\mu(D) : D \supset E,\ D\ \mathfrak{T}_q\text{-open}\right\} \leq \inf\left\{\mu(D) : D \supset E,\ D\ \text{ open}\right\} = \mu(E).$$

The assertion on compact sets is unchanged and the statement (i) follows.
(ii) The measure $\mu_t := u(t, .)dx$ converges to μ in the weak-$*$ topology. Hence we have

$$\begin{aligned} &\limsup_{t\to 0} \mu_t(E) \leq \mu(E) \text{ for any compact set } E \subset \mathbb{R}^N \\ &\liminf_{t\to 0} \mu_t(A) \geq \mu(A) \text{ for any open set } A \subset \mathbb{R}^N. \end{aligned} \tag{1.343}$$

If E is a $\mathfrak{T}_q$-closed set, there exists an increasing sequence of closed sets $\{K_m\}$ such that $cap_{\frac{2}{q},q'}(E \cap K_m^c) \to 0$ when $m \to \infty$. Then, for any open O containing E, one has

$$\limsup_{t\to 0} \mu_t(E) \leq \limsup_{t\to 0} \mu_t(K_m) + \limsup_{t\to 0} \mu_t(E \cap K_m^C) \leq \mu_t(O) + \limsup_{t\to 0} \mu_t(E \cap K_m^C).$$

We will prove by contradiction that

$$\lim_{m\to\infty} \limsup_{t\to 0} \mu_t(E \cap K_m^c) = 0. \tag{1.344}$$

Assume that (1.344) does not hold and let $\epsilon > 0$ be the value of the above limit. For fixed $m \in \mathbb{N}$, let $\{t_{n,m}\}$ be a decreasing sequence converging to 0 such that

$$\lim_{t_{n,m}\to 0} \mu_{t_{n,m}}(K_m) = \limsup_{t\to 0} \mu_t(K_m) = \epsilon_m.$$

The sequence $\{\epsilon_m\}$ is decreasing with limit ϵ when $m \to \infty$. Let $u_{n,m}$ be the sequence of solutions of (1.30) in Q_∞ such that $u_{n,m}(., 0) = \mathbf{1}_{E\cap K_m^c}\mu_{t_{n,m}}$. Clearly

$$u_{n,m}(x,t) \leq u(x, t + t_{n,m}) \quad \text{for all } (x,t) \in Q_T,$$

and

$$u_{n,m}(x,t) \leq V_{\widetilde{E\cap K_m^c}} \quad \text{for all } (x,t) \in Q_T,$$

where $V_{\widetilde{E\cap K_m^c}}$ is the maximal σ-moderate solution of (1.30) in Q_∞ with initial data ν where $\nu \in \mathfrak{M}_+(\mathbb{R}^N)$ vanishes in $E \cap K_m^c$ and is q-admissible (this notion is developed in the next section). Because

$$cap_{\frac{2}{q},q'}(\widetilde{E \cap K_m^c}) \leq \widetilde{C} cap_{\frac{2}{q},q'}(E \cap K_m^c) \to 0 \quad \text{as } m \to \infty,$$

it follows from Proposition 1.23 that

$$V_{\widetilde{E\cap K_m^c}} \to 0 \quad \text{as } m \to \infty.$$

This is a contradiction. Hence (1.344) holds. Thus the proof is complete if E is $\mathfrak{T}_q$-closed.
If E is $\mathfrak{T}_q$-open, then

$$\mu(E) = \mu(\widetilde{E}),$$

since μ is q-admissible and the proof follows.
Let ϕ be a quasi-continuous function. Without loss of generality, we can suppose that it is nonnegative since $\phi = \phi_+ - \phi_-$ and bounded above by 1. If $k \in \mathbb{N}$ and $m = 2^k - 1, 2^k - 2, ..., 0$, we denote by $a_{m,k}$ a real number in the interval $(m2^{-k}, (m+1)2^{-k}$ such that

$$\mu\left(\phi^{-1}(\{a_{m,k}\})\right) = 0.$$

Set

$$A_{m,k} = \phi^{-1}\left((a_{m,k}, a_{m+1,k}]\right) \ \text{ for } m = 1, 2, ..., 2^k - 1 \ \text{ and } A_{0,k} = \phi^{-1}\left((a_{0,k}, a_{1,k}]\right).$$

Since ϕ has compact support, all the above sets are bounded and

$$\lim_{t\to 0} \mu_t(A_{m,k}) = \mu(A_{m,k}). \tag{1.345}$$

If we denote by ϕ_k the step function

$$\phi_k = \sum_{m=0}^{2^k-1} m2^{-k}\mathbf{1}_{A_{m,k}},$$

then $\phi_k \uparrow \phi$ uniformly, and by (1.345),

$$\lim_{t\to 0} \int_{\mathbb{R}^N} u(x,t)\phi_k(x) = \int_{\mathbb{R}^N} \phi_k d\mu.$$

This implies that (ii) holds. □

1.5.4 Localisation

1.5.4.1 Vanishing Properties

Definition 1.16 A continuous function $u \in \mathcal{U}_+(Q_T)$ vanishes on a $\mathfrak{T}_q$-open set $G \subset \mathbb{R}^N$, if for any $\eta \in B^{\frac{2}{q},q'}(\mathbb{R}^N) \cap L^\infty(\mathbb{R}^N)$ with $\mathfrak{T}_q$-supp$(\eta) \subset^q G$, there holds

$$\lim_{t\to 0} \int_{\mathbb{R}^N} u(x,t)\eta_+^{2q'}(x)dx = 0. \tag{1.346}$$

We write $u \approx_G 0$. We denote by $\mathcal{U}_G(Q_T)$ the subset of $u \in \mathcal{U}_+(Q_T)$ which vanish in the previous sense on G.

The following result is obvious.

Proposition 1.21 *Let $A \subset \mathbb{R}^N$ be a $\mathfrak{T}_q$-open set, and $u_1, u_2 \in \mathcal{U}_+(Q_T)$. If $u_1 \approx_A 0$ and $0 \leq u_2 \leq u_1$, then $u_2 \approx_A 0$.*

Proposition 1.22 *Let $G, G' \subset \mathbb{R}^N$ be $\mathfrak{T}_q$-open set such that $G \sim^q G'$. If $u \in \mathcal{U}_G(Q_T)$, then $u \in \mathcal{U}_{G'}(Q_T)$.*

Proof If $\eta \in B^{\frac{2}{q},q'}(\mathbb{R}^N) \cap L^\infty(\mathbb{R}^N)$ with $\mathfrak{T}_q$-supp$(\eta) \subset^q G$, then $\mathfrak{T}_q$-supp$(\eta) \subset^q G'$. Since $|G' \cap G^c| = |G \cap G'^c|$, the result follows. □

If G is an open subset, this notion coincides with the usual definition of vanishing, since we can take a test function $\eta \in C_c^\infty(G)$. In that case $u \in C(Q_T \cup (G \times \{0\})$.

Lemma 1.31 *Assume that $u \in \mathcal{U}_G(Q_T)$. Then for any $\eta \in B^{\frac{2}{q},q'}(\mathbb{R}^N) \cap L^\infty(\mathbb{R}^N)$ with $\mathfrak{T}_q$-supp$(\eta) \subset^q G$, there holds*

$$\int_0^T \int_{\mathbb{R}^N} u^q \left(\mathbb{H}[\eta]_+\right)^{2q'} dxdt + \int_{\mathbb{R}^N} u(x,T) \left(\mathbb{H}[\eta]_+\right)^{2q'} dx \leq C \|\eta\|_{L^\infty}^{q'} \|\eta\|^{q'}_{B^{\frac{2}{q},q'}}. \tag{1.347}$$

Proof Let u and η be as in the statement of the lemma, $h = \mathbb{H}[\eta]$ and $\phi(r) : r_+^{2q'}$. Then

$$\int_0^T \int_{\mathbb{R}^N} \left(-u\left(\partial_t\phi(h) + \Delta\phi(h)\right) + u^q\phi(h)\right) dxdt + \int_{\mathbb{R}^N} u(x,T)\phi(h)dx = 0. \tag{1.348}$$

Inequality (1.347) is a consequence of (1.327). □

Lemma 1.32 *Let $G \subset \mathbb{R}^N$ be a $\mathfrak{T}_q$-open set. Then there exists a nondecreasing sequence $\{u_n\} \subset \mathcal{U}_G(Q_T)$ which converges to $\sup\{v : v \in \mathcal{U}_G(Q_T)\}$. In addition, the function $u := \sup\{v : v \in \mathcal{U}_G(Q_T)\}$ belongs to $\mathcal{U}_G(Q_T)$.*

Proof We recall that by definition, $u = \sup\{v : v \in \mathcal{U}_G(Q_T)\}$ is defined by

$$u(x,t) = \sup\{v : v \in \mathcal{U}_G(Q_T)\}(x,t) := \sup\{v(x,t) : v \in \mathcal{U}_G(Q_T)\} \quad \text{for all } (x,t) \in Q_T.$$

If u_1 and u_2 belong to $\mathcal{U}_G(Q_T)$, then $u_1 + u_2$ is a supersolution of (1.30) which vanishes on G. Hence $u_1 \vee u_2$ is a solution smaller than $u_1 + u_2$; hence $u_1 \vee u_2 \in \mathcal{U}_G(Q_T)$. By Proposition 1.11, there exists an increasing sequence $\{u_n\} \subset \mathcal{U}_G(Q_T)$ which converges to u. Then

$$\int_0^T \int_{\mathbb{R}^N} \left(-u_n\left(\partial_t\phi(h) + \Delta\phi(h)\right) + u_n^q\phi(h)\right) dxdt + \int_{\mathbb{R}^N} u_n(x,T)\phi(h)dx = 0. \tag{1.349}$$

As in (1.329) $\{u_n^q\phi(h)\}$ and $\{u_n(x,T)\phi(h)\}$ are uniformly bounded in $L^1(Q_T)$ and $L^1(\mathbb{R}^N)$ respectively, and by Fatou's theorem $u_n^q\phi(h) \uparrow u^q\phi(h)$ in $L^1(Q_T)$ and $u_n(x,T)\phi(h) \uparrow u(x,T)\phi(h)$ in $L^1(\mathbb{R}^N)$. Furthermore, if E is any Borel subset of Q_T, we have from (1.327) and Hölder's inequality

$$\left|\int_0^T \int_E u_n\left(\partial_t\phi(h) + \Delta\phi(h)\right) dxdt\right| \\ \leq C(T)\left(\int_0^T \int_E u_n^q\phi(h) dxdt\right)^{\frac{1}{q}} \|\eta\|_{L^\infty} \|\eta\|_{B^{\frac{2}{q},q'}},$$

and the right-hand side tends to 0 as $|E| \to 0$ since $u_n^q\phi(h) \leq u^q\phi(h) \in L^1(Q_T)$. By Vitali's convergence theorem, we infer that

$$\int_0^T \int_{\mathbb{R}^N} \left(-u\left(\partial_t\phi(h) + \Delta\phi(h)\right) + u^q\phi(h)\right) dxdt + \int_{\mathbb{R}^N} u(x,T)\phi(h)dx = 0. \tag{1.350}$$

Thus $u \in \mathcal{U}_G(Q_T)$. □

Definition 1.17 (i) Let $u \in \mathcal{U}_+(Q_T)$ and let A denote the union of all $\mathfrak{T}_q$-open sets on which u vanishes. Then $u \in \mathcal{U}_A(Q_T)$ and A^c is called the precise initial support of u, denoted by $\mathfrak{T}_q$-$supp(u)$.
(ii) Let $F \subset \mathbb{R}^N$ be a Borel set, we denote by U_F the maximal element of $\mathcal{U}_{\widetilde{F}^c}(Q_T)$.

Note that by definition

$$U_F = U_{\widetilde{F}}. \tag{1.351}$$

1.5.4.2 Maximal Solutions

If μ is a q-admissible measure, i.e. μ is absolutely continuous with respect to $cap_{\frac{2}{q},q'}$, u_μ denote the solution of (1.30) in Q_∞ with initial data μ.

Definition 1.18 If E is a Borel set with positive $cap_{\frac{2}{q},q'}$-capacity, we set

$$\mathcal{V}_{mod}(E) = \left\{u_\mu : \mu \in \mathfrak{M}_+^b(\mathbb{R}^N) \cap B^{-\frac{2}{q},q}(\mathbb{R}^N)\}, \mu(E^c) = 0\right\}$$

and

$$V_E = \sup\{u_\mu : u_\mu \in \mathcal{V}_{mod}(E)\} = \sup \mathcal{V}_{mod}(E).$$

We recall that we have proved in Sect. 1.3 the following result due to Marcus and Véron [34]. *If $F \subset \mathbb{R}^N$ is a closed set $U_F = V_F$.*

Proposition 1.23 *If $\{A_n\}$ is a collection of Borel sets such that $cap_{\frac{2}{q},q'}(A_n) \to 0$ as $n \to \infty$, then $U_{A_n} \to 0$.*

Proof Let O_n be an open set such that $A_n \subset O_n$ and $cap_{\frac{2}{q},q'}(O_n) \leq cap_{\frac{2}{q},q'}(A_n) + \frac{1}{n}$. By Kellogg's result in Proposition 1.12,

$$cap_{\frac{2}{q},q'}(\overline{O}_n) = cap_{\frac{2}{q},q'}(\widetilde{O}_n) \leq \widetilde{C} cap_{\frac{2}{q},q'}(O_n).$$

Therefore $cap_{\frac{2}{q},q'}(\overline{O}_n) \to 0$ when $n \to \infty$. Since

$$U_{A_n} \leq U_{\overline{O}_n},$$

the result follows. □

Corollary 1.4 *If $E \subset \mathbb{R}^N$ is a Borel set such that $cap_{\frac{2}{q},q'}(E) = 0$ then $\mathcal{U}_{\widetilde{E}^c} = \{0\}$.*

Proposition 1.24 *Let E and F be Borel sets.*
(i) If E and F are $\mathfrak{T}_q$-closed, then $U_E \wedge U_F = U_{E \cap F}$.
(ii) If E and F are $\mathfrak{T}_q$-closed, then

$$\begin{aligned} U_E < U_F &\Longleftrightarrow E \subset^q F \text{ and } cap_{\frac{2}{q},q'}(F \setminus E) > 0. \\ U_E = U_F &\Longleftrightarrow E \sim^q F. \end{aligned} \tag{1.352}$$

(iii) If $\{F_n\}$ is a decreasing sequence of $\mathfrak{T}_q$-closed subset of $\mathbb{R}^N$, then

$$\lim_{n\to\infty} U_{F_n} = U_{\cap F_n}.$$

(iv) Let A be a $\mathfrak{T}_q$-closed subset of $\mathbb{R}^N$ and $u \in \mathcal{U}_+(Q_T)$. Assume that for any $\sigma \in A$ there exists a $\mathfrak{T}_q$-open subset A_σ of $\mathbb{R}^N$ containing σ and contained in A such that

$$u \approx_{A_\sigma} 0.$$

Then u vanishes on A. In particular any $u \in \mathcal{U}_+(Q_T)$ vanishes on the complement of the $\mathfrak{T}_q$ support of u.

Proof (i) $U_E \wedge U_F$ is the largest solution below $\inf\{U_E, U_F\}$. Hence it is the largest solution which vanishes outside $E \cap F$.
(ii) By construction, $E \sim^q F$ implies $U_E = U_F$, and $U_E < U_F \Longleftrightarrow E \subset^q F$. Furthermore, if $cap_{\frac{2}{q},q'}(F \setminus E) > 0$ there exists a compact set $K \subset F \setminus E$ with $cap_{\frac{2}{q},q'}(K) > 0$. Hence $0 < U_K \leq U_F$. Consequently, $u_E = U_F$ implies $E \sim^q F$.
(iii) Let $V = \lim_{n\to\infty} U_{F_n}$. Since $F \subset F_n$, we have $F \subset^q F$, hence $U_F \leq U_{F_n}$ which implies $U_F \leq V$. But the $\mathfrak{T}_q$-support of V is included in F_n, and therefore is also included in $F = \cap_n F_n$, which implies $v \leq U_F$, and finally $V = U_F$.
(iv) First we assume that $A = \cup_n A_n$ where A_n is $\mathfrak{T}_q$-open and $u \approx_{A_n} 0$ for every n. Then, for every $k \in \mathbb{N}_*$, u vanishes on $\cup_{n=0}^k A_k$, and we can assume that the sequence $\{A_k\}$ is increasing. Set $F_n = A_n^c$. Then $u \leq U_{F_n^c}$ and by (iii) $U_{F_n^c} \downarrow U_F$, thus $u \leq U_F$. Equivalently $u \approx_A 0$.

In the general case, we use the quasi-Lindelöf property which is satisfied by the $\mathfrak{T}_q$-topology. From the covering of A by the family of $\mathfrak{T}_q$-open subsets of A indexed by the $\sigma \in A$, we can extract a countable subcovering A_{σ_n} such that $cap_{\frac{2}{q},q'}\left(A \setminus \cup_n A_{\sigma_n}\right) = 0$. Since $u \approx_{\cup_n A_{\sigma n}} 0$, the claim follows. □

Proposition 1.25 *(i) Let E be a $\mathfrak{T}_q$-closed set. Then*

$$U_E = \inf\{U_D : E \subset D, D \textit{ open}\} = \sup\{U_K : K \subset E, K \textit{ closed}\}. \quad (1.353)$$

(ii) Let E, F be Borel sets. Then

$$U_E = U_{F\cap E} \oplus U_{F\cap E^c}.$$

(iii) Let E, $\{F_n\}$ be a countable family of Borel sets. Assume either that $cap_{\frac{2}{q},q'}(E\Delta F_n)$ tends to 0, or $\widetilde{F}_n \downarrow \widetilde{E}$. Then

$$U_{F_n} \to U_E \quad \textit{as } n \to \infty.$$

Proof (i) Let $\{D_j\}$ be the decreasing sequence of open sets containing E already used in Lemma 1.26 and satisfying

$$\cap_j D_j = \cap_j \widetilde{D}_j = E' \sim^q E.$$

Then, by Proposition 1.24, there holds $U_{D_j} \to U_E$, which implies the first equality in (i).
For the second equality, let $\{F_n\}$ be a nondecreasing sequence of compact subsets of E such that $cap_{\frac{2}{q},q'}(E \setminus F_n) \to 0$. If $\{D_j\}$ is the decreasing sequence used above, then $cap_{\frac{2}{q},q'}\left(D_j \setminus E\right) \to 0$. Because $E \subset F_n \cup (D_n \cap F_n^c)$ we have

$$U_{F_n} \leq U_E \leq U_{F_n} + U_{D_n \setminus F_n}.$$

But

$$cap_{\frac{2}{q},q'}(D_n \setminus F_n) \leq cap_{\frac{2}{q},q'}(E \setminus F_n) + cap_{\frac{2}{q},q'}(D_n \setminus E) \to 0 \quad \text{as } n \to \infty.$$

By Proposition 1.23 $U_{D_n \setminus F_n} \to 0$. This implies the claim.
(ii) Using (1.353) we have

$$U_E \leq U_{E\cap F} + U_{E\cap F^c} \text{ hence } U_E \leq U_{E\cap F} \oplus U_{E\cap F^c}.$$

Since $U_{E\cap F}$ and $U_{E\cap F^c}$ vanish outside $\widetilde{E}$, it follows that $U_{E\cap F} \oplus U_{E\cap F^c}$ vanishes outside $\widetilde{E}$, hence

$$U_E = U_{\widetilde{E}} \geq U_{E\cap F} \oplus U_{E\cap F^c},$$

which is the claim.
(iii) Using (ii) we have

$$U_E \leq U_{E\cap F_n^c} + U_{E\cap F_n} \text{ and } U_{F_n} \leq U_{F_n\cap E^c} + U_{F_n\cap E}.$$

If $cap_{\frac{2}{q},q'}(E\Delta F_n) \to 0$, then $U_{E\Delta F_n} \to 0$ by Proposition 1.23.
If $\widetilde{F}_n \downarrow E$, the result follows by (iii). □

Theorem 1.22 *If E is a $\mathfrak{T}_q$-closed set, then V_E and U_E satisfy the same capacitary estimates as if E were a closed set. Hence $V_E = U_E$ and therefore U_E is σ-moderate.*

Proof The proof follows [33, Theorem 3.10] if $\{E_k\}$ is a proper q-stratification of E and μ is a bounded nonnegative measure belonging to $B^{-\frac{2}{q},q}(\mathbb{R}^N))\}$ and satisfying $\mu(E^c) = 0$, then

$$u_\mu = \sup\{u_{\mu_k} : \mu_k = \mathbf{1}_{E_k}\mu\}.$$

Therefore $V_E = \sup_n V_{E_n}$. By Marcus-Véron's theorem (Sect. 1.3), $V_{E_k} = U_{E_k}$, and by Proposition 1.25-(iii), $U_{E_k} \to U_E$, hence $U_E = V_E$.
Note also that if W_{E_k} is the capacitary potential defined by (1.152) with F replaced by E_k. Hence

$$cap_{\frac{2}{q},q'}\left(\frac{E_k \cap F_n(x,t)}{\sqrt{t(n+1)}}\right) \to cap_{\frac{2}{q},q'}\left(\frac{E \cap F_n(x,t)}{\sqrt{t(n+1)}}\right) \quad \text{as } k \to \infty.$$

Then by the Lebesgue convergence theorem (applied to series) $W_{E_n}(x,t) \to W_F(x,t)$. Hence if E is just a $\mathfrak{T}_q$-closed set, U_E satisfies the same capacitary quasi-representation as if it were closed and given in Theorems 1.15 and 1.18. □

1.5.4.3 The Local Restrictions

The local restrictions are key processes compatible with the supercritical range. They roughly consist in truncating a solution u of (1.30) outside a Borel set A^c. More precisely, we have the following.

Definition 1.19 Let A be a Borel subset of $\mathbb{R}^N$. We denote by $[u]_A$ the supremum of the $v \in \mathcal{U}_+(Q_T)$ which are smaller than u and vanish on $\widetilde{A}^c$. Equivalently $[u]_A = u \wedge U_A$, that is, the largest solution smaller than the subsolution $\inf\{u, U_A\}$.

The following result is an immediate consequence of the fact that $U_A = U_{\widetilde{A}}$ and $[u]_A = u \wedge U_A$.

Lemma 1.33 *For any Borel set $A \subset \mathbb{R}^N$, $[u]_A = [u]_{\widetilde{A}}$.*

Lemma 1.34 *If $G \subset \mathbb{R}^N$ is a $\mathfrak{T}_q$-open set and $u \in \mathcal{U}_G(Q_T)$, then*

$$u = \sup\left\{v \in \mathcal{U}_G(Q_T) : v \leq u,\ v \text{ vanishes in some open neighbourhood of } \widetilde{G}\right\}. \tag{1.354}$$

Proof Set $A = G^c$ and $\{A_n\}$ be a nondecreasing sequence of closed subsets of A such that $cap_{\frac{2}{q},q'}\left(A \cap A_n^c\right) \to 0$ as $n \to \infty$. By Proposition 1.25, there holds

$$U_A \leq U_{A_n} + U_{A\cap A_n^c}.$$

Hence

$$u = u \wedge U_A \leq u \wedge U_{A_n} + u \wedge U_{A\cap A_n^c}.$$

By Proposition 1.23, $U_{A\cap A_n^c} \to 0$ as $n \to \infty$. Therefore $u \wedge U_{A\cap A_n^c}$ converges also to 0, and

$$u = \lim_{n\to\infty} u \wedge U_{A_n},$$

which implies the claim. □

In the next result we analyse the regularity of the correspondence $E \mapsto [u]_E$.

Proposition 1.26 *Let $u \in \mathcal{U}_+(Q_T)$.*
(i) If E is $\mathfrak{T}_q$-closed, then,

$$[u]_E = \inf\{[u]_D : E \subset D,\ D \text{ open}\} = \inf\{[u]_F : F \subset E,\ F \text{ closed}\}. \tag{1.355}$$

(ii) If E and F are two $\mathfrak{T}_q$-closed sets then

$$[u]_E \leq [u]_{F\cap E} + [u]_{E\cap F^c}, \tag{1.356}$$

and

$$[[u]_E]_F = [[u]_F]_E = [u]_{F\cap E}. \tag{1.357}$$

(iii) Let E and F_n, $n = 1, 2, \ldots$ be Borel sets. If either $cap_{\frac{2}{q},q'}(E\Delta F_n) \to 0$, or $\widetilde{F}_n \downarrow \widetilde{E}$, then

$$[u]_{F_n} \to [u]_E.$$

Proof *Mutatis mutandis* the arguments we use are very similar to the ones in [33], but we keep them for the sake of completeness.
(i) Let $\mathcal{D} = \{D\}$ be the family of all open sets containing E as in (1.355). Using the first equality of (1.353), we have

$$\inf\{u, U_E\} = \inf\left\{u, \inf_{D\in\mathcal{D}} U_D\right\} = \inf_{D\in\mathcal{D}} \inf\{u, U_D\} \geq \inf_{D\in\mathcal{D}} [u]_D. \tag{1.358}$$

Clearly

$$[u]_{D_1} \wedge [u]_{D_2} \geq [u]_{D_1 \cap D_2},$$

then it is a consequence of Proposition 1.11 that $v = \inf_{D \in \mathcal{D}} [u]_D$ is a solution of (1.30). It follows from (1.358) that $[u]_E \geq v$. The reverse inequality is clear.
For the second equality, let $\{F_n\}$ be a nondecreasing sequence of compact subsets of E such that $cap_{\frac{2}{q},q'}(E \setminus F_n) \to 0$. If $\{D_j\}$ is the decreasing sequence used above, then $cap_{\frac{2}{q},q'}\left(D_j \setminus E\right) \to 0$. Because $E \subset F_n \cup (D_n \cap F_n^c)$ we have

$$U_{F_n} \leq U_E \leq U_{F_n} + U_{D_n \setminus F_n}.$$

But

$$cap_{\frac{2}{q},q'}(D_n \setminus F_n) \leq cap_{\frac{2}{q},q'}(E \setminus F_n) + cap_{\frac{2}{q},q'}(D_n \setminus E) \to 0 \quad \text{as } n \to \infty.$$

By Proposition 1.23 $U_{D_n \setminus F_n} \to 0$. This implies the claim.
(ii) Let $v \in \mathcal{U}_+(Q_T)$, dominated by u with $\mathfrak{T}_q$ support in E and let D and D' be open sets such that $\widetilde{E \cap F} \subset D$ and $\widetilde{E \cap F^c} \subset D'$. For any integer $j > \frac{1}{T}$, let v_j^1 be the solution of (1.30) in $\mathbb{R}^N \times (\frac{1}{j}, T)$ satisfying $v_j^1(.,\frac{1}{j}) = v(.,\frac{1}{j})\mathbf{1}_D$. We also denote by v_j^2 and v_j^3 the solution of (1.30) in $\mathbb{R}^N \times (\frac{1}{j}, T)$ with respective initial data $v_j^2(.,\frac{1}{j}) = v(.,\frac{1}{j})\mathbf{1}_{D'}$ and $v_j^3(.,\frac{1}{j}) = v(.,\frac{1}{j})\mathbf{1}_{(D \cup D')^c}$. Since v vanishes outside E, it vanishes in $(D \cup D')^c$, consequently $v(.,\frac{1}{j})\mathbf{1}_{(D \cup D')^c} \to 0$ when $j \to \infty$, which implies v_j^3 when $j \to \infty$. Therefore

$$v \leq \liminf_{j \to \infty}(v_j^1 + v_j^2) \leq [u]_D + [u]_{D'}.$$

Since $\widetilde{E \cap F} \subset D$ and $\widetilde{E \cap F^c} \subset D'$, it follows from (1.355)

$$v \leq [u]_{\widetilde{E \cap F}} + [u]_{\widetilde{E \cap F^c}} = [u]_{E \cap F} + [u]_{E \cap F^c}.$$

This implies (1.356).
For proving (1.357), we just have to notice that

$$[[u]_E]_F = [u]_E \vee U_F = (u \vee U_E) \vee U_F = [\max\{u, U_E, U_F\}]_\dagger = [[u]_F]_E.$$

(iii) By (1.356) there holds

$$[u]_E \leq [u]_{F_n \cap E} + [u]_{E \cap F_n^c} \quad \text{and } [u]_{F_n} \leq [u]_{F_n \cap E} + [u]_{F_n \cap E^c}.$$

If $cap_{\frac{2}{q},q'}(E \Delta F_n) \to 0$, then by Proposition 1.23 $U_{E \Delta F_n} \to 0$. Since

$$\max\{[u]_{E \cap F_n^c}, [u]_{F_n \cap E^c}\} \leq U_{E \Delta F_n},$$

we have that $U_{E\Delta F_n} \to 0$ when $n \to \infty$.
If $\widetilde{F}_n \downarrow \widetilde{E}$, then $U_{E_n} \to U_E$ by Proposition 1.23. Therefore

$$[u]_E \le \lim_{n\to\infty} U_{E_n} = \lim_{n\to\infty} u \vee U_{E_n} \le \lim_{n\to\infty} \{u, U_{E_n}\} \le \inf\{u, U_E\}.$$

Since $[u]_E$ is the largest solution dominated by $\inf\{u, U_E\}$ and the function $v = \lim_{n\to\infty} [u]_{F_n}$ is a solution, there holds $U_E \le v$. Thus (iii) follows. □

Definition 1.20 Let μ be a nonnegative Radon measure which is absolutely continuous with respect to the $cap_{\frac{2}{q},q'}$-capacity.
(i) The $\mathfrak{T}_q$-support of μ, denoted by $\mathfrak{T}_q$-supp(μ), is the intersection of all the $\mathfrak{T}_q$-closed sets F such that $\mu(F^c) = 0$.
(ii) We say that μ is concentrated on a Borel set E if $\mu(E^c) = 0$.

Proposition 1.27 *Let μ be a Radon measure as in Definition 1.20. Then*

$$\mathfrak{T}_q\text{-}supp\,(\mu) \sim^q \mathfrak{T}_q\text{-}supp\,(u_\mu).$$

Proof Set $F = \mathfrak{T}_q$-supp(u_μ). By Proposition 1.24-(iv), u_μ vanishes on F^c, and by Lemma 1.34 there exists an increasing sequence of positive solutions $\{u_n\}$ vanishing outside a closed subset F_n and converging to u. Set $S_n := \mathfrak{T}_q$-supp (u_n). Then $S_n \subset F_n$ and $S_n \subset S_{n+1}$. Thus $\{\overline{S}_n\}$ is an increasing sequence of closed subsets of F. If we set $\mu_n = \mathbf{1}_{\overline{S}_n}\mu$, we have that $u_n \le u_{\mu_n} \le u_\mu$. Hence the increasing sequence $\{u_{\mu_n}\}$ converges to u_μ as $n \to \infty$. Consequently

$$\mu_n \uparrow \mu \quad \text{and } \mathfrak{T}_q\text{-supp}\,(\mu) \subset^q \widetilde{\bigcup_{n=1}^{\infty} \overline{S}_n} \subset F.$$

If D is open and $\mu(D) = 0$, then u_μ vanishes in D. Therefore u_{μ_n} vanishes outside $\overline{S}_n$ and consequently it vanishes outside $\mathfrak{T}_q$-supp (μ). Hence u_μ vanishes outside $\mathfrak{T}_q$-supp (μ). This means $F \subset^q \mathfrak{T}_q$-supp (μ). □

Definition 1.21 Let $u \in \mathcal{U}_+(Q_T)$ and A be a Borel set. Then

$$[u]^A := \sup\left\{[u]_F : F \subset^q A,\ F\ \mathfrak{T}_q\text{-closed}\right\}.$$

Remark Note that since $[u]_E = [u]_{\widetilde{E}}$, if A is $\mathfrak{T}_q$-closed, we have $[u]_A = [u]^A$. In the general case, we have only $[u]^A \le [u]_A$.

Definition 1.22 Let $\beta > 0$ and $u \in C(Q_T)$, $u \ge 0$. For any Borel set $A \subset \mathbb{R}^N$, we denote by u_β^A the solution of

$$\begin{aligned} \partial_t v - \Delta v + |v|^{q-1}v = 0 \qquad & \text{in } \mathbb{R}^N \times (\beta, \infty) \\ v(., \beta) = \mathbf{1}_A u(., \beta) \text{ in } & \mathbb{R}^N. \end{aligned}$$

Proposition 1.28 *Let $u \in \mathcal{U}_+(Q_T)$ and E be $\mathfrak{T}_q$-supp (u).*
(i) If D is a $\mathfrak{T}_q$-open set such that $E \subset^q D$, then

$$[u]^D = \lim_{\beta\to 0} u_\beta^D = [u]_D = u. \tag{1.359}$$

(ii) If A is a $\mathfrak{T}_q$-open set, then

$$u \approx_A 0 \Longleftrightarrow u^Q = \lim_{\beta\to 0} u_\beta^Q = 0 \text{ for all } \mathfrak{T}_q\text{-open set s.t. } \widetilde{Q} \subset^q A. \tag{1.360}$$

(iii) Finally,

$$u \approx_A 0 \Longleftrightarrow [u]^A = 0. \tag{1.361}$$

Proof *Case 1: Assume first that E is closed.* Since u vanishes on E^c and is continuous in $Q_T \cup E^c \times \{0\}$, we have that $u = 0$ on E^c hence $u \in C(Q_T \cup E^c \times \{0\}$. If D is an open neighbourhood of E, then for all $\phi \in C_c(E^c)$ there holds

$$\lim_{t\to 0} \int_{E^c} u(x,t)\phi(x)dx = 0.$$

Therefore

$$\lim_{\beta\to 0} u_\beta^{E^c} = 0.$$

But

$$u_\beta^D(x,t) \le u(x,t) \le u_\beta^D(x,t) + u_\beta^{D^c}(x,t) \quad \text{for all } (x,t)) \in \mathbb{R}^N \times [\beta, T).$$

From this relation we deduce that

$$u = \lim_{\beta\to 0} u_\beta^D. \tag{1.362}$$

If we assume now that D is $\mathfrak{T}_q$-open and $E \subset^q D$, then for every $\epsilon > 0$ there exists an open set O_ϵ such that $D \subset O_\epsilon$, $E \subset O_\epsilon$ and $cap_{\frac{2}{q},q'}(O_\epsilon \cap D^c) < \epsilon$. Therefore

$$u_\beta^{O_\epsilon}(x,t) - u_\beta^D(x,t) \le U_{O'_\epsilon}(x, t-\beta) \quad \text{for all } t \ge \beta,$$

where $O'_\epsilon = O_\epsilon \cap D^c$. We observe that $\lim_{\epsilon\to 0} U_{O'_\epsilon}(x, t-\beta) = 0$ uniformly w.r.t. β. Since $\lim_{\beta\to 0} u_\beta^{O_\epsilon}(x,t) = u(x,t)$ for all $(x,t) \in Q_T$, it follows that $\lim_{\beta\to 0} u_\beta^D(x,t) = u(x,t)$. The same argument shows that $\lim_{\beta\to 0} u_\beta^{D^c}(x,t) = 0$ for all $(x,t) \in Q_T$. Combining all these results we obtain

$$\lim_{\beta\to 0} u_\beta^D \le [u]_D \le u,$$

hence $[u]_D = u$. By Proposition 1.17 there exists a $\mathfrak{T}_q$-open set Q such that $E \subset^q \subset Q \subset \widetilde{Q} \subset D$, therefore $u = [u]_Q \leq [u]^D$, hence $u = [u]^D$.

In addition there holds $E \subset^q A^c \subset^q \widetilde{Q}^c$. If we replace D by $\widetilde{Q}^c$ in the above argument, we have that $u \approx_A 0$ which implies $u^Q = \lim_{\beta\to 0} u^Q_\beta = 0$. For the opposite implication in equivalence (1.360) we use the fact that for any $\xi \in A$ there exists a $\mathfrak{T}_q$-open neighbourhood O_ξ of ξ such that $\widetilde{O}_\xi \subset^q A$. By (i) we have that $\lim_{\beta\to 0} u_\beta^{\widetilde{O}^c_\xi}$. Finally, since $u_\beta^{\widetilde{O}^c_\xi} \approx_{O_\xi} 0$ for all $\beta > 0$, we deduce that $u \approx_{O_\xi} 0$ by Proposition 1.23. Using Proposition 1.24 (iv) we deduce (1.360) in the case where E is closed.

Case 2: Assume next that E is $\mathfrak{T}_q$-closed. Let $\{E_n\}$ be a $\mathfrak{T}_q$-stratification of E such that $cap_{\frac{2}{q},q'}(E \cap E_n^c) \to 0$ as $n \to \infty$. If D is a $\mathfrak{T}_q$-open set such that $E \subset^q D$, then by Case 1,

$$\lim_{\beta\to 0}([u]_{E_n})^D_\beta = [u]_{E_n}. \tag{1.363}$$

By Proposition 1.26-(ii), using the definition of u^D_β and the fact that $[u]_E = u$,

$$u^D_\beta = ([u]_E)^D_\beta \leq ([u]_{E\cap E_n})^D_\beta + ([u]_{E\cap E_n^c})^D_\beta = ([u]_{E_n})^D_\beta + ([u]_{E\cap E_n^c})^D_\beta. \tag{1.364}$$

Let $\{\beta_k\}$ be a sequence decreasing to 0 such that there exists

$$w := \lim_{\beta_k\to 0} u^D_{\beta_k} \quad \text{and} \quad w_n := \lim_{\beta_k\to 0}([u]_{E\cap E_n^c})^D_{\beta_k} \quad \text{for } n = 1, 2, ...$$

Then, using the two previous inequalities

$$[u]_{E_n} \leq w \leq [u]_{E_n} + w_n \leq [u]_{E_n} + U_{E\cap E_n^c}.$$

Using (1.354) and the fact that $U_{E\cap E_n^c} \to 0$ and $U_{E_n} \to U_E$, we deduce that $w = u$. This implies (i).

In order to prove (ii), we apply (1.364) with D replaced by Q and get

$$([u]_E)^Q_\beta \leq ([u]_{E_n})^Q_\beta + ([u]_{E\cap E_n^c})^Q_\beta.$$

From Case 1 we have already proved that

$$\lim_{\beta\to 0}([u]_{E_n})^Q_\beta = 0.$$

There exists a decreasing sequence $\{\beta_k\}$ such that $\{u^Q_{\beta_k}\}$ and $\{([u]_{E\cap E_n^c})^Q_{\beta_k}\}$ admit a limit when $\beta_k \to 0$, for any $n = 1, 2, ...$ Therefore

$$\lim_{\beta_k\to 0} u^Q_{\beta_k} \leq \lim_{\beta_k\to 0}([u]_{E\cap E_n^c})^Q_{\beta_k} \leq U_{E\cap E_n}.$$

Since $U_{E\cap E_n} \to 0$ as $n \to \infty$, this implies the implication $\Longrightarrow$ in (1.360). The implication $\Longleftarrow$ in (1.360) is proved as in Case 1.

Proof of (iii). We assume first that $u \approx_A 0$. If F is a $\mathfrak{T}_q$-closed set such that $F \subset^q A$, then, by Lemma 1.25 there exists a $\mathfrak{T}_q$-open set Q such that $F \subset^q Q \subset \widetilde{Q} \subset^q A$. Applying (1.359) to $v := [u]_F$ and using (1.360) we obtain

$$v = \lim_{\beta\to 0} v_\beta^Q \leq \lim_{\beta\to 0} u_\beta^Q = 0.$$

It is thus a consequence of the definition of $[u]^A$ that $[u]^A = 0$.
If $[u]^A = 0$, then for any $\mathfrak{T}_q$-open set $Q \subset \widetilde{Q} \subset^q A$, there holds $[u]_Q = 0$. Because $\mathfrak{T}_q$-supp$(u_\beta^Q) \subset^q \widetilde{Q}$, there exists a subsequence β_k decreasing to 0 such that

$$\lim_{\beta_k\to 0} u_{\beta_k}^Q \leq [u]_Q = 0.$$

Therefore $u \approx_Q 0$ by (1.360). Applying again Lemma 1.25 and Proposition 1.24-(iv), we infer that $u \approx_A 0$. □

Definition 1.23 Let $u, v \in \mathcal{U}_+(Q_T)$ and A be a $\mathfrak{T}_q$-open set. We say that $u = v$ on A if both $u \ominus v$ and $v \ominus u$ vanish on A. This relation is denoted by $u \approx_A v$.

Proposition 1.29 *Let $u, v \in \mathcal{U}_+(Q_T)$ and A be a $\mathfrak{T}_q$-open set. Then,*
(i)

$$u \approx_A v \Longleftrightarrow \lim_{\beta\to 0} |u - v|_\beta^Q = 0, \tag{1.365}$$

for every $\mathfrak{T}_q$-open set Q such that $\widetilde{Q} \subset^q A$.
(ii)

$$u \approx_A v \Longleftrightarrow [u]_F = [v]_F, \tag{1.366}$$

for every $\mathfrak{T}_q$-closed set F such that $F \subset^q A$.

Proof The idea of the proof is the adaptation to the parabolic framework of the construction in the elliptic case performed in [33]. If $u \approx_A v$, then $u \ominus v \approx_A 0$ and $v \ominus u \approx_A 0$. Hence, by (1.360), we have that $w_\beta := (u \ominus v)_\beta^Q \to 0$ as $\beta \to 0$. We set $f_\beta = ((u - v)_+)_\beta^Q$ and consider the truncated problem in $B_j \times (\beta, \infty)$ for $j = 1, 2, ...,$

$$\begin{aligned} \partial_t w - \Delta w + |w|^{q-1} w = 0 \quad &\text{in } B_j \times (\beta, \infty) \\ w = 0 \quad &\text{on } \partial B_j \times (\beta, \infty) \\ w(., \beta) = \phi \ &\text{in } B_j, \end{aligned}$$

and denote by w_j and f_j respectively the solutions with initial data $\mathbf{1}_Q(u \ominus v)(., \beta)$ and $\mathbf{1}_Q(u - v)_+(., \beta)$. By the maximum principle, the sequences $\{w_j\}$ and $\{f_j\}$ are increasing. Since $u \ominus v$ is the smallest solution which dominates the subsolution $(u - v)_+$, we have $w_j \geq f_j$ for all $j \in \mathbb{N}_*$. When $j \to \infty$, $w_i \to w_\beta$ and $f_i \to f_\beta$. Then $w_\beta \geq f_\beta$. This implies

$$((u-v)_+)^Q_\beta \quad \text{as } \beta \to 0.$$

Similarly

$$((v-u)_+)^Q_\beta \quad \text{as } \beta \to 0.$$

This yields the implication $\Longrightarrow$ in (1.365).

For the reverse implication we introduce the problem

$$\begin{aligned} \partial_t w - \Delta w + |w|^{q-1} w = 0 \quad &\text{in } B_j \times (\beta, \infty) \\ w = h \quad &\text{on } \partial B_j \times (\beta, \infty) \\ w(.,\beta) = \phi \text{ in } B_j. \end{aligned}$$

Let $Q \subset \widetilde{Q} \subset^q A$ be a $\mathfrak{T}_q$-open. Denote by w_j the solution of the above problem with $h = \mathbf{1}_Q |u-v| \llcorner_{\partial B_j \times (\beta,\infty)}$ and $\phi = \mathbf{1}_Q |u-v|$, and f_j the solution with $h = \mathbf{1}_{Q^c} |u-v| \llcorner_{\partial B_j \times (\beta,\infty)}$ and $\phi = \mathbf{1}_{Q^c} |u-v|$. Then

$$|u-v| \leq w_j + f_j.$$

Up to some subsequence, w_j and f_j converge respectively to w and f which are solutions of (1.30) in $\mathbb{R}^N \times (\beta, \infty)$ with respective initial data $w(.,\beta) = \mathbf{1}_Q |u-v|(.,\beta)$ and $f(.,\beta) = \mathbf{1}_{Q^c} |u-v|(.,\beta)$. Because of uniqueness and Definition 1.22, $w = |u-v|^Q_\beta$ and $f = |u-v|^{Q^c}_\beta$. When $\beta \to 0$ we have by assumption

$$\lim_{\beta \to 0} |u-v|^Q_\beta = 0.$$

Let $\{\beta_k\}$ be a subsequence decreasing to 0 such that there exists $\lim_{\beta_k \to 0} |u-v|^{Q^c}_\beta$. Then

$$|u-v| \leq \lim_{\beta_k \to 0} |u-v|^{Q^c}_{\beta_k}.$$

But $|u-v|^{Q^c}_{\beta_k} \approx_Q 0$, hence $\lim_{\beta_k \to 0} |u-v|^{Q^c}_{\beta_k} \approx_Q 0$. Since $u \ominus v$ is the smallest solution which dominates the subsolution $(u-v)_+$ there holds

$$\max\{u \ominus v, v \ominus u\} \leq \lim_{\beta_k \to 0} |u-v|^{Q^c}_{\beta_k}.$$

The proof follows from Lemma 1.34 and Proposition 1.24-(i).

(ii) Let us assume that $u \approx_A v$, then

$$u + (u-v)_+ \leq v + (u-v)_+ \leq v + u \ominus v. \tag{1.367}$$

If F is a $\mathfrak{T}_q$-closed set and Q a $\mathfrak{T}_q$-open set such that $F \subset^q Q$, we claim that

$$[u]_F \leq [v]_Q + [u \ominus v]_Q. \tag{1.368}$$

This can be proved as follows: we first have

$$u = [u]_{\mathbb{R}^N} \leq [u]_Q + [u]_{Q^c}$$

by (1.356). Using (1.367) it infers

$$[u]_F \leq [u]_{\mathbb{R}^N} \leq v + u \ominus v \leq [v]_Q + [v]_{Q^c} + [u \ominus v]_Q + [u \ominus v]_{Q^c}.$$

The subsolution $w := \big([u]_F - ([v]_Q + [u \ominus v]_Q)\big)_+$ is dominated by $[u \ominus v]_{Q^c} + [v]_{Q^c}$ which is a supersolution. From the definition we have

$$w \leq [w]_\dagger \leq [u \ominus v]_{Q^c} \oplus [v]_{Q^c} \leq [u \ominus v]_{Q^c} + [v]_{Q^c}.$$

Therefore $[w]_\dagger \approx_Q 0$. Since $w \leq [u]_F$ we deduce $[w]_\dagger \leq [u]_F$, which means $\mathfrak{T}_q$-supp$([w]_\dagger) \subset^q F \subset^q Q$. As $[w]_\dagger \approx_Q 0$ we obtain that $w = [w]_\dagger = 0$ and (1.368) follows.

Let Q be a $\mathfrak{T}_q$-open set such that $F \subset^q Q \subset \widetilde{Q} \subset^q A$, and because $u \ominus v \approx_A 0$ implies $[u \ominus v]_F = 0$ by (1.361) and (1.368), we deduce that

$$[u]_F \leq [v]_Q.$$

By Lemma 1.26-I, there exists a decreasing sequence $\{Q_j\}$ of open sets such that $\cap_j Q_j \sim^q F$. Then by Proposition 1.25-(iii) there holds

$$[u]_F \leq \lim_{j\to\infty} [v]_{Q_j} \leq [v]_F.$$

Similarly $[v]_F \leq [u]_F$.

To prove the reverse implication, we assume that $[v]_F = [u]_F$ for any $\mathfrak{T}_q$-closed set $F \subset^q A$. If Q is a $\mathfrak{T}_q$-open set such that $F \subset^q Q \subset^q \widetilde{Q} \subset A$, we notice that

$$u \ominus v \leq [u]_Q \oplus [u]_{Q^c} \ominus [v]_Q,$$

since (for the last inequality)

$$u = [u]_{\mathbb{R}^N} \leq [u]_Q + [u]_{Q^c} \Longrightarrow u \leq [u]_Q \oplus [u]_{Q^c} \leq [u]_Q + [u]_{Q^c}.$$

Because $([u]_Q \oplus [u]_{Q^c}) \ominus [v]_Q$ is the smallest solution dominating $(([u]_Q \oplus [u]_{Q^c}) \ominus [v]_Q)_+$, we have, using the assumption that $[u]_Q = [v]_Q$,

$$\left(([u]_Q \oplus [u]_{Q^c}) \ominus [v]_Q\right)_+ \leq \left(([u]_Q + [u]_{Q^c}) \ominus [v]_Q\right)_+ = [u]_Q + [u]_{Q^c} - [v]_Q = [u]_{Q^c}.$$

Therefore

$$[u \ominus v]_F \leq u \ominus v \leq [u]_{Q^c}.$$

Hence $\mathfrak{T}_q$-supp$([u \ominus v]_F) \subset^q F$ and $[u \ominus v]_F \approx Q0$. This in turn implies that $[u \ominus v]_F = 0$. Using (1.361) in Proposition 1.28 we obtain $u \ominus v \approx_A 0$. Similarly $v \ominus u \approx_A 0$. □

As an immediate consequence of (1.365), we have the following.

Corollary 1.5 *If $A \subset \mathbb{R}^N$ is a $\mathfrak{T}_q$-open set, the relation $\approx_A$ is an equivalence relation in $\mathcal{U}_+(Q_T)$.*

1.5.5 The Regular Initial Trace

1.5.5.1 The Local Test

Lemma 1.35 *Let $u \in \mathcal{U}_+(Q_T)$ and Q be a $\mathfrak{T}_q$-open set. Then for any $\eta \in B^{\frac{2}{q},q'}(\mathbb{R}^N) \cap L^\infty(\mathbb{R}^N)$ with $\mathfrak{T}_q$-support in $\widetilde{Q}^c$, we have*

$$\int_0^T \int_{\mathbb{R}^N} (u \wedge U_Q)^q \, (\mathbb{H}[\eta]_+)^{2q'}(t,x)dxdt < \infty. \tag{1.369}$$

Proof By Proposition 1.21, there holds

$$\lim_{t \to 0} \int_Q \left(u \wedge U_Q\right)(x,t)\eta_+(x)dx = 0,$$

and the result follows by estimate (1.347) in Lemma 1.31. □

Proposition 1.30 *Let $u \in \mathcal{U}_+(Q_T)$ and Q be a $\mathfrak{T}_q$-open set such that $u \wedge U_Q$ is a moderate solution with initial data μ. Then for any $\xi \in Q$ there exists a $\mathfrak{T}_q$-open set $O_\xi \subset Q$ such that*

$$\int_0^T \int_{\mathbb{R}^N} u^q \left(\mathbb{H}[\mathbf{1}_{Q_\xi}]\right)^{2q'} (x,t)dxdt < \infty. \tag{1.370}$$

Furthermore, for any $\eta \in B^{\frac{2}{q},q'}(\mathbb{R}^N) \cap L^\infty(\mathbb{R}^N)$ with $\mathfrak{T}_q$-support in Q, we have

$$\lim_{t\to 0} \int_Q u(x,t)\eta_+^{2q'} dx = \int_Q \eta^{2q'} d\mu. \tag{1.371}$$

Proof If η is as above, the function $\eta_+^{2q'}$ is quasi-continuous and there holds by Theorem 1.21,

$$\lim_{t\to 0} \int_Q u \wedge U_Q(x,t)\eta_+^{2q'}(x)dx = \int_Q \eta_+^{2q'} d\mu,$$

and, by the properties of U_{Q^c},

$$\lim_{t\to 0} \int_Q u \wedge U_{Q^c}(x,t)\eta_+^{2q'}(x)dx = 0.$$

Since $u \wedge U_Q \leq u \leq u \wedge U_Q + u \wedge U_{Q^c}$, we get

$$\begin{aligned}\int_Q u \wedge U_Q(x,t)\eta_+^{2q'} dx &\leq \int_Q u(x,t)\eta_+^{2q'} dx \\ &\leq \int_Q u \wedge U_Q(x,t)\eta_+^{2q'} dx + \int_Q u \wedge U_{Q^c}(x,t)\eta_+^{2q'} dx.\end{aligned}$$

This implies

$$\lim_{t\to 0} \int_Q u(x,t)\eta_+^{2q'} dx = \int_Q \eta_+^{2q'} d\mu_Q.$$

By Propositions 1.18 and 1.19,

$$\int_0^T \int_{\mathbb{R}^N} (u \wedge U_Q)^q \,(\mathbb{H}[\eta]_+)^{2q'}(t,x)dxdt < \infty, \tag{1.372}$$

for any $\eta \in B^{\frac{2}{q},q'}(\mathbb{R}^N) \cap L^\infty(\mathbb{R}^N)$ with $\mathfrak{T}_q$-support in $\widetilde{Q}$. By Lemma 1.28, we can assume that the above function η has its values in [0, 1], with $\mathfrak{T}_q$-support in Q and value 1 on a $\mathfrak{T}_q$-open neighbourhood O_ξ of ξ. Then (1.372) implies

$$\int_0^T\int_{\mathbb{R}^N}(u\wedge U_Q)^q\left(\mathbb{H}[\mathbf{1}_{O_\xi}]\right)^{2q'}(t,x)dxdt<\infty. \tag{1.373}$$

□

Definition 1.24 If $E\subset\mathbb{R}^N)$ is a Borel set, we denote by $B^{\frac{2}{q},q'}(E)$ the closure for the norm of $B^{\frac{2}{q},q'}$ of the set of C^∞ functions with compact support in E.

If E is an open set, $B^{\frac{2}{q},q'}(E)$ coincides with $B_0^{\frac{2}{q},q'}(E)$.

Proposition 1.31 *Let $u\in\mathcal{U}_+(Q_T)$ and Q be a bounded $\mathfrak{T}_q$-open set such that*

$$\int_0^T\int_{\mathbb{R}^N}u^q\left(\mathbb{H}[\mathbf{1}_Q]\right)^{2q'}(x,t)dxdt<\infty. \tag{1.374}$$

(i) There exists an increasing sequence of $\mathfrak{T}_q$-open set $\{Q_n\}$ satisfying $Q_n\subset Q$, $\widetilde{Q}_n\subset^q Q_{n+1}$ and $Q_0:=\cup_{n=1}^\infty Q_n\sim^q\widetilde{Q}$ such that the solution $v_n:=u\wedge U_{Q_n}$ is moderate, $v_n\uparrow[u]_Q$ and there exists a nonnegative measure μ_Q on Q such that $tr\,(v_n)\to\mu_Q$ as $n\to\infty$.
(ii) For any $\eta\in B^{\frac{2}{q},q'}(Q)\cap L^\infty(Q)$, we have

$$\lim_{t\to0}\int_Q u(x,t)\eta_+^{2q'}(x)dx=\int_Q\eta_+^{2q'}d\mu_Q. \tag{1.375}$$

Proof Let $z\in Q$. By Lemma 1.28 there exist a $\mathfrak{T}_q$-open set V such that $z\in V\subset\widetilde{V}\subset Q$ and a function $\psi\in B^{\frac{2}{q},q'}(\mathbb{R}^N)$ such that $\psi=1$ q.a.e. on V, $\psi=0$ outside Q and $0\leq\psi\leq1$. By Lemma 1.25 there exists a $\mathfrak{T}_q$-open neighbourhood O_z of z such that $O_z\subset\widetilde{O}_z\subset V$.

We claim that the function

$$v_z=u\wedge U_{O_z} \tag{1.376}$$

is a moderate solution. Actually, let $R>0$ such that $Q\subset\bar{Q}\subset B_R$ and let $\eta\in C_c^\infty(B_{2R})$ with value 1 on B_R and $0\leq\eta\leq1$. Then the function $\zeta=(1-\psi)\eta$ belongs to $B^{\frac{2}{q},q'}(\mathbb{R}^N)\cap L^\infty(\mathbb{R}^N)$ and has compact support in $B_{2R}\cap\widetilde{V}^c$. Therefore

$$\begin{aligned}\int_0^T\int_{\mathbb{R}^N}v_z^q\left(\mathbb{H}[\mathbf{1}_{B_R}]\right)^{2q'}(x,t)dxdt&\leq\int_0^T\int_{\mathbb{R}^N}v_z^q\left(\mathbb{H}[\psi]\right)^{2q'}(x,t)dxdt\\&\quad+\int_0^T\int_{\mathbb{R}^N}v_z^q\left(\mathbb{H}[1-\psi]\right)^{2q'}(x,t)dxdt\\&\leq\int_0^T\int_{\mathbb{R}^N}v_z^q\left(\mathbb{H}[\psi]\right)^{2q'}(x,t)dxdt\\&\quad+\int_0^T\int_{\mathbb{R}^N}v_z^q\left(\mathbb{H}[\zeta]\right)^{2q'}(x,t)dxdt<\infty,\end{aligned}$$

because the first integral in the last inequality is finite by assumption and the second integral is finite by Lemma 1.35. As R is arbitrary, $u \wedge U_{O_z}$ is a moderate solution.

By the quasi-Lindelöf property there exists a nondecreasing sequence of $\mathfrak{T}_q$-open sets $\{O_n\}$ such that $\cup_n O_n \sim^q Q$ and, using the construction above, the solution $u \wedge U_{O_n}$ is moderate for any $n \in \mathbb{N}_*$. By Proposition 1.25-(II), for any n there exists a sequence $\{A_{n,j}\}$ of $\mathfrak{T}_q$-open sets such that $\widetilde{A}_{n,j} \subset^q A_{n,j+1} \subset^q E_n$ and $\cup_{j=1}^{\infty} A_{n,j} \sim^q E_n$. Put

$$Q_n = \bigcup_{k+j=n} A_{k,j}.$$

Then

$$\widetilde{Q}_n \subset \bigcup_{k+j=n} \widetilde{A}_{k,j} \subset^q \bigcup_{k+j=n} \widetilde{A}_{k,j+1} = Q_{n+1}.$$

Therefore

$$Q_0 := \bigcup_n Q_n \sim^q Q.$$

Next we prove that $v_n = u \wedge U_{Q_n} \to u \wedge U_Q$. By Proposition 1.25-(ii),

$$v_n = u \wedge U_{Q_n} \leq u \wedge U_Q \leq u \wedge U_{Q_n} + u \wedge U_{Q \cap Q_n^c}.$$

Since $Q \cap Q_n^c \downarrow F$ and $cap_{\frac{2}{q},q'}(F) = 0$, we infer from Proposition 1.25-(iii)

$$u \wedge U_{Q \cap Q_n^c} \to 0 \quad \text{as } n \to \infty.$$

Hence $v_n \uparrow u \wedge U_Q$. Again, by Proposition 1.25-(ii), $v_n = [v_{n+k}]_{Q_n}$. Therefore, with $\mu_n = \mathbf{1}_{Q_n} \mu_Q$,

$$\mu_n(Q_n) = \mu_{n+k}(Q_n) = \mu_Q(Q_n) < \infty \Longrightarrow tr\,(v_n) \to \mu_Q. \tag{1.377}$$

(iii) We assume at first that the function $\eta \in B^{\frac{2}{q},q'}(Q) \cap L^\infty(Q)$ is nonnegative (which is not a restriction) and has compact support in Q. By Lemma 1.29 there exists a function η_k with $\mathfrak{T}_q$ support included in Q_k such that $0 \leq \eta_k \leq \eta$, $\eta_{k+1} \leq \eta_k$,

$$\|\eta - \eta_k\|_{B^{\frac{2}{q},q'}} \leq \frac{1}{k}, \tag{1.378}$$

and, for k large enough,

$$\int_0^T \int_{\mathbb{R}^N} u^q \left(\mathbb{H}[\eta - \eta_k]\right)^{2q'}(x,t)dxdt \leq \left(\frac{1}{k}\right)^{2q'}.$$

Since

$$\lim_{t\to 0}\int_Q u(x,t)\eta_k dx = \int_Q \eta_k^{2q'} d\mu_Q \quad \text{and} \quad \lim_{k\to\infty}\int_Q \eta_k^{2q'} d\mu_Q = \int_Q \eta^{2q'} d\mu_Q,$$

by a standard limit theorem

$$\begin{aligned}\liminf_{t\to 0}\lim_{k\to\infty}\int_Q u(x,t)\eta_k^{2q'}dx &= \liminf_{t\to 0}\int_Q u(x,t)\eta^{2q'}dx\\ &\geq \lim_{k\to\infty}\lim_{t\to 0}\int_Q u(x,t)\eta_k^{2q'}dx = \int_Q \eta^{2q'}d\mu_Q.\end{aligned} \tag{1.379}$$

By (1.377) and Proposition 1.30 and Hölder's inequality

$$\begin{aligned}\left(\int_Q u(x,t)\eta^{2q'}dx\right)^{\frac{1}{2q'}} &\leq \left(\int_Q u(x,t)(\eta-\eta_k)^{2q'}dx\right)^{\frac{1}{2q'}} + \left(\int_Q u(x,t)\eta_k^{2q'}dx\right)^{\frac{1}{2q'}}\\ &\leq \left(\int_Q u(x,t)\eta_k^{2q'}dx\right)^{\frac{1}{2q'}} + C\left\|\eta-\eta_k\right\|_{B^{\frac{2}{q},q'}}\left\|\eta-\eta_k\right\|_{L^\infty}\\ &\quad + C\left(\int_0^T\int_{\mathbb{R}^N} u^q(x,t)\left(\mathbb{H}[\eta-\eta_k]\right)^{2q'}\right)^{\frac{1}{2q'}}\\ &\leq \left(\int_Q u(x,t)\eta_k^{2q'}dx\right)^{\frac{1}{2q'}} + \frac{C'}{k}.\end{aligned} \tag{1.380}$$

Hence

$$\limsup_{t\to 0}\left(\int_Q u(x,t)\eta^{2q'}dx\right)^{\frac{1}{2q'}} \leq \left(\int_Q \eta_k^{2q'}d\mu_Q\right)^{\frac{1}{2q'}} + \frac{C'}{k},$$

which implies, by letting $k\to\infty$,

$$\limsup_{t\to 0}\left(\int_Q u(x,t)\eta^{2q'}dx\right)^{\frac{1}{2q'}} \leq \left(\int_Q \eta^{2q'}d\mu_Q\right)^{\frac{1}{2q'}}. \tag{1.381}$$

Combining (1.379) and (1.381) we obtain (1.375).

In the general case, by Netrusov's approximation theorem [1, Theorem 10.1.1] there exists a function η_k with compact support in Q such that $0\leq\eta_k\leq\eta$ and (1.378) holds. The end of the proof is as above. □

Proposition 1.31 admits the following easy extension to the case where the set Q is non-necessarily bounded. An overview of the proof is given in Proposition 1.34.

Corollary 1.6 *Let Q be $\mathfrak{T}_q$-open set and $u\in\mathcal{U}_+(Q_T)$ satisfying (1.374) for any $\mathfrak{T}_q$-open and bounded subset of Q.*

(i) There exists an increasing sequence of $\mathfrak{T}_q$-open set $\{Q_n\}$ satisfying $Q_n \subset Q$, $\widetilde{Q}_n \subset^q Q_{n+1}$ and $Q_0 := \cup_{n=1}^{\infty} Q_n \sim^q \widetilde{Q}$ such that the solution $v_n := u \wedge U_{Q_n}$ is moderate, $v_n \uparrow [u]_Q$ and there exists a nonnegative measure μ_Q on Q such that $tr\,(v_n) \to \mu_Q$ as $n \to \infty$.
(ii) For any $\eta \in B^{\frac{2}{q},q'}(Q_n) \cap L^\infty(Q)$ for some $n \in \mathbb{N}$, we have

$$\lim_{t\to 0} \int_Q u(x,t)\eta_+^{2q'}(x)dx = \int_Q \eta_+^{2q'} d\mu_Q. \tag{1.382}$$

Proposition 1.32 *Let Q be a $\mathfrak{T}_q$-open set and $u \in \mathcal{U}_+(Q_T)$ satisfying (1.374). Then*
(i)

$$[u]_Q = \sup\{[u]_F : F \subset^q Q,\ F\ \mathfrak{T}_q\text{-closed}\}. \tag{1.383}$$

(ii) For every $\mathfrak{T}_q$-open set $O \subset \widetilde{O} \subset^q Q$ such that $[u]_O$ is a moderate solution, we have

$$\mathbf{1}_{\widetilde{O}}\mu_Q = tr\left(([[u]_Q]_O\right). \tag{1.384}$$

Furthermore the measure μ_Q defined in Proposition 1.31-(i) is $\mathfrak{T}_q$-locally finite on Q and σ-finite on $Q' := \cup_n Q_n$ where the sets Q_n form an increasing sequence $\{Q_n\}$ of $\mathfrak{T}_q$-open subsets of Q satisfying $\widetilde{Q}_n \subset^q Q_{n+1}$ and $Q' \sim^q Q$ as in Proposition 1.31-(i).
(iii) If $\{w_n\} \subset \mathcal{U}_+(Q_T)$ is a nondecreasing sequence of moderate solutions of (1.30) such that $\mathfrak{T}_q$-supp.$(w_n) \subset^q Q$ and $w_n \uparrow [u]_Q$, then $tr\,(w_n) \uparrow \mu_Q$.

Proof (i) Let u^* denote the right-hand side of (1.383). By Proposition 1.11 there exists a nondecreasing sequence $\{[u]_{F_n}\}$ such that F_n is $\mathfrak{T}_q$-closed and $[u]_{F_n} \uparrow u^*$. By Proposition 1.26 we have

$$[u]_{F_n} \leq [u]_{F_n \cap Q_m} + [u]_{F_n \cap Q_m^c}.$$

Notice that $F_n \cap Q_m^c$ is $\mathfrak{T}_q$-closed and $\cap_{m=1}^{\infty} F_n \cap Q_m^c = A_n$ and $cap_{\frac{2}{q},q'}(A_n) = 0$. Therefore, by Proposition 1.25 we have that $U_{F_n \cap Q_m^c} \to 0$ as $m \to \infty$, hence $[u]_{F_n \cap Q_m^c} \to 0$ as $m \to \infty$. Therefore $[u]_{F_n} \leq \lim_{m\to\infty}[u]_{F_m} = u_Q$. Letting $n \to \infty$ we infer $u^* \leq u_Q$. By the definition of u^* we have $u_Q \leq u^*$.
(ii) Set $\mu_O = tr\,([u]_O)$. If F is $\mathfrak{T}_q$-closed such that $F \subset^q O$, then by Proposition 1.26-(ii),

$$tr\,([u]_F) = tr\left([[u]_{\widetilde{O}}]_F\right) = \mathbf{1}_F\,\mu_O = \mathbf{1}_F\,\mu_O. \tag{1.385}$$

If $O' \subset \widetilde{O'} \subset^q Q$ is $\mathfrak{T}_q$-open such that $[u]_{O'}$ is a moderate solution, then clearly

$$\mu_{O \cap O'} = \mathbf{1}_{\widetilde{O} \cap \widetilde{O'}}\mu_O = \mathbf{1}_{\widetilde{O} \cap \widetilde{O'}}\mu_{O'}. \tag{1.386}$$

Since $[u]_F$ is moderate,

$$[[u]_{Q_n}]_F = [u]_{Q_n \cap F} \uparrow [u]_F \quad \text{as } n \to \infty. \tag{1.387}$$

In addition, $[u_Q]_F \geq \lim_{n\to\infty} [[u]_{Q_n}]_F = [u]_F$, jointly with $u_Q \leq u$, leads to

$$u_F = [u_Q]_F. \tag{1.388}$$

By (1.385) and (1.387), if F is a $\mathfrak{T}_q$-closed subset of $\mathcal{R}_q(u)$, and $[u]_F$ is moderate

$$tr\left([u]_F\right) = \lim_{n\to\infty} tr\left([[u]_{Q_n}]_F\right) = \mathbf{1}_F \mu_{\mathcal{R}_q(u)}, \tag{1.389}$$

and (1.384) follows.

Since $Q' = \cup_n Q_n$ and $\mu_Q(Q_n) < \infty$, μ_Q is σ-finite on $Q' \sim^q Q$. Since for any $\xi \in Q$ there exists a $\mathfrak{T}_q$-open neighbourhood O_ξ of ξ included in Q with the property that $\mu_Q(O_\xi) < \infty$, μ_Q is $\mathfrak{T}_q$-locally finite on Q.
(iii) If w is a moderate solution dominated by u_Q, with $\mathfrak{T}_q$-supp$\,(w) \subset^q Q$, then $tr(w) \leq \mu_Q$, since $[w]_{Q_n} \leq [u]_{Q_n}$ and $[w]_{Q_n} \uparrow w$ implies that

$$tr([w]_{Q_n}) \uparrow tr(w) \leq \lim_{n\to\infty} tr\left([u]_{Q_n}\right) = \mu_Q.$$

Let $\{w_n\}$ be an increasing sequence of moderate solutions with $F_n := \mathfrak{T}_q$-supp$\,(w) \subset^q Q$ and $w_n \uparrow u_Q$. We claim that if $\nu_n := tr(w_n)$, then

$$\nu := \lim_{n\to\infty} \nu_n = \mu_Q. \tag{1.390}$$

Clearly $\nu \leq \mu_Q$. To prove the reverse inequality, let D be a $\mathfrak{T}_q$-open set such that $[u]_D$ is moderate and $K \subset D$ a compact set such that $cap_{\frac{2}{q},q'}(K) > 0$. Then

$$w_n \leq [w_n]_D + [w_n]_{D^c} \longrightarrow u_Q = \lim_{n\to\infty} w_n \leq \lim_{n\to\infty} [w_n]_D + U_{D^c}.$$

Since $[u]_D$ is moderate, the sequence $\{[w_n]_D\}$ which is dominated by $[u]_D$ has an initial trace $tr\,(w_n]_D) := \mathbf{1}_{\tilde{D}}\nu_n$ which increases and converges to $\mathbf{1}_{\tilde{D}}\nu$. Hence, $\mathbf{1}_{\tilde{D}}\nu$ is a Radon measure which vanishes on sets with zero $cap_{\frac{2}{q},q'}$-capacity. Hence

$$[w_n]_D \uparrow u_{\mathbf{1}_{\tilde{D}}\nu}$$

where $u_{\mathbf{1}_{\tilde{D}}\nu}$ is the moderate solution with initial trace $\mathbf{1}_{\tilde{D}}\nu$. Therefore

$$u_Q = \lim_{n\to\infty} w_n \leq u_{\mathbf{1}_{\tilde{D}}\nu} + U_{D^c}.$$

This implies

$$\left([u_Q]_K - u_{\mathbf{1}_{\tilde{D}}\nu}\right)_+ \leq \inf\left\{U_{D^c}, U_K\right\}.$$

Notice that the left-hand side of the above inequality is a subsolution while the right-hand side is a supersolution. This implies

$$\left([u_Q]_K - u_{\mathbf{1}_{\tilde{D}}\nu}\right)_+ \leq U_{D^c} \wedge U_K = [[U]_{D^c}]_K = 0.$$

Therefore $u_Q]_K \leq u_{\mathbf{1}_{\tilde{D}}\nu}$ which implies $\mathbf{1}_K\mu_Q \leq \mathbf{1}_{\tilde{D}}\nu$. Moreover, if O is a $\mathfrak{T}_q$-open set such that $\widetilde{O} \subset^q D$, then, using the fact that

$$\mathbf{1}_{\tilde{O}}\mu_Q = \sup\left\{\mathbf{1}_K\mu_Q : K \subset O,\ K \text{ compact}\right\},$$

we obtain

$$\mathbf{1}_{\tilde{O}}\mu_Q \leq \mathbf{1}_{\tilde{D}}\nu. \tag{1.391}$$

Applying this series of inequalities to the sets Q_m, Q_{m+1}, ..., we infer

$$\mathbf{1}_{Q_m}\mu_Q \leq \mathbf{1}_{\tilde{Q}_{m+1}}\nu \leq \mathbf{1}_{Q_{m+2}}\nu.$$

Letting $m \to \infty$ we deduce that $\mu_Q = \nu$. □

1.5.5.2 $\mathfrak{T}_q$-Perfect Measures

Definition 1.25 Let μ be a positive Borel measure on $\mathbb{R}^N$.
(i) We say that μ is essentially absolutely continuous with respect to the $cap_{\frac{2}{q},q'}$-capacity if the following condition holds:
If Q is a $\mathfrak{T}_q$-open set and A a Borel subset such that $cap_{\frac{2}{q},q'}(A) = 0$, then

$$\mu(Q) = \mu(Q \cap A^c).$$

This relation is denoted by

$$\mu \lll cap_{\frac{2}{q},q'}.$$

(ii) We say that μ is regular with respect to the $\mathfrak{T}_q$-topology if, for every Borel set E, there holds

$$\begin{aligned}\mu(E) &= \inf\left\{\mu(D) : E \subset D,\ D\ \mathfrak{T}_q\text{-open}\right\}\\ &= \sup\left\{\mu(K) : K \subset E,\ K \text{ compact}\right\},\end{aligned} \tag{1.392}$$

and μ is outer regular with respect to the $\mathfrak{T}_q$-topology if there only holds

$$\mu(E) = \inf\left\{\mu(D) : E \subset D,\ D\ \mathfrak{T}_q\text{-open}\right\}. \tag{1.393}$$

(iii) A positive Borel measure is called $\mathfrak{T}_q$-perfect if it is essentially absolutely continuous with respect to the $cap_{\frac{2}{q},q'}$ and outer regular with respect to the $\mathfrak{T}_q$-topology. The space of $\mathfrak{T}_q$-perfect Borel measures is denoted by $\mathfrak{M}_q(\mathbb{R}^N)$.

Proposition 1.33 *If $\mu \in \mathfrak{M}_q(\mathbb{R}^N)$ and A is a non-empty Borel subset of $\mathbb{R}^N$ such that $cap_{\frac{2}{q},q'}(A) = 0$, then*

$$\mu = \begin{cases} \infty & \textit{if } \mu(Q \cap A^c) = \infty \textit{ for all } \mathfrak{T}_q\textit{-open neighbourhood } Q \textit{ of } A \\ 0 & \textit{otherwise.} \end{cases} \tag{1.394}$$

If μ_0 is an essentially absolutely continuous positive measure in $\mathbb{R}^N$ and Q is a $\mathfrak{T}_q$-open set such that $\mu_0(Q) < \infty$, then $\mu_0 \lfloor_Q$ is absolutely continuous with respect to the $cap_{\frac{2}{q},q'}$-capacity in the strong sense, that is, for any sequence of Borel subsets $\{A_n\}$ of $\mathbb{R}^N$,

$$cap_{\frac{2}{q},q'}(A_n) \to 0 \Longrightarrow \mu_0(Q \cap A) \to 0 \quad as\ n \to \infty.$$

If μ_0 is an essentially absolutely continuous positive Borel measure on $\mathbb{R}^N$ and if for every Borel subset of $\mathbb{R}^N$ we denote

$$\mu(E) = \inf\left\{\mu_0(D) : E \subset D\ D\ \mathfrak{T}_q\text{-}open\right\}, \tag{1.395}$$

then μ is a Borel measure and

$$\begin{aligned} &(i) \quad \mu_0 \leq \mu, \quad \mu_0(Q) = \mu(Q) \textit{ for all } \mathfrak{T}_q\textit{-open set } Q \\ &(ii) \quad \mu\lfloor_Q = \mu_0\lfloor_Q \textit{ for all } \mathfrak{T}_q\textit{-open set } Q \textit{ s.t. } \mu_0(Q) < \infty. \end{aligned} \tag{1.396}$$

Proof The first assertion follows from the definition of $\mathfrak{M}_q(\mathbb{R}^N)$. Next, if μ_0 is essentially absolutely continuous and $\mu_0(Q) < \infty$ where Q is $\mathfrak{T}_q$-open, then $\mathbf{1}_Q\mu_0$ is a bounded Borel measure which vanishes on Borel sets with zero $cap_{\frac{2}{q},q'}$-capacity. If $\{A_n\}$ is a sequence of Borel sets that we can assume it to be decreasing, such that $cap_{\frac{2}{q},q'}(A_n) \to 0$ when $n \to \infty$, and $\mu_n = \mathbf{1}_{Q\cap A_n}\mu_0$, then by [30, Lemma 2.8] there exists a unique moderate solution u_{μ_n} with initial trace μ_n. There holds $u_{\mu_n} \leq U_{Q\cap A_n}$. Since $cap_{\frac{2}{q},q'}(Q \cap A_n) \to 0$, $U_{Q\cap A_n}$ converges to 0 when $n \to \infty$, and so does u_{μ_n}. Then $\mu_n \to 0$ in the weak topology of Radon measures, which implies that $u_{\mu_n} \to 0$ locally uniformly in Q_T. Therefore $\mu(Q \cap A_n) \to 0$ which implies that $\mu_0\lfloor_Q$ is absolutely continuous with respect to the $cap_{\frac{2}{q},q'}$-capacity in the usual sense.

Assertion (1.396)-(i) follows from the definition (1.395). If Q is a $\mathfrak{T}_q$-open set such that $\mu_0(Q) < \infty$ then $\mu(Q) < \infty$. Since $\mu_0\lfloor_Q$ and $\mu\lfloor_Q$ are regular Borel measures which coincide on open sets, they coincide on all Borel sets. This implies (1.396)-(ii).

At the end, if A is a Borel set such that $cap_{\frac{2}{q},q'}(A) = 0$, then $cap_{\frac{2}{q},q'}(\widetilde{A}) = 0$. If Q is a $\mathfrak{T}_q$-open set, then $Q \cap \widetilde{A}^c$ is $\mathfrak{T}_q$-open. Therefore

$$\mu(Q) = \mu_0(Q) = \mu_0(Q \cap \widetilde{A}^c) = \mu(Q \cap \widetilde{A}^c).$$

Hence μ is essentially absolutely continuous. Using (1.396)-(i) and the definition of μ we infer that μ is outer regular with respect to the capacity $cap_{\frac{2}{q},q'}$. Hence $\mu \in \mathfrak{M}_q(\mathbb{R}^N)$. □

1.5.5.3 The Initial Trace on the Regular Set

In the next propositions we study the initial trace of a positive solution u of (1.30) on the regular initial set $\mathcal{R}_q(u)$ given in Definition 1.24, and we study the properties of the measure $\mu_{\mathcal{R}_q(u)}$ constructed by Proposition 1.31, Corollary 1.6 and Proposition 1.32.

Proposition 1.34 *Let* $u \in \mathcal{U}_+(Q_T)$.
(i) There exists an increasing sequence of $\mathfrak{T}_q$*-open sets* $\{Q_n\}$ *with the following properties:* $Q_n \subset \mathcal{R}_q(u)$, $\widetilde{Q}_n \subset^q Q_{n+1}$ *and* $\mathcal{R}_{q,0}(u) := \cup_{n=0}^{\infty} Q_n \sim^q \mathcal{R}_q(u)$, *such that*

$$v_n := u \wedge U_{Q_n} \text{ is moderate } \; v_n \uparrow v_{\mathcal{R}_q(u)} \text{ and } tr\,(v_n) \uparrow \mu_{\mathcal{R}_q(u)}. \tag{1.397}$$

(ii) There holds

$$v_{\mathcal{R}_q(u)} := \sup\left\{[u]_F : \; F \subset^q \mathcal{R}_q(u), \;\; F\ \mathfrak{T}_q\text{ -closed}\right\}. \tag{1.398}$$

Thus $v_{\mathcal{R}_q(u)}$ *is* σ*-moderate.*
(iii) If $[u]_F$ *is moderate and* $F \subset^q \mathcal{R}_q(u)$, *there exists a* $\mathfrak{T}_q$*-open set* $Q \subset \mathcal{R}_q(u)$ *such that* $F \subset^q Q$, $[u]_Q$ *is moderate.*
(iv) For every $\mathfrak{T}_q$*-open set* Q *such that* $[u]_Q$ *is a moderate solution, we have*

$$\mathbf{1}_{\widetilde{Q}}\mu_{\mathcal{R}_q(u)} = tr\,\left([u]_Q\right) = tr\,\left(\left[v_{\mathcal{R}_q(u)}\right]\right). \tag{1.399}$$

Finally $\mu_{\mathcal{R}_q(u)}$ *is* $\mathfrak{T}_q$*-locally finite on* $\mathcal{R}_q(u)$ *and* σ*-finite on* $\mathcal{R}_{q,0}(u)$.
(v) If $\{w_n\}$ *is a sequence of moderate solutions such that* $w_n \uparrow u_{\mathcal{R}_q(u)}$, *then*

$$\mu_{\mathcal{R}_q(u)} = \lim_{n\to\infty} tr\,(w_n) = \lim_{n\to\infty} tr(v_n). \tag{1.400}$$

(vi) The regularised measure $\bar{\mu}_{\mathcal{R}_q(u)}$ *defined for Borel sets* $E \subset \mathbb{R}^N$ *by*

$$\bar{\mu}_{\mathcal{R}_q(u)}(E) = \inf\left\{\mu_{\mathcal{R}_q(u)}(Q) : E \subset Q, \;\; Q\ \mathfrak{T}_q\text{-open}\right\} \tag{1.401}$$

is $\mathfrak{T}_q$*-perfect.*
(vii) There holds

$$u \approx_{\mathcal{R}_q(u)} v_{\mathcal{R}_q(u)}.$$

(viii) For every $\mathfrak{T}_q$*-closed set* $F \subset^q \mathcal{R}_q(u)$,

$$[u]_F = \left[v_{\mathcal{R}_q(u)}\right]_F. \tag{1.402}$$

Furthermore, if $\mu_{\mathcal{R}_q(u)}(F \cap K) < \infty$ *for every compact subset* $K \subset \mathbb{R}^N$, *then* $[u]_F$ *is moderate and*

$$tr\,([u]_F) = \mathbf{1}_F\mu_{\mathcal{R}_q(u)}. \tag{1.403}$$

(ix) If F is a $\mathfrak{T}_q$-closed set with positive $cap_{\frac{2}{q},q'}$-capacity, we have

$$\mu_{\mathcal{R}_q(u)}(F \cap K) < \infty \quad \text{for all compact set } K \subset \mathbb{R}^N \Longleftrightarrow [u]_F \quad \text{is moderate.} \tag{1.404}$$

Proof (i) For every $z \in \mathcal{R}_q(u)$ there exists a $\mathfrak{T}_q$-bounded open set $O_z \subset \mathcal{R}_q(u)$ such that $[u]_{O_z}$ is moderate. With the previous notations and the construction of the sequence $\{Q_n\}$ in Proposition 1.31, we recall that $v_n = [u]_{Q_n} = u \wedge U_{Q_n}$ satisfies $v_n = [v_{n+k}]_{Q_n}$ for any $k \in \mathbb{N}$ and

$$\mu_n(Q_n) = \mu_{n+k}(Q_n) = \mu_{\mathcal{R}_q(u)}(Q_n). \tag{1.405}$$

(ii) The proof has already been made in Proposition 1.32.

(iii) We assume firstly that F is bounded. Using the definition and (i), every point in $\mathcal{R}_q(u)$ possesses a $\mathfrak{T}_q$-open neighbourhood A such that $[u]_A$ is moderate. By Proposition 1.17-(II), for any $\epsilon > 0$ there exists a $\mathfrak{T}_q$-open set Q_ϵ such that $cap_{\frac{2}{q},q'}(F \cap Q_\epsilon^c) < \epsilon$ and $[u]_{Q_\epsilon}$ is moderate. Since F is bounded, we can assume that so is Q_ϵ. Let O_ϵ be an open set containing $F \cap Q_\epsilon^c$ and such that $cap_{\frac{2}{q},q'}(O_\epsilon) < 2\epsilon$. We define a $\mathfrak{T}_q$-closed set F_ϵ included in Q_ϵ by

$$F_\epsilon = F \cap O_\epsilon, \tag{1.406}$$

and $F_\epsilon \subset F$ with $cap_{\frac{2}{q},q'}(F \cap F_\epsilon^c) < 2\epsilon$.

Claim 1: Let E be a $\mathfrak{T}_q$-closed set, D a $\mathfrak{T}_q$-open set such that $[u]_D$ is moderate and $E \subset^q D$. There exists a decreasing sequence of $\mathfrak{T}_q$-open sets $\{G_n\}$ such that

$$E \subset^q G_{n+1} \subset \widetilde{G}_{n+1} \subset^q G_n \subset^q D, \tag{1.407}$$

and

$$[u]_{G_\epsilon} \to [u]_E \quad \text{in } L^q(K) \text{ for every compact set } K \subset \overline{Q}_T. \tag{1.408}$$

By Lemma 1.26 and Proposition 1.18-(iii), there exists a decreasing sequence of $\mathfrak{T}_q$-open sets $\{G_n\}$ satisfying (1.407) and such that $[u]_{G_n} \downarrow [u]_E$ locally uniformly in Q_T. Since $[u]_{G_n} \leq [u]_D$ which is a moderate solution, we deduce (1.408).

Next we assume that F is a non-necessarily bounded $\mathfrak{T}_q$-closed set. If $x \in F$ we set $B_n = B_n(x) \cap F$, $n \in \mathbb{N}_*$ and

$$E_n = \bigcup_{m=1}^{n} (F \cap B_n)_{2^{-m}},$$

where $(F \cap B_n)_{2^{-m}}$ is the set defined in (1.406) with F replaced by $F \cap B_n$ and ϵ replaced by 2^{-m}. We can also assume that the sequence $\{E_n\}$ is increasing. We set

$Q^n_{m^{-1}} = (F \cap B_n)_{m^{-1}}$ and

$$Q_n = \bigcup_{m=1}^{n} Q^n_{m^{-1}},$$

and as for $\{F_n\}$ we can assume that the sequence $\{Q_n\}$ is increasing. Therefore, we have that $E_n \subset E$, Q_n is $\mathfrak{T}_q$-open, $[u]_{Q_n}$ is moderate and $E_n \subset^q Q_n$. Furthermore $\cup_n E_n = E' \sim^q F$ since for any $n \in \mathbb{N}_*$ there holds

$$\begin{aligned} cap_{\frac{2}{q},q'}\left(F \setminus \bigcup_{j=1}^{\infty} E_j\right) &\leq \sum_{k=1}^{n} cap_{\frac{2}{q},q'}\left((F \cap B_k) \setminus \bigcup_{j=1}^{\infty} E_j\right) \\ &\quad + \sum_{k=n+1}^{\infty} cap_{\frac{2}{q},q'}\left((F \cap B_k) \setminus E_k\right) \\ &\leq \frac{1}{2^n} + \sum_{k=n+1}^{\infty} \frac{1}{2^k} = \frac{1}{2^{n-1}}. \end{aligned}$$

Thus, by Assertion 1, we can choose a sequence of $\mathfrak{T}_q$-open sets $\{V_n\}$ such that

$$E_n \subset^q V_n \subset \widetilde{V}_n \subset^q Q_n \quad \text{and} \quad \left\| [u]_{V_n} - [u]_{E_n} \right\|_{L^q(B_n(0)\times(0,T))} \leq 2^{-n}. \tag{1.409}$$

Notice that since E_n and Q_n are bounded sets, the functions $[u]_{V_n}$ and $[u]_{E_n}$ which are moderate belong to $L^q(\mathbb{R}^N \times (0,T))$.
Because $[u]_F$ is moderate, there exists a Radon measure $\mu_F = tr\,([u]_F)$ and $[u]_F = [u]_{E'}$ since $F \sim^q E'$. At the end, using (1.357) and the fact that $E_n \subset^q F$, we have

$$[u]_{E_n} = [u]_{F\cap E_n} = [[u]_{E_n}]_F.$$

Because $[u]_F$ is moderate we have $tr\left([u]_{E_n}\right) = \mathbf{1}_{E_n}\mu_F$. Since $E_n \uparrow E' \sim^q F$ we deduce that $[u]_{E_n} \uparrow [u]_F$ in $L^q_{loc}(\overline{Q}_T)$. Hence, we have from (1.409) that $[u]_{V_n} \to [u]_F$ in $L^q_{loc}(\overline{Q}_T)$ as $n \to \infty$.

Let $\{V_{n_k}\}$ be a subsequence such that

$$\left(\int_0^T \int_{B_k} \left|[u]_{V_{n_k}} - [u]_F\right|^q\right)^{\frac{1}{q}} \leq 2^{-k}. \tag{1.410}$$

If $K \subset \mathbb{R}^N$ is compact, it is included in B_k for $k \geq k_0$. We set $W = \cup_{k=1}^{\infty} V_{n_k}$, then

$$[u]_W \leq \sum_{k=1}^{\infty} [u]_{V_{n_k}}.$$

Therefore

$$\begin{aligned}\left(\int_0^T\int_K |[u]_W-[u]_F|^q\right)^{\frac{1}{q}} &\leq \sum_{k=1}^{k_0}\left(\int_0^T\int_{B_k}|[u]_{V_{n_k}}-[u]_F|^q\right)^{\frac{1}{q}}\\ &\quad+\sum_{k=k_0+1}^{\infty}\left(\int_0^T\int_K|[u]_{V_{n_k}}-[u]_F|^q\right)^{\frac{1}{q}}\\ &\leq \sum_{k=1}^{k_0}\left(\int_0^T\int_{B_k}|[u]_{V_{n_k}}-[u]_F|^q\right)^{\frac{1}{q}}+\sum_{k=k_0+1}^{\infty}2^{-k}\\ &<\infty.\end{aligned}$$

Because $F\subset^q W$, W is $\mathfrak{T}_q$-open, $[u]_F$ is moderate and K is arbitrary it follows from the above inequality that $[u]_W$ is moderate, therefore $W\subset\mathcal{R}_q(u)$ by Proposition 1.30.
(iv) Let Q be a $\mathfrak{T}_q$-open set such that $[u]_Q$ is a moderate solution, and $\mu_Q=tr\left([u]_Q\right)$. If F is a $\mathfrak{T}_q$-closed set such that $F\subset^q Q$, then by Proposition 1.18-(ii),

$$tr([u]_F)=tr\left(\left[[u]_Q\right]_F\right)=\mathbf{1}_F\mu_Q. \tag{1.411}$$

In particular, if Q and Q' are regular sets in the sense of Definition 1.14, then

$$\mu_{Q\cap Q'}=\mathbf{1}_{\tilde{Q}\cap\tilde{Q}'}\mu_Q=\mathbf{1}_{\tilde{Q}\cap\tilde{Q}'}\mu_{Q'}. \tag{1.412}$$

Using the notations of (i), we have $[v_{n+k}]_{Q_k}=v_k$ and hence $\mathbf{1}_{\tilde{Q}_k}\mu_{n+k}=\mu_k$ for every integer k.

Let F be a $\mathfrak{T}_q$-closed regular subset of $\mathcal{R}_q(u)$. Since $[u]_F$ is moderate we have by (1.412)

$$[v_n]_F=[u]_{F\cap\tilde{Q}_n}\uparrow[u]_F. \tag{1.413}$$

Furthermore, since we have

$$\left[v_{\mathcal{R}_q(u)}\right]_F\geq\lim_{n\to\infty}[v_n]_F=[u]_F,$$

and $v_{\mathcal{R}_q(u)}\leq u$, we infer

$$[u]_F=\left[v_{\mathcal{R}_q(u)}\right]_F. \tag{1.414}$$

It follows from (1.411) and (1.413) that if F is a $\mathfrak{T}_q$-closed subset of $\mathcal{R}_q(u)$ and $[u]_F$ is moderate that

$$tr\left([u]_F\right)=\lim_{n\to\infty}tr\left([v_n]_F\right)=\lim_{n\to\infty}\mathbf{1}_F\mu_n=\mathbf{1}_F\mu_{\mathcal{R}_q(u)}, \tag{1.415}$$

which yields (1.399).

Finally, since $\mathcal{R}_q(u)$ has a regular decomposition, $\mu_{\mathcal{R}_q(u)}$ is σ-finite on $\mathcal{R}_q(u)$. As for the claim that $\mu_{\mathcal{R}_q(u)}$ is $\mathfrak{T}_q$-locally finite on $\mathcal{R}_q(u)$, it is a consequence of the fact that every point $\xi \in \mathcal{R}_q(u)$ is contained in a $\mathfrak{T}_q$-open set $O_\xi \subset \mathcal{R}_q(u)$ such that $[u]_{O_\xi}$ is moderate and thus $\mu_{\mathcal{R}_q(u)}(O_\xi) < \infty$.

(v) If w is a moderate solution dominated by $v_{\mathcal{R}_q(u)}$ and the $\mathfrak{T}_q$-supp$(w) \subset^q \mathcal{R}_q(u)$ then $\tau := tr(w) \leq \mu_{\mathcal{R}_q(u)}$.

Now, let $\{w_n\}$ be an increasing sequence of moderate solutions such that $F_n := \mathfrak{T}_q$-supp$(w_n) \subset^q \mathcal{R}_q(u)$ and $w_n \uparrow v_{\mathcal{R}_q(u)}$. *We claim that*

$$\nu := \lim_{n\to\infty} \nu_n := \lim_{n\to\infty} tr(w_n) = \mu_{\mathcal{R}_q(u)}. \tag{1.416}$$

By the previous argument, $\nu \leq \mu_{\mathcal{R}_q(u)}$. In order to prove the opposite inequality, we proceed as follows: Let D be a $\mathfrak{T}_q$-open set such that $[u]_D$ is moderate and let K be a compact subset of D with positive $cap_{\frac{2}{q},q'}$-capacity. Then

$$w_n \leq [w_n]_D + [w_n]_{D^c} \Longleftarrow v_{\mathcal{R}_q(u)} := \lim_{n\to\infty} w_n \leq \lim_{n\to\infty} [w_n]_D + U_{D^c}.$$

The sequence $\{[w_n]_D\}$ is dominated by the moderate solution $[v_{\mathcal{R}_q(u)}]_D$. In addition $tr\left([w_n]_D\right) = \mathbf{1}_{\tilde{D}}\nu_n \uparrow \mathbf{1}_{\tilde{D}}\nu \leq \mathbf{1}_{\tilde{D}}v_{\mathcal{R}_q(u)}$. Hence $\mathbf{1}_{\tilde{D}}\nu$ is a Radon measure which vanishes on Borel sets with zero $cap_{\frac{2}{q},q'}$-capacity. Also $[w_n]_D \uparrow u_{\mathbf{1}_{\tilde{D}}\nu}$, with the usual notation. Consequently

$$v_{\mathcal{R}_q(u)} := \lim_{n\to\infty} w_n \leq u_{\mathbf{1}_{\tilde{D}}\nu} + U_{D^c}.$$

This implies

$$\left([v_{\mathcal{R}_q(u)}]_K - u_{\mathbf{1}_{\tilde{D}}\nu}\right)_+ \leq \inf\{U_{\tilde{D}^c}, U_K\}.$$

By the same argument as in the proof of Proposition 1.32-(ii) this yields

$$\left([v_{\mathcal{R}_q(u)}]_K - u_{\mathbf{1}_{\tilde{D}}\nu}\right)_+ \leq [[U_{D^c}]_K\} = 0.$$

Hence $[v_{\mathcal{R}_q(u)}]_K \leq u_{\mathbf{1}_{\tilde{D}}\nu}$ and hence $\mathbf{1}_K\mu_{\mathcal{R}_q(u)} \leq\leq \mathbf{1}_{\tilde{D}}\nu$. Next, if Q is a $\mathfrak{T}_q$-open set such that $Q \subset^q D$, we use the fact that

$$\sup\{\mathbf{1}_K\mu_{\mathcal{R}_q(u)} : K \subset Q,\ K \text{ compact}\} = \mathbf{1}_Q\mu_{\mathcal{R}_q(u)}$$

to obtain that

$$\mathbf{1}_Q\mu_{\mathcal{R}_q(u)} \leq \mathbf{1}_{\tilde{D}}\nu. \tag{1.417}$$

Applying this inequality to the couple of sets (Q_m, Q_{m+1}) we deduce that

$$\mathbf{1}_{Q_m}\mu_{\mathcal{R}_q(u)} \leq \mathbf{1}_{\tilde{Q}_{m+1}}\nu \leq \mathbf{1}_{Q_{m+2}}\nu.$$

Letting $m \to \infty$ implies $\mu_{\mathcal{R}_q(u)} \leq \nu$. This completes the proof of the claim (1.417) and assertion (v).

(vi) Since the measure $\mu_{\mathcal{R}_q(u)}$ is essentially absolutely continuous with respect to the $cap_{\frac{2}{q},q'}$-capacity, the claim follows from Proposition 1.33.

(vii) For any $n \in \mathbb{N}$, we have

$$u \leq [u]_{Q_n} + [u]_{Q_n^c}.$$

Since Q_n^c is $\mathfrak{T}_q$-closed and $\cap_n Q_n^c = \mathcal{R}_{q,0}^c(u)$, we have by Proposition 1.26-(iii)

$$[u]_{Q_n^c} \downarrow [u]_{\mathcal{R}_{q,0}^c(u)}.$$

Therefore

$$\lim_{n\to\infty} (u - [u]_{Q_n} = u - v_{\mathcal{R}_q(u)} \leq [u]_{\mathcal{R}_{q,0}^c(u)}.$$

It follows that $u \ominus v_{\mathcal{R}_q(u)} \approx_{\mathcal{R}_{q,0}(u)} 0$. Because $v_{\mathcal{R}_q(u)} \leq u$, this is equivalent to the statement $u \approx_{\mathcal{R}_{q,0}(u)} v_{\mathcal{R}_q(u)}$.

(viii) The fact that $[u]_F = [v_{\mathcal{R}_q(u)}]_F$ for every $\mathfrak{T}_q$-closed subset $F \subset^q \mathcal{R}_q(u)$ follows from assertion (vii). Next we assume that $\mu_{\mathcal{R}_q(u)}(F \cap K) < \infty$ for any compact set K and we set $F_n = F \cap \widetilde{Q}_n$. By relation (1.356) we have

$$[u]_{F_n} \leq [u]_F \leq [u]_{F_n} + [u]_{F\cap F_n^c} = [u]_{F_n} + [u]_{F\cap \widetilde{Q}_n^c} \leq [u]_{F_n} + [u]_{F\cap Q_n^c}.$$

Since $F \cap Q_n^c$ is $\mathfrak{T}_q$-closed and $\cap_n F \cap Q_n^c = G$ with $cap_{\frac{2}{q},q'}(G) = 0$, we deduce from Proposition 1.26-(iii) that $[u]_{F\cap Q_n^c} \to [u]_G = 0$ as $n \to \infty$. Hence $[u]_{F_n} \uparrow [u]_F$ and

$$tr\left([u]_{F_n}\right) = \mathbf{1}_{F_n}\mu_{\mathcal{R}_q(u)} \uparrow \mathbf{1}_{F_0}\mu_{\mathcal{R}_q(u)} = \mathbf{1}_F\mu_{\mathcal{R}_q(u)} \quad \text{as } n \to \infty,$$

since $\mathbf{1}_{F_0}\mu_{\mathcal{R}_q(u)} = \mathbf{1}_F\mu_{\mathcal{R}_q(u)}$ if $F_0 = \cap_n F_n$. Because $\mathbf{1}_F\mu_{\mathcal{R}_q(u)}$ is a Radon measure essentially absolutely continuous with respect to the $cap_{\frac{2}{q},q'}$-capacity, $[u]_F$ is moderate and (1.403) is verified.

(ix) If $\mu_{\mathcal{R}_q(u)}(F \cap K) < \infty$ for any compact set K, then by (viii) $[u]_F$ is moderate. Conversely, if $[u]_F$ is moderate, then by (iv), there holds $\mu_{\mathcal{R}_q(u)}(F \cap K) < \infty$ for any compact set $K \subset \mathbb{R}^N$. □

Example There exist functions $u \in \mathcal{U}_+(Q_T)$ such that $\mathcal{R}_q(u) = \mathbb{R}^N$ which are not moderate solutions. We construct one of them as follows. Let $\eta : [0, \infty) \mapsto [0, \infty)$ be a smooth function which is positive on $(0, \infty)$, $\eta^k(0) = 0$ for all $k \in \mathbb{N}$ (e.g. $\eta(r) = e^{-r^{-2}}$). We define the closed set $K \subset \mathbb{R}^N$ by

$$K = \{x = (x', x_N) \in \mathbb{R}^{N-1} \times \mathbb{R}_+ : |x'| \leq \eta(x_N)\}.$$

Then K is $\mathfrak{T}_q$-thin at 0 for the capacity $cap_{\frac{2}{q},q'}$. We set

$$f(x) = \begin{cases} \eta^{-N}(x_N) \text{ if } x \in K \\ 0 \qquad\quad \text{if } x \in \mathbb{R}^N \setminus K, \end{cases}$$

and define the measure

$$\mu = f dx.$$

Then the following properties hold:
1- μ is $\mathfrak{T}_q$-locally finite.
2- $\mu(Q_n) < \infty$ if $Q_n = B_{2n} \setminus \overline{B}_{\frac{1}{n}}$ and $\cup_n Q_n \sim^q \mathbb{R}^N$.
3- $\mu(F) = 0$ for any Borel set F such that $cap_{\frac{2}{q},q'}(F) = 0$.
4- There exists a nondecreasing sequence of bounded nonnegative Radon measures $\{\mu_n\}$ absolutely continuous with respect to the $cap_{\frac{2}{q},q'}$-capacity such that
(i) $\mathfrak{T}_q$-supp $(\mu_n) \subset \widetilde{Q}_n$, $\mu_n(A) = \mu_{n+k}(A)$ for any $A \subset \widetilde{Q}_n$ and any $n, k \in \mathbb{N}_*$.
(ii) $\lim_{n\to\infty} \mu_n = \mu$.
5- We can construct a solution u of (1.30) such that $\mathcal{R}_q(u) = \mathbb{R}^N$ and $\mu_{\mathcal{R}_q(u)} = \mu$.

We will prove later on that this solution is actually the unique one which has this initial trace since it is σ-moderate.

Lemma 1.36 *Let μ satisfy the conditions 1–4 above. Then there exists an open set $\mathcal{R}_q \sim^q \mathbb{R}^N$ such that the measure μ is a Radon measure on $\mathcal{R}_q$.*

Proof By [33, Lemma 2.5] for any $R > 1$ and $\epsilon > 0$ there exist a sequence of open sets $\{O_m\}$ and $n(m) \in \mathbb{N}_*$ such that $cap_{\frac{2}{q},q'}(O_m) < \epsilon 2^{-m}$ and

$$\overline{B}_R \setminus O_m \subset \bigcup_{j=1}^{n(m)} Q_j \quad \text{where } Q_j = B_{2j} \setminus \overline{B}_{\frac{1}{j}}. \tag{1.418}$$

Since $\overline{O}_m \subset \widetilde{O}$

$$cap_{\frac{2}{q},q'}\left(\overline{O}_m\right) \leq cap_{\frac{2}{q},q'}\left(\widetilde{O}\right) \leq \tilde{C} cap_{\frac{2}{q},q'}\left(O_m\right) \leq c\epsilon 2^{-m}.$$

If $x \in B_R \setminus \bigcap_{m=1}^{\infty} \overline{O}_m$ there exist $r_x > 0$ and $k \in \mathbb{N}$ such that

$$B_{r_x}(x) \subset B_R \setminus \bigcap_{m=1}^{k} \overline{O}_m.$$

Jointly with (1.418) it implies that

$$\mu(B_{r_x}(x)) < \infty.$$

We set

$$\mathcal{R}_q = \left\{x \in \mathbb{R}^N : \exists r_x > 0 \text{ such that } \mu(B_{r_x}(x)) < \infty\right\}.$$

The set $\mathcal{R}_q$ is open and by letting $R \to \infty$ and $\epsilon \to 0$, we obtain that $\mathcal{R}_q \sim^q \mathbb{R}^N$. By the definition of $\mathcal{R}_q$, for any compact set $K \subset \mathcal{R}_q$ there holds $\mu(K) < \infty$. Hence μ is a Radon measure in $\mathcal{R}_q$. □

1.5.6 The Precise Initial Trace

1.5.6.1 Definition and First Properties

We can now define the *precise initial trace* of an element of $\mathcal{U}_+(Q_T)$ in the super-critical case.

Definition 1.26 Let $q \geq q_c$ and $u \in \mathcal{U}_+(Q_T)$.
1- The function $v_{\mathcal{R}_q(u)}$ defined in (1.398) is called the regular component of u and will be denoted by u_{reg}.
2- Let $\{v_n\}$ be an increasing sequence of moderate solutions satisfying condition (1.397) and put $\mu_{\mathcal{R}_q(u)} := \lim_{n\to\infty} tr(v_n)$. Then, the regularised measure $\overline{\mu}_{\mathcal{R}_q(u)}$, defined by (1.401), is called the regular initial trace of u. It will be denoted by $tr_{\mathcal{R}_q(u)}(u)$.
3- The couple $(tr_{\mathcal{R}_q(u)}(u), \mathcal{S}_q(u))$ is called the precise initial trace of u and will be denoted by $tr^c(u)$.
4- Let ν be the Borel measure on $\mathbb{R}^N$ given by

$$\nu(E) = \begin{cases} tr_{\mathcal{R}_q(u)}(E) & \text{if } E \subset \mathcal{R}_q(u) \\ \infty & \text{if } E \cap \mathcal{S}_q(u) \neq \emptyset \end{cases} \tag{1.419}$$

for every Borel set $E \subset \mathbb{R}^N$. Then ν is the measure representation of the precise trace of u and it is denoted by $tr(u)$.

Remark In the definitions of $tr^c(u)$ and $tr(u)$, the exponent c stands for *couple*, but the two objects are the same in their respective classes. Thanks to Proposition 1.34 the measure $\mu_{\mathcal{R}_q(u)}$ is independent of the choice of the sequence $\{v_n\}$.

The next fundamental result is the parabolic version of the construction given in [33].

Theorem 1.23 *Assume that $u \in \mathcal{U}_+(Q_T)$ is a σ-moderate solution, and more precisely that there exists an increasing sequence $\{u_n\}$ of positive moderate solutions such that $u_n \uparrow u$ and $tr(u_n) = \mu_n$. Set $\mu_0 = \lim_{n\to\infty} \mu_n$ and define μ on Borel sets $E \subset \mathbb{R}^N$ by*

$$\mu(E) = \inf\left\{\mu_0(Q) : E \subset Q,\ Q\ \mathfrak{T}_q\text{-open}\right\}. \tag{1.420}$$

Then
(i) μ is the precise initial trace of u and μ is $\mathfrak{T}_q$-perfect. In particular μ is independent of the sequence $\{u_n\}$ which appears in its definition.

(ii) If A is a Borel set such that $\mu(A) < \infty$, then $\mu(A) = \mu_0(A)$.
(iii) A solution $u \in \mathcal{U}_+(Q_T)$ is σ-moderate if and only if

$$u = \sup\left\{v \in \mathcal{U}_+(Q_T),\ v \le u,\ v \text{ moderate}\right\}. \tag{1.421}$$

This statement is equivalent to

$$u = \sup\left\{u_\tau \in \mathcal{U}_+(Q_T) : \tau \in B^{-\frac{2}{q},q}(\mathbb{R}^N) \cap \mathfrak{M}_+^b(\mathbb{R}^N),\ \tau \le tr(u)\right\}. \tag{1.422}$$

(iv) If u and w are σ-moderate solutions,

$$tr(w) \le tr(u) \Longleftrightarrow w \le u. \tag{1.423}$$

Proof (i) Since the μ_n are Radon measures absolutely continuous with respect to $cap_{\frac{2}{q},q'}$, μ_0 which is the limit of the μ_n shares this property. By Proposition 1.33, μ is $\mathfrak{T}_q$-perfect.

Let $\{Q_n\}$ be the family of $\mathfrak{T}_q$-open sets of Proposition 1.34-(i). Set $Q'_n = \mathcal{R}_q(u) \setminus Q_n$. Since $\cup_n Q_n = \lim_{n\to\infty} Q_n \sim^q \mathcal{R}_q(u)$, then $Q'_n \downarrow E$ and $cap_{\frac{2}{q},q'}(E) = 0$. Consequently, for any $n \in \mathbb{N}$, we have

$$\lim_{m\to\infty} u_{\mathbf{1}_{Q'_m}\mu_n} = 0.$$

Therefore, there exists a subsequence still denoted by $\{Q'_n\}$ such that

$$\left(\int_0^T \int_{\mathbb{R}^N} u^q_{\mathbf{1}_{Q'_m}\mu_n} dxdt\right)^{\frac{1}{q}} \le \frac{1}{2^n}.$$

Since

$$\mathbf{1}_{\mathcal{R}_q(u)}\mu_n = \mathbf{1}_{Q_n}\mu_n + \mathbf{1}_{Q_n^c}\mu_n,$$

it follows that

$$\lim_{n\to\infty} |u_{\mathbf{1}_{\mathcal{R}_q(u)}\mu_n} - u_{\mathbf{1}_{Q_n}\mu_n}| = \lim_{n\to\infty} u_{\mathbf{1}_{Q_n^c}\mu_n} = 0.$$

Since we have also

$$u_n = u_{\mu_n} \le u_{\mathbf{1}_{\mathcal{R}_q(u)}\mu_n} + u_{\mathbf{1}_{\mathcal{S}_q(u)}\mu_n} \le u_{\mathbf{1}_{\mathcal{R}_q(u)}\mu_n} + [u]_{\mathcal{S}_q(u)},$$

we infer

$$0 \le u - [u]_{\mathcal{S}_q(u)} \le w := \lim_{n\to\infty} u_{\mathbf{1}_{\mathcal{R}_q(u)}\mu_n} = \lim_{n\to\infty} u_{\mathbf{1}_{D_n}\mu_n} \le u_{reg}.$$

This implies $u \ominus [u]_{\mathcal{S}_q(u)} \le u_{reg}$ and $u \le u_{reg} \oplus [u]_{\mathcal{S}_q(u)}$. For the opposite inequality, we have by Proposition 1.34-(iv)

$$[u]_{D_n} \uparrow u_{reg}.$$

By relation (1.368) in Proposition 1.29, using the fact that $\widetilde{D}_n \subset^q D_{n+1} \subset \widetilde{D}_{n+1} \subset^q \mathcal{R}_q(u)$, we have that $cap_{\frac{2}{q},q'}\left(\widetilde{D}_{n+1} \cap \mathcal{S}_q(u)\right) = 0$ and

$$[u]_{D_n} \leq [[u]_{\mathcal{S}_q(u)}]_{D_{n+1}} + [u \ominus [u]_{\mathcal{S}_q(u)}]_{D_{n+1}} = [u \ominus [u]_{\mathcal{S}_q(u)}]_{D_{n+1}} \leq u \ominus [u]_{\mathcal{S}_q(u)}.$$

Letting $n \to \infty$, we derive $u_{reg} \leq u \ominus [u]_{\mathcal{S}_q(u)}$. Therefore $\lim_{n\to\infty} u_{\mathbf{1}_{D_n}\mu_n} = u_{reg}$. Therefore the sequence $\left\{u_{\mathbf{1}_{D_n}\mu_n}\right\}$ satisfies condition (1.397) and by Proposition 1.34-(iv) and Definition 1.26 we obtain

$$\lim_{n\to\infty} \mathbf{1}_{D_n}\mu_n = \mu_{\mathcal{R}_q(u)} \quad \text{and} \quad tr_{\mathcal{R}_q(u)}(u) = \overline{\mu}_{\mathcal{R}_q(u)}. \tag{1.424}$$

Next we show that the q-singular set $\mathcal{S}_q(u)$ is singular for the sequences of measures $\{\mu_n\}$ in the sense that if $\xi \in \mathcal{S}_q(u)$, then for every $\mathfrak{T}_q$-open neighbourhood Q of ξ, $\mu_n(\widetilde{Q}) \to \infty$ when $n \to \infty$. Indeed, we can assume that Q is bounded and we consider a nonnegative function $\eta \in B^{\frac{2}{q},q'}(\mathbb{R}^N) \cap L^\infty(\mathbb{R}^N)$ with $\mathfrak{T}_q$-support included in Q. We put $h = \mathbb{H}\eta]$ and $\phi(r) = r_+^{2q'}$. Then, using Theorem 1.21, Proposition 1.19 and the computations in Proposition 1.18, we have

$$\int_0^T \int_{\mathbb{R}^N} \left(-u_n\left(\partial_t\phi(h) + \Delta\phi(h)\right) + u_n^q\phi(h)\right) dx d\tau + \int_{\mathbb{R}^N} (\phi(h)u_n)(x,T)dx$$
$$= \int_{\mathbb{R}^N} \eta^{2q'} d\mu_n,$$

and

$$\int_0^T \int_{\mathbb{R}^N} u_n^q \phi(h) dx d\tau \leq C(q)\left(\|\eta\|^{2q'}_{B^{\frac{2}{q},q'}} + \|\eta\|_{L^\infty} + \int_{\mathbb{R}^N} \eta^{2q'} d\mu_n\right).$$

We can assume that the function η has value 1 in some $\mathfrak{T}_q$-open set $D \subset Q$ and $0 \leq \eta \leq 1$ (see Lemma 1.28). If we let $n \to \infty$, we obtain from the above relations

$$\lim_{n\to\infty} \int_0^T \int_{\mathbb{R}^N} u_n^q \left(\mathbb{H}[\mathbf{1}_D]\right)^{2q'} dx d\tau$$
$$\leq C(q)\left(\|\eta\|^{2q'}_{B^{\frac{2}{q},q'}} + \|\eta\|_{L^\infty} + \lim_{n\to\infty} \int_{\mathbb{R}^N} \eta^{2q'} d\mu_n\right).$$

Then the assertion follows from Proposition 1.20.

In conclusion, we have proved that if $\xi \in \mathcal{S}_q(u)$ and Q is a $\mathfrak{T}_q$-open neighbourhood of ξ, then $\mu_0(\widetilde{Q}) = \infty$. By the outer regularity of μ with respect to the $\mathfrak{T}_q$-topology, it means that $\mu(\xi) = \infty$. Combined with (1.424) this implies that μ is the precise trace of u.

(ii) If $\mu(A) < \infty$, then A is contained in a $\mathfrak{T}_q$-open set D such that $\mu_0(D) < \infty$. By Proposition 1.33 we have that $\mu(A) = \mu_0(A)$.

(iii) Let $u \in \mathcal{U}_+(Q_T)$ be σ-moderate and denote

$$u^* := \sup\{v : v \text{ moderate } v \leq u\}. \tag{1.425}$$

By expression (1.425) $u^* \leq u$. Since u is σ-moderate there exists an increasing sequence $\{u_n\}$ of moderate solutions which converges to u. For any n we have proved in the beginning of the section on moderate solutions that given u_n there exists an increasing sequence $\{u_{n,m}\}_{m\in\mathbb{N}} = \{u_{\mu_{n,m}}\}_{m\in\mathbb{N}}$ of elements of $\mathcal{U}_+(Q_T)$ where $\mu_{n,m} \in B^{-\frac{2}{q},q}(\mathbb{R}^N) \cap \mathfrak{M}_+^b(\mathbb{R}^N)$ converging to u_n. Therefore

$$u_n \leq \sup\left\{u_\tau : \tau \in B^{-\frac{2}{q},q}(\mathbb{R}^N) \cap \mathfrak{M}_+^b(\mathbb{R}^N),\ \tau \leq tr(u)\right\} = u^{**}.$$

By letting $n \to \infty$ we infer $u \leq u^{**}$.

However, if u is σ-moderate, $\tau \in B^{-\frac{2}{q},q}(\mathbb{R}^N) \cap \mathfrak{M}_+^b(\mathbb{R}^N)$ and $\tau \leq tr(u)$, then we have that $tr(u_\tau \ominus u_n) = (\tau - \mu_n)_+$ and the corresponding sequence decreases to 0 when $n \to \infty$. Therefore $u_\tau \ominus u_n \downarrow 0$ which implies $u_\tau \leq u$ and thus $u^{**} \leq u$. Consequently, (1.421) implies (1.422). This shows that the two identities which define σ-moderate solutions are equivalent.

(iv) The implication $\Longrightarrow$ follows from (1.422). For proving the opposite implication, it is sufficient to show that if u is σ-moderate, w is moderate and $w \leq u$, then $tr(w) \leq tr(u)$. For this task, we consider an increasing sequence of moderate solutions $\{u_n\}$ which converges to u. Then $u_n \wedge w \leq u$ and consequently $u_n \leq u_n \wedge w \uparrow u$. This implies $tr(u_n \wedge w) \uparrow \mu' \leq tr(u)$. Hence $tr(w) \leq tr(u)$. □

This results extend Proposition 1.34 which deals with the regular initial trace.

Theorem 1.24 *Let $u \in \mathcal{U}_+(Q_T)$ and $\nu = tr(u)$.*
(i) u_{reg} is σ-moderate and $tr(u_{reg}) = tr_{\mathcal{R}_q(u)}(u)$.
(ii) If $v \in \mathcal{U}_+(Q_T)$, then

$$v \leq u \Longrightarrow tr(v) \leq tr(u), \tag{1.426}$$

and if F is a $\mathfrak{T}_q$-closed set, then

$$tr([u]_F) \leq \mathbf{1}_F \nu. \tag{1.427}$$

(iii) A singular point of the trace can be characterised in terms of the measure ν as follows:

$$\xi \in \mathcal{S}_q(u) \Longrightarrow \nu(Q) = \infty \quad \text{for all } \mathfrak{T}_q\text{-open set containing } \xi. \tag{1.428}$$

(iv) If Q is a $\mathfrak{T}_q$-open set then

$$[u]_Q \textit{ moderate } \iff \exists \textit{ Borel set } A \textit{ s.t. } cap_{\frac{2}{q},q'}(A) = 0 \textit{ and } \nu(A \cap \widetilde{Q} \setminus K) < \infty, \tag{1.429}$$

for any compact set $K \subset \mathbb{R}^N$.
(v) The singular set of u_{reg} *may not be empty. Actually*

$$\mathcal{S}_q(u) \setminus b_q(\mathcal{S}_q(u)) \subset \mathcal{S}_q(u_reg) \subset \mathcal{S}_q(u) \cap \widetilde{\mathcal{R}_q(u)}, \tag{1.430}$$

where $b_q(\mathcal{S}_q(u))$ *is the set of thick points of* $\mathcal{S}_q(u)$ *for the* $\mathfrak{T}_q$*-topology.*
(vi) Put

$$\mathcal{S}_{q,0}(u) := \left\{\xi \in \mathbb{R}^N : \nu(Q \setminus \mathcal{S}_q(u)) = \infty \textit{ for all } \mathfrak{T}_q\textit{-neighbourhood of } \xi\right\}. \tag{1.431}$$

Then

$$\mathcal{S}_q(u_{reg}) \setminus b_q(\mathcal{S}_q(u)) \subset \mathcal{S}_{q,0}(u) \subset \mathcal{S}_q(u_{reg}) \bigcup b_q(\mathcal{S}_q(u)). \tag{1.432}$$

Remark We will prove later on that any element of $\mathcal{U}_+(Q_T)$ is σ-moderate. Hence implication (1.426) is actually an equivalence.

Proof The first part of assertion (i) is proved in Proposition 1.34-(i) and the fact that $u_{reg} = v_{\mathcal{R}_q(u)}$. The second part follows from Definition 1.26 and Theorem 1.23-(i).
(ii) If $v \leq u$, then $\mathcal{R}_q(u) \subset \mathcal{R}_q(v)$ and by definition $v_{reg} \leq u_{reg}$. By Theorem 1.23-(iv) we have $tr(v_{reg}) \leq tr(u_{reg})$. This implies $tr(v) \leq tr(u)$. Inequality (1.427) is a consequence of (1.426).
(iii) If ξ is a regular point, there exists a $\mathfrak{T}_q$-open neighbourhood of ξ, say Q, such that $[u]_Q$. Therefore $\nu(Q) = tr_{\mathcal{R}_q(u)}(u)(Q) < \infty$. Conversely, if $\xi \in \mathcal{S}_q(u)$, it follows from the definition of the precise trace that $\nu(Q) = \infty$ for all $\mathfrak{T}_q$-open neighbourhood Q of ξ.
(iv) If Q is $\mathfrak{T}_q$-open and $[u]_Q$ is moderate, then $Q \subset \mathcal{R}_q(u)$. By Proposition 1.34-(ix) we obtain the implication $\Longrightarrow$ in (1.429). Conversely,

$$\nu(\widetilde{Q} \cap K \setminus A) < \infty, \text{ for all compact set } K \subset \mathbb{R}^N \Longrightarrow \widetilde{Q} \subset^q \mathcal{R}_q(u),$$

and $\mu_{\mathcal{R}_q(u)}(\widetilde{Q} \cap K) = \mu_{\mathcal{R}_q(u)}(\widetilde{Q} \cap K \setminus A) < \infty$. It follows by Proposition 1.34-(ix) that $[u]_Q$ is moderate.
(v) Because $\mathfrak{T}_q$-supp $(u_{reg}) \subset \widetilde{\mathcal{R}_q(u)}$ and $\mathcal{R}_q(u) \subset \mathcal{R}_q(u_{reg})$, we have

$$\mathcal{S}_q(u_{reg}) \subset \mathcal{S}_q(u) \cap \widetilde{\mathcal{R}_q(u)}.$$

Next we prove that $\mathcal{S}_q(u) \setminus b_q(\mathcal{S}_q(u)) \subset \mathcal{S}_q(u_{reg})$.
If $\xi \in \mathcal{S}_q(u) \setminus b_q(\mathcal{S}_q(u))$, then $\mathcal{R}_q(u) \cup \{\xi\}$ is a $\mathfrak{T}_q$-open neighbourhood of ξ. By (i) u_{reg} is σ-moderate and thus its trace is $\mathfrak{T}_q$-perfect (see Theorem 1.23)-(i)). Therefore if Q_0 is a $\mathfrak{T}_q$-open neighbourhood of ξ and $Q = Q_0 \cap \left(\{\xi\} \cup \mathcal{R}_q(u)\right)$, then

$$tr(u_{reg})(Q) = tr(u_{reg})(Q \setminus \{\xi\}) = tr(u)(Q \setminus \{\xi\}),$$

where, it the last inequality, we have used the fact that $Q \setminus \{\xi\} \subset \mathcal{R}_q(u)$. Now, let D be a $\mathfrak{T}_q$-open set such that $\xi \in D \subset \widetilde{D} \subset Q$. If $tr(u)(\widetilde{Q} \setminus \{\xi\}) < \infty$, then, by (iv), $[u]_D$ is moderate and $\xi \in \mathcal{R}_q(u)$, contrary to our assumption. Therefore $tr(u)(\widetilde{Q} \setminus \{\xi\}) = \infty$ which implies $tr(u_{reg})(Q_0 \setminus \{\xi\}) = \infty$ for every bounded $\mathfrak{T}_q$-open neighbourhood Q_0 of ξ, and consequently $\xi \in \mathcal{S}_q(u_{reg})$, which ends the proof of (v).

(vi) If $\xi \notin b_q(\mathcal{S}_q(u))$ there exists a $\mathfrak{T}_q$-open neighbourhood D of ξ such that $(D \setminus \{\xi\}) \cap \mathcal{S}_q(u) = \emptyset$, and thus

$$tr(u_{reg})(D \setminus \{\xi\} = tr(u_{reg})(D \setminus \mathcal{S}_q(u)) = tr(u)(D \setminus \mathcal{S}_q(u)). \tag{1.433}$$

Furthermore, if we assume that $\xi \in \mathcal{S}_{q,0}(u)$, then

$$tr(u)(D \setminus \mathcal{S}_q(u)) = tr(u_{reg})(D \setminus \mathcal{S}_q(u)) = \infty.$$

If Q is an arbitrary $\mathfrak{T}_q$-open neighbourhood of ξ, then the same relation holds if D is replaced by $D \cap Q$. Therefore $tr(u_{reg})(Q \setminus \{\xi\} = \infty$ for any such Q. This implies that $\xi \in \mathcal{S}_q(u_{reg})$ and $\mathcal{S}_{q,0}(u) \setminus b_q(\mathcal{S}_q(u)) \subset \mathcal{S}_q(u_{reg})$.

On the other hand, if $\xi \in \mathcal{S}_q(u_{reg}) \setminus b_q(\mathcal{S}_q(u))$, there exists a $\mathfrak{T}_q$-open neighbourhood D of ξ such that (1.433) holds and $tr(u_{reg})(D) = \infty$. Since u_{reg} is σ-moderate $tr(u_{reg})$ is $\mathfrak{T}_q$-perfect, which infers $tr(u_{reg})(D) = tr(u_{reg})(D \setminus \{\xi\}) = \infty$. Using (1.433) we obtain that $tr(u)(D \setminus \mathcal{S}_q(u)) = \infty$. At the end, if Q is $\mathfrak{T}_q$-open neighbourhood of ξ, then D can be replaced by $D \cap Q$, which yields $tr(u)(Q \setminus \mathcal{S}_q(u)) = \infty$. This proves that $\xi \in \mathcal{S}_{q,0}(u)$ and ends the proof of (1.433). □

Proposition 1.35 *Let F be a $\mathfrak{T}_q$-closed set. Then $\mathcal{S}_q(U_F) = b_q(F)$.*

Proof Let $\xi \in \mathbb{R}^N$ such that F is $\mathfrak{T}_q$-thin at ξ. Let Q be a $\mathfrak{T}_q$-open neighbourhood of ξ such that $\widetilde{Q} \subset^q F^c$. Then $[U_F]_Q = U_{F\cap\widetilde{Q}} = 0$. Then $\xi \in \mathcal{R}_q(u)$.

Conversely, if $\xi \in F \cap \mathcal{R}_q(U_F)$ there exists a $\mathfrak{T}_q$-open neighbourhood Q of ξ such that $[U_F]_Q$ is moderate. But the relation $[U_F]_Q = U_{F\cap\widetilde{Q}}$, combined with the previous assertion, implies that $cap_{\frac{2}{q},q'}(F \cap \widetilde{Q}) = 0$ and therefore $Q \subset \mathcal{R}_q(u)$. Since

$$cap_{\frac{2}{q},q'}(F) \leq cap_{\frac{2}{q},q'}(F \cap \widetilde{Q}) + cap_{\frac{2}{q},q'}(Q^c),$$

we conclude that F is $\mathfrak{T}_q$-thin at ξ. □

1.5.6.2 The Initial Value Problem

We introduce below some definitions and notations which will be useful in the sequel.

Definition 1.27 I- $\mathfrak{M}_+(\mathbb{R}^N)$ is the space of positive outer regular Borel measure in $\mathbb{R}^N$.

II- $C_q(\mathbb{R}^N)$ is the space of couples (τ, F) such that F is a $\mathfrak{T}_q$-closed subset of $\mathbb{R}^N$, $\tau \in \mathfrak{M}_+(\mathbb{R}^N)$ with $\mathfrak{T}_q$-supp $(\tau) \subset \widetilde{F^c}$ and $\mathbf{1}_{F^c}\tau$ is $\mathfrak{T}_q$-locally finite.

III- $\mathbb{T}$ denotes the mapping from $C_q(\mathbb{R}^N)$ into $\mathfrak{M}_+(\mathbb{R}^N)$ defined by $\nu = \mathbb{T}(\tau, F)$ where ν is defined as in (1.419) with $\mathcal{R}_q(u)$ and $\mathcal{S}_q(u)$ replaced respectively by F^c and F. In this setting ν is the measure representation of the couple (τ, F).

IV- If $(\tau, F) \in C_q(\mathbb{R}^N)$, the set

$$F_\tau = \{\xi \in \mathbb{R}^N : \tau(Q \setminus F) = \infty \text{ for all } \mathfrak{T}_q\text{-open neighbourhood of } \xi\} \quad (1.434)$$

is called the set of explosion points of τ.

Remark Since $\mathbf{1}_{F^c}\tau$ is locally finite, $F_\tau \subset F$. If F_τ is not included in $\widetilde{F^c}$, there would exist a $\mathfrak{T}_q$-open neighbourhood Q of ξ with an empty intersection with F^c, hence included in F, thus $Q \setminus F = \emptyset$ and $\tau(Q \setminus F) = 0$, a contradiction. Therefore $F_\tau \subset \widetilde{F^c}$ and consequently

$$F_\tau \subset \widetilde{F^c} \cap F = \left(F^c \bigcup b_q(F^c)\right) \cap F = b_q(F^c) \cap F. \quad (1.435)$$

This result has to be compared with Theorem 1.12 which deals with a necessary and sufficient conditions for the existence of a maximal solution u of (1.30) with a rough initial trace (S, μ).

The next result points out the crucial role of the set $\mathfrak{M}_q(\mathbb{R}^N)$ defined in Definition 1.25 for describing the link between $\mathcal{U}_+(Q_T)$ and $C_q(\mathbb{R}^N)$.

Proposition 1.36 *Let ν be a positive Borel measure in $\mathbb{R}^N$.*
(i) The initial value problem

$$\begin{aligned} \partial_t u - \Delta u + |u|^{q-1}u &= 0 \;\; in \;\; Q_\infty \\ u &\geq 0 \;\; in \;\; Q_\infty \\ tr(u) &= \nu \end{aligned} \quad (1.436)$$

possesses a solution if and only if $\nu \in \mathfrak{M}_q(\mathbb{R}^N)$.
(ii) Let $(\tau, F) \in C_q(\mathbb{R}^N)$ and set $\nu := \mathbb{T}(\tau, F)$. Then $\nu \in \mathfrak{M}_q(\mathbb{R}^N)$ if and only if there holds

$$\tau \in \mathfrak{M}_q(\mathbb{R}^N) \;\; and \;\; F = b_q(F) \bigcup F_\tau. \quad (1.437)$$

(iii) Let $\nu \in \mathfrak{M}_q(\mathbb{R}^N)$ and set

$$\begin{aligned} &\mathcal{E}_\nu := \{E : E \;\mathfrak{T}_q\text{-closed sets s.t. } \nu(E \cap K) < \infty \text{ for all compact } K \subset \mathbb{R}^N\} \\ &\mathcal{D}_\nu := \{D : D \;\mathfrak{T}_q\text{-open sets s.t. } \widetilde{D} \sim^q E \text{ for some } E \in \mathcal{E}_\nu\}. \end{aligned} \quad (1.438)$$

Then a solution of (1.436) is given by $u = v \oplus U_F$ *where*

$$G := \bigcup_{D \in \mathcal{D}_\nu} D, \quad F := G^c, \quad v := \sup\{u_{\mathbf{1}_E \nu} : E \in \mathcal{E}_\nu\}. \tag{1.439}$$

(iv) The solution $u := v \oplus U_F$ *is* σ*-moderate and it is the unique solution of problem (1.436) in the class of* σ*-moderate solutions. Furthermore* u *is the largest solution of this problem.*

Remark 1- We recall that if $E \in E_\nu$ then $\mathbf{1}_E \nu$ is a locally bounded Borel measure which does not charge sets of $cap_{\frac{2}{q},q'}$-capacity zero. Recall also that if μ is a positive measure possessing these properties, then u_μ denotes the moderate solution with initial trace μ.
2- We will see later on that $u := v \oplus U_F$ is the only solution to problem (1.436) since every solution happens to be σ-moderate.

Proof (I) If $u \in \mathcal{U}_+(Q_T)$,

$$tr(u) = \nu \Longrightarrow \nu \in \mathfrak{M}_q(\mathbb{R}^N). \tag{1.440}$$

By Proposition 1.34, u_{reg} is σ-moderate and $u \approx_{\mathcal{R}_q(u)} u_{reg}$. Therefore

$$\mathbf{1}_{\mathcal{R}_q(u)} tr(u) = \mathbf{1}_{\mathcal{R}_q(u)} tr(u_{reg}).$$

By Theorem 1.23 $\widetilde{\mu}_{\mathcal{R}_q} = tr(u_{reg}) \in \mathfrak{M}_q(\mathbb{R}^N)$. If v is defined by (1.438), then

$$v = \sup\left\{[u]_F : F\ \mathfrak{T}_q\text{-closed},\ F \subset^q \mathcal{R}_q(u)\right\} = u_{reg}, \tag{1.441}$$

where the second equality holds by definition. Actually, by Theorem 1.24, for every $\mathfrak{T}_q$-open set Q $[u]_Q$ is moderate if and only if $\nu(K \cap \widetilde{Q} \setminus A) < \infty$ for some set A with $cap_{\frac{2}{q},q'}(A) = 0$ and for every compact set $K \subset \mathbb{R}^N$. Hence, by Proposition 1.34-(ix), $\widetilde{E}$ is regular, in the sense that there exists a $\mathfrak{T}_q$-open regular set such that $E \subset^q Q$. Hence $u_{\mathbf{1}_E \nu} \leq [u]_Q$. This implies that $v \leq u_{reg}$, which proves (1.441). Furthermore, if $E \cap \mathcal{S}_q(u) \neq \emptyset$, then $\nu(E) = \infty$ by Definition 1.26. Therefore ν is outer regular with respect to the $\mathfrak{T}_q$-topology.

Next we prove that ν is essentially absolutely continuous (cf. Definition 1.25-(iii)). Let Q be a $\mathfrak{T}_q$-open set and A a non-empty $\mathfrak{T}_q$-closed subset of Q such that $cap_{\frac{2}{q},q'}(A) = 0$. Either $\nu(Q \setminus A) = \infty$ in which case $\nu(Q \setminus A) = \nu(Q) = \infty$, or $\nu(Q \setminus A) < \infty$. In that case $Q \setminus A \subset \mathcal{R}_q(u)$ and

$$\nu(Q \setminus A) = \widetilde{\mu}(Q \setminus A) = \widetilde{\mu}(Q)$$

since $cap_{\frac{2}{q},q'}(A) = 0 \Longrightarrow \widetilde{\mu}(A) = 0$.
Let $\xi \in A$ and D be a $\mathfrak{T}_q$-open subset of Q such that $\xi \in D \subset \widetilde{D} \subset^q Q$. Consider

now a sequence $\{B_n\}$ of $\mathfrak{T}_q$-open neighbourhoods of $A \cap \widetilde{D}$ such that $B_n \subset^q D$ and $cap_{\frac{2}{q},q'}(B_n) < 2^{-n}$. If we set $E_n = \widetilde{D} \setminus B_n$, we have

$$[u]_D \leq [u]_{E_n} + [u]_{B_n}.$$

Since $\lim_{n\to\infty}[u]_{B_n} = 0$, it follows that $[u]_D \leq \liminf_{n\to\infty}[u]_{E_n}$. Because $E_n \subset \mathcal{R}_q(u)$, $\nu(E_n) \leq \nu(Q \setminus A) < \infty$, we have by the definition of ν and Proposition 1.34-(ix) that $[u]_{E_n}$ is moderate. Using Lemmas 1.26, 1.25-(ii) and [30] there holds

$$\int_0^T \int_K [u]_{E_n}^q dxdt \leq C\nu(E_n) \leq C\nu(Q \setminus A) < \infty,$$

for any compact set $K \subset \mathbb{R}^N$. Therefore

$$\int_0^T \int_K [u]_D^q dxdt \leq C\nu(E_n) \leq C\nu(Q \setminus A) < \infty \quad \text{for all compact set } K \subset \mathbb{R}^N.$$

This implies that $[u]_D$ is moderate and thus $D \subset \mathcal{R}_q(u)$. Therefore, since every point A has a neighbourhood D as above, we conclude that $A \subset \mathcal{R}_q(u)$ and hence $\nu(A) = tr_{\mathcal{R}_q(u)}(u)(A) = 0$. If A is any a non-empty Borel subset of Q such that $cap_{\frac{2}{q},q'}(A) = 0$, we use the inequality $cap_{\frac{2}{q},q'}(\widetilde{A}) \leq \widetilde{C}cap_{\frac{2}{q},q'}(A)$ to conclude that ν is absolutely continuous and hence $\nu \in \mathfrak{M}_q(\mathbb{R}^N)$.

Next we prove the following:
(II) *Suppose that* $(\tau, F) \in C_q(\mathbb{R}^N)$ *satisfies* (1.437) *and put* $\nu = \mathbb{T}(\tau, F)$. *Then the solution* $u := v \oplus U_F$ *with* v *as in* (1.439) *satisfies* $tr(u) = \nu$. *Notice that this implies that* $\nu \in \mathfrak{M}_q(\mathbb{R}^N)$ *by* (1.439).

The solution v is σ-moderate by construction. Since τ is locally $\mathfrak{T}_q$-finite in F^c and essentially absolutely continuous with respect to $cap_{\frac{2}{q},q'}$ we have that

$$G := F^c \subset \mathcal{R}_q(u) \quad \text{and} \quad \mathbf{1}_G tr(v) = \tau_G. \tag{1.442}$$

Therefore, it follows from the definition of v that $F_\tau \subset S_q(v)$. By Proposition 1.35 and Theorem 1.24-(iv) we have that

$$F = b_q(F) \bigcup F_\tau \subset S_q(v) \bigcup S_q(U_F) \subset S_q(u) \subset F. \tag{1.443}$$

Hence $F = S_q(u)$, $v = u_{reg}$ and $\tau = tr(u_{reg})$. In turn, this implies $tr(u) = (\tau, F)$, which is equivalent to $\tau = tr(u)$.

Then we prove the following:
(III) *Suppose that* $(\tau, F) \in C_q(\mathbb{R}^N)$ *and that there exists a solution* u *such that* $tr^c(u) = (\tau, F)$ *(see Definition* 1.26 *for the definition of* tr^c*). Then*

$$\tau = tr_{\mathcal{R}_q(u)}(u) = tr(u_{reg}) \quad \text{and} \quad F = S_q(u). \tag{1.444}$$

If $U := u_{reg} \oplus U_F$, then $tr(U) = tr(u)$ and $u \le U$. U is the only σ-moderate solution of (1.436) and (τ, F) satisfies (1.437). Assertion (1.444) follows by Proposition 1.34-(i) and Definition 1.26. Since u_{reg} is σ-moderate, we have that $\tau \in \mathfrak{M}_q(\mathbb{R}^N)$ by Theorem 1.23.

By Proposition 1.34-(vi) there holds $u \approx_{\mathcal{R}_q(u)} u_{reg}$. Therefore the function $w := u \ominus u_{reg}$ which vanishes on $\mathcal{R}_q(u)$ is dominated by U. Note that $u - u_{reg} \le w$ and therefore

$$u \le u_{reg} \oplus w \le U. \tag{1.445}$$

By definition, $\mathcal{S}_{q,0}(u) = F_\tau$ and by Theorem 1.24-(vi) and Proposition 1.35 we have

$$\begin{aligned} \mathcal{S}_q(u) &= \mathcal{S}_q(u_{reg}) \bigcup \mathcal{S}_q(U_F) = \mathcal{S}_q(u_{reg}) \bigcup b_q(U_F) \\ &= \mathcal{S}_{q,0}(u) \bigcup b_q(U_F) = F_\tau \bigcup b_q(U_F). \end{aligned} \tag{1.446}$$

On the other hand $\mathcal{R}_q(U) \supset \mathcal{R}_q(u_{\mathcal{R}_q(u)}) = \mathcal{R}_q(u)$. As $u \le U$ we have $\mathcal{R}_q(U) \subset \mathcal{R}_q(u)$. Hence $\mathcal{R}_q(U) = \mathcal{R}_q(u)$ and $\mathcal{S}_q(U) = \mathcal{S}_q(u)$. Therefore, by (1.443), (1.445),

$$F = \mathcal{S}_q(U) = F_\tau \cup b_q(U_F).$$

This implies that (τ, F) satisfies (1.437) and $tr^c(u) = (\tau, F)$. That U is the maximal solution with this trace follows from (1.445).

The solution U is σ-moderate because u_{reg} and U_F are σ-moderate (see Theorem 1.22).

Finally we prove

(IV) *If $\nu \in \mathfrak{M}_q(\mathbb{R}^N)$ then the couple (τ, F) defined by*

$$v := \sup \left\{ u_{1_E \nu} : E \in \mathcal{E}_\nu \right\}, \; \tau = tr(v), \; F = \mathcal{R}_q^c(v), \tag{1.447}$$

(see (1.438) for the definition of $\mathcal{E}_\nu$) satisfies (1.437). This is the only couple belonging to $C_q(\mathbb{R}^N)$ satisfying $\nu = \mathbb{T}(\tau, F)$. The solution v is σ-moderate so that $\tau \in \mathfrak{M}_q(\mathbb{R}^N)$.

We first prove that $u := v \oplus U_F$ is a solution with initial trace $tr(u) = (\tau, F)$. Actually $u \ge v$, so that $\mathcal{R}_q(u) \subset \mathcal{R}_q(v)$. On the other hand, since τ is $\mathfrak{T}_q$-locally finite in $\mathcal{R}_q(v) = F^c$, it follows that $\mathcal{S}_q(u) \subset F$. Therefore $\mathcal{R}_q(v) \subset \mathcal{R}_q(u)$, and finally $\mathcal{R}_q(u) = \mathcal{R}_q(v)$ and $F = \mathcal{S}_q(u)$. This also implies $v = u_{reg}$.

At the end

$$\mathcal{S}_q(u) = \mathcal{S}_q(v) \bigcup b_q(\mathcal{S}_q(U_F)) = b_q(F) \bigcup F,$$

which means that (1.437) holds.

That for $\nu \in \mathfrak{M}_q(\mathbb{R}^N)$ the couple (τ, F) defined by (1.437) is the only one couple belonging to $C_q(\mathbb{R}^N)$ satisfying $\nu = \mathbb{T}(\tau, F)$ is a mere consequence of their expression in Definition 1.27.

Finally, statements (i)–(iv) follow from (I)–(IV). □

Remark If $\nu \in \mathfrak{M}_q(\mathbb{R}^N)$ then G and υ as defined by (1.439) have the following alternative representation:

$$G := \bigcup_{\mathcal{E}_\nu} E = \bigcup_{Q \in \mathcal{F}_\nu} Q, \qquad \upsilon := \sup\left\{u_{1_Q \nu} : Q \in \mathcal{F}_\nu\right\}), \tag{1.448}$$

where

$$\mathcal{F}_\nu = \left\{Q : Q \ \mathfrak{T}_q\text{-open}, \ \nu(Q \cap K) < \infty \text{ for all compact set } K \subset \mathbb{R}^N\right\} r. \tag{1.449}$$

In order to prove (1.448) we first observe that if A is a $\mathfrak{T}_q$-open set, then there exists an increasing sequence of $\mathfrak{T}_q$-quasi-closed sets $\{E_n\}$ such that $A = \cup_1^\infty E_n$. This follows from Lemma 1.26-(II-i-ii) with $E_n = F_n \setminus L$ where $L = A' \setminus A$ and $cap_{\frac{2}{q},q'}(L) = 0$. Thus

$$\bigcup_{\mathcal{D}_\nu} D \subset \bigcup_{\mathcal{F}_\nu} Q \subset \bigcup_{\mathcal{E}_\nu} E = H.$$

On the other hand, if $E \in \mathcal{E}_\nu$, then $\mu_{\mathcal{R}_q(u)}(K \cap \widetilde{E}) = \mu_{\mathcal{R}_q(u)}(K \cap E) = \nu(E \cap K) < \infty$ for any compact set $K \subset \mathbb{R}^N$. By Proposition 1.34 (ix), $\widetilde{E}$ is regular in the sense that there exists a $\mathfrak{T}_q$-open regular set Q such that $E \subset^q Q$, therefore $H = \bigcup_{\mathcal{D}_\nu} D$.

If D is a $\mathfrak{T}_q$-open regular set, then $D = \cup_{n=1}^\infty E_n$ where $\{E_n\}$ is an increasing sequence of $\mathfrak{T}_q$-quasi-closed sets. This implies

$$u_{1_D \nu} = \lim_{n \to \infty} u_{1_{E_n} \nu}.$$

Hence

$$\sup\left\{u_{1_Q \nu} : Q \in \mathcal{D}_\nu\right\} \leq \sup\left\{u_{1_Q \nu} : Q \in \mathcal{F}_\nu\right\} \leq \sup\left\{u_{1_Q \nu} : Q \in \mathcal{E}_\nu\right\}.$$

However, if $E \in \mathcal{E}_\nu$ there exists a $\mathfrak{T}_q$-open regular set Q such that $E \subset^q Q$. This implies the inequality in (1.448).

1.5.7 Representation of Positive Solutions of $\partial_t u - \Delta u + V u = 0$

In this section we prove a general representation theorem for positive solutions of

$$\partial_t u - \Delta u + V u = 0 \quad \text{in } Q_T := \mathbb{R}^N \times (0, T) \tag{1.450}$$

where $V : Q_T \to \mathbb{R}$ is a Borel function satisfying

$$0 \leq V(x,t) \leq \frac{C}{t} \qquad \text{for all } (x,t) \in Q_T, \tag{1.451}$$

for some positive constant. Our results are the parabolic counterpart of Ancona's results [3] concerning representation of positive solutions of

$$-\Delta u + V(x)u = 0 \qquad \text{in } \Omega, \tag{1.452}$$

by means of a Martin operator when $\Omega \subset \mathbb{R}^N$ is a bounded Lipschitz domain and V a Borel function defined in Ω satisfying

$$0 \leq V(x) \leq \frac{C}{(\rho(x))^2} \qquad \text{for almost all } x \in \Omega, \tag{1.453}$$

with $\rho(x) = \operatorname{dist}(x, \partial\Omega)$ and $C > 0$.

We recall first some well-known facts concerning weak solutions of (1.450).

Definition 1.28 Let $\mu \in \mathfrak{M}(\mathbb{R}^N)$. We say that u is a weak solution of

$$\begin{aligned} \partial_t u - \Delta u + Vu &= 0 \text{ in } Q_T \\ u(.,0) &= \mu \text{ in } \mathbb{R}^N, \end{aligned} \tag{1.454}$$

if $u, Vu \in L^1_{loc}(\overline{Q}_T)$ and there holds

$$\iint_{Q_T} u\,(-\partial_t\zeta - \Delta\zeta + V\zeta)\,dxdt = \int_{\mathbb{R}^N} \zeta(x,0)d\mu(x) \quad \text{for all } \zeta \in \mathbb{X}(Q_T), \tag{1.455}$$

where

$$\mathbb{X}(Q_T) = \left\{\zeta \in C_c(\mathbb{R}^N \times [0,T)) : \zeta + \Delta\zeta \in L^\infty_{loc}(\overline{Q}_T)\right\}.$$

Note that this definition implies that the function u admits the measure μ as an initial trace as described in Sect. 1.2. The next result is an easy adaptation of the techniques developed in Sect. 1.2.

Lemma 1.37 *Let $\mu \in \mathfrak{M}_+(\mathbb{R}^N)$ and assume that there exists a positive weak solution u of problem (1.454) where V satisfies (1.453). Then for any smooth bounded domain Ω there exists a unique positive weak solution $v = v_\Omega$ of problem*

$$\begin{aligned} \partial_t v - \Delta v + Vv &= 0 && \text{in } Q_T^\Omega := \Omega \times (0,T) \\ v &= 0 && \text{in } \partial_\ell Q_T^\Omega := \partial\Omega \times (0,T) \\ v(.,0) &= \mathbf{1}_\Omega \mu && \text{in } \Omega. \end{aligned} \tag{1.456}$$

Furthermore $0 \leq v_\Omega \leq u$ and the mapping $\Omega \mapsto v_\Omega$ is nondecreasing.

Proof Let ϵ_n be a sequence converging to 0 and v_n the solution of

$$\begin{array}{rll} \partial_t v - \Delta v + Vv = 0 & \text{in } \Omega \times (\epsilon_n, T) \\ v = 0 & \text{in } \partial\Omega \times (\epsilon_n, T) \\ v(., \epsilon_n) = \mathbf{1}_\Omega u(., \epsilon_n) & \text{in } \Omega. \end{array} \tag{1.457}$$

Such a solution exists since $u(., \epsilon_n) \in L^1(\Omega)$ and it satisfies $0 \leq v_n \leq u$ in $\Omega \times (\epsilon_n, T)$. By classical parabolic regularity estimates we may assume that the sequence $\{v_n\}$ converges locally uniformly in Q_T^Ω to a nonnegative function v dominated by u. Let $\zeta \in C^{1,1;1}(\overline{Q}_\Omega^T)$ vanish on $\partial_\ell Q_T^\Omega$ and for $t \geq T - \delta$ for some $\delta > 0$. Set $\zeta_n(x, t) = \zeta(x, t - \epsilon_n)$, then from (1.456), and assuming that $\epsilon_n \leq \delta$ we have

$$\int_{\epsilon_n}^T \int_\Omega v_n \left(-\partial_t \zeta_n - \Delta \zeta_n + V\zeta_n\right) dxdt = \int_\Omega \zeta(., 0) u(., \epsilon_n) dx. \tag{1.458}$$

Because $v_n V \leq uV \in L^1_{loc}(\overline{Q}_T)$ we deduce from the dominated convergence theorem that the left-hand side of (1.458) converges to $\int_0^T \int_\Omega v \left(-\partial_t \zeta - \Delta \zeta + V\zeta_n\right)$ $dxdt$ while the right-hand side converges to $\int_\Omega \zeta(., 0) d\mu(x)$. The final assertion on the monotonicity of $\Omega \mapsto v_\Omega$ is a consequence of the maximum principle. This ends the proof. □

Lemma 1.38 *Let the assumptions on μ, V and u of Lemma 1.37 be satisfied and denote by $v_R := v_{B_R}$ the solution of (1.455) with $\Omega = B_R$. Then*

$$v_R \uparrow u \qquad \textit{as } R \to \infty.$$

Furthermore this convergence is uniform on compact subsets of Q_T.

Proof Since the mapping $R \mapsto v_R$ is increasing and v_R is dominated by u, there exists a function w such that

$$v_R \uparrow w \leq u \qquad \text{as } R \to \infty,$$

and this convergence is locally uniformly in Q_T^Ω. Because for any $\zeta \in C_c(\overline{Q}_T)$,

$$\lim_{R \to \infty} \int_{B_R} \zeta(., 0) d\mu(x) = \int_{\mathbb{R}^N} \zeta(., 0) d\mu(x),$$

we infer that w is a weak solution of problem (1.454). Therefore the function $\widetilde{w} = u - w$ is nonnegative and satisfies

$$\begin{array}{rll} \partial_t \widetilde{w} - \Delta \widetilde{w} \leq 0 & \text{in } Q_T \\ \widetilde{w} \geq 0 & \text{in } Q_T \\ \widetilde{w}(., 0) = 0 & \text{in } \mathbb{R}^N. \end{array} \tag{1.459}$$

Moreover $\widetilde{w}$ belongs to $L^1_{loc}(\overline{Q}_T)$. We extend it by 0 in $\mathbb{R}^N \times (-T, 0)$ and the resulting function $\widetilde{w}^*$ is a nonnegative sub-caloric function in $\mathbb{R}^N \times (-T, T)$ that we can suppose to be C^∞ by replacing it by $J_\epsilon * \widetilde{w}^*$ where J_ϵ is a sequence of mollifiers in $\mathbb{R}^{N+1}$. By the maximum principle $J_\epsilon * \widetilde{w}^* = 0$. Hence $\widetilde{w}^* = 0$ which yields $u = w$. □

The next result is the extension of the initial trace theorem for nonnegative caloric functions to nonnegative solutions of (1.450).

Lemma 1.39 *Let V satisfy (1.451) and $u \in C^{2;1}(Q_T)$ be a nonnegative function satisfying (1.450). Assume that for any $x \in \mathbb{R}^N$ there exists a bounded open set $U \subset \mathbb{R}^N$ such that*

$$\int_0^T \int_U u(y,t)V(y,t)dydt < \infty.$$

Then $u \in L^1(\times(0,T))$ and there exists a nonnegative Radon measure μ in $\mathbb{R}^N$ such that

$$\lim_{t\to 0} \int_{\mathbb{R}^N} u(x,t)\zeta(x)dx = \int_{\mathbb{R}^N} \zeta d\mu \quad \textit{for all } \zeta \in C_c^\infty(\mathbb{R}^N).$$

Proof Without loss of generality we can assume that ∂U is smooth and since $Vu \in L^1(U \times (0,T))$ it is classical that there exists a solution v to the problem

$$\begin{aligned} \partial_t v - \Delta v &= Vu && \text{in } Q_T^U \\ v &= 0 && \text{in } \partial_\ell Q_T^U \\ v(.,0) &= 0 && \text{in } U. \end{aligned}$$

The function v is nonnegative and $w = u + v$ is a positive solution of the heat equation. Hence w admits an initial trace on U which is a nonnegative Radon measure. This implies that u admits the same initial trace on U. We end the proof by using a partition of unity. □

Now we can prove our fundamental *Representation Theorem*.

We assume that V satisfies (1.451) and let $u \in C^{2;1}(Q_T)$ be a nonnegative solution of (1.450). If $\psi \in C^{2;1}(Q_T)$ we define $v \in C^{2;1}(Q_T)$ by $v(x,t) = e^{-\psi(x,t)}u(x,t)$. Then

$$\partial_t v - \Delta v - 2\nabla v.\nabla\psi - |\nabla\psi|^2 v - 2v\Delta\psi + (\partial_t\psi + \Delta\psi + V)v = 0 \quad \text{in } Q_T. \tag{1.460}$$

We choose ψ to be the solution of

$$\begin{aligned} -\partial_t\psi - \Delta\psi &= V \text{ in } Q_T \\ \psi(.,T) &= 0 \text{ in } \mathbb{R}^N. \end{aligned} \tag{1.461}$$

Then

$$\psi(x,t) = \int_t^T \int_{\mathbb{R}^N} \frac{e^{-\frac{|x-y|^2}{4(s-t)}}}{(4\pi(s-t))^{\frac{N}{2}}} V(y,s)dyds. \tag{1.462}$$

Because of (1.451) the following estimates hold:

$$\begin{array}{ll} (i) & 0 \le \psi(x,t) \le C_0(T)\ln\left(\frac{T}{t}\right) \\ (ii) & |\nabla\psi(x,t)| \le C_1(T) + C_2(T)\ln\left(\frac{T}{t}\right). \end{array} \tag{1.463}$$

With this choice of ψ, Eq. (1.460) becomes

$$\partial_t v - \Delta v - 2\sum_{i=1}^{n}(v\psi_{x_i})_{x_i} - |\nabla\psi|^2 v = 0 \quad \text{in } Q_T. \tag{1.464}$$

Because $\ln t \in L^p(0,1)$ for all $p \in [1,\infty)$, it follows that for any $p \in [1,\infty)$ there exists $M_j := M_j(p) > 0$, $j = 1, 2$ such that

$$\begin{array}{ll} (i) & \displaystyle\int_0^T \sup_{x\in\mathbb{R}^N} |\psi(x,t)|^p dt \le M_1 \\ (ii) & \displaystyle\int_0^T \sup_{x\in\mathbb{R}^N} |\nabla\psi(x,t)|^p dt \le M_2. \end{array} \tag{1.465}$$

Using Aronson's estimates [4] with $A_{ij} = \delta_{ij}$, $A_i = 2\psi_{x_i}$, $B_i = 0$, $C = |\nabla\psi|^2$ and $p = \infty$ with the notations of this article, then the condition H therein is satisfied. Therefore there exists a kernel $\Gamma(x,t;y,s)$ defined in $Q_T \times Q_T$ satisfying

$$\begin{aligned} C_1(T,N,M_2)\frac{e^{-a_1\frac{|x-y|^2}{4(s-t)}}}{(4\pi(s-t))^{\frac{N}{2}}} &\le \Gamma(x,t;y,s) \\ &\le C_2(T,N,M_2)\frac{e^{-a_2\frac{|x-y|^2}{4(s-t)}}}{(4\pi(s-t))^{\frac{N}{2}}}, \end{aligned} \tag{1.466}$$

for all $(x,t,y,s) \in Q_T \times Q_T$ with $a_1 \ge a_2 > 0$ depending on T, N and M_2, and such that v admits the following representation:

$$v(x,t) = \int_{\mathbb{R}^N} \Gamma(x,t;y,0)d\mu(y), \tag{1.467}$$

where μ is the initial trace of u obtained in Lemma 1.39. Furthermore there holds

$$\lim_{t\to 0}\int_{\mathbb{R}^N}\int_{\mathbb{R}^N} \Gamma(x,t;y,0)\zeta(x)d\mu(y)dx = \int_{\mathbb{R}^N} \zeta d\mu \quad \text{for all } \zeta \in C_c^\infty(\mathbb{R}^N).$$

Note that if the initial trace of u is a function u_0 such that $e^{-\gamma|x|^2}u_0(.) \in L^2(\mathbb{R}^N)$ for some $\gamma > 0$ and u_0 is continuous at some $x \in \mathbb{R}^N$ then

$$\lim_{t\to 0}\int_{\mathbb{R}^N}\Gamma(x,t;y,0)u_0(y)dy = u_0(x). \tag{1.468}$$

Finally, we have the representation

$$u(x,t) = e^{\psi(x,t)}\int_{\mathbb{R}^N}\Gamma(x,t;y,0)d\mu(y). \tag{1.469}$$

1.5.8 σ-Moderate Solutions

1.5.8.1 The Marcus Approach

In this paragraph we adapt to the parabolic framework the construction in [28] used for characterising, by mean of their precise boundary trace, all the positive solutions of

$$-\Delta u + u^q = 0 \quad \text{in } \Omega \tag{1.470}$$

in a smooth bounded domain Ω.

Proposition 1.37 *Let $u \in \mathcal{U}_+(Q_T)$, then*

$$\max\left\{u_{\mathcal{R}_q(u)}, [u]_{\mathcal{S}_q(u)}\right\} \le u \le u_{\mathcal{R}_q(u)} + [u]_{\mathcal{S}_q(u)}. \tag{1.471}$$

Proof From Proposition 1.34-(ii) the function $v = u \ominus u_{\mathcal{R}_q(u)}$ has its $\mathfrak{T}_q$-support included in $\mathcal{S}_q(u)$ since its vanishes on $\mathcal{R}_q(u)$. Furthermore $v \le u$, hence $v \le [u]_{\mathcal{S}_q(u)}$. Therefore $u - u_{\mathcal{R}_q(u)} \le v$, which implies $u \le u_{\mathcal{R}_q(u)} + [u]_{\mathcal{S}_q(u)}$. The left-hand side inequality in (1.471) follows by the construction of $u_{\mathcal{R}_q(u)}$ and the definition of $[u]_{\mathcal{S}_q(u)}$. □

Proposition 1.38 *Let $u \in \mathcal{U}_+(Q_T)$ and A, B be two disjoint $\mathfrak{T}_q$-closed Borel subsets of $\mathbb{R}^N$. If the $\mathfrak{T}_q$-support of u is included in $A \cup B$ and $[u]_A$ and $[u]_B$ are σ-moderate, then u is σ-moderate. Furthermore*

$$u = [u]_A \oplus [u]_B = [u]_A \vee [u]_B. \tag{1.472}$$

Proof Because $[u]_A$ and $[u]_B$ are σ-moderate there exist two increasing sequences $\{\tau_n\}$ and $\{\tau'_n\}$ included in $B^{-\frac{2}{q},q'}(\mathbb{R}^N) \cap \mathfrak{M}^b_+(\mathbb{R}^N)$ such that

$$u_{\tau_n} \uparrow [u]_A \quad \text{and} \quad u_{\tau'_n} \uparrow [u]_B \quad \text{as } n \to \infty,$$

and $\mathfrak{T}_q$-supp $(\tau_n) \subset^q A$ while $\mathfrak{T}_q$-supp $(\tau_n') \subset^q B$ (see Proposition 1.27). Thus

$$cap_{\frac{2}{q},q'}\left(\mathfrak{T}_q\text{-supp}\,(\tau_n) \cap \mathfrak{T}_q\text{-supp}\,(\tau_n')\right) = 0,$$

and

$$u_{\tau_n} \wedge u_{\tau_n'} = u_{\tau_n} \oplus u_{\tau_n'} = u_{\tau_n+\tau_n'}.$$

Moreover, by Proposition 1.25-(ii) and Definition 1.21,

$$\max\{[u]_A, [u]_B\} \leq u \leq [u]_A + [u]_B. \tag{1.473}$$

Therefore

$$\max\{u_{\tau_n}, u_{\tau_n'}\} \leq u \Longrightarrow u_{\tau_n+\tau_n'} \leq u.$$

On the other hand

$$u - u_{\tau_n+\tau_n'} \leq [u]_A - u_{\tau_n} + [u]_B - u_{\tau_n'}.$$

Since the right-hand side tends to 0 as $n \to \infty$ we obtain

$$\lim_{n\to\infty} u_{\tau_n+\tau_n'} = u, \tag{1.474}$$

which means that u is σ-moderate.

By definition of the operations $\oplus$ and $\vee$, identity (1.472) admits the following equivalent formulation;
(a) u is the largest solution dominated by $[u]_A + [u]_B$,
(b) u is the smallest solution dominating by $\max\{[u]_A, [u]_B\}$.
Set $w := [u]_A \oplus [u]_B$, then

$$u \leq w \leq [u]_A + [u]_B.$$

Clearly $[u]_A \leq [w]_A$. Since $[w]_A \leq w \leq [u]_A + [u]_B$ implies $[w]_A - [u]_A \leq [u]_B$. This implies

$$v := [([w]_A - [u]_A)_+]_\dagger \leq [u]_B \quad v \leq [w]_A,$$

where $[([w]_A - [u]_A)_+]_\dagger$ is defined in the notations (e) in Sect. 4.1. This implies that

$$\mathfrak{T}_q\text{-supp}\,(\nu) \subset A \quad \text{and } \mathfrak{T}_q\text{-supp}\,(\nu) \subset B.$$

Since $A \cap B = \emptyset$ we obtain $v = 0$ and $[w]_A \leq [u]_A$. In a similar way $[w]_B \leq [u]_B$. Using (1.473) and the fact that for any Borel set E we have $[u]_E \leq [u]_{\widetilde{E}\cap A} + [u]_{\widetilde{E}\cap B}$ we infer

$$\mathcal{S}_q(u) = \mathcal{S}_q(w).$$

As a consequence any regular $\mathfrak{T}_q$-open regular subset $Q \subset \mathcal{R}_q(w)$ is included into $\mathcal{R}_q(u)$. Using now Proposition 1.25-(ii) and the fact that the $\mathfrak{T}_q$-support of w is included into $A \cap B$ we deduce

$$[w]_Q \leq [w]_{\tilde{Q}\cap A} + [w]_{\tilde{Q}\cap B} = [[w]_A]_{\tilde{Q}} + [[w]_B]_{\tilde{Q}} = [[u]_A]_{\tilde{Q}} + [[u]_B]_{\tilde{Q}w}.$$

Now $[w]_Q$ and $[u]_Q$ are moderate solutions. Because $A \cap B = \emptyset$ there also holds $[u]_{\tilde{Q}\cap A} \oplus [u]_{\tilde{Q}\cap B} \leq [u]_Q$, which implies in turn $[u]_Q = [w]_Q$. Therefore, by Proposition 1.34-(ii), $w_{\mathcal{R}_q(u)} = u_{\mathcal{R}_q(u)}$. Using Proposition 1.36 and the remark hereafter we derive

$$u \leq w \leq u_{\mathcal{R}_q(u)} + U_F,$$

where F is defined in (1.439). Since σ-moderate solutions are uniquely defined, w and u coincide. Hence the result follows from (1.473) and (1.474) by letting $n \to \infty$. □

1.5.8.2 Characterisation of Positive Solutions of $\partial_t u - \Delta u + u^q = 0$

If $u \in \mathcal{U}_+(Q_T)$ we set

$$V(x,t) = u^{q-1}. \tag{1.475}$$

Then u is a solution of

$$\partial_t u - \Delta u + Vu = 0 \quad \text{in } Q_T, \tag{1.476}$$

where V satisfies estimate (1.451) with $C = (q-1)^{-\frac{1}{q-1}}$. The function u belongs to $C^{2;1}(Q_T)$ and there exists a nonnegative Radon measure μ in $\mathbb{R}^N$ such that the following representation formula holds:

$$u(x,t) = e^{\psi(x,t)} \int_{\mathbb{R}^N} \Gamma(x,t;y,0)d\mu(y) \quad \text{for all } (x,t) \in Q_T, \tag{1.477}$$

where ψ is the solution of (1.461) expressed by (1.462). The measure μ is called the *extended initial trace of* u.

If $E \subset \mathbb{R}^N$ is a Borel set we put

$$\mu_E = \mathbf{1}_E \mu \quad \text{and} \quad (u)_E := e^{\psi(x,t)} \int_{\mathbb{R}^N} \Gamma(x,t;y,0)d\mu_E(y) \quad \text{in } Q_T.$$

The next result is fundamental and points out the importance of the function $(u)_E$.

Lemma 1.40 *Let $F \subset \mathbb{R}^N$ be compact, then*

$$(u)_E \leq [u]_E \quad \textit{in } Q_T. \tag{1.478}$$

Proof Let $A \subset \mathbb{R}^N$ be a Borel set and $0 < \beta \leq \frac{T}{2}$. We denote by v_β^A the solution of

$$\begin{aligned} \partial_t v - \Delta v + Vv &= 0 && \text{in } \mathbb{R}^N \times (\beta, T) \\ v(.,\beta) &= \mathbf{1}_A u(.,\beta) && \text{in } \mathbb{R}^N, \end{aligned} \tag{1.479}$$

and by w_β^A the one of

$$\begin{aligned} \partial_t w - \Delta w + w^q &= 0 && \text{in } \mathbb{R}^N \times (\beta, T) \\ v(.,\beta) &= \mathbf{1}_A u(.,\beta) && \text{in } \mathbb{R}^N. \end{aligned} \tag{1.480}$$

Since $u^{q-1} \leq V$, there holds $0 \leq w_\beta^A \leq v_\beta^A \leq u$. For any sequence $\{\beta_k\}$ decreasing to 0 one can extract a subsequence still denoted by $\{\beta_k\}$ such that $\{v_{\beta_k}^A\}$ and $\{w_{\beta_k}^A\}$ converges locally uniformly to v^A and w^A respectively. Clearly $w^A \in \mathcal{U}_+(Q_T)$ while v^A is a solution of (1.476). Since the $\mathfrak{T}_q$-support of $w_\beta^A(.,\beta)$ is included into $\widetilde{Q}$ for any open set Q which contains A, we have

$$v^A \leq w^A \leq [u]_{\widetilde{Q}}. \tag{1.481}$$

Next we set $\widetilde{v}_k = e^{-\psi} v_{\beta_k}^A$, then $\widetilde{v}_k$ satisfies

$$\begin{aligned} \partial_t v - \Delta v - 2\nabla v.\nabla\psi - |\nabla\psi|^2 v - 2v\Delta\psi + (\partial_t\psi + \Delta\psi + V)v &= 0 && \text{in } Q_T \\ v(.,\beta_k) &= \mathbf{1}_A \int_{\mathbb{R}^N} \Gamma(.,\beta_k,y,0)d\mu(y) && \text{in } \mathbb{R}^N. \end{aligned} \tag{1.482}$$

Using Duhamel's formula (see [4] in a similar case), we have

$$\begin{aligned} \widetilde{v}_k(x,t) &= \int_{\mathbb{R}^N} \mathbf{1}_A(x)\Gamma(x,t-\beta_k;y,0)\left(\int_{\mathbb{R}^N} \Gamma(x,\beta_k;y,0)d\mu(y)\right)dx \\ &= \int_{\mathbb{R}^N}\left(\int_{\mathbb{R}^N} \mathbf{1}_A(x)\Gamma(x,t-\beta_k;y,0)\Gamma(x,\beta_k;y,0)dx\right)d\mu(y) \\ &\leq \int_{\mathbb{R}^N}\left(\int_{\mathbb{R}^N} \mathbf{1}_Q(x)\Gamma(x,t-\beta_k;y,0)\Gamma(x,\beta_k;y,0)dx\right)d\mu(y). \end{aligned}$$

Using the estimates on Γ (see (1.466)) the continuity and the property (1.468) we can let $k \to \infty$ and obtain by the dominated convergence theorem

$$\lim_{k\to\infty} \widetilde{v}_k(x,t) \leq \int_{\mathbb{R}^N} \Gamma(x,t_k;y,0)d\mu_Q(y).$$

This implies

$$v^A \leq (u)_{\widetilde{Q}}. \tag{1.483}$$

Then we can proceed in the same way with A^c. Extracting a subsequence from the previous subsequence (and denoting it still by $\{k\}$) we obtain limits v^{A^c} and w^{A^c} and they satisfy

$$v^{A^c} \leq w^{A^c} \leq [u]_{\widetilde{Q'^c}} \quad \text{for all open sets } Q' \supset A^c.$$

Since $v^A_{\beta_k} + v^{A^c}_{\beta_k}$ satisfies (1.30) in $(\beta_k, T) \times \mathbb{R}^N$ with initial data $u(., \beta_k)$, we have

$$v^A + v^{A^c} = u\,, \quad v^A \leq (u)_{\widetilde{Q}}\,, \quad v^{A^c} \leq (u)_{\widetilde{Q'^c}}$$

from what we derive

$$v^A = u - v^{A^c} \geq (u)_{\widetilde{Q'^c}}. \tag{1.484}$$

Next, if F is a compact subset of $\mathbb{R}^N$, O an open set such that $F \subset O \subset A$, we obtain from (1.484) with $Q' = A^c$ (and thus $A^c \cap F = Q' \cap F = \emptyset$),

$$v^A \geq (u)_O.$$

By (1.483),

$$v \leq w \leq [u]_{\widetilde{Q}} \quad \text{for all open set } Q \supset A,$$

which implies

$$(u)_F \leq (u)_O \leq [u]_{\widetilde{Q}}. \tag{1.485}$$

By Lemma 1.26 we can fix a sequence of open sets $\{Q_n\}$ such that $\cap_n \widetilde{Q}_n = E' \sim^q F$. This implies $[u]_{Q_n} \downarrow [u]_F$ (see Proposition 1.312-(iii)). The result follows from (1.485). □

In the next result we prove that the extended initial trace of a positive solution of (1.30) is absolutely continuous with respect to the $cap_{\frac{2}{q},q'}$-capacity.

Proposition 1.39 *Let $u \in \mathcal{U}_+(Q_T)$ and μ be its extended initial trace as defined in (1.477). Then $\mu(E) = 0$ for any Borel set $E \subset \mathbb{R}^N$ such that $cap_{\frac{2}{q},q'}(E) = 0$.*

Proof If K is a compact set satisfying $cap_{\frac{2}{q},q'}(K) = 0$, then $U_K = 0$ by Corollary 1.4. Therefore $[u]_K = u \vee U_K = 0$. Consequently, by Lemma 1.40 $(u)_K = 0$ and $\mu(K) = 0$. Since this holds for any such K, it also holds for E by outer regularity. □

We recall that for any $\nu \in B^{-\frac{2}{q},q}(\mathbb{R}^N) \cap \mathfrak{M}^b_+(\mathbb{R}^N)$ and any $T > 0$ $C(T) > 0$ such that

$$C^{-1} \|\nu\|_{B^{-\frac{2}{q},q}(\mathbb{R}^N)} \leq \|\mathbb{H}[\nu]\|_{L^q(Q_T)} \leq C \|\nu\|_{B^{-\frac{2}{q},q}(\mathbb{R}^N)}. \tag{1.486}$$

Proposition 1.40 *Let $u \in \mathcal{U}_+(Q_T)$, μ be its extended initial trace and $\nu \in B^{-\frac{2}{q},q}(\mathbb{R}^N) \cap \mathfrak{M}^b_+(\mathbb{R}^N)$. Suppose that there exists no positive solution of (1.30) dominated by $v = \inf\{u, \mathbb{H}[\nu]\}$. Then μ and ν are mutually singular that we denote by $\mu \perp \nu$.*

Proof Set $V' = v^{q-1}$. Then v is a supersolution of

$$\partial_t w - \Delta w + V' w = 0 \quad \text{in } Q_T. \tag{1.487}$$

We first prove by contradiction that there exists no positive solution of (1.487) dominated by v. Indeed, if such a solution w of this equation does exist, there holds

$$\partial_t w - \Delta w + w^q \leq \partial_t w - \Delta w + V' w = 0.$$

Because of (1.486) the function w is a moderate solution of (1.30) dominated by v, contrary to the assumption. Next, we have a representation formula valid in Q_T where we use Aronson's estimates 1.465 and the constants a_2 and C from this inequality,

$$\begin{aligned}
\inf\{u, \mathbb{H}[\nu]\} &= \inf\left\{e^{\psi} \int_{\mathbb{R}^N} \Gamma(x,t;y,0)d\mu(y), \mathbb{H}[\nu]\right\} \\
&\geq \inf\left\{\int_{\mathbb{R}^N} \Gamma(x,t;y,0)d\mu(y), \mathbb{H}[\nu]\right\} \\
&\geq C \inf\left\{\mathbb{H}[\mu]\left(\tfrac{t}{a_2}, x\right), \mathbb{H}[\nu](t,x)\right\} \\
&\geq C \inf\left\{\mathbb{H}[\mu]\left(\tfrac{t}{\max\{a_2,1\}}, x\right), \mathbb{H}[\nu]\left(\tfrac{t}{\max\{a_2,1\}}, x\right)\right\}.
\end{aligned}$$

We notice that

$$(t,x) \mapsto \inf\left\{\mathbb{H}[\mu]\left(\tfrac{t}{\max\{a_2,1\}}, x\right), \mathbb{H}[\nu]\left(\tfrac{t}{\max\{a_2,1\}}, x\right)\right\},$$

is a supersolution of the equation $\partial_t w - \frac{1}{\max\{a_2,1\}}\Delta w = 0$; therefore there exists a nonnegative Radon measure $\widetilde{\nu}$ in $\mathbb{R}^N$ such that

$$\begin{aligned}
\lim_{t\to 0} \int_{\mathbb{R}^N} \phi(x) \inf\left\{\mathbb{H}[\mu]\left(\tfrac{t}{\max\{a_2,1\}}, x\right), \mathbb{H}[\nu]\left(\tfrac{t}{\max\{a_2,1\}}, x\right)\right\} dx \\
= \int_{\mathbb{R}^N} \phi(x) d\widetilde{\nu}(x).
\end{aligned} \tag{1.488}$$

By Lemmas 1.38 and 1.39 there exists a positive solution $\tilde{v} \leq v$ of the initial value problem

$$\begin{aligned}
\partial_t w - \Delta w + V w &= 0 \text{ in } Q_T \\
w(.,0) &= \widetilde{\nu} \text{ in } \mathbb{R}^N.
\end{aligned}$$

By the first claim it yields $\widetilde{\nu} = 0$.

By the Radon-Nikodym theorem there exists a positive measure σ and a Borel function $\theta \in L^1(\mathbb{R}^N, \mu)$ such that $\sigma \perp \mu$ and $\nu = \theta\mu + \sigma$. Therefore if H is the heat kernel in Q_∞, we obtain

$$
\begin{aligned}
0 &= \lim_{t\to 0}\int_{\mathbb{R}^N}\phi(x)\inf\left\{\mathbb{H}[\mu]\left(\tfrac{t}{\max\{a_2,1\}},x\right),\mathbb{H}[\nu]\left(\tfrac{t}{\max\{a_2,1\}},x\right)\right\}dx\\
&\geq \lim_{t\to 0}\int_{\mathbb{R}^N}\phi(x)H\left(\tfrac{t}{\max\{a_2,1\}},x,y\right)\min\{1,\theta(y)\}d\mu(y)dx\\
&\geq \lim_{t\to 0}\int_{\mathbb{R}^N}\phi(y)\min\{1,\theta(y)\}d\mu(y)\\
&= 0.
\end{aligned}
$$

Hence $\theta = 0$ and $\nu \perp \mu$. □

Lemma 1.41 *Let $u \in U_+(Q_T)$, μ its extended initial trace and suppose that for every $\nu \in \mathfrak{M}^b_+(\mathbb{R}^N) \cap B^{-\frac{2}{q},q'}(\mathbb{R}^N)$ there exists no positive solution of (1.30) dominated by $v = \inf\{u, \mathbb{H}[\nu]\}$. Then $u = 0$.*

Proof As in the previous lemma, the proof is an adaptation to the parabolic framework of the construction in [28]. By the previous lemma,

$$\mu \perp \nu \quad \text{for all } \nu \in \mathfrak{M}^b_+(\mathbb{R}^N) \cap B^{-\frac{2}{q},q'}(\mathbb{R}^N).$$

Suppose now that $\mu \neq 0$, then by Lemma 1.40 μ vanishes on Borel sets $E \subset \mathbb{R}^N$ such that $cap_{\frac{2}{q},q'}(E) = 0$. Therefore, there exists an increasing sequence $\{\nu_k\} \subset \mathfrak{M}^b_+(\mathbb{R}^N) \cap B^{-\frac{2}{q},q'}(\mathbb{R}^N)$ which converges to μ. Therefore $\mu \perp \nu_k$ and for every $k \in \mathbb{N}$ there exists a Borel set $A_k \subset \mathbb{R}^N$ such that

$$\mu(A_k) = 0 \quad \text{and} \quad \nu_k(A_k^c) = 0.$$

If we denote $A = \cup_k A_k$, then

$$\mu(A) = 0 \quad \text{and for all integer } k,\ \nu_k(A^c) = 0.$$

But since $\nu_k \leq \mu$ we have also $\nu_k(A) = 0$ and thus $\nu_k = 0$ for all k, a contradiction. □

The next result is fundamental.

Proposition 1.41 *Let $u \in U_+(Q_T$, then $[u]_{\mathcal{S}_q(u)}$ is σ-moderate.*

Proof We simplify the notations in setting $u_{\mathcal{S}} = [u]_{\mathcal{S}_q(u)}$ (there will be no ambiguity), and we denote $F = \mathfrak{T}_q\text{-supp}\,(u_{\mathcal{S}})$. Then $F \subset \mathcal{S}_q(u)$. We know that if $\mathcal{S}_q(u)$ is thin at ξ, then $\mathcal{S}_q(u)^c \cup \{\xi\}$ is $\mathfrak{T}_q$-open and $\mathcal{S}_q(u)^c \cup \{\xi\} \sim^q \mathcal{S}_q(u)^c$. Since F is the $\mathfrak{T}_q$-support of $u_{\mathcal{S}}$ we see that F consists exactly in the set of $cap_{\frac{2}{q},q'}$-thick points of $\mathcal{S}_q(u)$, and therefore $\mathcal{S}_q(u) \setminus F$ is contained in the singular set of $u_{\mathcal{R}_q(u)}$.

If $\nu \in \mathfrak{M}^b_+(\mathbb{R}^N) \cap B^{-\frac{2}{q},q'}(\mathbb{R}^N)$ and u_ν is the solution of (1.30) with initial trace ν we put

$$u^* := \sup\left\{u_\nu : \nu \in \mathfrak{M}_+^b(\mathbb{R}^N) \cap B^{-\frac{2}{q},q'}(\mathbb{R}^N),\ u_\nu \leq u_S\right\}. \tag{1.489}$$

By the previous lemma, u^* do exist since some elements u_ν of this family exist. Also u^* is σ-moderate by Theorem 1.23-(iii). Therefore u^* is the largest σ-moderate moderate solution of (1.30) dominated by u_S. Let $\{\nu_k\} \subset \mathfrak{M}_+^b(\mathbb{R}^N) \cap B^{-\frac{2}{q},q'}(\mathbb{R}^N)$ be an increasing sequence such that $u_{\nu_k} \uparrow u^*$.

Let F^* be the $\mathfrak{T}_q$ support of u^*, then F^* is $\mathfrak{T}_q$-closed and included in F. Let us assume that

$$cap_{\frac{2}{q},q'}(F \setminus F^*) > 0,$$

then there exists a compact set $E \subset F \setminus F^*$ such that $cap_{\frac{2}{q},q'}(E) > 0$ and $(F^*)^c := Q^*$ is $\mathfrak{T}_q$-open and contains E. By Lemma 1.25 there exists a $\mathfrak{T}_q$-open set Q' such that $E \subset^q Q' \subset \widetilde{Q}' \subset^q Q^*$. Because $Q' \subset^q \mathfrak{T}_q$-supp (u_S), $[U_S]_{Q'} > 0$ and by Proposition 1.39 there exists a positive bounded measure $\tau \in B^{-\frac{2}{q},q'}(\mathbb{R}^N)$ with support in $\widetilde{Q}'$ such that $u_\tau \leq u_S$. As the $\mathfrak{T}_q$-support of τ is a $\mathfrak{T}_q$-closed set disjoint from F^*, the inequality $u^* \geq u_\tau$ cannot hold. However since $\tau \in \mathfrak{M}_+^b(\mathbb{R}^N) \cap B^{-\frac{2}{q},q'}(\mathbb{R}^N$ is such that $u_\tau \leq u_S$), it follows that $u \leq u^*$, which is a contradiction. Hence

$$cap_{\frac{2}{q},q'}(F \setminus F^*) = 0.$$

Since $u_{\nu_k} \uparrow u^*$, the $\mathfrak{T}_q$ support of ν_k is contained into the $\mathfrak{T}_q$ support of u^* which is F^*. Therefore there exists a $\mathfrak{T}_q$-closed set F_0^* contained into F such that $\mathcal{S}_q(u^*) = F_0^*$ and $\mathcal{R}_q(u^*) = (F_0^*)^c$. Suppose now that

$$cap_{\frac{2}{q},q'}(F \setminus F_0^*) > 0,$$

and let Q' be a $\mathfrak{T}_q$-open set contained into $\mathcal{R}_q(u^*)$ such that $[u_S]_{Q'}$ is a moderate solution of (1.30). Then $\widetilde{Q}' \subset^q \mathcal{R}_q(u^*)$ and $[u^*]_{\widetilde{Q}'}$ is a moderate solution too, thus

$$\int\int_{Q_T} [u^*]^q_{\widetilde{Q}'} \phi(x) dx dt < \infty \quad \text{for all } \phi \in C_0^\infty(\mathbb{R}^N),\ \phi \geq 0.$$

On the other hand Q' is a $\mathfrak{T}_q$-open subset of F which is the $\mathfrak{T}_q$ support of u_S. Consequently the initial trace of $[u^*]_{\widetilde{Q}'}$ has no regular part, that is

$$\mathcal{R}_q([u^*]_{\widetilde{Q}'}) = \emptyset \quad \text{and } \mathcal{S}_q([u^*]_{\widetilde{Q}'}) = \mathfrak{T}_q\text{-support}([u^*]_{\widetilde{Q}'}).$$

In such a case we call $[u^*]_{\widetilde{Q}'}$ a *purely singular solution of* (1.30). It implies that

$$v = \left[[u_S]_{\widetilde{Q}'} - [u^*]_{\widetilde{Q}'}\right]_+$$

is a purely singular solution too.

Let v^* be defined as in expression (1.489) with u_S replaced by v. Then v^* is a singular σ-moderate solution of (1.30). As it is dominated by u and σ-moderate, it is smaller than u^*. Now, $\mathfrak{T}_q$-supp$(v^*) \subset^q \widetilde{Q}' \subset^q \mathcal{R}_q(u^*)$, therefore u^* cannot be larger or equal to v^*, hence $(v^* - u^*)_+$ is not identically zero. Since both u^* and v^* are σ-moderate, it follows that there exists a nonnegative bounded measure $\tau \in B^{-\frac{2}{q},q'}(\mathbb{R}^N)$ such that $u_\tau \leq v^*$ and $(u_\tau - u^*)_+$ is not identically zero, and obviously that $u^* \leq \max\{u_\tau, v^*\}$. The function $\max\{u^*, u_\tau\}$ is a nontrivial subsolution of (1.30) and there exists a smallest solution Z above it, which is also strictly larger than u^*. However $u_\tau \leq v^* \leq u^*$ and thus $u^* = Z$, a contradiction. As a consequence $cap_{\frac{2}{q},q'}(Q') = 0$ for any $\mathfrak{T}_q$-open set included in $\mathcal{R}_q(u^*)$ such that $[u^*]_{Q'}$ is a moderate solution. Hence

$$cap_{\frac{2}{q},q'}(F \setminus F_0^*) = 0. \tag{1.490}$$

In conclusion u^* is σ-moderate, $\mathfrak{T}_q$-supp $(u^*) \subset F$ and $F_0^* = \mathcal{S}_q(u^*) \sim^q F$. Therefore, by Proposition 1.36 and the remark which follows $u^* = U_F$. Since by definition (1.489) $u^* \leq u_S \leq U_F$ it follows that $u^* = u_S$ and thus u_S is σ-moderate. □

The following result is the *icing on the cake* of the precise trace theory.

Theorem 1.25 *Every positive solution of (1.30) is σ-moderate.*

Proof Let $u \in \mathcal{U}_+(Q_T)$. By Proposition 1.34-(i), $\mathcal{R}_q(u)$ has a regular decomposition $\{Q_n\}$ and

$$v_n := [u]_{Q_n} \uparrow u_{\mathcal{R}_q(u)}.$$

Then $u_{\mathcal{R}_q(u)}$ is σ-moderate and

$$u \ominus u_{\mathcal{R}_q(u)} \leq [u]_{\mathcal{S}_q(u)}.$$

Set

$$u_n = v_n \oplus [u]_{\mathcal{S}_q(u)}.$$

By Proposition 1.41 $[u]_{\mathcal{S}_q(u)}$ is σ-moderate. Using the fact that $\widetilde{Q}_n \cap \mathcal{S}_q(u)$, it follows by Proposition 1.38 that u_n is σ-moderate. The sequence $\{u_n\}$ is increasing and converges to some $\overline{u}$ of (1.30) which is σ-moderate too. Furthermore

$$v_n \vee [u]_{\mathcal{S}_q(u)} = u_n = v_n \oplus [u]_{\mathcal{S}_q(u)} \Longrightarrow \max\{u_{\mathcal{R}_q(u)}, [u]_{\mathcal{S}_q(u)}\} \leq \overline{u} \leq u_{\mathcal{R}_q(u)} \oplus [u]_{\mathcal{S}_q(u)}.$$

This implies that $\mathcal{S}_q(u) = \mathcal{S}_q(\overline{u})$. Now, by construction we have

$$v_n = [u]_{Q_n} \leq [\overline{u}]_{Q_n}.$$

then letting $n \to \infty$ we obtain by Proposition 1.34

$$u_{\mathcal{R}_q(u)} \leq \overline{u}_{\mathcal{R}_q(u)} \Longrightarrow u_{\mathcal{R}_q(u)} = \overline{u}_{\mathcal{R}_q(u)},$$

therefore $tr(u) = tr(\overline{u})$. But since $\overline{u} \leq u$, it follows by Proposition 1.36 and the uniqueness of σ-moderate solutions that $\overline{u} = u$. □

1.5.9 Further Studies and Open Problems

1.5.9.1 Lateral Boundary Trace

Let Ω be either a C^2 open subset or $\mathbb{R}^N_+$. The problem is to analyse the trace on the lateral boundary of $\Omega \times (0, T)$ of any positive solution of (1.30). It is proved in [29] that there exists a lateral trace in the class of outer regular Borel measures in $\partial_\ell \Omega \times (0, T) := \partial\Omega \times (0, T)$. The critical value for the exponent q is $\tilde{q} - c = \frac{N+3}{N+1}$ since above this value the boundary isolated singularities are removable (see. e.g. [34]). The geometry of the cylindrical domain makes much more difficult the study of the supercritical case. A similar study was performed by Kuznetsov [25, 26] in the framework of superprocesses and with the restriction that $1 < q < 2$.

1.5.9.2 Full Trace Problem

It is an extension of the initial trace problem treated in this survey and the lateral boundary trace. In the paper [31] the initial trace is considered for solutions in a cylinder Q^Ω_T and the existence and uniqueness theorem in the subcritical case is proved provided the lateral boundary value is integrable in $\partial\Omega \times (0, T)$ and the initial measure has compact support in Ω, or at least is bounded near $\partial\Omega$. This is due to the fact that for general measure μ in $\mathfrak{M}_+(\Omega)$, the behaviour of μ near $\partial\Omega$ is fundamental. A general study of the trace of positive solutions of (1.30) on $\partial Q^\Omega_T := (\Omega \times \{0\}) \bigcup (\partial\Omega \times [0, T))$ would be a great interest. In this direction we can mention the work [24] dealing with the lateral boundary trace of positive solutions of

$$\frac{\partial u}{\partial t} - \Delta u - u^q = 0 \qquad \text{in } \mathbb{R}^N_+ \times (0, T). \tag{1.491}$$

Extension to general domain are expected. See also [35] for related problems.

1.5.9.3 Equations of General Absorption-Convection

$$\partial_t u - \Delta u + u^p |\nabla u|^q = 0. \tag{1.492}$$

Since this is an equation with absorption, the construction of an initial trace should be tractable. To our knowledge, the study of the self-similar solutions and isolated singularities has not yet been done. This study needs a preliminary analysis of the problem

$$\begin{aligned} \partial_t u - \Delta u + u^p|\nabla u|^q &= 0 \quad \text{in } Q_T \\ u(.,0) &= \mu \quad \text{in } \mathcal{D}'(\mathbb{R}^N), \end{aligned} \tag{1.493}$$

where μ is a nonnegative Radon measure.

1.5.9.4 Equations of Hamilton Jacobi Type

$$\partial_t u - \Delta u + m|\nabla u|^q = 0 \quad \text{in } Q_T. \tag{1.494}$$

The subcritical case has been treated by Bidaut-Véron-Dao [10, 11]. They prove the existence of a critical exponent $q^* = \frac{N+2}{N+1}$. When $1 < q < q^*$, they obtain the existence of solutions u with a Dirac mass as initial data and the existence and uniqueness of a positive very singular solution. When $q \geq q^*$ they prove that isolated singularities at $t = 0$ are removable. The detailed analysis of the initial trace in the supercritical case seems open.

1.5.9.5 Equations of Mixed Absorption-Reaction-Convection

$$\partial_t u - \Delta u + u^p - m|\nabla u|^q = 0, \tag{1.495}$$

or

$$\partial_t u - \Delta u + m|\nabla u|^q - u^p = 0. \tag{1.496}$$

For these two types of equations, the existence of an initial trace seems open except in some specific cases. The study has to be put in parallel with the ones dealing with the boundary value problem and the boundary trace for the elliptic equations

$$-\Delta u + u^p - m|\nabla u|^q = 0 \qquad \text{in } \Omega, \tag{1.497}$$

obtained in [12] or

$$-\Delta u + m|\nabla u|^q - u^p = 0 \qquad \text{in } \Omega, \tag{1.498}$$

obtained in [13]. In these two papers, a method is developed which associates some specific supersolutions and subsolutions namely

$$-\Delta u - m|\nabla u|^q = 0 \text{ in } \Omega \text{ and } -\Delta u + u^p = 0 \text{ in } \Omega \tag{1.499}$$

for (1.497) in [12] and

$$-\Delta u - u^p = 0 \text{ in } \Omega \text{ and } -\Delta u + m|\nabla u|^q = 0 \text{ in } \Omega \tag{1.500}$$

for (1.498) in [13]. It appears that this could be adapted to the study of (1.495) and (1.496). We also refer to the book of Quittner and Souplet [36] which contains an impressive quantity of results concerning semilinear heat equations with reaction terms of the type $-u^p$ or $-|\nabla u|^q$.

References

1. Adams, D.R., Hedberg, L.I.: Function spaces and potential theory. Grundlehren der mathematischen Wissenschaften, vol. 145. Springer (1967)
2. Aikawa, H., Borichev, A.A.: Quasiadditivity and measure property of capacity and the tangential boundary behavior of harmonic functions. Trans. Am. Math. Soc. **348**, 1013–1030 (1996)
3. Ancona, A.: Negatively curved manifolds, elliptic operators and the Martin boundary. Ann. Math. Second. Ser. **125**, 495–536 (1987)
4. Aronson, D.G.: Non-negative solutions of linear parabolic equations. Ann. Sc. Norm. Sup. Pisa, Classe di Sci. **22**, 607-694 (1968)
5. Aronson, D.G.: Widder's inversion theorem and the initial distribution problems. SIAM J. Math. Anal. **12**, 639–651 (1981)
6. Baras, P., Pierre, M.: Singularités éliminables pour des équations semilinéaires. Ann. Inst. Fourier **34**, 185–206 (1984)
7. Baras, P., Pierre, M.: Problèmes paraboliques semi-linéaires avec données mesures. Appl. Anal. **18**, 111–149 (1984)
8. Berens, H., Butzer, P.: Semigroups of operators and approximations. Grundlehren der mathematischen Wissenschaften, vol. 314. Springer, New York (1996)
9. Bidaut-Véron, M.F., Chasseigne, E., Véron, L.: Initial trace of solutions of some quasilinear parabolic equations with absorption. J. Funct. Anal. **193**, 140–205 (2002)
10. Bidaut-Véron, M.F., Dao, A.N.: Initial trace of solutions of Hamilton-Jacobi parabolic equation with absorption. Adv. Nonlinear. Stud. **15**, 889–921 (2015)
11. Bidaut-Véron, M.F., Dao, A.N.: Isolated initial singularities for the viscous Hamilton-Jacobi equation. Adv. Differ. Eqs. **17**, 903–934 (2012)
12. Bidaut-Véron, M.F., Garcia-Huidobro, M., Véron, L.: Boundary singular solutions of a class of equations with mixed absorption-reaction. Calc. Var. Partial Differ. Eqs. **61**, Paper No. 2150051, 37 pp (2022)
13. Bidaut-Véron, M.F., Véron, L.: Trace and boundary singularities of solutions of a class of quasilinear equations. Discr. Cont. Dyn. Syst. **43**, 1026–1051 (2023)
14. Brezis, H.: Nonlinear elliptic equation in $\mathbb{R}^N$ without condition at infinity. Appl. Math. Opt. **12**, 271–282 (1985)
15. Brezis, H., Friedman, A.: Nonlinear parabolic equations involving measures as initial conditions. J. Math. Pures Appl. **62**, 73–97 (1983)
16. Brezis, H., Peletier, L.A., Terman, D.: A very singular solution of the heat equation with absorption. Arch. Rational Mech. Anal. **95**, 185–209 (1986)
17. Dynkin, E.B.: Superdiffusions and Partial Differential Equations, American Mathematical Society Colloquium Publications, vol. 50, vi+236 pp. American Mathematical Society, Providence (2002)
18. Dynkin, E.B.: Superdiffusions and positive solutions of nonlinear partial differential equations. University Lecture Series, vol. 34, vi+120 pp. American Mathematical Society, Providence (2004)
19. Dynkin, E.B., Kuznetsov, S.E.: Superdiffusions and removable singularities for quasilinear partial differential equations. Commun. Pure Appl. Math. **49**, 125–176 (1996)
20. Dynkin, E.B., Kuznetsov, S.E.: Fine topology and fine trace on the boundary associated with a class of quasilinear differential equations. Commun. Pure Appl. Math. **51**, 897–936 (1998)

21. Feyel, D., de la Pradelle, A.: Topologies fines et compactifications associées à certains espaces de Dirichlet. Ann. Inst. Fourier Grenoble **27**, 121–146 (1977)
22. Gkikas, K., Véron, L.: Initial value problems for diffusion equations with singular potential. Contemporary Math. **594**, 201–230 (2013)
23. Gkikas, K., Véron, L.: Complete classification of the positive solutions of heat equation with super critical absorption. Adv. Nonlinear Stud. **14**, 47–113 (2014)
24. Hisa, K., Ishige, K., Takahashi, J.: Initial traces and solvability for a semilinear heat equation on a half space of $\mathbb{R}^N$. Trans. Am. Math. Soc., to appear
25. Kuznetsov, S.: On removable lateral singularities for quasilinear parabolic PDE'. C. R. Acad. Sci. Paris t. **325**, 627–632 (1997)
26. Kuznetsov, S.: Polar boundary sets for superdiffusions and removable lateral singularities for non-linear parabolic PDE's. Commun. Pure Appl. Math. **51**, 303–340 (1998)
27. Le Gall, J.F.: A probabilistic approach to the trace at the boundary for solutions of a semilinear parabolic partial differential equation. J. Appl. Math. Stochastic Anal. **9**, 399–414 (1996)
28. Marcus, M.: Complete classification of the positive solutions of $-\Delta u + u^q = 0$. J. Anal. Mathématique **117**, 187–220 (2012)
29. Marcus, M., Véron, L.: Trace au bord lat 'éral des solutions positives d'équations paraboliques non-linéaires. C. R. Acad. Sci. Paris t. **324**, 783–788 (1997)
30. Marcus, M., Véron, L.: Initial trace of positive solutions of some nonlinear parabolic equations. Commun. P.D.E. **24**, 1445–1499 (1999)
31. Marcus, M., Véron, L.: Initial trace of positive solutions to semilinear parabolic inequalities. Adv. Nonlinear Stud. **4**, 395–436 (2002)
32. Marcus, M., Véron, L.: Capacitary estimates of positive solutions of semilinear elliptic equations with absorbtion. J. Eur. Math. Soc. **6**, 483–527 (2004)
33. Marcus, M., Véron, L.: The precise boundary trace of positive solutions of the equation $\Delta u = u^q$ in the supercritical case. Perspectives in nonlinear partial differential equations. Contemporary Mathematics, vol. 446, pp. 345–383. American Mathematical Society, Providence, RI (2007)
34. Marcus, M., Véron, L.: Capacitary estimates of positive solutions of semilinear parabolic equations with absorbtion. Calc. Var. Partial Differ. Eqs. **48**, 131–183 (2013)
35. Marcus, M., Véron, L.: Semilinear parabolic equations with measure boundary data and isolated singularities. J. Anal. Mathématique **80**, 870–900 (2001)
36. Quittner, P., Souplet, Ph.: Superlinear parabolic problems. Blow-up, global existence and steady states, 2nd edition. Birkhäuser Advanced Texts, Birkhäuser Verlag, Basel, xvi+725 pp (2007)
37. Tartar, L.: Sur un lemme d'équivalence utilisé en analyse Numérique. Calcolo **24**, 129–140 (1987)
38. Triebel, H.: Interpolation Theory, Function Spaces, Differential Operators, vol. 18. North-Holland Publishing Company, North-Holland Mathematical Library. Amsterdam-New-York-Oxford (1978)
39. Widder, D.V.: Positive temperature on an infinite rod. Trans. Am. Math. Soc. **55**, 85–95 (1944)
40. Whittaker, E.T., Watson, G.N.: A Course of Modern Analysis, 4th edn. Cambridge University Press, Chapter XXI (1927)

Chapter 2
Graph Gradient Flows: From Discrete to Continuum

Yves van Gennip, Yoshikazu Giga, and Jun Okamoto

Abstract This paper gives a framework to study a continuum limit of a gradient flow on a graph where the number of vertices increases in an appropriate way. As examples, we prove the convergence of a discrete total variation flow and a discrete Allen–Cahn flow on discretised tori to their respective continuum limits.

2.1 Introduction

We prove the convergence of two graph-based discrete gradient flows, the total variation flow, and the Allen–Cahn flow, to a continuum limiting problem when the graph approximates a Euclidean space. Although characterisation of continuum limits has been studied variationally at the energy level quite extensively in the past decade, mainly Γ-convergence results, the convergence problem for discrete flows has only been attracting more attention fairly recently.

If $\Phi : H \to \mathbb{R}$ is a Fréchet differentiable function on a Hilbert space $(H, \langle\cdot,\cdot\rangle)$, with gradient $\nabla_H \Phi$, then the gradient flow of Φ (w.r.t. the inner product on H) is a function $u : (0, \infty) \to H$ that satisfies $\dot{u} = -\nabla_H \Phi(u)$. If Φ is not differentiable, but convex, the differential *equation* is replaced by the differential *inclusion* $\dot{u} \in -\partial\Phi(u)$, where $\partial\Phi(u)$ denotes the subdifferential of Φ at u. The function Φ is often called the *energy*, even when there is no direct physical interpretation as

Y. van Gennip
Delft Institute of Applied Mathematics, Technische Universiteit Delft, Delft, The Netherlands
e-mail: y.vangennip@tudelft.nl

Y. Giga (✉)
Graduate School of Mathematical Sciences, The University of Tokyo, Meguro-ku, Tokyo, Japan
e-mail: labgiga@ms.u-tokyo.ac.jp

J. Okamoto
Faculty of Medicine Bldg. B, Institute for the Advanced Study of Human Biology, Kyoto University, Sakyo-ku, Kyoto, Japan
e-mail: okamoto.jun.8n@kyoto-u.ac.jp

M. Jleli et al. (eds.), *Advances in Nonlinear Evolution Equations*, Infosys Science Foundation Series, https://doi.org/10.1007/978-981-97-9812-4_2

such. An equivalent formulation can be given in terms of an *evolution variational inequality* (see Definition 2.1; recent in-depth overviews are given in [4, 52]). Such a formulation offers more flexibility when generalising the notion of gradient flow to metric spaces that lack a Hilbert space structure. Additionally, the variational setting can be exploited when comparing gradient flows that are formulated on different underlying spaces. We make crucial use of this advantage to prove convergence of gradient flows that are formulated on discrete graphs to a continuum gradient flow.

In the current work, we consider two families of gradient flows: total variation gradient flows with the energy Φ being the graph-based or continuum total variation functional (Sect. 2.4.1) and Allen–Cahn gradient flows with the energy Φ being the graph-based or continuum Allen–Cahn functional (also known as the Ginzburg–Landau functional or Modica–Mortola functional; Sect. 2.5.1).

To be able to prove convergence results, and thus to compare flows formulated on a graph with flows formulated on a continuum, we require a way to embed the discrete flows in continuum space. In this paper we restrict ourselves to graphs that are obtained as a regular (square or (hyper)cubic) discretisation of the flat torus. In that setting, we can associate functions on the graph with functions on the torus via piecewise-constant or piecewise-linear interpolation. We use the former for our convergence results for total variation flow and the latter for Allen–Cahn flow.

To avoid additional technical complexities, for the Allen–Cahn flows we only consider the one-dimensional case.

The two main results in this paper are Theorems 2.4 and 2.6. The former states that (under assumptions that will be given in detail later) the L^2-difference between the (piecewise-constant interpolant of the) graph-based total variation flow and the continuum total variation flow is controlled by the L^2-difference of their initial conditions. In particular, if both flows start from the same initial condition, they are equal. It has been already noted in [42] that if a function is constant on rectangles, it remains constant on those rectangles under the anisotropic ℓ^1-total variation flow. The relation to the piecewise-constant interpolant of the graph-based total variation flow, however, was not mentioned there. Analysis of the space-discrete total variation flow by means of mode-decomposition methods is presented in the recent preprint [17].

The second main result, Theorem 2.6, states that (again under assumptions that will be specified in due time) on any finite time interval the difference between the (piecewise-linear interpolant of the) graph-based and continuum Allen–Cahn flows converges to zero, uniformly in time, as the mesh size of the discretisation of the one-dimensional torus goes to zero.

We start the paper with an abstract general framework based on evolution variational inequalities [5] for gradient flows. Within this framework we can derive Theorem 2.1 and Corollary 2.1 which, after carefully casting the total variation flow problem in the language of the abstract setting and confirming that the required assumptions are satisfied in Sects. 2.3 and 2.4, immediately lead to Theorem 2.4. Proving our result for the Allen–Cahn flows requires more work, because in this case it is not possible to work with piecewise-constant embeddings of the graph-based functions (since the continuum Allen–Cahn functional is not well-defined for functions with jumps) and the embedding via linear interpolation that we use instead is not an isometry, which means the results from our abstract framework are not directly applicable. Moreover, inequality (2.6) between the graph-based functions Φ_h and continuum function Φ does not hold when Φ_h and Φ are the graph-based and continuum Allen–Cahn functionals, respectively. Since this inequality is one of the requirements for Theorem 2.1, we need to resort to using Theorem 2.2 instead, which requires a weaker condition (2.9) to be satisfied, but at the cost of leading to a weaker conclusion, namely only an asymptotic comparison (when the mesh size approaches zero), rather than an explicit estimate at positive mesh size.

In this paper we prove the convergence of the flows using the evolution variational inequality formulation of the flows directly. An alternative approach is possible based on [21, Theorem 4.11], which requires Γ-convergence of the discrete energies Φ_h to the continuum energy Φ, asymptotic local equicoercivity of the energies Φ_h [21, Definition 4.4], and well-preparedness of the initial conditions. The Γ-convergence results may be obtained in a way similar to the discrete-to-continuum Γ-convergence results in [61]. Alternatively, we see that condition (2.6) or condition (2.9) allows for the construction of recovery sequences in the proof of Γ-convergence, and lower semicontinuity of Φ together with (2.11) implies the limit inferior inequality that is required for Γ-convergence. The (asymptotically local) equicoercivity condition follows from condition (2.5) together with compactness properties of sublevel sets of Φ. We conclude that the a priori alternative approach from [21] corresponds to our approach using Theorem 2.2, or at the very least, can be employed after establishing the same conditions we use. We note that Theorem 2.1 does not require the limit inferior condition, but can only be expected to be useful in those special occasions where the discrete gradient flows are also continuum gradient flows (after embedding)—a situation which is not expected to be typical.

We note that the ideas from [21] (and relatedly [53]) have been generalised to functions that are not geodesically λ-convex (see Sect. 2.2 for a definition) by Sandier and Serfaty and others [8, 31, 52, 56–58]. A general framework provided by Mielke requires convergence of metrics as well as Γ-convergence of functionals to conclude convergence of gradient flows [46, Theorem 3.4.3]. It is assumed that the sets underlying the metric spaces in the sequence are independent of the limit parameter but the metrics can depend on it. We suspect that a suitable modification of such an idea can lead to similar convergence results as ours, but we do not pursue that course in the present paper. For other literature about this energy-dissipation-principle (EDP) convergence, which can be seen as an extension of the work by Sandier and Serfaty, we refer to [22, 47–50]. A key ingredient is the characterisation of gradient flows as minimisers (at value zero) of a De Giorgi functional. For example, in [26, Theorem 3.14] this formulation is used to establish the continuum limit of nonlocal-interaction equations on graphs as gradient flows in a probability space equipped with Wasserstein distance.

As examples of other recent literature that is interested in the convergence of graph-based gradient flows to continuum gradient flows, we mention [35, 36], in which the limit of the graph-based p-Laplacian evolution problem is established.

The paper is structured as follows. The general abstract setting is introduced in Sect. 2.2, and the setup of the graphs that we use, with embeddings and projections into and from the continuum space, is given in Sect. 2.3. The results from this section will be applied to obtain our two main convergence results, which are given in Sects. 2.4 (total variation flow) and 2.5 (Allen–Cahn flow). The main part of the paper ends with conclusions and suggestions for future work in Sect. 2.6, after which follow appendices that give detailed proofs for some properties of total variation for piecewise-constant functions (Appendix A2.7), properties of the operator Γ which is important in Sect. 2.5 (Appendix B2.8), and two Poincaré–Wirtinger-type inequalities that we need in the proof of Lemma 2.7 (Appendix C2.9).

2.2 An Abstract Framework

Let (M, d) be a complete metric space with a distance d. Let γ be a continuous function from an interval I to M, i.e., $\gamma \in C(I, M)$. Following [5] we say that γ is an absolutely continuous curve if there is an integrable function[1] f defined on I such that, for all $t, s \in I$,

$$d(\gamma(t), \gamma(s)) \leq \int_s^t f(r)\, dr.$$

For any pair of metric spaces M and $\tilde{M}$, by $C(M; \tilde{M})$ we will denote the space of continuous functions with domain M and codomain $\tilde{M}$. If additionally M and $\tilde{M}$ are differentiable manifolds and k is a positive integer k, then $C^k(M; \tilde{M})$ will be the space of k times continuously differentiable functions with domain[2] M and codomain $\tilde{M}$. If $\tilde{M} = \mathbb{R}$, we write $C^k(M)$. Moreover, $C^\infty(M; \tilde{M}) = \bigcup_{k \in \mathbb{N}} C^k(M; \tilde{M})$.

A curve $\gamma : (0, \infty) \to M$ is locally absolutely continuous if it is absolutely continuous in any bounded interval $(a, b) \subset (0, \infty)$.

In this paper we use the notation Φ for a functional $\Phi : M \to \mathbb{R} \cup \{+\infty\}$. For $\lambda \in \mathbb{R}$, such a functional is called λ-convex on a curve $\gamma \in C([0, 1], M)$ if

$$\Phi(\gamma(t)) \leq (1-t)\Phi(\gamma(0)) + t\Phi(\gamma(1)) - \frac{1}{2}\lambda t(1-t) d^2(\gamma(0), \gamma(1))$$

for all $t \in [0, 1]$. We note that, if $\lambda_2 \leq \lambda_1$ and Φ is λ_1-convex on γ, Φ is also λ_2-convex on γ.

A curve $\gamma \in C([0, 1], M)$ is a constant-speed geodesic (from $\gamma(0)$ to $\gamma(1)$) if

$$d(\gamma(s), \gamma(t)) = d(\gamma(0), \gamma(1))(t - s) \quad \text{for all} \quad s, t \in [0, 1] \quad \text{with} \quad s \leq t.$$

[1] We note that if such an f exists, it necessarily is nonnegative almost everywhere on I.

[2] In a slight and common abuse of notation, we will sometimes write $u \in C^k(M; \tilde{M})$ if $u : \hat{M} \to \tilde{M}$ with $\hat{M} \supset M$ and $u|_M \in C^k(M; \tilde{M})$.

A functional Φ is said to be geodesically λ-convex if for any $z_0, z_1 \in \mathcal{D}(\Phi)$ there is a constant-speed geodesic $\gamma \in C([0,1], M)$ such that $\gamma(0) = z_0, \gamma(1) = z_1$, and such that Φ is λ-convex on the curve γ. Here $\mathcal{D}(\Phi) := \{z \in M \mid \Phi(z) < \infty\}$ is the effective domain of Φ.

Remark 2.1 In a Hilbert space, geodesic λ-convexity of Φ is nothing but[3] convexity of $\Phi(\cdot) - \frac{\lambda\|\cdot\|^2}{2}$. This follows from the fact that in a (real or complex) Hilbert space for each pair of distinct points a and b, there is a unique constant-speed geodesic from a to b, given by the straight line of the form $\gamma(t) = a + t(b-a)$. It can be checked directly that this is indeed a constant-speed geodesic. To prove uniqueness, let z be a point on a constant-speed geodesic $\tilde{\gamma}$ from $a := \tilde{\gamma}(0)$ to $b := \tilde{\gamma}(1)$, not equal to a or to b. Then, by the triangle inequality, $\|x+y\| \le \|x\| + \|y\|$ for $x := z - a$, and $y := b - z$. By the definition of a constant-speed geodesic, if $t \in (0,1)$ is such that $\tilde{\gamma}(t) = z$, we require

$$\begin{aligned} \|x\| &= \|z-a\| = \|b-a\|t = \|x+y\|t \quad \text{and} \\ \|y\| &= \|b-z\| = \|b-a\|(1-t) = \|x+y\|(1-t), \end{aligned} \tag{2.1}$$

so we require equality in the triangle inequality. In the case of a Hilbert space, this holds if and only if $(\|x\| + \|y\|)^2 - \|x+y\|^2 = 2\left(\|x\|\|y\| - \langle x, y\rangle\right) = 0$, where $\langle x, y\rangle$ denotes the inner product. Hence we require equality in the Cauchy–Schwarz inequality $\langle x, y\rangle \le \|x\|\|y\|$, which is equivalent to x and y being linearly dependent, i.e., in geometric terms, x and y are parallel. Thus there exists a scalar c such that $y = cx$ and thus $b - z = c(z-a)$. We note that $c \neq -1$, since $a \neq b$. It follows that $z = \frac{1}{1+c}b + \frac{c}{1+c}a$. Moreover, by (2.1), $\|x+y\| = |1+c|\|x\| = \frac{1}{t}\|x\|$ and thus $\left|\frac{1}{1+c}\right| = t \in (0,1)$. Hence we can write $\frac{1}{1+c} = te^{\iota\vartheta}$ where $\vartheta := \operatorname{Arg}\left(\frac{1}{1+c}\right)$. We write ι for the imaginary unit. Thus

$$\|\tilde{\gamma}(t) - b\| = \|z - b\| = \left|1 - te^{\iota\vartheta}\right| \|a-b\|.$$

By continuity of $\tilde{\gamma}$, if we take $t \uparrow 1$, we find $\left|1 - te^{\iota\vartheta}\right| \|a-b\| = 0$. Since $a \neq b$, this implies $e^{\iota\vartheta} = 1$. Hence $t = \frac{1}{1+c}$ and thus $\tilde{\gamma}(t) = z = tb + (1-t)a = a + t(b-a)$.

We note that the uniqueness may not be true for a Banach space even if it has finite dimension. A simple example is $\mathbb{R}^2$ equipped with the ℓ^1-norm,[4] so that the unit ball is a square.

If $\lambda = 0$, then a geodesically λ-convex Φ is called geodesically convex; if $\lambda \le 0$, then such Φ is often called geodesically semiconvex.[5]

---

[3] Thus in particular, in a Hilbert space setting geodesic 0-convexity is equivalent to convexity in the sense of vector spaces.

[4] See Sect. 2.3.1.

[5] This terminology is not consistent in the literature: sometimes $\lambda < 0$ is required for semiconvexity, while some sources even allow any $\lambda \in \mathbb{R}$.

Definition 2.1 (*Evolution variational inequality formulation of gradient flow*) Assume that Φ is geodesically λ-convex. A gradient flow of Φ is a locally absolutely continuous curve $u : (0, \infty) \to M$, which satisfies the evolution variational inequality

$$\frac{1}{2}\frac{d}{dt}d^2(u(t), v) + \frac{1}{2}\lambda d^2(u(t), v) \leq \Phi(v) - \Phi(u(t))$$

for almost every $t > 0$ and for all $v \in \mathcal{D}(\Phi)$. Formally we write $\dot{u} \in -\partial\Phi(u)$. If we wish to emphasise the space or metric in which the gradient flow takes place, we speak of an M-gradient flow.

Remark 2.2 Consider the setting of Definition 2.1. For a fixed $v \in \mathcal{D}(\Phi)$, the function $w \mapsto d^2(w, v)$ is Lipschitz continuous on any bounded subset of M. By continuity of u, if $0 < T_1 < T_2$, then $u([T_1, T_2]) \subset M$ is bounded. Since the composition of a Lipschitz-continuous and absolutely continuous function is again absolutely continuous, on any interval $[T_1, T_2]$, $t \mapsto d^2(u(t), v)$ is absolutely continuous and thus its derivative exists for almost every $t \in [T_1, T_2]$. As $T_1 > 0$ is arbitrary, the derivative $\frac{d}{dt}d^2(u(t), v)$ exists for almost every $t \in (0, \infty)$.

Lemma 2.1 *Let Φ be as in Definition 2.1. If $c \in M$ is such that, for all $v \in \mathcal{D}(\Phi)$, $\Phi(v) \geq \Phi(c)$, then the constant curve $u = c$ is a gradient flow of Φ.*

Proof Let $v \in \mathcal{D}(\Phi)$. If γ is a constant-speed geodesic from c to v, then by the geodesic λ-convexity of Φ we have for all $t \in [0, 1]$,

$$\Phi(\gamma(t)) \leq (1-t)\Phi(c) + t\Phi(v) - \frac{1}{2}\lambda t(1-t)d^2(c, v).$$

By assumption Φ has a global minimum at c, hence

$$t\Phi(c) \leq \Phi(\gamma(t)) - \Phi(c) + t\Phi(c) \leq t\Phi(v) - \frac{1}{2}\lambda t(1-t)d^2(c, v).$$

Let $t \neq 0$ and divide by t to find $\Phi(c) \leq \Phi(v) - \frac{1}{2}\lambda(1-t)d^2(c, v)$. Taking the limit $t \downarrow 0$, we conclude that $\Phi(c) \leq \Phi(v) - \frac{1}{2}\lambda d^2(c, v)$ and thus, since $\frac{d}{dt}d^2(c, v) = 0$, $\frac{1}{2}\frac{d}{dt}d^2(c, v) + \frac{1}{2}\lambda d^2(c, v) \leq \Phi(v) - \Phi(c)$. Therefore the constant curve $u = c$ satisfies the inequality from Definition 2.1 and hence is a gradient flow of Φ. $\square$

Lemma 2.2 *Let Φ be as in Definition 2.1 and additionally assume that Φ is lower semicontinuous. Let u be a gradient flow of Φ with $u((0, \infty)) \subset \mathcal{D}(\Phi)$ and continuous on $[0, \infty)$. Then $t \mapsto \Phi(u(t))$ is a non-increasing function on $[0, \infty)$.*

Moreover, if $I \subset [0, \infty)$ is a nondegenerate interval, Φ is as in Definition 2.1 (not necessarily lower semicontinuous) and u is as above in the first part of this lemma, then for almost every $t \in I$, $\Phi(u(t)) = \Phi(u(\inf I)) < +\infty$ if and only if, for all $t \in I$, $u(t) = u(\inf I)$.

Proof First we prove the following claim: If $v \in M$, then $t \mapsto d^2(u(t), v)$ is locally absolutely continuous on $(0, \infty)$. To prove this, let $(a, b) \subset (0, \infty)$, since u is locally absolutely continuous on $(0, \infty)$, if $a < b < \infty$, there exists an integrable function f on (a, b) such that, for all $s, t \in (a, b)$,

$$d(u(t), u(s)) \leq \int_s^t f(r)\, dr.$$

Since u is continuous on $[a, b]$, so is $t \mapsto d(u(t), v)$; hence this function is bounded on (a, b). In particular, there is a $C > 0$ such that, for all $s, t \in (a, b)$, $|d(u(t), v) + d(u(s), v)| \leq C$; hence, for all $s, t \in (a, b)$,

$$\begin{aligned} |d^2(u(t), v) - d^2(u(s), v)| &= |d(u(t), v) + d(u(s), v)||d(u(t), v) - d(u(s), v)| \\ &\leq Cd(u(t), u(s)) \\ &\leq \int_s^t Cf(r)\, dr, \end{aligned}$$

where we used the reverse triangle inequality for the first inequality. Hence $t \mapsto d^2(u(t), v)$ is absolutely continuous on (a, b) and thus locally absolutely continuous on $(0, \infty)$.

Next we prove that $\Phi \circ u$ is non-increasing. Let $t^* \in [0, \infty)$. If $\Phi(u(t^*)) = +\infty$, then it is trivially true that, for all $t \geq t^*$, $\Phi(u(t)) \leq \Phi(u(t^*))$. Now assume that $\Phi(u(t^*)) < +\infty$, so that $u(t^*) \in \mathcal{D}(\Phi)$. Define

$$S := \{t > t^* : \Phi(u(t)) > \Phi(u(t^*))\}.$$

Since Φ is lower semicontinuous and u is (absolutely) continuous, $\Phi \circ u$ is lower semicontinuous; hence S is an open set. For a proof by contradiction, assume that $S \neq \emptyset$, then S is an (at most) countable union of pairwise disjoint open intervals [54, Theorem 2.24]. Since S is bounded below, so is each of these intervals. Let $J \subset S$ be such an interval. Since S is open, $a := \inf S \in [t^*, \infty) \setminus S$, thus $\Phi(u(a)) \leq \Phi(u(t^*))$. In particular, $a \notin J$.

For notational convenience, define $g : J \to \mathbb{R}$ by $g(t) := d^2(u(t), u(a))$. Recall that g is locally absolutely continuous on J.

In the inequality in Definition 2.1, choose $v = u(a)$. Then, for almost every $t \in J$,

$$\frac{1}{2}\left[\frac{d}{dt}g(t) + \lambda g(t)\right] \leq \Phi(u(a)) - \Phi(u(t)) \leq \Phi(u(t^*)) - \Phi(u(t)) < 0.$$

We apply a Grönwall argument: We have, for almost every $t \in J$,

$$\frac{d}{dt}\left(e^{\lambda t} g(t)\right) = e^{\lambda t}\left[\frac{d}{dt}g(t) + \lambda g(t)\right] < 0.$$

Let $t \in J$ and integrate the inequality over (a, t), using $g(a) = 0$, to find

$$e^{\lambda t} g(t) - e^{\lambda a} g(a) = e^{\lambda t} g(t) < 0$$

and hence

$$g(t) < 0,$$

which contradicts $g(t) = d^2(u(t), u(a)) \geq 0$. Therefore $S = \emptyset$ and thus, for all $t \geq t^*$, $\Phi(u(t)) \leq \Phi(u(t^*))$, which proves the first claim of the lemma.

To prove the second claim, let $I \subset [0, \infty)$ be a nondegenerate interval such that, for almost every $t \in I$, $\Phi(u(t)) = \Phi(u(b)) < +\infty$, where $b := \inf I$. The "if" statement follows directly since by assumption, if $t \in I \setminus \{0\}$, then $\Phi(u(t)) < +\infty$. For the "only if" claim, note that, by the inequality in Definition 2.1 with $v = u(b)$, we have for almost every $t \in I$, $h'(t) + \lambda h(t) \leq 0$, where $h(t) := d^2(u(t), u(b))$. By a similar argument as above for g, we find that $h(t) \leq 0$, hence $h(t) = 0$, and thus $u(t) = u(b)$ for all $t \in I$. □

Remark 2.3 In the first part of Lemma 2.2 above, we assume that Φ is lower semicontinuous. In fact, in the proof we only use the fact that $\Phi \circ u$ is lower semicontinuous and thus the conclusions of the lemma remain true under this weaker assumption.

Remark 2.4 Definition 2.1 is a natural extension of a gradient flow in a Hilbert space with an inner product $\langle \cdot, \cdot \rangle$ and corresponding norm $\|\cdot\|$. By [5, Remark 1.1.3], in the Hilbert space setting the derivative $\dot{u}(t)$ of an absolutely continuous curve u exists for a.e. t. Assume t is such that $\dot{u}(t)$ exists. If Φ is convex ($\lambda = 0$), then by definition of the subdifferential of a convex function [24, Chap. I, Sect. 5.1] we see that $\dot{u}(t) \in -\partial\Phi(u(t))$ is equivalent to, for all $v \in \mathcal{D}(\Phi)$,

$$\langle -\dot{u}(t), v - u(t) \rangle \leq \Phi(v) - \Phi(u(t)).$$

The left-hand side equals $\frac{d}{dt}\frac{\|u(t)-v\|^2}{2}$ and thus u satisfies the evolution variational inequality (Definition 2.1) for $\lambda = 0$ if and only if $\dot{u}(t) \in -\partial\Phi(u(t))$.

For $\lambda \neq 0$ we interpret $\dot{u} \in -\partial\Phi(u)$ (for those t for which $\dot{u}(t)$ exists) as

$$\dot{u}(t) \in -\partial\left(\Phi - \lambda\frac{\|\cdot\|^2}{2}\right)(u(t)) - \lambda u(t), \tag{2.2}$$

which is equivalent to, for all $v \in \mathcal{D}(\Phi)$,

$$\langle -\dot{u}(t) - \lambda u(t), v - u(t) \rangle \leq \Phi(v) - \lambda\frac{\|v\|^2}{2} - \left(\Phi(u(t)) - \lambda\frac{\|u(t)\|^2}{2}\right).$$

Since

$$\langle \lambda u(t), u(t) - v \rangle + \lambda\frac{\|v\|^2}{2} - \lambda\frac{\|u(t)\|^2}{2} = \lambda\frac{\|u(t) - v\|^2}{2},$$

the function u satisfies the evolution variational inequality if and only if (2.2) is satisfied.

We recall a basic unique solvability result. The next result is a special case of [5, Theorem 4.0.4], where more general Φ are allowed.

Proposition 2.1 *Assume that $\Phi : M \to \mathbb{R} \cup \{+\infty\}$ is geodesically λ-convex and lower semicontinuous with $\Phi \not\equiv +\infty$. Assume furthermore that Φ is locally bounded below at some point $u_* \in \mathcal{D}(\Phi)$, i.e., there exists an $r_* > 0$ such that*

$$m_* := \inf \{\Phi(v) \mid v \in M \text{ and } d(v, u_*) < r_*\} > -\infty.$$

Then for all $u_0 \in \overline{\mathcal{D}(\Phi)}$ (the closure of $\mathcal{D}(\Phi)$ in the topology of M), there exists a solution $u \in C([0, \infty), M)$ to $\dot{u} \in -\partial\Phi(u)$ satisfying $u(0) = u_0$. Furthermore, for all $t > 0$, $u(t) \in \mathcal{D}(\Phi)$. Moreover, such a solution is unique among all locally absolutely continuous curves u for which $u(t) \to u_0$ as $t \downarrow 0$.

Remark 2.5 Without the assumption of lower semicontinuity of Φ in Proposition 2.1 above, global-in-time existence with $u(t) \in \mathcal{D}(\Phi)$ for all $t > 0$ is not guaranteed, even though a local-in-time solution may exist. Consider for example the function $\Phi : \mathbb{R} \to \mathbb{R} \cup \{+\infty\}$, defined by

$$\Phi(v) := \begin{cases} -v, & \text{if } v < 0, \\ +\infty, & \text{if } v \geq 0, \end{cases}$$

which is convex and locally bounded (at all points in its effective domain), but not lower semicontinuous. From Definition 2.1—with $\lambda = 0$ and the Euclidean distance on $\mathbb{R}$ for d—it follows that, for any $a > 0$, $u(t) = t - a$ satisfies the evolution variational inequality for $t < a$. However, at $t = a$ the curve leaves the effective domain of Φ, i.e., $u(a) \notin \mathcal{D}(\Phi)$. Moreover, any function $\tilde{u}$ that satisfies the evolution variational inequality locally in time must be of the form $\tilde{u}(t) = t - b$, for some $b > 0$. Indeed, if $\Phi(\tilde{u}(t)) = +\infty$, the evolution variational inequality does not hold, thus $\tilde{u}(t) < 0$ has to hold for almost every $t > 0$ and thus, by continuity, for all $t > 0$. By Remark 2.4, the evolution variational inequality in this case is equivalent to the differential inclusion $\dot{\tilde{u}}(t) \in \partial\Phi(\tilde{u}(t)) = \{1\}$, where the equality follows from differentiability of Φ on $(-\infty, 0)$ with $\Phi'(v) = -1$ [24, Proposition 5.3]. Thus $\tilde{u}(t) = t - b$, as claimed and no global-in-time solutions exist. On the other hand, if we define $\Phi(0) := 0$ instead, then Φ is lower semicontinuous and indeed the function $u(t) = \min(t - a, 0)$ is a global solution.

Definition 2.2 If u is a gradient flow of Φ and $u(t) \to u_0 \in \overline{\mathcal{D}(\Phi)}$ as $t \downarrow 0$, then u is called a gradient flow of Φ starting from u_0.

If the assumptions from Proposition 2.1 are satisfied, uniqueness allows us to call u *the* gradient flow of Φ starting from u_0.

In the setting of Proposition 2.1, $\Phi(u(t)) < +\infty$ for all $t > 0$. Hence the evolution variational inequality from Definition 2.1, which was assumed to hold for all $v \in \mathcal{D}(\Phi)$, trivially also holds for all $v \in M \setminus \mathcal{D}(\Phi)$.

The results when M is a Hilbert space go back to the theory of maximal monotone operators [9] pioneered by Kōmura [41], where the first existence result has been established.

If M is a normed vector space, in particular if M is a Banach space, the assumption $m_* > -\infty$ for some u_* and r_* is automatically fulfilled. Indeed, by the Hahn–Banach theorem, any lower semicontinuous convex function that is not identically equal to $+\infty$ is bounded below by an affine continuous function [10, Proposition 1.10] and thus in particular locally bounded below. Since $\Phi - \lambda\|\cdot\|^2/2$ is convex for λ-convex functions Φ, such Φ are locally bounded from below if they are lower semicontinuous and not identically equal to $+\infty$. In specific instances, still we may note local boundedness below explicitly to emphasise this condition is satisfied.

Returning to the case of a general complete metric space M, to observe the uniqueness part of Proposition 2.1, let u and v be two solutions and let $s, t \in (0, \infty)$ be two "times" at which the evolution variational inequality holds, then

$$\frac{1}{2}\frac{d}{dt}d^2(v(t), u(s)) + \frac{1}{2}\lambda d^2(v(t), u(s)) \leq \Phi(u(s)) - \Phi(v(t))$$

and

$$\frac{1}{2}\frac{d}{dt}d^2(u(t), v(s)) + \frac{1}{2}\lambda d^2(u(t), v(s)) \leq \Phi(v(s)) - \Phi(u(t)).$$

Adding both sides and observing that

$$\frac{d}{dt}d^2(u(t), v(t)) = \frac{d}{dt}\, d^2(u(t), v(s))\big|_{s=t} + \frac{d}{dt}\, d^2(v(t), u(s))\big|_{s=t},$$

we take $s = t$ to get

$$\frac{1}{2}\frac{d}{dt}d^2(u(t), v(t)) + \lambda d^2(u(t), v(t)) \leq 0.$$

Via an integration, this implies a generalised contraction principle: for all $t \geq 0$,

$$d^2(u(t), v(t)) \leq e^{-2\lambda t} d^2(u(0), v(0)). \tag{2.3}$$

If the initial data $u(0)$ and $v(0)$ are the same, then we conclude that $u \equiv v$. This idea is useful to define our convergence result.

In what follows, we consider a family of complete metric spaces $(M_h, d_h)_{h \in \mathfrak{H}}$ approximating (M, d) as $h \to 0$ in some sense which is to be made precise. Here $\mathfrak{H}$ is a nonempty subset of $(0, 1)$ of which 0 is a limit point. In particular, we allow $\mathfrak{H}$ to be a proper subset of $(0, 1)$. In fact, from Sect. 2.3 onwards we only consider $h \in (0, 1)$ for which $\frac{1}{h}$ is an integer.

Let $i_h : M_h \to M$ be an embedding map which is an isometry, i.e., $d_h(v_1, v_2) = d(i_h v_1, i_h v_2)$, for all $v_1, v_2 \in M_h$. In particular, i_h is continuous. Let $p_h : M \to M_h$ be a mapping such that $p_h \circ i_h$ is the identity map on M_h. This is (in spirit[6]) a projection, in the sense that $p_h \circ i_h \circ p_h = p_h$ on M. We assume the following non-strict contraction property (i.e., 1-Lipschitz continuity) holds:

$$d_h(p_h w_1, p_h w_2) \leq d(w_1, w_2) \qquad \text{for all} \quad w_1, w_2 \in M. \tag{2.4}$$

In particular, p_h is continuous.

We want to consider functionals $\Phi : M \to \mathbb{R} \cup \{+\infty\}$ and, for all $h \in \mathfrak{H}$, $\Phi_h : M_h \to \mathbb{R} \cup \{+\infty\}$, for which gradient flows exist. Hence we will require that these functionals satisfy the conditions of Proposition 2.1, i.e., they are geodesically λ-convex, lower semicontinuous, not identically equal to $+\infty$, and locally bounded below at some point in their respective domains. If, for all $h \in \mathfrak{H}$, the functionals Φ_h and Φ satisfy these conditions with the same λ, we will say Φ_h and Φ satisfy condition **(λ-AGS)**, in deference to [5, Theorem 4.0.4]. We note that we do not require the points in the effective domain at which local lower boundedness is satisfied to be the same for all functionals, nor the values of the lower bounds or the radii of the local neighbourhoods on which the bounds hold.

Additionally, we sometimes assume any or all of the following. When we do so, we will explicitly state this:

$$\Phi(i_h v) \leq \Phi_h(v) \qquad \text{for all } h \in \mathfrak{H} \text{ and all } v \in M_h \tag{2.5}$$

$$\Phi_h(p_h w) \leq \Phi(w) \qquad \text{for all } h \in \mathfrak{H} \text{ and all } w \in M. \tag{2.6}$$

$$d_h^2(v, p_h w) + d^2(i_h p_h w, w) = d^2(i_h v, w) \qquad \text{for all } h \in \mathfrak{H}, \text{ all } v \in M_h, \text{ and all } w \in M. \tag{2.7}$$

The last assumption is similar to the Pythagorean theorem and thus says that as a projection p_h is "orthogonal", in a sense. A typical example where this Pythagorean theorem holds is the case where M and all M_h are inner product spaces with inner products $\langle \cdot, \cdot \rangle$ and $\langle \cdot, \cdot \rangle_h$, respectively, which satisfy $\langle i_h u, i_h v \rangle = \langle u, v \rangle_h$ and

$$\langle i_h p_h w - w, i_h u \rangle = 0, \tag{2.8}$$

for all $h \in \mathfrak{H}$, all $w \in M$, and all $u, v \in M_h$.

Inner product structures are not necessary for (2.7) to hold, as the following example shows. If (M, d) is a complete metric space such that d cannot be generated by an inner product and, for all $h \in \mathfrak{H}$, $(M_h, d_h) = (M, d)$, then we can define, for all $h \in \mathfrak{H}$, both i_h and p_h to be equal to the identity operator on M. In that case i_h

[6] We will call p_h a projection, even though $p_h \circ p_h$ is not necessarily well-defined, let alone equal to p_h.

is an isometry, p_h satisfies (2.4), and $p_h \circ i_h$ is the identity operator. Moreover, (2.7) holds, since, for all $h \in \mathfrak{H}$ and for all $v, w \in M_h = M$,

$$d_h^2(v, p_h w) + d^2(i_h p_h w, w) = d^2(v, w) + d^2(w, w) = d^2(v, w) = d^2(i_h v, w).$$

Condition (2.5) implies that $i_h(\mathcal{D}(\Phi_h)) \subset \mathcal{D}(\Phi)$. Similarly, condition (2.6) implies that $p_h(\mathcal{D}(\Phi) \subset \mathcal{D}(\Phi_h)$.

Since $p_h \circ i_h$ is the identity map on M_h, we note that (2.5) and (2.6) together imply that, for all $v \in M_h$,

$$\Phi(i_h v) \le \Phi_h(v) = \Phi_h(p_h i_h v) \le \Phi(i_h v)$$

and thus $\Phi(i_h v) = \Phi_h(v)$.

On the other hand, if $\Phi(i_h v) = \Phi_h(v)$ holds for all $v \in M_h$, this is not sufficient to imply (2.6), as the latter condition has to hold for all[7] $w \in M$ and while we can conclude that $\Phi_h(p_h i_h v) = \Phi(i_h v)$ for all $v \in M_h$, i_h is typically not surjective.

Theorem 2.1 *Assume there exists a $\lambda \le 0$ such that Φ_h and Φ satisfy condition (λ-AGS). Assume that $\overline{\mathcal{D}(\Phi_h)} = M_h$. Let u^h be the gradient flow of Φ_h starting from $u_0^h \in M_h$. Assume moreover* (2.5), (2.6), *and* (2.7). *Then $i_h u^h$ is the gradient flow of Φ starting from $i_h u_0^h$.*

Corollary 2.1 *Assume there exists a $\lambda \le 0$ such that Φ_h, M_h, Φ, M, and u^h satisfy all the hypotheses from Theorem 2.1. Let u be the gradient flow of Φ starting from $u_0 \in M = \overline{\mathcal{D}(\Phi)}$. Then*

$$d^2(i_h u^h(t), u(t)) \le e^{-2\lambda t} d^2(i_h u_0^h, u_0), \qquad \text{for all} \quad t > 0.$$

If $i_h u_0^h \to u_0$ as $h \to 0$, this yields, for all $T > 0$, uniform convergence of $i_h u^h$ to u on $(0, T)$ as $h \to 0$. For example if $i_h p_h u_0 \to u_0$ in M as $h \to 0$ and $u_0^h = p_h u_0$, then $i_h u^h$ converges to u in M uniformly on $(0, T)$ as $h \to 0$, for any $T > 0$. Corollary 2.1 follows from Theorem 2.1 by the generalised contraction principle (2.3).

Proof (of Theorem 2.1) By definition our assumptions (2.5) and (2.6) yield

$$\begin{aligned}\frac{1}{2}\frac{d}{dt} d_h^2(u^h, p_h w) + \frac{\lambda}{2} d_h^2(u^h, p_h w) &\le \Phi_h(p_h w) - \Phi_h(u^h)\\ &\le \Phi(w) - \Phi(i_h u^h) \qquad \text{for all} \quad w \in M.\end{aligned}$$

By (2.7) we see

$$\frac{d}{dt} d_h^2(u^h, p_h w) = \frac{d}{dt} d^2(i_h u^h, w),$$

[7] This is related to the requirement in Definition 2.1 that the evolution variational inequality holds for all $v \in \mathcal{D}(\Phi)$.

since $d^2(i_h p_h w, w)$ does not depend on t. We thus conclude that

$$\frac{1}{2}\frac{d}{dt}d^2(i_h u^h, w) + \frac{\lambda}{2}d^2(i_h u^h, w) \leq \Phi(w) - \Phi(i_h u^h), \qquad \text{for all} \quad w \in M,$$

since $\lambda \leq 0$ and $d_h(u^h, p_h w) = d_h(p_h i_h u^h, p_h w) \leq d(i_h u^h, w)$ by the non-strict contraction property (2.4). This says that $i_h u^h$ is a gradient flow of Φ. Since $u^h(t) \to u_0^h$ as $t \downarrow 0$ and since i_h is continuous, we have $i_h u^h(t) \to i_h u_0^h$ as $t \downarrow 0$. Thus $i_h u^h$ is a gradient flow starting at u_0^h. □

The next theorem gives convergence under a weaker assumption. Instead of (2.6) we consider the weaker assumption

$$\overline{\lim_{h\to 0}}\, \Phi_h(p_h w) \leq \Phi(w) \qquad \text{for all} \quad w \in M \tag{2.9}$$

and the bound in (2.10). Instead of (2.5) we assume (2.11).

Theorem 2.2 *Assume there exists a $\lambda \leq 0$ such that Φ_h and Φ satisfy condition (λ-AGS) and let conditions* (2.7) *and* (2.9) *be satisfied. Let u^h be the gradient flow of Φ_h starting from $u_0^h \in M_h = \overline{\mathcal{D}(\Phi_h)}$ and let u be the gradient flow of Φ starting from $u_0 \in M = \overline{\mathcal{D}(\Phi)}$. Assume further that there exists a $T \in (0, \infty]$, a $\delta > 0$, and a nonnegative function $\Psi : M \to \mathbb{R} \cup \{+\infty\}$ such that the function $t \mapsto e^{2\lambda t}\Psi(u(t))$ is in $L^1(0, T)$*[8] *and, for all $w \in M$ and for all $h \in (0, \delta) \cap \mathfrak{H}$,*

$$\Phi_h(p_h w) \leq \Psi(w). \tag{2.10}$$

Moreover, let there be a function $\varepsilon : (0, 1) \to (0, \infty)$ such that $\varepsilon(h) \to 0$ as $h \to 0$ and such that, for all $t \in [0, T]$ and for all $h \in (0, \delta) \cap \mathfrak{H}$,

$$\Phi\left(i_h u^h(t)\right) \leq \Phi_h\left(u^h(t)\right) + \varepsilon(h). \tag{2.11}$$

If $i_h u_0^h \to u_0$ and

- *$T \in (0, \infty)$ or*
- *$T = \infty$, $\lambda < 0$, and $\underline{\lim}_{t\to\infty} \Phi(u(t)) > -\infty$,*

then

$$\lim_{h\to 0} \sup_{t\in[0,T]} e^{2\lambda t} d(i_h u^h(t), u(t)) = 0.$$

If $T = \infty$, we interpret $[0, T]$ as $[0, \infty)$.

---

[8] By $L^1(0, T)$ we denote the set of Lebesgue integrable functions on $(0, T)$.

Proof As in the proof of uniqueness, leading up to (2.3), we observe that

$$\frac{1}{2}\frac{d}{dt}d_h^2(u^h(t), p_h u(s)) + \frac{\lambda}{2}d_h^2(u^h(t), p_h u(s)) \le \Phi_h(p_h u(s)) - \Phi_h(u^h(t)), \tag{2.12}$$

$$\frac{1}{2}\frac{d}{dt}d^2(u(t), i_h u^h(s)) + \frac{\lambda}{2}d^2(u(t), i_h u^h(s)) \le \Phi(i_h u^h(s)) - \Phi(u(t)), \tag{2.13}$$

for all s and t at which the evolution variational inequality is satisfied. By the chain rule and (2.7), we see that

$$\begin{aligned}\frac{d}{dt}d^2(u(t), i_h u^h(t)) &= \frac{d}{dt}\, d^2(u(t), i_h u^h(s))\big|_{s=t} + \frac{d}{dt}\, d^2(i_h u^h(t), u(s))\big|_{s=t} \\ &= \frac{d}{dt}\, d^2(u(t), i_h u^h(s))\big|_{s=t} + \frac{d}{dt}\, d_h^2(u^h(t), p_h u(s))\big|_{s=t}.\end{aligned}$$

Adding (2.12) and (2.13) we now observe that

$$\begin{aligned}&\frac{1}{2}\frac{d}{dt}d^2(u(t), i_h u^h(t)) + \lambda d^2(u(t), i_h u^h(t)) \\ &\le \frac{1}{2}\frac{d}{dt}d^2(u(t), i_h u^h(t)) + \frac{\lambda}{2}d^2(u(t), i_h u^h(t)) + \frac{\lambda}{2}d_h^2(u^h(t), p_h u(t)) \\ &\le \Phi(i_h u^h(t)) - \Phi_h(u^h(t)) + \Phi_h(p_h u(t)) - \Phi(u(t)) \\ &\le \varepsilon(h) + \Phi_h\left(p_h u(t)\right) - \Phi\left(u(t)\right),\end{aligned}$$

where for the first inequality we used that, by the non-strict contraction property (2.4), $d_h^2(u^h(t), p_h u(t)) = d_h^2(p_h i_h u^h(t), p_h u(t)) \le d^2(i_h u^h(t), p_h u(t))$ and that $\lambda \le 0$. The final inequality follows from (2.11).

For any real-valued function f that is differentiable at t, we have

$$e^{-2\lambda t}\frac{d}{dt}\left(e^{2\lambda t} f(t)\right) = \frac{df}{dt}(t) + 2\lambda f(t).$$

Using this identity for $f(t) := \frac{d}{dt}d^2(u(t), i_h u^h(t))$, we multiply the inequality above by $2e^{2\lambda t}$ and integrate both sides of $[0, t]$, for a fixed $t \in [0, T]$, to get

$$\begin{aligned}e^{2\lambda t} d^2\left(u(t), i_h u^h(t)\right) &\le d^2(u_0, i_h u_0^h) \\ &\quad + 2\int_0^t e^{2\lambda\tau}\left[\Phi_h\left(p_h u(\tau)\right) - \Phi\left(u(\tau)\right) + \varepsilon(h)\right] d\tau.\end{aligned} \tag{2.14}$$

Next we claim that if a family of measurable functions f_h on $[0, T)$ satisfies, for almost every $t \in (0, T)$, $\overline{\lim}_{h\to 0} f_h(t) \le 0$ and there exists a $g \in L^1(0, T)$ with $g \ge 0$

such that, for all h, $f_h \le g$ on $(0, T)$, then

$$\varlimsup_{h\to 0} \sup_{0<t<T} \int_0^t f_h(\tau)\, d\tau \le 0. \tag{2.15}$$

We emphasise that the case $T = \infty$ is allowed. To prove (2.15), first we note that, for all $\eta > 0$, there exists a $t_{h,\eta} \in (0, T)$ such that

$$\sup_{0<t<T} \int_0^t f_h(\tau)\, d\tau \le \int_0^{t_{h,\eta}} f_h(\tau)\, d\tau + \eta.$$

For all h and for all $\eta > 0$, we set

$$\bar{f}_h(\tau) := \begin{cases} f_h(\tau) & \text{if } 0 \le \tau \le t_{h,\eta}, \\ 0 & \text{if } t_{h,\eta} < \tau \le T. \end{cases}$$

Since $\bar{f}_h \le g$, applying Fatou's lemma [28, Sect. 1.3, Theorem 1] for $g - \bar{f}_h \ge 0$ yields

$$\int_0^T \varliminf_{h\to 0} (g - \bar{f}_h)(\tau)\, d\tau \le \varliminf_{h\to 0} \int_0^T (g - \bar{f}_h)(\tau)\, d\tau.$$

Since $g \in L^1(0, T)$, this implies

$$\int_0^T \varliminf_{h\to 0} \left(-\bar{f}_h(\tau)\right) d\tau \le \varliminf_{h\to 0} \left(-\int_0^T \bar{f}_h(\tau)\, d\tau\right)$$

and thus

$$\varlimsup_{h\to 0} \int_0^T \bar{f}_h(\tau)\, d\tau \le \int_0^T \varlimsup_{h\to 0} \bar{f}_h(\tau)\, d\tau.$$

We observe that, for almost all $\tau \in (0, T)$, $\varlimsup_{h\to 0} \bar{f}_h(\tau) \le 0$ and hence

$$\varlimsup_{h\to 0} \int_0^{t_{h,\eta}} f_h(\tau)\, d\tau = \varlimsup_{h\to 0} \int_0^T \bar{f}_h(\tau)\, d\tau \le 0.$$

Thus we obtain that

$$\varlimsup_{h\to 0} \sup_{0<t<T} \int_0^t f_h(\tau)\, d\tau \le 0 + \eta.$$

Sending η to zero, we establish our claim in (2.15).

Writing

$$f_h(\tau) := e^{2\lambda\tau} \left[\Phi_h\left(p_h u(\tau)\right) - \Phi\left(u(\tau)\right) + \varepsilon(h)\right]$$

and taking the supremum over $(0, T)$ on both sides of inequality (2.14), we get

$$\sup_{0<t<T} e^{2\lambda t} d^2\left(u(t), i_h u^h(t)\right) \leq d^2(u_0, i_h u_0^h) + 2 \sup_{0<t<T} \int_0^t f_h(\tau)\, d\tau. \tag{2.16}$$

We define

$$\tilde{g}(\tau) := e^{2\lambda\tau} \left[\Psi\left(u(\tau)\right) - \Phi\left(u(T)\right) + \varepsilon(h)\right] \quad \text{and} \quad g(\tau) := \left(\tilde{g}(\tau)\right)^+,$$

where, for $x \in \mathbb{R}$, $x^+ := \max(0, x)$ denotes the positive part of x. In the case that $T = \infty$, we set $\Phi\left(u(\infty)\right) := \underline{\lim}_{t\to\infty} \Phi\left(u(t)\right) > -\infty$.

By construction $g \geq 0$. By (2.10) we have that, for all $h \in (0, \delta) \cap \mathfrak{H}$ and all $\tau \in [0, T]$, $\Phi_h\left(p_h u(\tau)\right) \leq \Psi\left(u(\tau)\right)$. Furthermore, by Lemma 2.2, the function $t \mapsto \Phi\left(u(t)\right)$ is non-increasing on $[0, \infty)$. Therefore the functions f_h satisfy, for all $h \in (0, \delta) \cap \mathfrak{H}$, $f_h \leq \tilde{g} \leq g$ on $(0, T)$.

Moreover, by assumption the function $t \mapsto e^{2\lambda t}\Psi\left(u(t)\right)$ is in $L^1(0, T)$ and hence, if $T < \infty$, then $\tilde{g} \in L^1(0, T)$ and thus $g \in L^1(0, T)$. In the case where $T = \infty$, we also have assumed that $\lambda < 0$ and $\Phi\left(u(\infty)\right) > -\infty$. Hence it follows that $e^{2\lambda t}\left[-\Phi\left(u(\infty)\right) + \varepsilon(h)\right]^+$ is integrable on $(0, \infty)$. Since $g(\tau) \leq e^{2\lambda t}\left\{\left[\Psi\left(u(\tau)\right)\right]^+ + \left[-\Phi\left(u(\infty)\right) + \varepsilon(h)\right]^+\right\}$, we again deduce that $g \in L^1(0, T)$.

By Proposition 2.1 we know that, for all $t > 0, u(t) \in M$ and therefore, by assumption (2.9) and $\lim_{h\to 0} \varepsilon(h) = 0$, we have that, for all $t \in (0, T)$, $\overline{\lim}_{h\to 0} f_h(t) \leq 0$. Thus all the required assumptions for (2.15) are satisfied. Applying (2.15) to the inequality in (2.16), we obtain

$$\overline{\lim_{h\to 0}} \sup_{0<t<T} e^{2\lambda t} d^2\left(u(t), i_h u^h(t)\right) \leq \overline{\lim_{h\to 0}}\, d^2(u_0, i_h u_0^h) + 0.$$

Since $\underline{\lim}_{h\to 0} \sup_{0<t<T} d^2\left(u(t), i_h u^h(t)\right) \geq 0$, the result follows. □

Remark 2.6 If $T < \infty$ in Theorem 2.2, we may replace the assumption of integrability of the function $t \mapsto e^{2\lambda t}\Psi(u(t))$ on $(0, T)$ by an assumption of integrability of $t \mapsto \Psi(u(t))$ on $(0, T)$, since the latter implies the former. Moreover, in that case the conclusion of the theorem,

$$\lim_{h\to 0} \sup_{t\in[0,T]} e^{2\lambda t} d(i_h u^h(t), u(t)) = 0,$$

implies that

$$\lim_{h\to 0} \sup_{t\in[0,T]} d(i_h u^h(t), u(t)) = 0,$$

since, by nonnegativity of λ,

$$\sup_{t\in[0,T]} e^{2\lambda t} d(i_h u^h(t), u(t)) \leq \sup_{t\in[0,T]} d(i_h u^h(t), u(t))$$

$$\leq e^{-2\lambda T} \sup_{t\in[0,T]} e^{2\lambda t} d(i_h u^h(t), u(t)).$$

We use Theorem 2.1 and Corollary 2.1 to prove convergence of discrete total variation flows to a continuum total variation flow in Theorem 2.4. These results are not applicable in our proof of convergence of discrete Allen–Cahn flows to a continuum Allen–Cahn flow in Theorem 2.6, as explained in detail in Sect. 2.5.4. Instead, we first approximate the continuum flow by different discrete flows and use Theorem 2.2 to prove convergence of those new flows to the continuum flow in Theorem 2.7, after which Theorem 2.8 establishes that the new flows approximate the discrete flows, resulting in Theorem 2.6.

2.3 Setting of the Problem

We consider an anisotropic total variation flow and an Allen–Cahn flow on "cubic" graphs and their convergence to their continuum versions when the mesh size of the graphs tends to zero. In this section we define the graphs, the corresponding metric spaces M_h and their continuum counterpart M, the embeddings i_h and "projections" p_h, and the functionals which generate the gradient flows.

2.3.1 *The Graphs*

We construct a graph G_h by discretising a flat torus $\mathbb{T}^n := (\mathbb{R}/\mathbb{Z})^n$ with mesh size $h \in (0, 1)$. In the remainder of this paper, we assume that h is such that $1/h$ is an integer, i.e.,

$$\mathfrak{H} := \left\{ h \in (0,1) \;\middle|\; \exists k \in \mathbb{N}\; h = \frac{1}{k} \right\}.$$

For any $n \in \mathbb{N}$, we denote the Euclidean norm on $\mathbb{R}^n$ by $|\cdot|$. We use the same notation for the induced Euclidean norms on $\mathbb{T}^n$.

Define the graphs $G_h = (V_h, E_h)$ via their node sets V_h and edge sets E_h:

$$\begin{aligned} V_h &:= \left(h(\mathbb{Z}/h^{-1}\mathbb{Z})\right)^n = \{0, h, 2h, \ldots, 1-h\}^n, \\ E_h &:= \left\{ (z, \tilde{z}) \in V_h \times V_h \;\middle|\; |z - \tilde{z}|_{\ell^1} = h \right\} \\ &\quad \cup \left\{ (z, \tilde{z}) \in V_h \times V_h \;\middle|\; \exists j\; |z_j - \tilde{z}_j| = 1 - h \text{ and } \forall i \neq j\; z_i = \tilde{z}_i \right\}. \end{aligned}$$

Here $|x|_{\ell^1} := \sum_{i=1}^n |x_i|$. We parametrise the torus $\mathbb{T}^n$ by the hypercube $\Omega_h := [-h/2, 1-h/2)^n$ with periodic boundary conditions (see Fig. 2.1 for the case with $n = 2$). Then the first set in E_h contains the edges within $\overline{\Omega_h}$ and the second set the

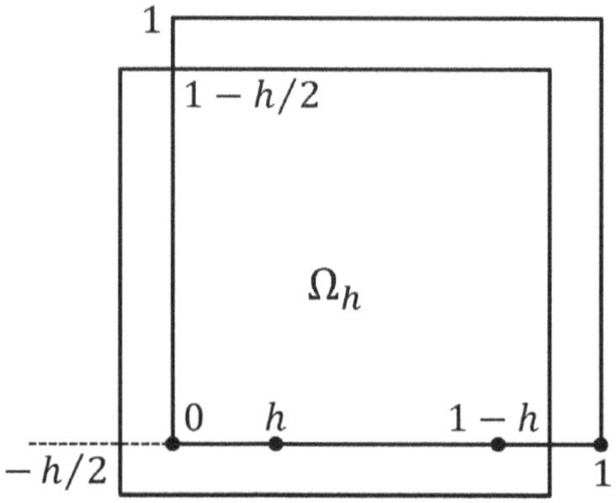

Fig. 2.1 The square $\Omega_h := [-h/2, 1-h/2)^2$ with periodic boundary conditions parametrises the torus $\mathbb{T}^2$. The black circles indicate some of the nodes in the corresponding node set V_h, where the node at $(0, 1)$ is identified with the one at $(0, 0)$

edges that "go over" the boundary of Ω_h and capture the periodicity on $\mathbb{T}^n$. We note that $(z, \tilde{z}) \in E_h$ if and only if $(\tilde{z}, z) \in E_h$. We identify these edges with each other so that the graphs are undirected. When it is clear from the context what h is, we may also write $z \sim \tilde{z}$ to indicate $(z, \tilde{z}) \in E_h$.

We note that elsewhere in the literature (e.g., in [61] and some of the references therein) the graph G_h is viewed as an edge-weighted graph, in which a weight h^{n-1} is assigned to each edge (and to each pair of unconnected nodes a weight of zero). In this paper, we have chosen to directly incorporate the dependence on the mesh size into the definitions of our inner products and functionals which we give below, e.g., the factors h^{n-1} and h^n in $\|\nabla u\|_1$ and $\langle \cdot, \cdot \rangle_h$, respectively (see Remark 2.9).

Using the notation $[n] := \{1, \ldots, n\}$, for $z \in V_h$ we define

$$Q_z^h := \left\{ hy + z \in \mathbb{R}^n \;\middle|\; \forall i \in [n] \; -1/2 \le y_i < 1/2 \right\},$$

which is a "half-open" n-cube centred at the vertex z with edges of length h. With

$$I_{z_i}^h := [z_i - h/2, z_i + h/2), \tag{2.17}$$

we have $Q_z^h = \prod_{i=1}^n I_{z_i}^h := I_{z_1}^h \times \ldots \times I_{z_n}^h$. We note that, if f is integrable on $\mathbb{T}^n$, then

$$\int_{\mathbb{T}^n} f(x)\, dx = \sum_{z \in V_h} \int_{Q_z^h} f(x)\, dx.$$

2.3.2 *Function Spaces, Embedding, and Projection*

Let $h \in \mathfrak{H}$. We define the L^2 inner product of functions $v_1, v_2 : V_h \to \mathbb{R}$ as

$$\langle v_1, v_2 \rangle_h := \sum_{z \in V_h} v_1(z) v_2(z) h^n.$$

We denote the inner product space of all functions $v : V_h \to \mathbb{R}$ equipped with $\langle\cdot,\cdot\rangle_h$ by L_h^2. If we want to consider the set of real-valued vertex functions $v : V_h \to \mathbb{R}$ without assuming a priori it is equipped with a specific inner product, we denote this by $\mathcal{V}_h$. For functions $v \in \mathcal{V}_h$ we also define the norm $\|v\|_\infty := \max_{z\in V} |v(z)|$.

Since L_h^2 is a finite (n-)dimensional inner product space, it is complete (and thus a Hilbert space) and the corresponding norm $\|v\|_h := \sqrt{\langle v, v\rangle_h}$ is finite for all $v \in L_h^2$. In Sect. 2.4, L_h^2 will play the role that M_h played in Sect. 2.2. In Sect. 2.5 we will introduce a different inner product on $\mathcal{V}_h$ and the resulting Hilbert space $\overline{L}_h^2$ will play the role of M_h in that section instead (see Sect. 2.5.4).

The role of M will be played by the Lebesgue space $L^2(\mathbb{T}^n)$ consisting of all functions $w : \mathbb{T}^n \to \mathbb{R}$ which have finite norm $\|w\|_{L^2(\mathbb{T}^n)} := \sqrt{\langle w, w\rangle_{L^2(\mathbb{T}^n)}}$, where

$$\langle w_1, w_2\rangle_{L^2(\mathbb{T}^n)} := \int_{\mathbb{T}^n} w_1(x)w_2(x)\,dx = \int_{\Omega_h} w_1(x)w_2(x)\,dx,$$

for functions $w_1, w_2 \in L^2(\mathbb{T}^n)$. We emphasise that the value of the integral on the right does not depend on the choice of h. For later use, for $w : \mathbb{T}^n \to \mathbb{R}$ and $p \in \mathbb{N}$ we also define the norms $\|w\|_{L^p(\mathbb{T}^n)} := \left(\int_{\mathbb{T}^n} |w(x)|^p\,dx\right)^{1/p}$ and $\|w\|_{L^\infty(\mathbb{T}^n)} := \operatorname{esssup}_{x\in\mathbb{T}^n} |w(x)|$.

We define the embedding operator $i_h : L_h^2 \to L^2(\mathbb{T}^n)$. If $v \in L_h^2$ and $x \in \mathbb{T}^n$, we set

$$i_h v(x) := v(z),$$

where $z \in V_h$ is such that $x \in Q_z^h$. We note that, by our identification of $\mathbb{T}^n$ with Ω_h (with periodic boundary conditions) for each x, z is uniquely determined.

Next we define the projection operator $p_h : L^2(\mathbb{T}^n) \to L_h^2$. If $w \in L^2(\mathbb{T}^n)$ and $z \in V_h$, then

$$p_h w(z) := \frac{1}{|Q_z^h|}\int_{Q_z^h} w(x)\,dx = h^{-n}\int_{Q_z^h} w(x)\,dx,$$

where $|\cdot|$ denotes the Lebesgue measure and the second equality follows since $|Q_z^h| = h^n$, for all $z \in V_h$.

We end this subsection with some useful properties of i_h and p_h. First we introduce some new notation which we will require in the proof of Theorem 2.3 (viii) and at later points in this paper. For $i \in [n]$, we define

$$\tilde{Q}_{z,i}^h := \prod_{j\in[n]\setminus\{i\}} I_{z_j}^h, \tag{2.18}$$

where again $\prod$ denotes the Cartesian product of sets. We refer to (2.17) for the definition of $I_{z_j}^h$.

Theorem 2.3 *Let $h \in \mathfrak{H}$.*

(i) *If $v_1, v_2 \in L^2_h$, then $\langle i_h v_1, i_h v_2\rangle_{L^2(\mathbb{T}^n)} = \langle v_1, v_2\rangle_h$. In particular, it follows that $i_h(L^2_h) \subset L^2(\mathbb{T}^n)$ as required, and that i_h is an isometry.*
(ii) *The composition $p_h \circ i_h$ is the identity on L^2_h.*
(iii) *If $w \in L^2(\mathbb{T}^n)$ and $x \in Q^h_z$, we have*

$$i_h p_h w(x) = h^{-n} \int_{Q^h_z} w(y)\, dy.$$

(iv) *If $w_1, w_2 \in L^2(\mathbb{T}^n)$, then*

$$\langle i_h p_h w_1, w_2\rangle_{L^2(\mathbb{T}^n)} = \langle p_h w_1, p_h w_2\rangle_h = \langle w_1, i_h p_h w_2\rangle_{L^2(\mathbb{T}^n)}.$$

In particular the operator $i_h \circ p_h$ is self-adjoint on $L^2(\mathbb{T}^n)$.
(v) *Condition* (2.8) *is satisfied and thus the Pythagorean equality* (2.7) *holds.*
(vi) *If $f \in C(\mathbb{T}^n)$, then $i_h p_h f \to f$ uniformly, as $h \to 0$.*
(vii) *For all $w \in L^2(\mathbb{T}^n)$, $i_h p_h w \to w$ in $L^2(\mathbb{T}^n)$ as $h \to 0$.*
(viii) *For all $w \in L^2(\mathbb{T}^n)$ and for all $z \in V_h$,*

$$\int_{Q^h_z} (i_h p_h w)(x)\, dx = \int_{Q^h_z} w(x)\, dx \quad \textit{and}$$
$$\int_{Q^h_z} |i_h p_h w|(x)\, dx \le \int_{Q^h_z} |w|(x)\, dx.$$

Proof (i) A direct computation shows

$$\langle i_h v_1, i_h v_2\rangle_{L^2(\mathbb{T}^n)} = \sum_{z\in V_h} \int_{Q^h_z} i_h v_1(x) i_h v_2(x)\, dx = \sum_{z\in V_h} v_1(z) v_2(z) h^n = \langle v_1, v_2\rangle_h.$$

(ii) Let $v \in L^2_h$ and $z \in V_h$, then

$$p_h i_h v(z) = h^{-n} \int_{Q^h_z} i_h v(x)\, dx = h^{-n} \int_{Q^h_z} v(z)\, dx = v(z).$$

(iii) This follows directly from $i_h p_h w(x) = p_h w(z)$ and the definition of p_h.
(iv) We compute

$$\begin{aligned}\langle i_h p_h w_1, w_2\rangle_{L^2(\mathbb{T}^n)} &= \sum_{z\in V_h} \int_{Q^h_z} (i_h p_h w_1)(x) w_2(x)\, dx \\ &= \sum_{z\in V_h} \int_{Q^h_z} \left(h^{-n} \int_{Q^h_z} w_1(y)\, dy \right) w_2(x)\, dx\end{aligned}$$

$$= \sum_{z \in V_h} \left(h^{-n} \int_{Q_z^h} w_1(y)\,dy \right) \left(h^{-n} \int_{Q_z^h} w_2(x)\,dx \right) h^n$$
$$= \langle p_h w_1, p_h w_2 \rangle_h,$$

which proves the first equality. Continuing the computation to prove the second equality, we find

$$\sum_{z \in V_h} \int_{Q_z^h} \left(h^{-n} \int_{Q_z^h} w_1(y)\,dy \right) w_2(x)\,dx = \sum_{z \in V_h} \int_{Q_z^h} w_1(y) \left(h^{-n} \int_{Q_z^h} w_2(x)\,dx \right) dy$$
$$= \sum_{z \in V_h} \int_{Q_z^h} w_1(y)(i_h p_h w_2)(y)\,dy = \langle w_1, i_h p_h w_2 \rangle_{L^2(\mathbb{T}^n)}.$$

Since $i_h \circ p_h$ is a bounded operator, as follows from part (iii), the symmetry just proven implies self-adjointness.

(v) Since $i_h \circ p_h$ is self-adjoint on $L^2(\mathbb{T}^n)$ by part (iv) and $p_h \circ i_h$ is the identity on L_h^2, we have, for $u \in L_h^2$ and $w \in L^2(\mathbb{T}^n)$,

$$\langle i_h p_h w - w, i_h u \rangle_{L^2(\mathbb{T}^n)} = \langle w, i_h p_h i_h u - i_h u \rangle_{L^2(\mathbb{T}^n)} = 0.$$

(vi) By compactness of $\mathbb{T}^n$, f is uniformly continuous. Let $\varepsilon > 0$, then there exists a $\delta > 0$ such that, if $|x - y| < \delta$, then $|f(x) - f(y)| < \varepsilon$. Let $h < n^{-1/2}\delta$, $x \in \mathbb{T}^n$, and let $z \in V_h$ be such that $x \in Q_z^h$. If $y \in Q_z^h$, then $|x - y| < \sqrt{n}h = \delta$. Thus

$$|i_h p_h f(x) - f(x)| = \left| h^{-n} \int_{Q_z^h} f(y)\,dy - f(x) \right| = \left| h^{-n} \int_{Q_z^h} (f(y) - f(x))\,dy \right|$$
$$\leq h^{-n} \int_{Q_z^h} |f(y) - f(x)|\,dy < h^{-n} \int_{Q_z^h} \varepsilon\,dy = \varepsilon.$$

(vii) Let $f \in C(\mathbb{T}^n)$. By the previous point, $i_h p_h f \to f$ uniformly as $h \to 0$. Since $i_h \circ p_h$ is self-adjoint, we have, for $w \in L^2(\mathbb{T}^n)$,

$$\langle i_h p_h w, f \rangle_{L^2} = \langle w, i_h p_h f \rangle_{L^2} \to \langle w, f \rangle_{L^2} \quad \text{as} \quad h \to 0.$$

By the Cauchy–Schwarz inequality, applied on a single cube Q_z^h, we get, for all $\tilde{w} \in L^2(\mathbb{T}^n)$,

$$\left(\int_{Q_z^h} \tilde{w}(y)\,dy \right)^2 \leq \left(\int_{Q_z^h} \tilde{w}(y)\,dy \right) \left(\int_{Q_z^h} 1^2\,dy \right) = \left(\int_{Q_z^h} \tilde{w}(y)\,dy \right) h^n.$$

If $\|\tilde{w}\|_{L^2(\mathbb{T}^n)} = 1$, then

$$\|i_h p_h \tilde{w}\|^2_{L^2(\mathbb{T}^n)} = \sum_{z\in V_h} \int_{Q_z^h} \left(h^{-n} \int_{Q_z^h} \tilde{w}(y)\, dy \right)^2 dx = \sum_{z\in V_h} h^{-n} \left(\int_{Q_z^h} \tilde{w}(y)\, dy \right)^2$$
$$\leq \sum_{z\in V_h} \int_{Q_z^h} \tilde{w}(y)\, dy = \int_{\mathbb{T}^n} \tilde{w}(y)\, dy = 1.$$

Hence the operator norm of $i_h \circ p_h$ satisfies

$$\|i_h \circ p_h\|_{op} := \sup\left\{ \|i_h p_h \tilde{w}\|_{L^2(\mathbb{T}^n)} \;\middle|\; \tilde{w} \in L^2(\mathbb{T}^n) \text{ and } \|\tilde{w}\|_{L^2(\mathbb{T}^n)} = 1 \right\} \leq 1.$$

Let $\tilde{w} \in L^2(\mathbb{T}^n)$. Since $C(\mathbb{T}^n)$ is dense in $L^2(\mathbb{T}^n)$, there exists a sequence (f_m) in $C(\mathbb{T}^n)$ which converges to $\tilde{w}$ in $L^2(\mathbb{T}^n)$ as $m \to \infty$. Combining the Cauchy–Schwarz inequality with the bound on the operator norm, we find

$$\begin{aligned}
\langle i_h p_h w, \tilde{w}\rangle_{L^2(\mathbb{T}^n)} &= \langle i_h p_h w, f_m\rangle_{L^2(\mathbb{T}^n)} + \langle i_h p_h w, \tilde{w} - f_m\rangle_{L^2(\mathbb{T}^n)} \\
&\leq \langle i_h p_h w, f_m\rangle_{L^2(\mathbb{T}^n)} + \|i_h p_h w\|_{L^2(\mathbb{T}^n)} \|\tilde{w} - f_m\|_{L^2(\mathbb{T}^n)} \\
&\leq \langle i_h p_h w, f_m\rangle_{L^2(\mathbb{T}^n)} + \|w\|_{L^2(\mathbb{T}^n)} \|\tilde{w} - f_m\|_{L^2(\mathbb{T}^n)}.
\end{aligned}$$

Taking first $h \to 0$ and then $m \to \infty$, the right-hand side vanishes and thus $i_h p_h w \rightharpoonup w$ (i.e., weakly) in $L^2(\mathbb{T}^n)$. By lower semicontinuity of the $L^2(\mathbb{T}^n)$ norm with respect to weak $L^2(\mathbb{T}^n)$ convergence and again the estimate on the operator norm, we have that

$$\|w\|_{L^2(\mathbb{T}^n)} \leq \underline{\lim}_{h\to 0} \|i_h p_h w\|_{L^2(\mathbb{T}^n)} \leq \overline{\lim}_{h\to 0} \|w\|_{L^2(\mathbb{T}^n)} = \|w\|_{L^2(\mathbb{T}^n)}.$$

Therefore $\|i_h p_h w\|_{L^2(\mathbb{T}^n)} \to \|w\|_{L^2(\mathbb{T}^n)}$. Together with the weak convergence $p_h w \rightharpoonup w$ in $L^2(\mathbb{T}^n)$, this implies that $i_h p_h w \to w$ (i.e., strongly) in $L^2(\mathbb{T}^n)$.

(viii) For all $j \in [n]$, define the operator $q_h^j : L^2(\mathbb{T}^n) \to L^2(\mathbb{T}^n)$ by, for all $w \in L^2(\mathbb{T}^n)$ and all $x \in Q_z^h$,

$$q_h^j w(x) := h^{-1} \int_{I_{z_j}^h} w(\tilde{x}_j, x_j)\, dx_j = h^{-1} \int_{I_{z_j}^h} w(x)\, dx_j.$$

Assume $\{\alpha_1, \dots, \alpha_n\} = [n]$. Let $i, j \in [n]$ with $i \neq j$ and assume that $\alpha_{n-1} = i$ and $\alpha_n = j$. Let $x \in Q_z^h$, then

$$q_h^{\alpha_{n-1}} q_h^{\alpha_n} w(x) = q_h^i q_h^j w(x) = h^{-1} \int_{I_{z_i}^h} q_h^j w(x)\, dx_i = h^{-2} \int_{I_{z_i}^h} \int_{I_{z_j}^h} w(x)\, dx_i dx_j.$$

Repeating this argument for $\alpha_1, \dots, \alpha_{n-2}$, we find that

$$q_h^{\alpha_1} \dots q_h^{\alpha_n} w = i_h p_h w. \tag{2.19}$$

Furthermore, we compute for all $j \in [n]$ and all $z \in V_h$,

$$\begin{aligned}
\int_{Q_z^h} |q_h^j w(x)|\, dx &= \int_{Q_z^h} \left| h^{-1} \int_{I_{z_j}^h} w(\tilde{x}_j, y)\, dy \right| dx \\
&= \int_{I_{z_j}^h} \int_{\tilde{Q}_{z,j}^h} \left| h^{-1} \int_{I_{z_j}^h} w(\tilde{x}_j, y)\, dy \right| dx_j\, d\tilde{x}_j \\
&= \int_{\tilde{Q}_{z,j}^h} \left| \int_{I_{z_j}^h} w(\tilde{x}_j, y)\, dy \right| d\tilde{x}_j \leq \int_{\tilde{Q}_{z,j}^h} \int_{I_{z_j}^h} \left| w(\tilde{x}_j, y) \right| dy\, d\tilde{x}_j \\
&= \int_{Q_z^h} |w(x)|\, dx.
\end{aligned}$$

In the absence of the absolute value function in the integrals above, the inequality becomes an equality. Repeatedly applying this (in)equality according to (2.19) gives the desired result.

□

As mentioned before, in Sect. 2.5.4 we introduce a different inner product on $\mathcal{V}_h$ which gives rise to the Hilbert space $\overline{L}_h^2$. We postpone the introductions of a corresponding new embedding operator I_h (Sect. 2.5.4) and projection operator P_h (Sect. 2.5.7) to Sect. 2.5, where they will be needed for our proofs for the Allen–Cahn gradient flows.

2.4 Continuum Limit of a Discrete Total Variation Flow

In this section we consider the L_h^2-gradient flow of the discrete total variation φ_{TV}^h and want to claim convergence to anisotropic total variation flow, i.e., the $L^2(\mathbb{T}^n)$-gradient flow of the continuum anisotropic total variation φ_{TV}. First we define these functionals in Sect. 2.4.1.

2.4.1 *Discrete and Continuum Total Variation Functionals*

Let $h \in \mathfrak{H}$. The discrete (anisotropic) graph total variation functional $\varphi_{TV}^h : L_h^2 \to \mathbb{R}$ is defined by

$$\varphi_{TV}^h(u) := \frac{1}{2} \sum_{z;\tilde{z}\sim z} h^{n-1} |u(z) - u(\tilde{z})|,$$

where we have used the shorthand notation $\sum_{z;\tilde{z}\sim z} := \sum_{z\in V_h} \sum_{\tilde{z}\in V_h:\tilde{z}\sim z}$. Since L_h^2 is a Hilbert space and thus closed, we note that $\overline{\mathcal{D}\left(\varphi_{TV}^h\right)} = L_h^2$.

The anisotropic total variation functional for functions $u : \mathbb{T}^n \to \mathbb{R}$ is

$$\int_{\mathbb{T}^n} |Du|_{l^1} := \sup \left\{ \int_{\mathbb{T}^n} u(x) \operatorname{div} g(x)\, dx \;\middle|\; g \in C^1(\mathbb{T}^n; \mathbb{R}^n),\ \forall x \in \mathbb{T}^n\ |g(x)|_\infty \le 1 \right\},$$

where $|g(x)|_\infty := \max_{1\le i\le n} |g_i(x)|$. If u is regular enough to admit a weak partial derivative in each direction, the notation $\int_{\mathbb{T}^n} |Du(x)|_{l^1}\, dx$ is compatible with the earlier definition of $|\cdot|_{l^1}$.

If $g = (g_1, \ldots, g_n)$, the condition $|g(x)|_\infty \le 1$ is equivalent to, for all $i \in [n]$, $|g_i(x)| \le 1$, hence

$$\sup_g \int_{\mathbb{T}^n} u(x) \operatorname{div} g(x)\, dx = \sup_g \sum_{i=1}^n \int_{\mathbb{T}^n} u(x) \frac{\partial g_i}{\partial x_i}(x)\, dx = \sum_{i=1}^n \sup_{g_i} \int_{\mathbb{T}^n} u(x) \frac{\partial g_i}{\partial x_i}(x)\, dx, \tag{2.20}$$

where the supremum on the right-hand side is taken over all $g_i \in C^1(\mathbb{T}^n; \mathbb{R})$ that satisfy, for all $x \in \mathbb{T}^n$, $|g_i(x)| \le 1$. (We allow ourselves the slight misuse of notation, to stress that the set of admissible functions for each supremum depends on i.)

We denote the space of functions of bounded variation, i.e., the space of all $u \in L^1(\mathbb{T}^n)$ (i.e., Lebesgue integrable functions on $\mathbb{T}^n$) for which $\int_{\mathbb{T}^n} |Du|_{l^1} < +\infty$, by $BV(\mathbb{T}^n)$. It can be shown that $\int_{\mathbb{T}^n} |Du|_{l^1}$ and the isotropic total variation (as in, for example, [34]) are equivalent seminorms and thus $BV(\mathbb{T}^n)$ as defined above agrees with the standard definition based on the isotropic total variation. It is worth noting that $BV(\mathbb{T}^n)$ is not the same as $BV(\operatorname{int} \Omega_h)$ (where int denotes the topological interior), since the anisotropic total variation on $\mathbb{T}^n$ also measures variations on the boundary of Ω_h (taking into account the periodic boundary conditions), whereas the anisotropic total variation on int Ω_h does not.

An important property of the anisotropic total variation is that, for $u \in L^2_h$,

$$\int_{\mathbb{T}^n} |Di_h u|_{l^1} = \varphi^h_{TV}(u). \tag{2.21}$$

We provide a proof of (2.21) in Appendix A2.7.

We are now ready to define the continuum counterpart of the discrete anisotropic total variation functional, $\varphi_{TV} : L^2(\mathbb{T}^n) \to \mathbb{R} \cup \{+\infty\}$, by

$$\varphi_{TV}(u) := \begin{cases} \int_{\mathbb{T}^n} |Du|_{l^1}, & \text{if } u \in BV(\mathbb{T}^n) \cap L^2(\mathbb{T}^n), \\ +\infty, & \text{if } u \in L^2(\mathbb{T}^n) \setminus BV(\mathbb{T}^n). \end{cases} \tag{2.22}$$

Because $BV(\mathbb{T}^n) \cap L^2(\mathbb{T}^n)$ is dense in $L^2(\mathbb{T}^n)$ (as follows from the fact that the space of infinitely often continuously differentiable real-valued functions on $\mathbb{T}^n$, $C(\mathbb{T}^n)$, is dense in $L^2(\mathbb{T}^n)$ [1, Corollary 2.30]), we have $\overline{\mathcal{D}(\varphi_{TV})} = L^2(\mathbb{T}^n)$.

We note that, for all $h \in \mathfrak{H}$, φ^h_{TV} is convex (geodesically 0-convex), as is φ_{TV}. These functionals are nonnegative and thus bounded below. Furthermore, φ^h_{TV} is continuous. The lower semicontinuity of φ_{TV} is less obvious but it is standard [34], [28, Sect. 5.2, Theorem 1], [15, Appendix A] (note that these results survive the move from the $L^1(\mathbb{T}^n)$ topology to the $L^2(\mathbb{T}^n)$ topology). Since these functionals are also not identically equal to $+\infty$, φ^h_{TV} and φ_{TV} satisfy condition (0-AGS). Moreover, equation (2.21), which can be written as $\varphi_{TV}(i_h u) = \varphi^h_{TV}(u)$, implies that condition (2.5) is satisfied with $\Phi_h = \varphi^h_{TV}$ and $\Phi = \varphi_{TV}$. By Theorem 2.3(v) we know that (2.7) is satisfied.

Formally (ignoring issues when $\frac{\partial u}{\partial x_j} = 0$), the $L^2(\mathbb{T}^n)$-gradient flow of φ_{TV} is a solution to

$$u_t = \operatorname{div}\left(\sum_{j=1}^n \frac{\partial}{\partial x_j} \Big/ \left|\frac{\partial u}{\partial x_j}\right|\right).$$

2.4.2 Convergence of Flows

We will establish the following result.

Theorem 2.4 *Let $h \in \mathfrak{H}$ and let u^h be the L^2_h-gradient flow of φ^h_{TV} with initial data $u^h_0 \in L^2_h$. Let u be the $L^2(\mathbb{T}^n)$-gradient flow of φ_{TV} with initial data $u_0 \in L^2(\mathbb{T}^n)$. Then, for all $t \geq 0$,*

$$\|i_h u^h(t) - u(t)\|_{L^2(\mathbb{T}^n)} \leq \|i_h u^h_0 - u_0\|_{L^2(\mathbb{T}^n)}.$$

We will apply our abstract results (Theorem 2.1 and Corollary 2.1) to $\Phi = \varphi_{TV}$ on $M = L^2(\mathbb{T}^n)$ and $\Phi_h = \varphi^h_{TV}$ on $M_h = L^2_h$. As established in Sect. 2.4.1, these functionals satisfy condition (0-AGS). Moreover, $L^2_h = \mathcal{D}(\varphi^h_{TV})$, $L^2(\mathbb{T}^n) = \overline{\mathcal{D}(\varphi_{TV})}$, and conditions (2.7) and (2.5) hold. In order to apply Theorem 2.1 and Corollary 2.1, it remains to show that condition (2.6) is satisfied as well. This is the content of the following lemma.

Lemma 2.3 *Let $h \in \mathfrak{H}$. For all $w \in L^2(\mathbb{T}^n)$,*

$$\varphi^h_{TV}(p_h w) \leq \varphi_{TV}(w).$$

Proof If $w \in L^2(\mathbb{T}^n) \setminus BV(\mathbb{T}^n)$, the inequality is trivially true. Now assume that $w \in BV(\mathbb{T}^n)$.

By (2.21), (2.60), and Corollary 2.2, we have that

$$\begin{aligned}\varphi^h_{TV}(p_h w) &= \int_{\mathbb{T}^n} |D i_h p_h w|_{l^1} = \sum_{i=1}^n \int_{\mathbb{T}^n} |D_{x_i} i_h p_h w| \\ &= \sum_{i=1}^n \sup_{g \in H_{h,i}} \int_{\mathbb{T}^n} i_h p_h w(x) \frac{\partial g}{\partial x_i}(x)\, dx,\end{aligned}$$

where we recall the definition of the sets $H_{h,i}$ from (2.66). In particular, if $g \in H_{h,i}$, then $\frac{\partial g}{\partial x_i}$ is constant a.e. on each set int Q^h_z. Thus, if we denote its value on int Q^h_z by $c_z \in \mathbb{R}$, then

$$\begin{aligned}\int_{\mathbb{T}^n} i_h p_h w(x)\frac{\partial g}{\partial x_i}(x)\,dx &= \sum_{z\in V_h}\int_{Q^h_z} i_h p_h w(x)\frac{\partial g}{\partial x_i}(x)\,dx = \sum_{z\in V_h} c_z \int_{Q^h_z} i_h p_h w(x)\,dx \\ &= \sum_{z\in V_h} c_z \int_{Q^h_z} w(x)\,dx = \sum_{z\in V_h}\int_{Q^h_z} w(x)\frac{\partial g}{\partial x_i}(x)\,dx \\ &= \int_{\mathbb{T}^n} w(x)\frac{\partial g}{\partial x_i}(x)\,dx,\end{aligned}$$

where we used Theorem 2.3 (viii) to establish the third equality above.

We deduce that

$$\sup_{g\in H_{h,i}}\int_{\mathbb{T}^n} i_h p_h w(x)\frac{\partial g}{\partial x_i}(x)\,dx = \sup_{g\in H_{h,i}}\int_{\mathbb{T}^n} w(x)\frac{\partial g}{\partial x_i}(x)\,dx \le \int_{\mathbb{T}^n} |D_{x_i} w|,$$

where the inequality follows from the fact that each set $H_{h,i}$ is a subset of the corresponding set of admissible functions in the supremum in (2.61). Applying (2.60) once more, we conclude from (2.22) that

$$\varphi^h_{TV}(p_h w) \le \sum_{i=1}^n \int_{\mathbb{T}^n} |D_{x_i} w| = \int_{\mathbb{T}^n} |Dw|_{l^1} = \varphi_{TV}(w).$$

□

Proof (of Theorem 2.4) By Eq. (2.21), Lemma 2.3, and Theorem 2.3, we know that $\Phi = \varphi_{TV}$ and $\Phi_h = \varphi^h_{TV}$ satisfy conditions (2.5), (2.6), and (2.7). Moreover, by our discussion in Sect. 2.4.1 they also satisfy condition (0-AGS). We are able to apply Theorem 2.1 and Corollary 2.1 to get the desired result. □

Remark 2.7 In [42] it is proved that u^h is the gradient flow of φ_{TV} in $L^2(\mathbb{T}^n)$ by calculating the minimal section of the subdifferential of φ_{TV}, i.e., the element of the subdifferential (with respect to the $L^2(\mathbb{T})$ inner product) $\partial\varphi_{TV}(u^h(t))$ that is smallest in $L^2(\mathbb{T}^n)$ norm, at each time t. Our result gives a direct proof that u^h is indeed the gradient flow of φ_{TV} in $L^2(\mathbb{T}^n)$.

2.4.3 Convergence of Energy

We are interested in the convergence of the "energy" $\varphi^h_{TV}(u^h)$ to $\varphi_{TV}(u)$.

Theorem 2.5 *Let $h \in \mathfrak{H}$. Let u^h and u be as in Theorem 2.4 and let $T > 0$. Then*

$$\left|\int_0^T \left(\varphi_{TV}^h(u^h(s)) - \varphi_{TV}(u(s))\right) ds\right| \le \left(\|i_h u_0^h\|_{L^2(\mathbb{T}^n)} + \|u_0\|_{L^2(\mathbb{T}^n)}\right) \|i_h u_0^h - u_0\|_{L^2(\mathbb{T}^n)}.$$

Proof By definition $\dot{u}^h \in -\partial\varphi_{TV}^h(u^h)$ and $\dot{u} \in -\partial\varphi_{TV}(u)$. Since φ_{TV}^h and φ_{TV} are positively one-homogeneous, we see that,[9] for all $v \in \partial\varphi_{TV}^h(u^h)$ and for all $w \in \partial\varphi_{TV}(u)$ we have the Euler identity (see, for example, [6, Lemma 1.7])

$$\langle u^h, v\rangle_h = \varphi_{TV}^h(u^h), \qquad \langle u, w\rangle_{L^2(\mathbb{T}^n)} = \varphi_{TV}(u).$$

Choosing $v = \dot{u}^h$ and $w = \dot{u}$ yields

$$\varphi_{TV}^h(u^h) = \frac{1}{2}\frac{d}{dt}\|u^h\|_h^2, \qquad \varphi_{TV}(u) = \frac{1}{2}\frac{d}{dt}\|u\|_{L^2(\mathbb{T}^n)}^2.$$

These equalities hold almost everywhere on $(0, \infty)$. From this observation it follows that

$$\begin{aligned}\int_0^T \left(\varphi_{TV}^h(u^h(s)) - \varphi_{TV}(u(s))\right) ds =& \frac{1}{2}\left(\|i_h u^h(T)\|_{L^2(\mathbb{T}^n)}^2 - \|u(T)\|_{L^2(\mathbb{T}^n)}^2\right) \\ &- \frac{1}{2}\left(\|i_h u_0^h\|_{L^2(\mathbb{T}^n)}^2 - \|u_0\|_{L^2(\mathbb{T}^n)}^2\right), \end{aligned} \tag{2.23}$$

where we have also used the isometry property of i_h (Theorem 2.3).

By the contraction property from Theorem 2.4 we see that

$$\|i_h u^h(T) - 0\|_{L^2(\mathbb{T}^n)} \le \|i_h u_0^h - 0\|_{L^2(\mathbb{T}^n)}, \qquad \|u(T) - 0\|_{L^2(\mathbb{T}^n)} \le \|u_0 - 0\|_{L^2(\mathbb{T}^n)} \text{ and}$$
$$\|i_h u^h(T) - u(T)\|_{L^2(\mathbb{T}^n)} \le \|i_h u_0^h - u_0\|_{L^2(\mathbb{T}^n)}.$$

For the first two inequalities we used that the constant solutions $u^h = 0$ and $u = 0$ are gradient flows of φ_{TV}^h starting from 0 and of φ_{TV} starting from 0, respectively (Lemma 2.1). Since, for all $a, b \in \mathbb{R}$, $a^2 - b^2 = (a+b)(a-b)$, we find, using the inequalities above and the reverse triangle inequality, that

$$\begin{aligned}&\left|\|i_h u^h(T)\|_{L^2(\mathbb{T}^n)}^2 - \|u(T)\|_{L^2(\mathbb{T}^n)}^2\right| \\ &= \left|\|i_h u^h(T)\|_{L^2(\mathbb{T}^n)} + \|u(T)\|_{L^2(\mathbb{T}^n)}\right| \left|\|i_h u^h(T)\|_{L^2(\mathbb{T}^n)} - \|u(T)\|_{L^2(\mathbb{T}^n)}\right| \\ &\le \left|\|i_h u^h(T)\|_{L^2(\mathbb{T}^n)} + \|u(T)\|_{L^2(\mathbb{T}^n)}\right| \left\|i_h u^h(T) - u(T)\right\|_{L^2(\mathbb{T}^n)} \\ &\le \left|\|i_h u_0^h\|_{L^2(\mathbb{T}^n)} + \|u_0\|_{L^2(\mathbb{T}^n)}\right| \left\|i_h u_0^h - u_0\right\|_{L^2(\mathbb{T}^n)}\end{aligned}$$

[9] Assume φ is a convex function on a (real or complex) vector space V. This function is positively one-homogeneous if, for all $u \in V$ and for all $r > 0$, $\varphi(ru) = r\varphi(u)$. If $v \in \partial\varphi(u)$, then, for all $w \in V$, $\varphi(w) - \varphi(u) \ge \langle w - u, v\rangle$. Choosing $w = 0$ and $w = 2u$ leads to $\varphi(u) = \langle u, v\rangle$.

and

$$\begin{aligned}
&\left| \|i_h u_0^h\|_{L^2(\mathbb{T}^n)}^2 - \|u_0\|_{L^2(\mathbb{T}^n)}^2 \right| \\
&= \left| \|i_h u_0^h\|_{L^2(\mathbb{T}^n)} + \|u_0\|_{L^2(\mathbb{T}^n)} \right| \left| \|i_h u_0^h\|_{L^2(\mathbb{T}^n)} - \|u_0\|_{L^2(\mathbb{T}^n)} \right| \\
&\leq \left| \|i_h u_0^h\|_{L^2(\mathbb{T}^n)} + \|u_0\|_{L^2(\mathbb{T}^n)} \right| \left\| i_h u_0^h - u_0 \right\|_{L^2(\mathbb{T}^n)} .
\end{aligned}$$

By (2.23) we get the desired estimate. □

2.5 Continuum Limit of a Discrete Allen–Cahn Flow

In this section we will study the convergence of discrete Allen–Cahn gradient flows to a continuum Allen–Cahn gradient flow. These flows are determined, in the sense of Definition 2.1, by the discrete and continuum Allen–Cahn functionals, φ_{AC}^h and φ_{AC}, respectively. We start by defining these functionals in Sect. 2.5.1. From Sect. 2.5.3 onwards we restrict ourselves to the case $n = 1$.

2.5.1 Discrete and Continuum Allen–Cahn Functionals

Let $h \in \mathfrak{H}$. The discrete Allen–Cahn functional consists of two terms, the Dirichlet energy $\varphi_D^h : \mathcal{V}_h \to \mathbb{R}$ and the double-well potential energy $\varphi_W^h : \mathcal{V}_h \to \mathbb{R}$, which we define separately:

$$\varphi_D^h(u) := \frac{1}{4} \sum_{z;\tilde{z}\sim z} h^{n-2} \left(u(z) - u(\tilde{z})\right)^2 ,$$

$$\varphi_W^h(u) := \sum_{z\in V_h} h^n W\left(u(z)\right) .$$

Here W is a double-well potential with wells of equal depth; in this paper, we shall fix $W(x) = \alpha(x^2 - 1)^2/4$ for a given $\alpha > 0$. Now we define the discrete Allen–Cahn functional $\varphi_{AC}^h : \mathcal{V}_h \to \mathbb{R}$ as

$$\varphi_{AC}^h := \varphi_D^h + \varphi_W^h .$$

We note here that we have chosen $\mathcal{V}_h$ as the domain of φ_D^h, φ_W^h, and φ_{AC}^h. At various times we will want to emphasise a specific inner product structure (and induced topology) on the domain, in which case we will write L_h^2 or $\overline{L}_h^2$ for the domain. The latter of these two Hilbert spaces will be defined in Sect. 2.5.4.

The continuum versions of the Dirichlet energy, $\varphi_D : L^2(\mathbb{T}^n) \to \mathbb{R} \cup \{+\infty\}$, and the double-well energy, $\varphi_W : L^2(\mathbb{T}^n) \to \mathbb{R} \cup \{+\infty\}$, are given by

$$\varphi_D(u) := \begin{cases} \frac{1}{2}\int_{\mathbb{T}^n} |\nabla u(x)|^2\, dx, & \text{if } u \in H^1(\mathbb{T}^n), \\ +\infty, & \text{otherwise,} \end{cases} \quad \text{and}$$

$$\varphi_W(u) := \begin{cases} \int_{\mathbb{T}^n} W(u(x))\, dx, & \text{if } W \circ u \in L^1(\mathbb{T}^n), \\ +\infty, & \text{otherwise.} \end{cases}$$

Here $|\nabla u|$ denotes the Euclidean norm of the (weak) gradient of u. We remind ourselves that $H^m(\mathbb{T}^n)$ is[10] the space of functions in $L^2(\mathbb{T}^n)$ that are $m \in \mathbb{N}$ times weakly differentiable with each of their weak derivatives also being in $L^2(\mathbb{T}^n)$. It is equipped with the (squared) norm $\|w\|^2_{H^m(\mathbb{T}^n)} := \|w\|^2_{L^2(\mathbb{T}^n)} + \sum_{l=1}^m \|w^{(l)}\|^2_{L^2(\mathbb{T}^n)}$, where $w^{(l)}$ denotes the lth derivative of w.

We note that φ_W is $L^2(\mathbb{T}^n)$-coercive since, by Jensen's inequality [27, Appendix B Theorem 2],

$$\int_{\mathbb{T}^n} W(u)\, dx = \int_{\mathbb{T}^n} (u^2(x) - 1)^2\, dx \geq \left(\int_{\mathbb{T}^n} (u^2(x) - 1)\, dx\right)^2 = \left(\|u\|^2_{L^2(\mathbb{T}^n)} - 1\right)^2.$$

The continuum Allen–Cahn functional $\varphi_{AC} : L^2(\mathbb{T}^n) \to \mathbb{R} \cup \{+\infty\}$ is then defined to be

$$\varphi_{AC}(u) := \varphi_D(u) + \varphi_W(u).$$

Neither φ^h_{AC} nor φ_{AC} are convex, but they are geodesically $(-\alpha)$-convex, uniformly in h (for φ^h_{AC}). This follows since φ^h_D and φ_D are convex and since $\left(W(x) - \frac{(-\alpha)}{2}x^2\right)'' = \alpha(3x^2 - 1) + \alpha \geq 0$, so that W is geodesically $(-\alpha)$-convex, and thus so are φ^h_W and φ_W. We emphasise that α does not depend on h.

The functional φ^h_{AC} is continuous, if $\mathcal{V}_h$ is equipped with any norm-induced topology (as those topologies are all equivalent since $\mathcal{V}_h$ is finite-dimensional). Moreover, since the integrand of $\varphi_{AC}(u)$, $\frac{1}{2}|\nabla u|^2 + W(u)$, is smooth as a function of $(\nabla u, u)$ and convex in ∇u, φ_{AC} is lower semicontinuous with respect to weak $H^1(\mathbb{T}^n)$ convergence [27, §8.2 Theorem 1] and thus, by the compact Rellich–Kondrachov embedding [1, Theorem 6.3] also with respect to (strong) $L^2(\mathbb{T}^n)$ convergence. Since φ^h_{AC} and φ_{AC} are not identically equal to $+\infty$ and nonnegative (and thus bounded below), they satisfy the ($-\alpha$-AGS) condition.

The Allen–Cahn functional (discrete or continuum) also goes by various other names in the literature, such as Ginzburg–Landau functional [7] or Modica–Mortola functional [51].

---

[10] And similarly if $\mathbb{T}^n$ is replaced by an open subset of itself or of $\mathbb{R}^n$. Recall that $w \in L^2(\mathbb{T}^n)$ is the weak derivative of $u \in L^2(\mathbb{T}^n)$ with respect to $\mathbb{T}^n$-coordinate x_i, if, for all $\varphi \in C^\infty(\mathbb{T}^n)$, $\langle w, \varphi\rangle_{L^2(\mathbb{T}^n)} = \langle u, \frac{\partial}{\partial x_i}\varphi\rangle_{L^2(\mathbb{T}^n)}$.

2.5.2 Additional Setup

We are interested in discrete Allen–Cahn flows, i.e., solutions of $\dot{u} \in -\partial\varphi^h_{AC}(u)$, and in particular in their convergence to a continuum Allen–Cahn flow, i.e., a solution of $\dot{u} \in -\partial\varphi_{AC}(u)$ (Theorem 2.6). An explicit form of the equation for the Allen–Cahn flow is

$$\dot{u} = \Delta u - W'(u) \quad \text{in } \mathbb{T}^n \times (0, \infty), \tag{2.24}$$

which is often called the Allen–Cahn equation [3]. As usual Δ denotes the (continuum) Laplacian $\sum_{i=}^n \frac{\partial^2}{\partial x_i^2}$ where x_i are the coordinates on $\mathbb{T}^n$. This explicit form of the gradient flow equation can be computed directly since the subdifferential in (2.2) is a singleton. Similarly an explicit form of the equation for the discrete Allen–Cahn flow can be computed:

$$\dot{u} = \Delta_h u - W'(u) \quad \text{in } V_h \times (0, \infty), \tag{2.25}$$

where

$$(\Delta_h u)(z) := h^{-2} \sum_{\tilde{z} \in V_h : \tilde{z} \sim z} (u(\tilde{z}) - u(z)) \tag{2.26}$$

defines the graph Laplacian [16, 63] for $u \in L^2_h$ and $z \in V_h$. This discrete flow equation has been studied in more detail in [11–13, 62]. Details of the derivation in the one-dimensional ($n = 1$) case follow in Lemma 2.6.

It will also be useful to define a graph gradient $\nabla_h : V_h \times V_h \to \mathbb{R}$; for $u \in L^2_h$ and $z, \tilde{z} \in V_h$,

$$\nabla_h u(z, \tilde{z}) = \begin{cases} h^{-1}\big(u(\tilde{z}) - u(z)\big), & \text{if } \tilde{z} \sim z, \\ 0, & \text{otherwise.} \end{cases}$$

We note that $\nabla_h u$ is defined on $V_h \times V_h$. For such functions $\chi, \phi : V_h \times V_h \to \mathbb{R}$ we introduce the inner product

$$\langle \chi, \phi \rangle_{h,h} := \frac{1}{2} \sum_{z;\tilde{z} \sim z} h^n \chi(z, \tilde{z}) \phi(z, \tilde{z})$$

and corresponding norm $\|\chi\|_{h,h} := \sqrt{\langle \chi, \chi \rangle_{h,h}}$. Then we have, for $u, v \in \mathcal{V}_h$,

$$\begin{aligned} \langle -\Delta_h u, v \rangle_h &= \sum_{z;\tilde{z} \sim z} h^{n-2} (u(z) - u(\tilde{z}))\, v(z) = \frac{1}{2} \sum_{z;\tilde{z} \sim z} h^{n-2} (u(z) - u(\tilde{z}))\, (v(z) - v(\tilde{z})) \\ &= \langle \nabla_h u, \nabla_h v \rangle_{h,h} = \langle u, -\Delta_h v \rangle_h. \end{aligned} \tag{2.27}$$

and $\varphi^h_D(u) = \frac{1}{2} \|\nabla_h u\|^2_{h,h}$.

Remark 2.8 Two different sign conventions for the (graph) Laplacian appear in the literature. Our choice in (2.26) is in line with the typical definition in the partial differential equations' literature which has the Laplacian as negative semidefinite. This differs by an overall minus sign from the common definition in the spectral graph literature, which has the graph Laplacian as positive semidefinite.

Remark 2.9 It is also interesting to compare how the choices we have made in this paper regarding the scaling with h correspond to the scaling choices in [62, Sect. 2] which are common in the literature. In [62] the scaling of the inner products, operators, and summands in the functionals depends on the edge weights ω, the node degrees $d \sim \omega$, and two parameters, r and q, as follows:

$$\begin{gathered}\langle\cdot,\cdot\rangle_h \sim d^r \sim \omega^r, \quad \langle\cdot,\cdot\rangle_{h,h} \sim \omega^{2q-1},\\ \nabla_h \sim \omega^{1-q}, \quad \Delta_h \sim \omega^{1-r},\\ \text{summands in } \varphi_D^h \sim \omega, \quad \text{summands in } \varphi_{TV}^h \sim \omega^q.\end{gathered}$$

We note that for the functionals φ_D^h and φ_{TV}^h, we consider the scaling of the summands only, i.e., we do not consider the fact that the number of summands is $|V_h| = h^{-n}$. According to our choices in this paper, we require the summands of φ_D^h to be proportional to h^{n-2}, which suggests the choice $\omega = h^{n-2}$ for the edge weights between neighbouring nodes. Comparisons for the other objects lead to

$$\omega^r = h^n, \quad \omega^{2q-1} = h^n, \quad \omega^{1-q} = h^{-1}, \quad \omega^{1-r} = h^{-2}, \quad \text{and} \quad \omega^q = h^{n-1}.$$

It follows that, when $n \neq 2$,

$$q = \frac{n-1}{n-2} \quad \text{and} \quad r = 2q - 1 = \frac{n}{n-2}.$$

This implies that $q > 1$ and $r > 1$, which are not choices that are typically considered. When $n = 2$, we have $\omega = 1$, which cannot be made compatible with the other requirements.

2.5.3 Restriction to One-Dimensional Case ($n = 1$)

To avoid additional technical complexity, we consider the convergence problem only for the one-dimensional case, i.e., $n = 1$. In this case the vertex set is $V_h = \{0, h, 2h, \ldots, 1 - h\}$. We remind ourselves that $h \in \mathfrak{H}$ and thus $|V_h| = h^{-1}$ is an integer. It will be useful to write $\left[\frac{1}{h} - 1\right]_0 := \{0\} \cup \left[\frac{1}{h} - 1\right] = \{0, 1, \ldots, \frac{1}{h} - 1\}$, so that to each $k \in \left[\frac{1}{h} - 1\right]_0$ corresponds one and only one $kh \in V_h$.

Remember that $\mathcal{V}_h$ denotes the set of real-valued vertex functions $v : V_h \to \mathbb{R}$, without assuming a specific inner product structure.

To simplify notation, if $v \in \mathcal{V}_h$, we write $v_k := v(kh)$. Similarly, if $\chi : V_h \times V_h \to \mathbb{R}$, then we write $\chi_{k,l} := \chi(kh, lh)$. By periodicity $v_{\frac{1}{h}} = v_0$, $v_{\frac{1}{h}+1} = v_1$, $v_{\frac{1}{h}-1} = v_{-1}$, etc. Periodicity imposes similar relationships on $\chi_{k,l}$.

For future reference it is convenient to explicitly state to what some of the relevant definitions reduce in this one-dimensional case:

$$(\Delta_h u)_k = h^{-2}(u_{k+1} + u_{k-1} - 2u_k), \qquad (\nabla_h u)_{k,k\pm 1} = h^{-1}(u_{k\pm 1} - u_k),$$

$$\langle u, v\rangle_h = \sum_{k=0}^{\frac{1}{h}-1} h u_k v_k, \qquad \langle \chi, \phi\rangle_{h,h} = \frac{1}{2}\sum_{k=0}^{\frac{1}{h}-1} h\left(\chi_{k,k+1}\phi_{k,k+1} + \chi_{k,k-1}\phi_{k,k-1}\right),$$

$$\varphi_D^h(u) = \frac{1}{2}\sum_{k=0}^{\frac{1}{h}-1} h^{-1}(u_{k+1} - u_k)^2.$$

We note that if χ and ϕ are both symmetric (i.e., $\chi_{k,l} = \chi_{l,k}$ and $\phi_{k,l} = \phi_{l,k}$) or both skew-symmetric (i.e., $\chi_{k,l} = -\chi_{l,k}$ and $\phi_{k,l} = -\phi_{l,k}$), then, by periodicity,

$$\langle \chi, \phi\rangle_{h,h} = \sum_{k=0}^{\frac{1}{h}-1} h \chi_{k,k+1}\phi_{k,k+1}. \tag{2.28}$$

In particular, this holds if χ and ϕ are both graph gradients of a vertex function (and thus skew-symmetric).

We also observe that $\Delta_h u$ is equal to the second order central difference approximation of the second derivative of a function $u : \mathbb{T} \to \mathbb{R}$.

Finally, we note that by the Sobolev embedding [1, Theorem 4.12] and the finite measure of $\mathbb{T}$, $H^1(\mathbb{T}) \subset C(\mathbb{T}) \subset L^4(\mathbb{T})$. Hence, if $u \in H^1(\mathbb{T})$, then $W \circ u \in L^1(\mathbb{T})$ and thus $\varphi_W(u) < +\infty$. It follows that $H^1(\mathbb{T}) \subset \mathcal{D}(\varphi_{AC})$. Hence, by density of $H^1(\mathbb{T})$ in $L^2(\mathbb{T})$ (as follows from the density of $C^\infty(\mathbb{T})$ in $L^2(\mathbb{T})$), we have $\overline{\mathcal{D}(\varphi_{AC})} = L^2(\mathbb{T})$.

2.5.4 Extension Operator and Induced Inner Product

In our study of the total variation flow in Sect. 2.4 we used an embedding operator $i_h : L^2_{\mathfrak{h}} \to L^2(\mathbb{T}^n)$, which preserved the total variation in the sense of formula (2.21). This, in turn, ensured that condition (2.5) was satisfied as an equality. In the present case, the situation is more complicated as we cannot find an embedding operator that preserves the Allen–Cahn functional in a similar way. We define a new embedding operator.

Let $h \in \mathfrak{H}$. We define $I_h : \mathcal{V}_h \to L^2(\mathbb{T})$ via piecewise-linear interpolation

$$I_h u(x) := \frac{u_{k+1} - u_k}{h}(x - kh) + u_k,$$

where $k \in \left[\frac{1}{h} - 1\right]_0$ is such that $kh \leq x < (k+1)h$, i.e., $x \in I^h_{kh} = Q^h_{kh}$ in the notation of Sect. 2.3.1.

A direct computation shows that, for all $u \in \mathcal{V}_h$,

$$\begin{aligned}\varphi_D(I_h u) &= \frac{1}{2}\int_{\mathbb{T}} \left|(I_h u)'(x)\right|^2 dx = \frac{1}{2}\sum_{k=0}^{\frac{1}{h}-1}\int_{kh}^{(k+1)h} \left|(I_h u)'(x)\right|^2 dx \\ &= \frac{1}{2}\sum_{k=0}^{\frac{1}{h}-1} h\left(\frac{u_{k+1}-u_k}{h}\right)^2 dx = \frac{1}{2}\sum_{k=0}^{\frac{1}{h}-1} h^{-1}\left(u_{k+1}-u_k\right)^2 dx \\ &= \varphi_D^h(u),\end{aligned} \tag{2.29}$$

where we used that $u_{\frac{1}{h}} - u_{\frac{1}{h}-1} = u_0 - u_{\frac{1}{h}-1}$ by periodicity. We note that this does not suffice to conclude that (2.5) holds for the full functionals φ_{AC} and φ^h_{AC}. In fact, when we prove Theorem 2.7 below, we will use Theorem 2.2 rather than Theorem 2.1, to avoid needing requirement (2.5). The identity in (2.29) above, will be very useful in the proof of Theorem 2.7.

We note that $I_h : \mathcal{V}_h \to L^2(\mathbb{T})$ is a linear operator, but unlike i_h, when the domain is equipped with our usual inner product, $I_h : L^2_h \to L^2(\mathbb{T})$ is not an isometry. This is problematic, as we required our embedding to be an isometry in Sect. 2.2. Therefore we introduce the inner product $(\cdot,\cdot)_h$ that is induced by I_h, i.e., for $v_1, v_2 \in \mathcal{V}_h$,

$$(v_1, v_2)_h := \langle I_h v_1, I_h v_2\rangle_{L^2(\mathbb{T})}.$$

To differentiate the Hilbert space of vertex functions equipped with this new inner product from L^2_h, we denote the set $\mathcal{V}_h$ equipped with $(\cdot,\cdot)_h$ by $\overline{L}^2_h$. We note that, by definition, this ensures that $I_h : \overline{L}^2_h \to L^2(\mathbb{T})$ *is* an isometry. We denote the norm for $v \in \overline{L}^2_h$ by $\|v\|_{\overline{h}} := \sqrt{(v,v)_h}$. We emphasise that, as sets, $\mathcal{V}_h$, L^2_h, and $\overline{L}^2_h$ are equal, i.e., $v \in \mathcal{V}_h$ if and only if $v \in L^2_h$, which in turn is true if and only if $v \in \overline{L}^2_h$.

Lemma 2.4 *The operator I_h satisfies, for all $u, v \in \mathcal{V}_h$,*

$$\langle I_h u, I_h v\rangle_{L^2(\mathbb{T})} = \frac{h}{3}\sum_{k=0}^{\frac{1}{h}-1}\left[2u_k v_k + \frac{1}{2}\left(u_k v_{k+1} + u_{k+1}v_k\right)\right].$$

Moreover, $\frac{1}{3}\|u\|^2_h = \frac{1}{3}\|i_h u\|^2_{L^2(\mathbb{T})} \leq \|I_h u\|^2_{L^2(\mathbb{T})} = \|u\|^2_{\overline{h}} \leq \|i_h u\|^2_{L^2(\mathbb{T})} = \|u\|^2_h$.

Proof For $a, b \in \mathbb{R}$, a direct computation shows that

$$\int_0^h \left|\frac{b-a}{h}x + a\right|^2 dx = \frac{h}{3}(a^2 + b^2 + ab).$$

This gives for $u \in \mathcal{V}_h$,

$$
\begin{aligned}
\|I_h u\|^2_{L^2(\mathbb{T})} &= \sum_{k=0}^{\frac{1}{h}-1} \int_{kh}^{(k+1)h} \left| \frac{u_{k+1} - u_k}{h}(x - kh) + u_k \right|^2 dx \\
&= \sum_{k=0}^{\frac{1}{h}-1} \int_0^h \left| \frac{u_{k+1} - u_k}{h} x + u_k \right|^2 dx \\
&= \frac{h}{3} \sum_{k=0}^{\frac{1}{h}-1} \left[u_k^2 + u_{k+1}^2 + u_k u_{k+1} \right]. \qquad (2.30)
\end{aligned}
$$

Let $u, v \in \mathcal{V}_h$. By the polarisation identity, we now get

$$
\begin{aligned}
4\langle I_h u, I_h v\rangle_{L^2(\mathbb{T})} &= \|I_h(u+v)\|^2_{L^2(\mathbb{T})} - \|I_h(u-v)\|^2_{L^2(\mathbb{T})} \\
&= \frac{h}{3} \sum_{k=0}^{\frac{1}{h}-1} \big[(u_k + v_k)^2 + (u_{k+1} + v_{k+1})^2 + (u_k + v_k)(u_{k+1} + v_{k+1}) \\
&\qquad - (u_k - v_k)^2 - (u_{k+1} - v_{k+1})^2 - (u_k - v_k)(u_{k+1} - v_{k+1}) \big] \\
&= \frac{h}{3} \sum_{k=0}^{\frac{1}{h}-1} \big[4u_k v_k + 4u_{k+1} v_{k+1} + 2u_k v_{k+1} + 2u_{k+1} v_k \big],
\end{aligned}
$$

from which the first desired identity follows, by periodicity.

Using periodicity in (2.30), we also find that

$$
\|I_h u\|^2_{L^2(\mathbb{T})} = \frac{h}{3} \sum_{k=0}^{\frac{1}{h}-1} (2u_k^2 + u_k u_{k+1}).
$$

Using Young's inequality $|u_k u_{k+1}| \leq (u_k^2 + u_{k+1}^2)/2$ and periodicity again yields

$$
\begin{aligned}
\frac{1}{3}\|u\|_h^2 &= \frac{h}{3} \sum_{k=0}^{\frac{1}{h}-1} u_k^2 = \frac{h}{3} \sum_{k=0}^{\frac{1}{h}-1} \left[2u_k^2 - \frac{u_k^2 + u_{k+1}^2}{2} \right] \leq \frac{h}{3} \sum_{k=0}^{\frac{1}{h}-1} (2u_k^2 + u_k u_{k+1}) \\
&= \|I_h u\|^2_{L^2(\mathbb{T})} \leq \frac{h}{3} \sum_{k=0}^{\frac{1}{h}-1} \left[2u_k^2 + \frac{u_k^2 + u_{k+1}^2}{2} \right] \\
&= \|u\|_h^2.
\end{aligned}
$$

By the isometry property of Theorem 2.3 the required inequalities now follow. □

Remark 2.10 Let $a, b \in \mathbb{R}$ and $m \in \mathbb{N}$. A direct computation shows

$$\int_0^h \left|\frac{b-a}{h}x + a\right|^m dx = \begin{cases} \frac{h}{(b-a)(m+1)}\left[b^{m+1} - a^{m+1}\right], & \text{if } m \text{ is even,} \\ \frac{h}{(b-a)(m+1)}\left[(\operatorname{sgn} b)b^{m+1} - (\operatorname{sgn} a)a^{m+1}\right], & \text{if } m \text{ is odd.} \end{cases}$$

A proof by induction shows that $b^{m+1} - a^{m+1} = (b-a)\sum_{l=0}^{m} a^l b^{m-l}$, hence if m is even we have

$$\int_0^h \left|\frac{b-a}{h}x + a\right|^m dx = \frac{h}{m+1}\sum_{\ell=0}^{m} a^\ell b^{m-\ell}.$$

This implies in particular that, for m even and for $u \in \mathcal{V}_h$,

$$\begin{aligned} \|I_h u\|^m_{L^m(\mathbb{T})} := \int_{\mathbb{T}} |I_h u|^m \, dx &= \sum_{k=0}^{\frac{1}{h}-1} \frac{h}{m+1} \sum_{l=0}^{m} u_k^l u_{k+1}^{m-l} \\ &\le \sum_{k=0}^{\frac{1}{h}-1} \frac{h}{m+1} \sum_{l=0}^{m} \left(l\frac{|u_k|^m}{m} + (m-l)\frac{|u_k|^m}{m} \right) \\ &= \sum_{k=0}^{1/h-1} h\,|u_k|^m := \|u\|^m_{L^m_h}. \end{aligned} \tag{2.31}$$

The inequality follows from Young's inequality, $|u_k^l u_{k+1}^{m-l}| \le \frac{l}{m}|u_k|^m + \frac{m-l}{m}|u_{k+1}|^m$, and periodicity in k.

On the other hand, when m is even, l is odd, and $1 \le l \le m-1$, Young's inequality $|u_k u_{k+1}| \le (u_k^2 + u_{k+1}^2)/2$ tells us that

$$\begin{aligned} u_k^l u_{k+1}^{m-l} = u_k^{l-1} u_{k+1}^{m-l-1}(u_k u_{k+1}) &\ge u_k^{l-1} u_{k+1}^{m-l-1}\left(-\frac{1}{2}u_k^2 - \frac{1}{2}u_{k+1}^2\right) \\ &= -\frac{1}{2}u_k^{l+1} u_{k+1}^{m-l-1} - \frac{1}{2}u_k^{l-1} u_{k+1}^{m-l+1}. \end{aligned}$$

Using periodicity in k, it follows that

$$\begin{aligned} \sum_{l=0}^{m} u_k^l u_{k+1}^{m-l} &= \frac{1}{2}u_k^m + \frac{1}{2}u_{k+1}^m + \sum_{\substack{l=1 \\ l \text{ is odd}}}^{m-1} \left(u_k^l u_{k+1}^{m-l} + \frac{1}{2}u_k^{l+1} u_{k+1}^{m-l-1} + \frac{1}{2}u_k^{l-1} u_{k+1}^{m-l+1} \right) \\ &\ge \frac{1}{2}u_k^m + \frac{1}{2}u_{k+1}^m + \sum_{\substack{l=1 \\ l \text{ is odd}}}^{m-1} \left(-\frac{1}{2}u_k^{l+1} u_{k+1}^{m-l-1} - \frac{1}{2}u_k^{l-1} u_{k+1}^{m-l+1} \right. \end{aligned}$$

$$+\frac{1}{2}u_k^{l+1}u_{k+1}^{m-l-1}+\frac{1}{2}u_k^{l-1}u_{k+1}^{m-l+1}\Bigg)$$

$$=\frac{1}{2}u_k^m+\frac{1}{2}u_{k+1}^m.$$

Summing over k and using periodicity again, we thus find that, for m even,

$$\begin{aligned}\|I_h u\|_{L^m(\mathbb{T})}^m &= \frac{h}{m+1}\sum_{k=0}^{\frac{1}{h}-1}\sum_{l=0}^{m} u_k^l u_{k+1}^{m-l} \geq \frac{h}{m+1}\sum_{k=0}^{\frac{1}{h}-1}\left(\frac{1}{2}|u_k|^m+\frac{1}{2}|u_{k+1}|^m\right)\\ &= \frac{h}{m+1}\sum_{k=0}^{\frac{1}{h}-1}|u_k|^m\\ &= \frac{1}{m+1}\|u\|_{L_h^m}^m. \end{aligned} \tag{2.32}$$

We note that $\|\cdot\|_h = \|\cdot\|_{L_h^2}$, so that the inequalities from Lemma 2.4 correspond to the case $m = 2$.

Lemma 2.5 *Let $u \in \mathcal{V}_h$, then*

$$\|I_h u - i_h u\|_{L^2(\mathbb{T})}^2 = \frac{h^2}{2}\|\nabla_h u\|_{h,h}^2 = h^2\varphi_D^h(u) \leq h^2\varphi_{AC}^h(u).$$

Proof We compute

$$\begin{aligned}\|I_h u - i_h u\|_{L^2(\mathbb{T})}^2 &= \sum_{k=0}^{\frac{1}{h}-1}\int_{kh}^{(k+1)h}\left|\frac{u_{k+1}-u_k}{h}(x-kh)+u_k-u_k\right|^2 dx\\ &= \sum_{k=0}^{\frac{1}{h}-1}\int_0^h\left(\frac{u_{k+1}-u_k}{h}\right)^2 x^2\,dx = \frac{h}{3}\sum_{k=0}^{\frac{1}{h}-1}(u_{k+1}-u_k)^2\\ &= \frac{h^2}{2}\|\nabla_h u\|_{h,h}^2 = h^2\varphi_D^h(u) \leq h^2\varphi_{AC}^h(u). \end{aligned}$$

□

As a consequence of Lemma 2.4 we have, for $u, v \in \mathcal{V}_h$,

$$\begin{aligned}(u, v)_h &= \frac{h}{3}\sum_{k=0}^{1/h-1}\left[2u_k v_k+\frac{1}{2}\left(u_k v_{k+1}+u_{k+1}v_k\right)\right]\\ &= \frac{h}{3}\sum_{k=0}^{1/h-1}\left[2u_k+\frac{u_{k-1}+u_{k+1}}{2}\right]v_k = \langle\Gamma u, v\rangle_h, \end{aligned} \tag{2.33}$$

where we used periodicity and defined the operator Γ by

$$(\Gamma u)_k := \frac{1}{3}\left(2u_k + \frac{u_{k-1}+u_{k+1}}{2}\right). \tag{2.34}$$

We can consider Γ as an operator from L^2_h to L^2_h or from $\overline{L}^2_h$ to $\overline{L}^2_h$. We note that $\langle \Gamma u, v\rangle_h = (u,v)_h = \langle I_h u, I_h v\rangle_{L^2(\mathbb{T})}$ and thus $\langle \Gamma u, u\rangle_h = \|u\|^2_{\overline{h}} = \|I_h u\|^2_{L^2(\mathbb{T})}$. Hence, by Lemma 2.4,

$$\frac{1}{3}\|u\|^2_h \leq \langle \Gamma u, u\rangle_h \leq \|u\|^2_h. \tag{2.35}$$

The operator Γ can be identified with a circulant $1/h$ by $1/h$-matrix, if one regards u as a $1/h$-dimensional vector. This identification is helpful in deriving some interesting properties of Γ and its matrix exponential in Appendix B2.8.

Defining the operators τ_+ and τ_- by

$$(\tau_\pm u)_k := u_{k\pm 1}, \tag{2.36}$$

we have $\Gamma = \frac{2}{3}\mathrm{Id} + \frac{1}{6}\tau_+ + \frac{1}{6}\tau_-$.

By direct computation we find that $\tau_\pm$ and Δ_h commute: $\tau_\pm \Delta_h u = \Delta_h \tau_\pm u$. It follows that Γ and Δ_h also commute. It is also useful to observe that

$$(\nabla_h \tau_\pm u)_{k,k+1} = h^{-1}\big((\tau_\pm u)_{k+1} - (\tau_\pm u)_k\big) = h^{-1}\left(u_{k+1\pm 1} - u_{k\pm 1}\right) = (\nabla_h u)_{k\pm 1, k+1\pm 1},$$

and thus by (2.28), periodicity, and Young's inequality,

$$\begin{aligned}
\langle \nabla_h u, \nabla_h \Gamma u\rangle_{h,h} &= h\sum_{k=0}^{\frac{1}{h}-1}\left[(\nabla_h u)_{k,k+1}\left(\nabla_h\left(\frac{2}{3}u + \frac{1}{6}\tau_- u + \frac{1}{6}\tau_+ u\right)\right)_{k,k+1}\right] \\
&= h\sum_{k=0}^{\frac{1}{h}-1}\left[\frac{2}{3}(\nabla_h u)^2_{k,k+1} + \frac{1}{6}(\nabla_h u)_{k,k+1}(\nabla_h u)_{k-1,k}\right. \\
&\qquad\left. + \frac{1}{6}(\nabla_h u)_{k,k+1}(\nabla_h u)_{k+1,k+2}\right] \\
&= h\sum_{k=0}^{\frac{1}{h}-1}\left[\frac{2}{3}(\nabla_h u)^2_{k,k+1} + \frac{1}{3}(\nabla_h u)_{k,k+1}(\nabla_h u)_{k-1,k}\right] \\
&\geq h\sum_{k=0}^{\frac{1}{h}-1}\left[\frac{2}{3}(\nabla_h u)^2_{k,k+1} - \frac{1}{6}(\nabla_h u)^2_{k,k+1} - \frac{1}{6}(\nabla_h u)^2_{k-1,k}\right] \\
&= h\sum_{k=0}^{\frac{1}{h}-1}\left[\frac{2}{3}(\nabla_h u)^2_{k,k+1} - \frac{1}{3}(\nabla_h u)^2_{k,k+1}\right] = h\sum_{k=0}^{\frac{1}{h}-1}\frac{1}{3}(\nabla_h u)^2_{k,k+1} \\
&= \frac{1}{3}\|\nabla_h u\|^2_{h,h}.
\end{aligned} \tag{2.37}$$

We note that by periodicity we also have

$$\|u - \tau_+ u\|_h^2 = \|u - \tau_- u\|_h^2 = h^2 \|\nabla_h u\|_{h,h}^2. \tag{2.38}$$

Again by periodicity, we have that $\langle \tau_\pm u, v\rangle_h = \langle u, \tau_\mp v\rangle_h$ and thus Γ is a self-adjoint operator on L_h^2. It follows that $\Gamma - \frac{1}{3}\mathrm{Id}$ is also self-adjoint; moreover, by Lemma 2.4 it is positive semidefinite on L_h^2:

$$\left\langle \left(\Gamma - \frac{1}{3}\mathrm{Id}\right) u, u \right\rangle_h = \|I_h u\|_{L^2(\mathbb{T})}^2 - \frac{1}{3}\|u\|_h^2 \geq 0.$$

As an aside, a direct computation also shows that $\langle (\Gamma - \mathrm{Id}/3)\, u, u\rangle_h$ is proportional to the *signless* graph Dirichlet functional[11] on $\mathcal{V}_h$ (see, for example, [40, Sect. 2.1]):

$$\begin{aligned}\left\langle \left(\Gamma - \frac{1}{3}\mathrm{Id}\right) u, u \right\rangle_h &= \sum_{k=0}^{\frac{1}{h}-1} h \left(\frac{1}{3}u_k^2 + \frac{1}{6}u_{k-1}u_k + \frac{1}{6}u_{k+1}u_k\right) = \frac{h}{3}\sum_{k=0}^{\frac{1}{h}-1} \left(u_{k+1}u_k + u_k^2\right) \\ &= \frac{h}{3}\sum_{k=0}^{\frac{1}{h}-1} \left(u_{k+1}u_k + \frac{1}{2}u_k^2 + \frac{1}{2}u_{k+1}^2\right) = \frac{h}{6}\sum_{k=0}^{\frac{1}{h}-1} (u_k + u_{k+1})^2 \geq 0,\end{aligned} \tag{2.39}$$

where we used periodicity twice. We point out that the computation above also proves that Γ itself is positive definite on L_h^2, since $\langle \Gamma u, u\rangle_h \geq \frac{1}{3}\langle u, u\rangle_h$. In particular, since L_h^2 has finite dimension, the (linear) inverse operator $\Gamma^{-1} : L_h^2 \to L_h^2$ exists.

By Lemma 2.4 it also follows that $\mathrm{Id} - \Gamma$ is self-adjoint. It is also positive semidefinite on L_h^2, since, by (2.33), $\langle (\mathrm{Id} - \Gamma)u, u\rangle_h = \|u\|_h^2 - \|I_h u\|_{L^2(\mathbb{T})}^2 \geq 0$. It will be useful to have an explicit expression for this difference of norms:

$$\begin{aligned}\|u\|_h^2 - \|I_h u\|_{L^2(\mathbb{T})}^2 &= \sum_{k=0}^{\frac{1}{h}-1} h \left[\frac{1}{2}u_k^2 + \frac{1}{2}u_{k+1}^2 - \frac{1}{3}u_k^2 - \frac{1}{3}u_{k+1}^2 - \frac{1}{3}u_k u_{k+1}\right] \\ &= \frac{1}{6}\sum_{k=0}^{\frac{1}{h}-1} h(u_{k+1} - u_k)^2 \\ &= \frac{h^2}{6}\|\nabla_h u\|_{h,h}^2,\end{aligned} \tag{2.40}$$

where we used (2.30) and periodicity.

[11] The plus sign in $u_k + u_{k+1}$ on the penultimate right-hand side of (2.39), rather than a minus sign, makes it (proportional to) the *signless* graph Dirichlet functional.

Since Γ is a positive definite operator on L_h^2, the square root $\Gamma^{\frac{1}{2}} : L_h^2 \to L_h^2$ exists and is positive definite. Moreover, since Γ is symmetric, so[12] is $\Gamma^{\frac{1}{2}}$. By (2.35), it follows that

$$(\Gamma u, u)_h = \langle \Gamma^2 u, u\rangle_h = \langle \Gamma^{\frac{1}{2}}\Gamma u, \Gamma^{\frac{1}{2}} u\rangle_h = \langle \Gamma\Gamma^{\frac{1}{2}} u, \Gamma^{\frac{1}{2}} u\rangle_h \geq \frac{1}{3}\langle \Gamma^{\frac{1}{2}} u, \Gamma^{\frac{1}{2}} u\rangle_h$$
$$= \frac{1}{3}\langle \Gamma u, u\rangle_h = \frac{1}{3}(u, u)_h,$$

thus Γ is also positive definite as operator on $\overline{L}_h^2$. Hence, since $\overline{L}_h^2$ has finite dimension, the (linear) inverse operator $\Gamma^{-1} : \overline{L}_h^2 \to \overline{L}_h^2$ exists. Furthermore, since $(\Gamma u, u)_h = \langle \Gamma^2 u, u\rangle_h = (u, \Gamma u)_h$, Γ is also symmetric as operator on $\overline{L}_h^2$.

We end this subsection with the computation of the subdifferential of φ_{AC}^h with respect to the L_h^2 and $\overline{L}_h^2$ inner products. For the definition of the subdifferential of a convex function, we refer to [24, Chap. I, Sect. 5.1]. To be able to distinguish the subdifferential with respect to $\langle\cdot,\cdot\rangle_h$ (which is the one we have been using until now) from the subdifferential with respect to $(\cdot,\cdot)_h$, from now on we will write $\partial_{L_h^2}$ for the former and $\partial_{\overline{L}_h^2}$ for the latter.

We recall from (2.2) that, since φ_{AC}^h is geodesically $(-\alpha)$-convex, we are interested in the shifted subdifferentials $-\partial_{L_h^2}\left(\varphi_{AC}^h + \frac{\alpha}{2}\|\cdot\|_h^2\right)(u) + \alpha u$ and

$$-\partial_{\overline{L}_h^2}\left(\varphi_{AC}^h + \frac{\alpha}{2}\|\cdot\|_{\overline{h}}^2\right)(u) + \alpha u.$$

Lemma 2.6 *Let $u \in \mathcal{V}_h$. The subdifferential of φ_D^h at u with respect to the L_h^2 inner product is the singleton $\partial_{L_h^2}\varphi_D^h(u) = \{-\Delta_h u\} \subset L_h^2$. Moreover,*

$$-\partial_{L_h^2}\left(\varphi_{AC}^h + \frac{\alpha}{2}\|\cdot\|_h^2\right)(u) + \alpha u = \{\Delta_h u - W' \circ u\} \subset L_h^2.$$

---

[12] Since Γ is self-adjoint, there is a spectral decomposition $\Gamma = \sum_j \lambda_j P_j$ where the P_j are the orthogonal projections onto the eigenspaces corresponding to the eigenvalues λ_j, which all are real. The summation is over finitely many j, since L_h^2 has finite dimension. Since Γ is positive definite, we have, for all j, $\lambda_j > 0$. Then $\Gamma^{1/2}$ is defined as

$$\Gamma^{1/2} := \sum_j \lambda_j^{1/2} P_j.$$

Since P_j is self-adjoint, this $\Gamma^{1/2}$ must be self-adjoint. This argument can be generalised to self-adjoint operators on a Hilbert space. For any real-valued continuous function f and self-adjoint operator A on a Hilbert space, $f(A)$ is defined by the spectral decomposition [64, Sect. XI.6, Theorem 1] and $f(A)$ is self-adjoint [64, Sect. XI.12, Theorem 3]. We also refer to [55, Chap. 10] for further details about square roots of positive-semidefinite self-adjoint operators.

The subdifferential of φ_D^h at u with respect to the $\overline{L}_h^2$ inner product is the singleton $\partial_{\overline{L}_h^2}\varphi_D^h(u) = \{-\Gamma^{-1}\Delta_h u\} \subset \overline{L}_h^2$. Furthermore,

$$-\partial_{\overline{L}_h^2}\left(\varphi_{AC}^h + \frac{\alpha}{2}\|\cdot\|_h^2\right)(u) + \alpha u = \{\Gamma^{-1}\Delta_h u - \Gamma^{-1}(W' \circ u)\} \subset L_h^2.$$

Proof First we note that, for all $v \in \mathcal{V}_h$,

$$\begin{aligned}\varphi_D^h(u+v) - \varphi_D^h(u) &= h^{-1}\sum_{k=0}^{\frac{1}{h}-1}\left[(u_{k+1}-u_k)(v_{k+1}-v_k) + \frac{1}{2}(v_{k+1}-v_k)^2\right]\\ &= \langle\nabla_h u, \nabla_h v\rangle_{h,h} + \frac{1}{2}\|\nabla_h v\|_{h,h}^2 \geq \langle -\Delta_h u, v\rangle_h,\end{aligned}$$

where we used (2.27) and (2.28). Since φ_D^h is convex, it follows from the definition of subdifferential that $-\Delta_h u \in \partial_{L_h^2}\varphi_D^h(u)$. Because the domain V_h of φ_D^h has finite dimension and since the functional is smooth, the subdifferential is in fact a singleton (and agrees with the gradient of the functional), thus $\partial_{L_h^2}\varphi_D^h(u) = \{-\Delta_h u\}$.

Moreover, by (2.33) we have $(-\Delta_h u, v)_h = \langle -\Gamma\Delta_h u, v\rangle_h$ and thus

$$\varphi_D^h(u+v) - \varphi_D^h(u) \geq (-\Gamma^{-1}\Delta_h u, v)_h.$$

As above, it follows that $-\Gamma^{-1}\Delta_h u \in \partial_{\overline{L}_h^2}\varphi_D^h(u)$ and, since the subdifferential is a singleton, in fact $\partial_{\overline{L}_h^2}\varphi_D^h(u) = \{-\Gamma^{-1}\Delta_h u\}$.

Since $W(\cdot) + \frac{\alpha}{2}\|\cdot\|_h^2$ is convex and differentiable, we have, for all $v \in \mathcal{V}_h$,

$$W(u+v) + \frac{\alpha}{2}\|u+v\|_h^2 - W(u) - \frac{\alpha}{2}\|u\|_h^2 \geq \langle W' \circ u + \alpha u, v\rangle_h.$$

Hence, for all $v \in \mathcal{V}_h$,

$$\varphi_{AC}^h(u+v) + \frac{\alpha}{2}\|u+v\|_h^2 - \varphi_{AC}^h(u) - \frac{\alpha}{2}\|u\|_h^2 \geq \langle -\Delta_h u + W' \circ u + \alpha u, v\rangle_h$$

and similarly

$$\begin{aligned}\varphi_{AC}^h(u+v) + \frac{\alpha}{2}\|u+v\|_{\overline{h}}^2 - \varphi_{AC}^h(u) - \frac{\alpha}{2}\|u\|_{\overline{h}}^2 &\geq \langle -\Delta_h u + W' \circ u + \alpha\Gamma u, v\rangle_h\\ &= -(\Gamma^{-1}\Delta_h u - \Gamma^{-1}(W' \circ u) - \alpha u, v)_h,\end{aligned}$$

where the equality follows from (2.33).

The remaining results now follow, using again the fact that the subdifferentials are singletons. □

2.5.5 Convergence Result

Our main result will be the convergence of discrete Allen–Cahn gradient flows to a continuum Allen–Cahn gradient flow, which we formalise in Theorem 2.6.

We remind ourselves that a discrete Allen–Cahn gradient flow is a solution of $\dot{u} \in -\partial_{L_h^2}\varphi_{AC}^h(u)$. It satisfies equation (2.25), which (by Lemma 2.6) for $n = 1$ is explicitly given by

$$\dot{u}_k = (\Delta_h u)_k - W'(u_k) \tag{2.41}$$

for $k \in \left[\frac{1}{h} - 1\right]_0$. A continuum Allen–Cahn flow is an $L^2(\mathbb{T})$-gradient flow of φ_{AC} in $L^2(\mathbb{T})$ and is given by (2.24) (with $n = 1$). Given initial data, the unique existence of such a flow is guaranteed by Proposition 2.1, since φ_{AC} is $(-\alpha)$-convex. The same proposition, or the Picard–Lindelöf theorem for ordinary differential equations (ODE) [37, Theorem I.3.1], also guarantees the unique existence of a discrete Allen–Cahn gradient flow, i.e., a solution of (2.41), given initial data.

Theorem 2.6 *Let $u_0 \in L^2(\mathbb{T})$ and, for all*[13] *$h \in \mathfrak{H}$, $u_0^h \in L_h^2$ (for $n = 1$). Assume that there exists an $h_* > 0$, such that $h \mapsto \|u_0^h\|_\infty$ is bounded on $(0, h_*) \cap \mathfrak{H}$. Furthermore, assume that $h\|\nabla u_0^h\|_{h,h} \to 0$ and $I_h u_0^h \to u_0$ in $L^2(\mathbb{T})$ as $h \to 0$ in $\mathfrak{H}$. Let u^h be the solution of (2.41) starting from u_0^h. Let u be the solution of (2.24) starting from u_0. Then, for all $T \in (0, \infty)$,*

$$\lim_{h\to 0} \sup_{t\in[0,T]} \|I_h u^h(t) - u(t)\|_{L^2(\mathbb{T})} = 0.$$

Remark 2.11 By Lemma 2.4, $\|\cdot\|_{L^2(\mathbb{T})} \circ i_h$ and $\|\cdot\|_{L^2(\mathbb{T})} \circ I_h$ are equivalent norms on $\mathcal{V}_h$. Hence the result of Theorem 2.6 also holds if the operator I_h is replaced by i_h in the assumptions of Theorem 2.6. Moreover, using Lemma 2.5 we have

$$\begin{aligned}\|i_h u^h(t) - u(t)\|_{L^2(\mathbb{T})} &\le \|i_h u^h(t) - I_h u(t)\|_{L^2(\mathbb{T})} + \|I_h u^h(t) - u(t)\|_{L^2(\mathbb{T})}\\ &\le h\left(\varphi_{AC}^h(u(t))\right)^{1/2} + \|I_h u^h(t) - u(t)\|_{L^2(\mathbb{T})}.\end{aligned}$$

By Lemma 2.2

$$\varphi_{AC}^h(u(t)) \le \varphi_{AC}^h(u_0^h) \le \frac{1}{2}\|\nabla_h u_0^h\|_{h,h}^2 + C_h,$$

with $C_h := \sup\left\{W(s) \mid |s| \le \|u_0^h\|_\infty\right\} = \max\left(W(0), W(\|u_0^h\|_\infty)\right)$. (The last equality follows from the specific definition of W that we are using.) Hence, for $h \in (0, h_*) \cap \mathfrak{H}$,

$$h\left(\varphi_{AC}^h(u(t))\right)^{1/2} \le \frac{h}{\sqrt{2}}\|\nabla_h u_0^h\|_{h,h} + Ch,$$

where $C := \max_{h\in(0,h_*)\cap\mathfrak{H}} C_h$. Thus, by the assumptions on the initial condition u_0^h,

[13] We can also restrict ourselves to a proper subset of $\mathfrak{H}$, as long as it contains a sequence converging to zero.

$$\lim_{h\to 0} h\left(\varphi_{AC}^h(u(t))\right)^{1/2} = 0.$$

Hence we can also replace I_h by i_h in the conclusion of Theorem 2.6 and the result still holds.

Unfortunately, we cannot apply our abstract Theorem 2.2 immediately since the embedding mapping I_h is not an isometry from L_h^2 to $L^2(\mathbb{T})$. To circumvent this difficulty, we proceed in two steps.

First we consider the gradient flow of φ_{AC}^h with respect to the $\overline{L}_h^2$ inner product. By Lemma 2.6 its explicit form is

$$\dot{U}_k = \left(\Gamma^{-1}\left(\Delta_h U - W' \circ U\right)\right)_k, \tag{2.42}$$

for $k \in \left[\frac{1}{h} - 1\right]_0$. As for (2.41), standard ODE techniques guarantee that, given initial data, a unique solution to (2.42) exists. Since $I_h : \overline{L}_h^2 \to L^2(\mathbb{T})$ *is* an isometry, we can apply Theorem 2.2 to compare a solution of (2.42) with a solution of (2.24). The following theorem makes this precise.

Theorem 2.7 *Assume the same hypotheses concerning u_0, u_0^h, and u as in Theorem 2.6. Let U^h be the solution of* (2.42) *starting from u_0^h. Then, for all $T \in (0, \infty)$,*

$$\lim_{h\to 0} \sup_{t\in[0,T]} \|I_h U^h(t) - u(t)\|_{L^2(\mathbb{T})} = 0.$$

We shall prove Theorem 2.7 as an application of Theorem 2.2 in Sect. 2.5.8 with $M_h = \overline{L}_h^2$ and $M = L^2(\mathbb{T})$.

The second step in our proof of Theorem 2.6 is a comparison between u^h, the solution of (2.41), and U^h, the solution of (2.42).

Theorem 2.8 *Assume the same hypotheses concerning u_0^h and u^h as in Theorem 2.6 and the same hypothesis concerning U^h as in Theorem 2.7. Then, for all $T \in (0, \infty)$,*

$$\lim_{h\to 0} \sup_{t\in[0,T]} \|I_h u^h(t) - I_h U^h(t)\|_{L^2(\mathbb{T})} = 0.$$

We defer the proof of Theorem 2.8 to Sect. 2.5.6.

Proof (Proof of Theorem 2.6) Our main convergence result Theorem 2.6 follows from Theorems 2.7 and 2.8. □

Remark 2.12 Since we allow the choice $T = \infty$ in Theorem 2.2, one may wonder why we do not allow this choice in Theorems 2.7 and 2.8 and consequently also not in Theorem 2.6.

This choice would introduce nontrivial difficulties, such as "infinite" constants N and C_N in (2.49) and (2.50), respectively, in the proof of Theorem 2.9 (which in turn is needed in the proof of Theorem 2.8) and the unboundedness of $t \mapsto e^{\alpha t}$ in (2.57) in the proof of Theorem 2.7.

The remaining parts of Sect. 2.5 are organised as follows. We actually prove our "second step" (Theorem 2.8) first, in Sect. 2.5.6. The reason for this reordering of our steps is that one of the results from that section, Lemma 2.42, is also required in our proof of Theorem 2.7. In Sect. 2.5.6 we prove Theorem 2.9 and in Sect. 2.5.8 Theorem 2.7, which also completes the proof of Theorem 2.6. This proof is an application of Theorem 2.2. We thus require a projection P_h as a counterpart to the embedding I_h. We introduce P_h in Sect. 2.5.7 and prove some of its properties.

2.5.6 Comparison of Two Discrete Flows (proof of Theorem 2.8)

In this subsection, we prove Theorem 2.8. In fact we consider equations more general than (2.42) and (2.41). Theorem 2.8 will then follow as a special case of Theorem 2.9.

As usual, if there is no further specification, we assume that $h \in \mathfrak{H}$. Up to the proof of Theorem 2.8 at the end of this section, h is fixed; therefore we have suppressed the dependence on h in the notation u and U for solutions of (2.43) and (2.44), respectively.

Let $F : \mathbb{R} \to \mathbb{R}$ be a continuously differentiable function with $F(0) = 0$ and, for all $x \in \mathbb{R}$, $F(x)x \geq 0$. Let $\mu \in \mathbb{R}$. We consider, for all $k \in \left[\frac{1}{h} - 1\right]_0$,

$$\dot{u}_k = (\Delta_h u)_k + \mu u_k - F(u_k), \tag{2.43}$$

as a generalisation of (2.41). We recover (2.41) if $F(x) = W'(x) + \mu x$ and $\mu > \alpha$. We note that $W'(x) = \alpha x(x^2 - 1)$ and $W''(x) = \alpha(3x^2 - 1)$, so that in this case indeed $F(0) = W'(0) = 0$ and $F'(x) = W''(x) + \mu > 3\alpha x^2 \geq 0$.

As a generalisation of (2.42), we consider, for all $k \in \left[\frac{1}{h} - 1\right]_0$,

$$\Gamma \dot{U}_k = (\Delta_h U + \mu U - F \circ U)_k . \tag{2.44}$$

By the Picard–Lindelöf theorem [37, Theorem I.3.1], the initial value problems corresponding to (2.43) and (2.44) are guaranteed to have unique continuously differentiable solutions locally in time, i.e., at least up to some time $T^* > 0$. By $C^1([0, T^*]; \mathcal{V}_h)$ we denote the set of functions $u : [0, T^*] \to \mathcal{V}_h$ that are continuously differentiable (considering one-sided derivatives at the endpoints of the interval). Any operator (such as Δ_h or Γ) which is defined on $\mathcal{V}_h$ has a natural action on $C^1([0, T^*]; \mathcal{V}_h)$, defined by $(\Delta_h u)(t) = \Delta_h u(t)$, $(\Gamma u)(t) = \Gamma u(t)$, etc.

We start by proving a minimum principle for supersolutions and a comparison principle for solutions of (2.43). If U solves equation (2.44), then it satisfies the required assumptions of part (i) of Proposition 2.2 below, since, for all $k \in \left[\frac{1}{h} - 1\right]_0$,

$$b_k = -\mu + \int_0^1 F'(\theta U_k)\, d\theta$$

is bounded below if U_k is bounded. This is the case for continuous solutions U_k on the compact interval $[0, T]$.

Proposition 2.2

(i) (Minimum principle). Let $T > 0$ and assume that there exist a function $b : [0, T] \times V_h \to \mathbb{R}$ that is bounded below and a function $v \in C^1([0, T]; \mathcal{V}_h)$ such that, for all $k \in \left[\frac{1}{h} - 1\right]_0$,

$$\dot{v}_k \geq (\Delta_h v)_k - b_k v_k.$$

If, for all $k \in \left[\frac{1}{h} - 1\right]_0$, $v_k(0) \geq 0$, then, for all $k \in \left[\frac{1}{h} - 1\right]_0$ and for all $t \in [0, T]$, $v_k(t) \geq 0$.

(ii) (Growth estimate). Let $T > 0$ and assume that $u \in C^1([0, T]; \mathcal{V}_h)$ is a solution of (2.43) starting from $u_0 \in \mathcal{V}_h$. Then, for all $t \in [0, T]$, $\|u(t)\|_\infty \leq \|u_0\|_\infty e^{\mu t}$.

Proof (i) Without loss of generality we may assume that $b < 0$, for if it is not, let $\tilde{\mu} \in \mathbb{R}$ be such that

$$\tilde{\mu} > - \inf_{(k,t) \in \left[\frac{1}{h} - 1\right]_0 \times [0,T]} b_k(t).$$

Since b is bounded below, this is well-defined. Then we consider the following inequality for $\tilde{v}(t) := e^{-\tilde{\mu}t} v(t)$:

$$\begin{aligned}\dot{\tilde{v}}(t) = -\tilde{\mu} e^{-\tilde{\mu}t} v(t) + e^{-\tilde{\mu}t} \dot{v}(t) &\geq -\tilde{\mu}\tilde{v}(t) + e^{-\tilde{\mu}t} \Delta_h v(t) - e^{-\tilde{\mu}t} b(t) v(t) \\ &= \Delta_h \tilde{v}(t) - \big(b(t) + \tilde{\mu}\big)\tilde{v}(t).\end{aligned}$$

By definition of $\tilde{\mu}$, we have $b + \tilde{\mu} > 0$. Moreover, $\tilde{v}_k(t) \geq 0$ if and only if $v_k(t) \geq 0$.

Suppose that the conclusion of part (i) were false. Since v is continuous in t, the minimum of $v_k(t)$ over all $(k, T) \in \left[\frac{1}{h} - 1\right]_0 \times [0, T]$ exists and is negative (i.e., < 0).

Take a minimiser $(k_0, t_0) \in \left[\frac{1}{h} - 1\right]_0 \times (0, T]$. Then $\dot{v}_{k_0}(t_0) \leq 0$ and

$$(\Delta_h v)_{k_0}(t_0) = \frac{(v_{k_0+1}(t_0) - v_{k_0}(t_0)) + (v_{k_0-1}(t_0) - v_{k_0}(t_0))}{h^2} \geq 0,$$

and thus $0 \geq -b_{k_0}(t_0) v_{k_0}(t_0)$. This contradicts $v_{k_0}(t_0) < 0$.

(ii) We will prove that, for all $t \in [0, T]$,

$$\max_k u_k(t) \leq \|u_0\|_\infty e^{\mu t}.$$

A symmetric argument, of which we will not give the details, yields the estimate from below by $-\|u_0\|_\infty e^{\mu t}$.
Define $\overline{u} \in C^1(\mathbb{R}; \mathcal{V}_h)$ by $\overline{u}(t) := \|u_0\|_\infty e^{\mu t}$. We observe that, for all $t \in \mathbb{R}$, $\overline{u}(t)$ is constant on V_h and thus $\Delta_h \overline{u} = 0$. Moreover, for all $t \in \mathbb{R}$ and all $k \in \left[\frac{1}{h} - 1\right]_0$, $(F \circ \overline{u})_k(t) = F\left(\|u_0\|_\infty e^{\mu t}\right) \geq 0$, since $F(0) = 0$ and $F(x) \geq 0$ for strictly positive[14] x. Thus $\overline{u}$ is a supersolution of (2.43), i.e., for all $t \in \mathbb{R}$,

$$\dot{\overline{u}}(t) = \mu\overline{u}(t) = \Delta_h\overline{u}(t) + \mu\overline{u}(t) \geq \Delta_h\overline{u}(t) + \mu\overline{u}(t) - F \circ \overline{u}(t).$$

Thus $v := \overline{u} - u$ solves, for all $t \in [0, T]$,

$$\dot{v} \geq \Delta_h v + \mu v - F \circ \overline{u} + F \circ u = \Delta_h v - bv,$$

where we defined, for all $k \in \left[\frac{1}{h} - 1\right]_0$ and for all $t \in [0, T]$,

$$b_k(t) := -\mu + \int_0^1 F'\left(\theta\overline{u}_k(t) + (1-\theta)u_k(t)\right) d\theta.$$

Since u and $\overline{u}$ are continuous, the sets $u([0, T])$ and $\overline{u}([0, T])$ are compact and thus, by Tychonoff's theorem so is $[0, 1] \times u([0, T]) \times \overline{u}([0, T])$. Since the function $\mathbb{R}^3 \to \mathbb{R}$, $(\theta, x, y) \mapsto \theta x + (1-\theta)y$ is continuous, the set $\Theta := \left\{\theta x + (1-\theta)y \in \mathbb{R} \;\middle|\; (\theta, x, y) \in [0, 1] \times u([0, T]) \times \overline{u}([0, T])\right\}$ is compact. Because F' is continuous, the set $F'(\Theta)$ is compact and in particular bounded. This implies that the function b is bounded and thus we can apply part (i) of the current proposition. Doing so yields, for all $t \in [0, T]$ and for all $k \in \left[\frac{1}{h} - 1\right]_0$, $v_k(t) \geq 0$. The proof is now complete.

□

Remark 2.13 As a corollary to Proposition 2.2, the results from that proposition can be extended to include the case $T = \infty$, if we then interpret $t \in [0, T]$ to mean $t \in [0, \infty)$. For a given $t \in [0, \infty)$, if we let $\tilde{T} > 0$ be large enough such that $t \in [0, \tilde{T}]$, then the results of both parts of Proposition 2.2 hold for that t.

Remark 2.14 We note that the growth estimate from Proposition 2.2 (ii) allows us to extend our earlier local-in-time existence results for the initial value problems associated with (2.43) to global-in-time existence results.[15] Given $u_0^h \in \mathcal{V}_h$, by the

---

[14] To avoid ambiguities regarding zero, we call numbers from $(0, \infty)$ and $(-\infty, 0)$ strictly positive and strictly negative, respectively.

[15] We remark that Proposition 2.1 also gives us global-in-time existence—based on an energy principle—but without guarantee of differentiability.

local-in-time result there exists $T^* > 0$ such that a solution $u^h \in C^1([0, T^*]; \mathcal{V}_h)$ exists of (2.43) with initial condition $u^h(0) = u_0^h$. Since the right-hand side of (2.43) is defined for all $u \in \mathcal{V}_h$ and continuous with respect to u, by standard ODE results (e.g., [37, Theorem I.2.1]), the maximal right-interval of existence of the solution is $[0, \infty)$ unless there exists a time T^{**} such that the solution can be extended to $[0, T^{**})$ and $\lim_{t \to T^{**}} u(t) = \pm\infty$. By the growth estimate it follows that such a T^{**} does not exist and thus the solution exists for all $t > 0$.

In the following proposition we compare a solution u of (2.43) with Γu.

Proposition 2.3 *Let $h \in \mathfrak{H}$ and let u be a solution of* (2.43) *with initial data $u_0 \in \mathcal{V}_h$. Then, for all t in the domain of u,*

$$e^{-2\mu t} \|u(t) - \Gamma u(t)\|_h^2 + 2 \int_0^t e^{-2\mu s} \|\nabla_h (u(s) - \Gamma u(s))\|_{h,h}^2 \, ds \le \frac{h^2}{9} \|\nabla_h u_0\|_{h,h}^2.$$

Proof To simplify notation, we suppress the explicit dependence on t (which is assumed to be in the domain of u) in the first half of this proof.

Remembering the definition of $\tau_\pm$ from (2.36), we have

$$3\,(u - \Gamma u) = \frac{1}{2} \left[(u - \tau_+ u) + (u - \tau_- u) \right]. \tag{2.45}$$

Hence

$$\|u - \Gamma u\|_h^2 = \frac{1}{6^2} \|(u - \tau_+ u) + (u - \tau_- u)\|_h^2 \le \frac{1}{18} \left(\|u - \tau_+ u\|_h^2 + \|u - \tau_- u\|_h^2 \right) = \frac{1}{18} \left(\|v_+\|_h^2 + \|v_-\|_h^2 \right),$$

where we used the notation $v_\pm := u - \tau_\pm u$. We also used that $\|a + b\|^2 \le (\|a\| + \|b\|)^2 = \|a\|^2 + \|b\|^2 + 2\|a\|\|b\| \le 2\|a\|^2 + 2\|b\|^2$, which holds for any norm. Because ∇_h is a linear operator, we thus also have

$$\|\nabla_h (u - \Gamma u)\|_{h,h}^2 = \frac{1}{6^2} \|\nabla_h v_+ + \nabla_h v_-\|_{h,h}^2 \le \frac{1}{18} \left(\|\nabla_h v_+\|_{h,h}^2 + \|\nabla_h v_-\|_{h,h}^2 \right).$$

Since $\tau_\pm u$ solves (2.43), calculating the inner product $\langle u - \tau_\pm u, \dot{u} - \tau_\pm \dot{u} \rangle_h = \frac{1}{2} \frac{d}{dt} \|v_\pm\|_h^2$ yields

$$\frac{1}{2} \frac{d}{dt} \|v_\pm\|_h^2 - \mu \|v_\pm u\|_h^2 + \|\nabla_h v_\pm\|_{h,h}^2 = -\langle u - \tau_\pm u, F \circ u - F \circ \tau_\pm u \rangle_h.$$

Here we used (2.27). The last term is nonpositive since F is nondecreasing. This implies

$$\frac{d}{dt}\left(e^{-2\mu t}\|v_\pm(t)\|_h^2\right) + 2e^{-2\mu t}\,\|\nabla_h v_\pm(t)\|_{h,h}^2$$
$$= e^{-2\mu t}\left(\frac{d}{dt}\|v_\pm(t)\|_h^2 - 2\mu\|v_\pm(t)\|_h^2 + 2\,\|\nabla_h v_\pm(t)\|_{h,h}^2\right) \le 0.$$

Integrating this inequality over $(0, t)$ and using (2.38), we find

$$\begin{aligned} 0 &\ge e^{-2\mu t}\|v_\pm(t)\|_h^2 - \|v_\pm(0)\|_h^2 + 2\int_0^t e^{-2\mu s}\,\|\nabla_h v_\pm(s)\|_{h,h}^2\,ds \\ &= e^{-2\mu t}\|v_\pm(t)\|_h^2 - h^2\|\nabla_h u_0\|_{h,h}^2 + 2\int_0^t e^{-2\mu s}\,\|\nabla_h v_\pm(s)\|_{h,h}^2\,ds. \end{aligned} \tag{2.46}$$

Recalling the estimates for $\|u - \Gamma u\|_h^2$ and $\|\nabla_h(u - \Gamma u)\|_{h,h}^2$ earlier in this proof, the desired inequality follows:

$$\begin{aligned} &e^{-2\mu t}\|u(t) - \Gamma u(t)\|_h^2 + 2\int_0^t e^{-2\mu s}\,\|\nabla_h(u(s) - \Gamma u(s))\|_{h,h}^2\,ds \\ &\le \frac{1}{18}\Big(e^{-2\mu t}\|v_+(t)\|_h^2 + 2\int_0^t e^{-2\mu s}\,\|\nabla_h v_+(s))\|_{h,h}^2\,ds \\ &\qquad + e^{-2\mu t}\|v_-(t)\|_h^2 + 2\int_0^t e^{-2\mu s}\,\|\nabla_h v_-(s))\|_{h,h}^2\,ds\Big) \\ &\le \frac{h^2}{9}\|\nabla_h u_0\|_{h,h}^2. \end{aligned}$$

□

Theorem 2.9 *Let $h \in \mathfrak{H}$, $\mu \ge 0$, and $T \in (0, \infty)$. Assume $u \in C^1([0, T]; \mathcal{V}_h)$ solves (2.43) starting from $u_0 \in \mathcal{V}_h$ and $U \in C^1([0, T]; \mathcal{V}_h)$ solves (2.44) starting from $U_0 \in \mathcal{V}_h$. Then there exists a constant C_N, depending only on μ, T, and $\|u_0\|_\infty$ (details in (2.49) and (2.50)), such that, for all $t \in [0, T]$ and for all $\delta > 0$,*

$$\|\Gamma^{1/2}(u(t) - U(t))\|_h^2 \le \left[h^2 e^{2\mu T}\left(\frac{1}{18} + \frac{C_N}{\mu\delta}\right)\|\nabla_h u_0\|_{h,h}^2 + \Gamma^{1/2}(u_0 - U_0)\|_h^2\right] e^{6(\mu+\delta)t}.$$

In particular, if $u_0 = U_0$, then, for all $t \in [0, T]$ and all $\delta > 0$, there exists a $C_* > 0$ depending on μ, δ, T, and $\|u_0\|_\infty$ (details in (2.51)), such that

$$\|u(t) - U(t)\|_h^2 \le C_* h^2 \|\nabla_h u_0\|_{h,h}^2 e^{6(\mu+\delta)t}.$$

Proof Let $h \in \mathfrak{H}$. In most lines of the proof, to simplify the notation we will not write the explicit dependence on $t \in [0, T]$.

Similarly as in the proof of Proposition 2.3, we shall calculate $(v, v)_h = \langle v, \Gamma v\rangle_h = \|\Gamma^{1/2} v\|_h^2$ with $v := u - U$. Using equations (2.43) for u and (2.44) for U, together with the self-adjointness of Γ and the fact that Γ and $\frac{d}{dt}$ commute, we obtain that

$$\begin{aligned}\frac{1}{2}\frac{d}{dt}\|\Gamma^{1/2}v\|_h^2 &= \langle v, \Gamma \dot{v}\rangle_h \\ &= \underbrace{\langle v, \Gamma\Delta_h u - \Delta_h U\rangle_h}_{=:I_1} \underbrace{- \langle v, \Gamma F\circ u - F\circ U\rangle_h}_{=:I_2} + \underbrace{\mu\langle v, \Gamma u - U\rangle_h}_{=:I_3}. \end{aligned} \tag{2.47}$$

We proceed by computing I_1, using (2.27):

$$\begin{aligned} I_1 &= \langle v, \Delta_h v\rangle_h + \langle v, \Gamma\Delta_h u - \Delta_h u\rangle_h \\ &= -\langle \nabla_h v, \nabla_h v\rangle_{h,h} - \langle \nabla_h v, \nabla_h(\Gamma u - u)\rangle_{h,h} \end{aligned}$$

since $\Delta_h\Gamma = \Gamma\Delta_h$. Applying Young's inequality $ab \le (a^2+b^2)/2$, we obtain

$$\begin{aligned} I_1 &\le -\|\nabla_h v\|_{h,h}^2 + \frac{1}{2}\|\nabla_h v\|_{h,h}^2 + \frac{1}{2}\|\nabla_h(\Gamma u - u)\|_{h,h}^2 \\ &= -\frac{1}{2}\|\nabla_h v\|_{h,h}^2 + \frac{1}{2}\|\nabla_h(u - \Gamma u)\|_{h,h}^2 . \end{aligned}$$

We next calculate I_2. Since F is nondecreasing, we see that

$$\begin{aligned} I_2 &= -\langle v, F\circ u - F\circ U\rangle_h - \langle v, \Gamma F\circ u - F\circ u\rangle_h \le -\langle v, \Gamma F\circ u - F\circ u\rangle_h \\ &\le \|v\|_h\, \|\Gamma F\circ u - F\circ u\|_h, \end{aligned} \tag{2.48}$$

where we used the Cauchy–Schwarz inequality to obtain the last inequality. Because

$$(\Gamma F\circ u)_k = \frac{2}{3}F(u_k) + \frac{1}{6}F(u_{k-1}) + \frac{1}{6}F(u_{k+1}),$$

a similar observation as the one for $u - \Gamma u$ in (2.45) shows that

$$F\circ u - \Gamma F\circ u = \frac{1}{6}\big[(F\circ u - F\circ\tau_+ u) + (F\circ u - F\circ\tau_- u)\big].$$

By Proposition 2.2 (ii), we know $\|u\|_\infty(t) \le \|u_0\|_\infty e^{\mu t}$. Thus, pointwise on V_h,

$$\begin{aligned} |F\circ u(t) - F\circ\tau_\pm u(t)| &\le \left|\int_0^1 F'(\theta u(t) + (1-\theta)\tau_\pm u(t))\, d\theta\right| |u(t) - \tau_\pm u(t)| \\ &\le N\,|u(t) - \tau_\pm u(t)| \end{aligned}$$

with

$$N := \sup\left\{F'(s) \mid |s| \le \|u_0\|_\infty e^{\mu T}\right\}. \tag{2.49}$$

This estimate, combined with (2.48) and the triangle inequality, yields

$$\begin{aligned} I_2 &\le \frac{1}{6}\|v\|_h \left(\|F\circ u - F\circ \tau_+ u\|_h + \|F\circ u - F\circ \tau_- u\|_h\right) \\ &\le \frac{N}{6}\|v\|_h \left(\|u - \tau_+ u\|_h + \|u - \tau_- u\|_h\right). \end{aligned}$$

Similarly, using (2.45) and the Cauchy–Schwarz and triangle inequalities again, we estimate

$$\begin{aligned} I_3 &= \mu\|v\|_h^2 + \mu\langle v, \Gamma u - u\rangle_h \\ &\le \mu\|v\|_h^2 + \frac{\mu}{6}\|v\|_h \left(\|u - \tau_+ u\|_h + \|u - \tau_- u\|_h\right). \end{aligned}$$

Let $\delta > 0$. Applying Young's inequality, $ab \le \frac{\delta}{2}a^2 + \frac{1}{2\delta}b^2$, to the estimates for I_2 and I_3 above, we find

$$\begin{aligned} I_2 &\le \frac{\delta}{2}\|v\|_h^2 + \frac{1}{2\delta}\left(\frac{N}{6}\right)^2 \left(\|u - \tau_+ u\|_h + \|u - \tau_- u\|_h\right)^2, \\ I_3 &\le \mu\|v\|_h^2 + \frac{\delta}{2}\|v\|_h^2 + \frac{1}{2\delta}\left(\frac{\mu}{6}\right)^2 \left(\|u - \tau_+ u\|_h + \|u - \tau_- u\|_h\right)^2. \end{aligned}$$

Combining our estimates for I_1, I_2, and I_3 with (2.47) we find

$$\begin{aligned} &\frac{1}{2}\frac{d}{dt}\|\Gamma^{1/2} v\|_h^2(t) + \frac{1}{2}\|\nabla_h v\|_{h,h}^2 \\ &\quad \le \frac{1}{2}\|\nabla_h(u - \Gamma u)\|_{h,h}^2 + (\mu + \delta)\|v\|_h^2 + \frac{C_N}{2\delta}\left(\|u - \tau_+ u\|_h^2 + \|u - \tau_- u\|_h^2\right), \end{aligned}$$

with

$$C_N := \frac{N^2 + \mu^2}{36}. \tag{2.50}$$

Integrating over $(0, t)$ and multiplying by 2, we get, by rearranging terms,

$$\begin{aligned} &\|\Gamma^{1/2} v(t)\|_h^2 + \int_0^t \|\nabla_h v(s)\|_{h,h}^2\, ds \le \int_0^t \|\nabla_h(u - \Gamma u)(s)\|_{h,h}^2\, ds \\ &+ 2(\mu + \delta)\int_0^t \|v(s)\|_h^2\, ds \\ &+ \frac{C_N}{\delta}\left(\int_0^t \|u(s) - \tau_+ u(s)\|_h^2\, ds + \int_0^t \|u(s) - \tau_- u(s)\|_h^2\, ds\right) + \|\Gamma^{1/2}(u_0 - U_0)\|_h^2. \end{aligned}$$

By Proposition 2.3 we have

$$\int_0^t \|\nabla_h(u(s)-\Gamma u(s))\|_{h,h}^2\, ds \le e^{2\mu T}\int_0^t e^{-2\mu s}\,\|\nabla_h(u(s)-\Gamma u(s))\|_{h,h}^2\, ds$$
$$\le \frac{h^2}{18}e^{2\mu T}\,\|\nabla_h u_0\|_{h,h}^2.$$

From (2.46) in the proof of Proposition 2.3, we see that

$$e^{-2\mu t}\,\|u(t)-\tau_{\pm}u(t)\|_h^2 + 2\int_0^t e^{-2\mu s}\,\|\nabla_h(u-\tau_{\pm}u)(s)\|_{h,h}^2\, ds \le h^2\|\nabla_h u_0\|_{h,h}^2.$$

This yields

$$\int_0^t \|u(s)-\tau_{\pm}u(s)\|_h^2\, ds \le h^2\|\nabla_h u_0\|_{h,h}^2\int_0^t e^{2\mu s}\, ds = \frac{h^2}{2\mu}\|\nabla_h u_0\|_{h,h}^2\,(e^{2\mu t}-1)$$
$$\le \frac{h^2}{2\mu}e^{2\mu T}\,\|\nabla_h u_0\|_{h,h}^2.$$

Applying these estimates, we deduce that

$$\|\Gamma^{1/2}v(t)\|_h^2 + \int_0^t \|\nabla_h v(s)\|_{h,h}^2\, ds$$
$$\le h^2 e^{2\mu T}\left(\frac{1}{18}+\frac{C_N}{\mu\delta}\right)\|\nabla_h u_0\|_{h,h}^2 + \|\Gamma^{1/2}(u_0-U_0)\|_h^2$$
$$+\,2(\mu+\delta)\int_0^t \|v(s)\|_h^2\, ds.$$

After applying $\|v\|_h^2 \le 3\|\Gamma^{1/2}v\|_h^2$ from (2.35) to the integrand on the right-hand side, Grönwall's inequality [37, Corollary I.6.6] yields, for all $t \in [0, T]$,

$$\|\Gamma^{1/2}v(t)\|_h^2 \le \left[h^2 e^{2\mu T}\left(\frac{1}{18}+\frac{C_N}{\mu\delta}\right)\|\nabla_h u_0\|_{h,h}^2 + \|\Gamma^{1/2}(u_0-U_0)\|_h^2\right]e^{6(\mu+\delta)t}.$$

This proves the first result.

In the special case where $u_0 = U_0$, we find

$$\|v(t)\|_h^2 \le C_* h^2\|\nabla_h u_0\|_{h,h}^2 e^{6(\mu+\delta)t},$$

with

$$C_* := e^{2\mu T}\left(\frac{1}{6}+\frac{3C_N}{\mu\delta}\right). \tag{2.51}$$

This proves the second desired estimate. □

We can now prove Theorem 2.8.

Proof (of Theorem 2.8) Let $h \in \mathfrak{H}$, $\delta > 0$, and $t \in [0, T]$. By Lemma 2.4 and Theorem 2.9 (with $u = u^h$, $U = U^h$, $\mu > \alpha$, and $F(x) = W'(x) + \mu x$) we have

$$\|I_h U^h(t) - I_h u^h(t)\|_{L^2(\mathbb{T})} = \|\Gamma^{1/2}(U^h(t) - u^h(t))\|_h \leq C_*^{1/2} h \|\nabla_h u_0^h\|_{h,h} e^{3(\mu+\delta)t}.$$

By assumption we also have that $\|u_0^h\|_\infty$ is bounded uniformly in h, for h small enough. Consequently so is $C_*^{1/2}$ (which depends on h only through $\|u_0^h\|_\infty$).

First taking the supremum over $t \in [0, T]$ in the inequality above and subsequently the limit for $h \to 0$ yields Theorem 2.8, since, by assumption, $h\|\nabla_h u_0^h\|_{h,h} \to 0$ as $h \to 0$. □

2.5.7 Projections

Let $h \in \mathfrak{H}$. We define H_h to be the space of real-valued continuous piecewise-linear functions on $\mathbb{T}$ whose derivatives can only have jumps at the points $kh \in \mathbb{T}$ with $k \in \{0, \ldots, 1/h - 1\}$. (As an aside we observe that, if $f \in H_{h,i}$—see (2.66)—then, keeping all coordinates but x_i fixed, the function $x_i \mapsto f(x)$ is in H_h.) We note that $I_h \overline{L}_h^2 = H_h$. Moreover, H_h is a linear subspace of $L^2(\mathbb{T})$ and, having finite dimension, it is also closed.

To be able to use Theorem 2.2 to prove Theorem 2.7, we require a projection P_h corresponding to the embedding $I_h : \overline{L}_h^2 \to L^2(\mathbb{T})$, as explained in Sect. 2.2. We will define this projection using a few intermediary functions, which will prove useful in their own right.

We define $\overline{P}_h : L^2(\mathbb{T}) \to H_h$ to be the orthogonal projection from $L^2(\mathbb{T})$ to H_h. This is well-defined, since H_h is a finite-dimensional, and thus closed, subspace of the Hilbert space $L^2(\mathbb{T})$. As one of the standard properties of orthogonal projections, it follows immediately that $\overline{P}_h$ satisfies, for all $w \in L^2(\mathbb{T})$,

$$\|\overline{P}_h w\|_{L^2(\mathbb{T})} \leq \|w\|_{L^2(\mathbb{T})}. \tag{2.52}$$

Moreover, for all $w \in L^2(\mathbb{T})$, there exists a unique $w^\perp \in H_h^\perp$ (where $H_h^\perp$ denotes the orthogonal complement of H_h in $L^2(\mathbb{T})$), such that $w = \overline{P}_h w + w^\perp$. Since $1 \in H_h$, we have

$$\int_{\mathbb{T}} \overline{P}_h w(x)\, dx = \int_{\mathbb{T}} w(x) \cdot 1\, dx - \int_{\mathbb{T}} w^\perp(x) \cdot 1\, dx = \int_{\mathbb{T}} w(x)\, dx. \tag{2.53}$$

With $C(\mathbb{T})$ being the set of continuous functions on $\mathbb{T}$, in this one-dimensional case we define the projection $\overline{\pi}_h : C(\mathbb{T}) \to \overline{L}_h^2$ by, for all $w \in C(\mathbb{T})$,

$$(\overline{\pi}_h w)(kh) = w(kh), \qquad \text{for } k \in \left[\frac{1}{h} - 1\right]_0.$$

Now we define $P_h : L^2(\mathbb{T}) \to \overline{L}_h^2$ as $P_h := \overline{\pi}_h \circ \overline{P}_h$. Since the restricted operator $\overline{P}_h|_{H_h}$ is the identity map Id_{H_h} on H_h, we have that $P_h \circ I_h = \overline{\pi}_h \circ \overline{P}_h \circ I_h = \overline{\pi}_h \circ \mathrm{Id}_{H_h} \circ I_h = \mathrm{Id}_{\overline{L}_h^2}$, the identity map on $\overline{L}_h^2$, as required by our discussion in Sect. 2.2.

Furthermore, we note that $I_h \circ \overline{\pi}_h|_{H_h} = \mathrm{Id}_{H_h}$. Hence $I_h \circ P_h = I_h \circ \overline{\pi}_h \circ \overline{P}_h = \overline{P}_h$, and thus for all $w \in L^2(\mathbb{T})$ we have, by (2.52),

$$\|P_h w\|_{\overline{h}} = \|I_h \circ P_h w\|_{L^2(\mathbb{T})} = \|\overline{P}_h w\|_{L^2(\mathbb{T})} \leq \|w\|_{L^2(\mathbb{T})}.$$

Hence the non-strict contraction requirement from (2.4) is also satisfied. Moreover, using again that $\overline{P}_h$ is an orthogonal projection, we have, for all $v \in \overline{L}_h^2$ and for all $w \in L^2(\mathbb{T})$,

$$\begin{aligned}\|v - P_h w\|_{\overline{h}}^2 + \|I_h P_h w - w\|_{L^2(\mathbb{T})}^2 &= \|I_h v - I_h P_h w\|_{L^2(\mathbb{T})}^2 + \|I_h P_h w - w\|_{L^2(\mathbb{T})}^2 \\ &= \|I_h v - \overline{P}_h w\|_{L^2(\mathbb{T})}^2 + \|\overline{P}_h w - w\|_{L^2(\mathbb{T})}^2 \\ &= \|I_h v - w\|_{L^2(\mathbb{T})}^2.\end{aligned}$$

For the final equality, we used that $I_h v \in H_h$. This establishes that P_h satisfies condition (2.7). We postpone our check of condition (2.9) to the proof of Theorem 2.7 in Sect. 2.5.8.

Our goal in the remainder of Sect. 2.5.7 is to establish some other useful properties of $\overline{P}_h$, which we will need in our proof of Theorem 2.7. We collect these results in Theorem 2.10: in particular, Theorem 2.10 parts (i) and (iv) are used in the proof of Theorem 2.7, while parts (i) and (ii) are used in the proof of part (iii), which in turn is used to prove part (iv). In numerical analysis, the properties in parts (ii) and (iii) combined sometimes go by the name of stability in $H^1(\mathbb{T})$.

Theorem 2.10

(i) For all $w \in H^1(\mathbb{T})$ the estimate

$$\left\|(\overline{P}_h w)'\right\|_{L^2(\mathbb{T})} \leq C\|w'\|_{L^2(\mathbb{T})}$$

holds with $C := 4\sqrt{3}/\pi + 2$.

(ii) For all $w \in H^1(\mathbb{T})$, $\lim_{h\to 0} \left\|(\overline{P}_h w - w)'\right\|_{L^2(\mathbb{T})} = 0$.

(iii) For all $w \in L^2(\mathbb{T})$, $\lim_{h\to 0} \left\|\overline{P}_h w - w\right\|_{L^2(\mathbb{T})} = 0$.

(iv) For all $w \in H^1(\mathbb{T})$, $\overline{P}_h w \to w$ uniformly, as $h \to 0$.

In [32, Theorem 1.5], this is proved using the Ritz operator, i.e., the projection operator from $H^1(\mathbb{T})$ to H_h, and elliptic regularity in a general domain. A more general estimate in $L^1(\mathbb{T})$ and $W^{1,p}(\mathbb{T})$ has been established in [18, Theorem 2], where the two-dimensional case is also discussed.

Here we prove Theorem 2.10 by other means. To do so, we first require a different projection operator, π_h, and its properties which we present in Lemma 2.7. We present a proof of Theorem 2.10 after we have proven this lemma. We define $\pi_h : C(\mathbb{T}) \to H_h \subset C(\mathbb{T})$, by following $\overline{\pi}_h$ with linear interpolation on each interval $\big(kh, (k+1)h\big)$, i.e., $\pi_h := I_h \circ \overline{\pi}_h$.

In the following lemma we collect some basic properties of the error $e_h w := \pi_h w - w \in C(\mathbb{T})$. Recall that $H^2(\mathbb{T}) \subset H^1(\mathbb{T}) \subset C(\mathbb{T})$ by the Sobolev embedding [1] and so π_h is well-defined on $H^1(\mathbb{T})$ and $H^2(\mathbb{T})$. We also note that $H_h \subset H^1(\mathbb{T})$.

In the proof of the following lemma, we require two Poincaré–Wirtinger(-type) inequalities. In order not to interrupt the flow of the main argument, we defer the statement and proof of these inequalities to Lemma 2.14 in Appendix C2.9.

Lemma 2.7 *The following estimates are valid.*

(i) $\|e_h w\|^2_{L^2(\mathbb{T})} \le (2h/\pi)^2 \|w'\|^2_{L^2(\mathbb{T})}$, *for all* $w \in H^1(\mathbb{T})$.
(ii) $\|e_h w\|^2_{L^2(\mathbb{T})} \le (h/\pi)^4 \|w''\|^2_{L^2(\mathbb{T})}$, *for all* $w \in H^2(\mathbb{T})$.
(iii) $\big\|(e_h w)'\big\|^2_{L^2(\mathbb{T})} \le \frac{1}{3} h^2 \|w''\|^2_{L^2(\mathbb{T})}$, *for all* $w \in H^2(\mathbb{T})$.
(iv) $\big\|(\pi_h w)'\big\|_{L^2(\mathbb{T})} \le \|w'\|_{L^2(\mathbb{T})}$, *for all* $w \in H^1(\mathbb{T})$.
(v) $\|w'\|^2_{L^2(\mathbb{T})} \le (12/h^2) \|w\|^2_{L^2(\mathbb{T})}$, *for all* $w \in H_h$.

Proof We first prove (iv). Let $w \in H^1(\mathbb{T})$. By definition

$$\int_0^h \big|(\pi_h w)'(x)\big|^2 \, dx = \left|\frac{w(h) - w(0)}{h}\right|^2 h = \left|\frac{1}{h}\int_0^h w'(x)\, dx\right|^2 \cdot h.$$

By Hölder's inequality or Jensen's inequality [27, Appendix B Theorem 2], this is dominated by

$$\frac{1}{h}\int_0^h \big|w'(x)\big|^2 \, dx \cdot h = \int_0^h |w'|^2 \, dx.$$

Similarly,

$$\int_{kh}^{(k+1)h} |(\pi_h w)'(x)|^2 \, dx \le \int_{kh}^{(k+1)} |w'(x)|^2 \, dx.$$

Adding these inequalities from $k = 0$ to $k = 1/h - 1$ and taking square roots yields (iv).

We next prove (i). Let $w \in H^1(\mathbb{T})$. Since $e_h w(0) = e_h w(h) = 0$, by the Poincaré–Wirtinger inequality in Lemma 2.14 (i) we have

$$\int_0^h |e_h w(x)|^2 \, dx \le (h/\pi)^2 \, \big\|(e_h w)'\big\|^2_{L^2((0,h))} \, .$$

By (iv) we have $\|(\pi_h w)'\|_{L^2(\mathbb{T})} \leq \|w'\|_{L^2(\mathbb{T})}$, so that $\|(e_h w)'\|_{L^2(\mathbb{T})} \leq 2\|w'\|_{L^2(\mathbb{T})}$. Hence

$$\int_0^h |e_h w(x)|^2 \, dx \leq (2h/\pi)^2 \|w'\|^2_{L^2((0,h))}.$$

Similar inequalities can be derived on the intervals $(kh, (k+1)h)$. Adding these from $k = 0$ to $k = 1/h - 1$ yields (i).

For the proof of (ii), assume that $w \in H^2(\mathbb{T})$. Since $e_h w(0) = e_h w(h) = 0$, by the Poincaré–Wirtinger-type inequality in Lemma 2.14 (ii) we have

$$\int_0^h |e_h w(x)|^2 \, dx \leq (h/\pi)^4 \int_0^h \left|(e_h w)''(x)\right|^2 \, dx.$$

Since $(\pi_h w)'' = 0$ a.e. on $(0, h)$, this implies

$$\int_0^h |e_h w(x)|^2 \, dx \leq (h/\pi)^4 \int_0^h |w''(x)|^2 \, dx.$$

Similarly as before, this yields

$$\|e_h w\|^2_{L^2(\mathbb{T})} \leq (h/\pi)^4 \|w''\|^2_{L^2(\mathbb{T})}.$$

To prove (iii), we again let $w \in H^2(\mathbb{T})$ and consider the integral

$$\int_0^h \left|(e_h w)'(x)\right|^2 \, dx = \int_0^h \left| w'(x) - \frac{w(h) - w(0)}{h} \right|^2 \, dx.$$

Setting $u = w'$, this integral becomes

$$\int_0^h \left| u(x) - \frac{1}{h} \int_0^h u(y) \, dy \right|^2 \, dx = \int_0^h \left| \frac{1}{h} \int_0^h (u(x) - u(y)) \, dy \right|^2 \, dx.$$

This is dominated by

$$\int_0^h \left(\frac{1}{h} \int_0^h |u(y) - u(x)|^2 \, dy \right) \, dx,$$

by the Hölder or Jensen inequality. Either of these inequalities also implies that

$$|u(y) - u(x)|^2 = \left| \int_x^y u'(s) \, ds \right|^2 \leq |x - y| \int_x^y |u'(s)|^2 \, ds \leq |x - y| \int_0^h |u'(s)|^2 \, ds,$$

where the final inequality holds for $x, y \in (0, h)$. Combining these observations, we have

$$\int_0^h \left|(e_h w)'(x)\right|^2 dx \le \frac{1}{h}\int_0^h \int_0^h |x-y|\,dx\,dy \int_0^h |w''(s)|^2\,ds$$
$$= \tilde{C}h^2 \int_0^h |w''(s)|^2\,ds,$$

with, as required,

$$\tilde{C} = \int_0^1 \int_0^1 |x-y|\,dx\,dy = \int_0^1 \left[\int_0^y (y-x)\,dx + \int_y^1 (x-y)\,dx\right] dy$$
$$= \int_0^1 \left[\left(y^2 - \frac{1}{2}y^2\right) + \left(\frac{1}{2} - y - \frac{1}{2}y^2 + y^2\right)\right] dy = \int_0^1 \left(y^2 - y + \frac{1}{2}\right) dy = \frac{1}{3}.$$

We thus obtain (iii).

Finally, we shall prove the reverse Poincaré–Wirtinger-type estimate (v) for $w \in H_h$. It suffices to prove this inequality in $(0, h)$ for a linear function w, as the full result on $\mathbb{T}$ will then follow by summing the contributions over all the intervals $(kh, (k+1)h)$ as in earlier parts of this proof. We may assume that $w(x) = ax - b$ for $a > 0, b \in \mathbb{R}$, by the symmetry provided by the square in the integrand. By overall multiplication of w by a constant, we may assume that $a = 1$. Then

$$\int_0^h (w'(x))^2\,dx = h,$$

while

$$\int_0^h w^2(x)\,dx = \int_0^h (x-b)^2\,dx$$
$$\ge \int_0^h (x-h/2)^2\,dx = 2\cdot\frac{1}{3}\left(\frac{h}{2}\right)^3 = \frac{1}{12}h^3.$$

The inequality follows since $b \mapsto \int_0^h (x-b)^2\,dx$ achieves its minimum value at $b = \frac{h}{2}$. Thus we have

$$\int_0^h \left(w'(x)\right)^2 dx \le (12/h^2) \int_0^h w^2(x)\,dx.$$

As is common practice in this proof by now, adding similar inequalities for each interval $(kh, (k+1)h)$ yields (iv). □

Proof ***(of Theorem 2.10)***

(i) Let $w \in H^1(\mathbb{T})$. This implies that also $e_h w \in H^1(\mathbb{T})$. Since $\pi_h w \in H_h$ and $\overline{P}_h$ is a linear operator, we observe that

$$\left((\overline{P}_h - I)w\right)' = -\left((\overline{P}_h(e_h w)\right)' + (e_h w)'.$$

From Lemma 2.7 (v) we see that

$$\left\|\left(\overline{P}_h(e_h w)\right)'\right\|_{L^2(\mathbb{T})} \le \frac{2\sqrt{3}}{h}\left\|\overline{P}_h(e_h w)\right\|_{L^2(\mathbb{T})} \le \frac{2\sqrt{3}}{h}\|e_h w\|_{L^2(\mathbb{T})},$$

where for the last inequality we used (2.52). By Lemma 2.7 (iv), we have

$$\left\|(e_h w)'\right\|_{L^2(\mathbb{T})} \le 2\|w'\|_{L^2(\mathbb{T})}.$$

Collecting these inequalities, we deduce that

$$\left\|\left((\overline{P}_h - I)w\right)'\right\|_{L^2(\mathbb{T})} \le \frac{2\sqrt{3}}{h}\|e_h w\|_{L^2(\mathbb{T})} + 2\|w'\|_{L^2(\mathbb{T})}.$$

Estimating the first term of the right-hand side by Lemma 2.7 (i) yields

$$\left\|\left((\overline{P}_h - I)w\right)'\right\|_{L^2(\mathbb{T})} \le \left(\frac{4\sqrt{3}}{\pi} + 2\right)\|w'\|_{L^2(\mathbb{T})}.$$

(ii) First we assume that $w \in H^2(\mathbb{T})$. As in (i) we have

$$\left\|\left((\overline{P}_h - I)w\right)'\right\|_{L^2(\mathbb{T})} \le \frac{2\sqrt{3}}{h}\|(e_h w)\|_{L^2(\mathbb{T})} + \left\|(e_h w)'\right\|_{L^2(\mathbb{T})}.$$

By Lemma 2.7 (ii) and (iii), this estimate yields

$$\left\|\left((\overline{P}_h - I)w\right)'\right\|_{L^2(\mathbb{T})} \le \frac{2\sqrt{3}}{h}(h/\pi)^2\|w''\|_{L^2(\mathbb{T})} + \frac{h}{\sqrt{3}}\|w''\|_{L^2(\mathbb{T})}.$$

Thus, by part (i) of this theorem,

$$\left\|\left(\overline{P}_h - I)w\right)'\right\|_{L^2(\mathbb{T})} \to 0, \quad \text{as } h \to 0.$$

Now assume $w \in H^1(\mathbb{T})$ instead and $\delta > 0$. By density, there exists a $w_\delta \in H^2(\mathbb{T})$ such that $\left\|(w - w_\delta)'\right\|_{L^2(\mathbb{T})} < \delta$. As observed above,

$$\left\|\left(\overline{P}_h w_\delta - w_\delta\right)'\right\|_{L^2(\mathbb{T})} \to 0, \quad \text{as } h \to 0.$$

By part (i) of this theorem and the triangle inequality, we have

$$\begin{aligned}\left\|\left(\overline{P}_h w - w\right)'\right\|_{L^2(\mathbb{T})} &\leq \left\|\left(\overline{P}_h w_\delta - w_\delta\right)'\right\|_{L^2(\mathbb{T})} + \left\|\left(\overline{P}_h (w - w_\delta)\right)'\right\|_{L^2(\mathbb{T})} \\ &\quad + \left\|(w - w_\delta)'\right\|_{L^2(\mathbb{T})} \\ &\leq \left\|\left(\overline{P}_h w_\delta - w_\delta\right)'\right\|_{L^2(\mathbb{T})} + (C+1)\left\|(w - w_\delta)'\right\|_{L^2(\mathbb{T})}.\end{aligned}$$

Sending h to zero yields

$$\overline{\lim_{h\to 0}}\left\|\left(\overline{P}_h w - w\right)'\right\|_{L^2(\mathbb{T})} \leq (C+1)\left\|(w - w_\delta)'\right\|_{L^2(\mathbb{T})} < (C+1)\delta.$$

Taking $\delta \to 0$ gives the desired convergence.

(iii) First we assume that $w \in H^1(\mathbb{T})$. From (2.52) and part (i) of this theorem, we know that also $\overline{P}_h w \in H^1(\mathbb{T})$. By (2.53) we have that $\int_{\mathbb{T}} \left(\overline{P}_h w - w\right) dx = 0$, so by the Poincaré–Wirtinger inequality for functions with zero "mass" [44, Theorem 13.27 and Exercise 13.13], part (ii) of this theorem implies that $\overline{P}_h w \to w$ strongly in $L^2(\mathbb{T})$.
Now let $w \in L^2(\mathbb{T})$. By density of $H^1(\mathbb{T})$ in $L^2(\mathbb{T})$, for all $\delta > 0$ there exists a $w_\delta \in H^1(\mathbb{T})$ such that $\|w - w_\delta\|_{L^2(\mathbb{T})} < \delta$. From (2.52) and the triangle inequality it follows that

$$\begin{aligned}\left\|\overline{P}_h w - w\right\|_{L^2(\mathbb{T})} &\leq \left\|\overline{P}_h w_\delta - w_\delta\right\|_{L^2(\mathbb{T})} + \left\|\overline{P}_h w_\delta - \overline{P}_h w\right\|_{L^2(\mathbb{T})} \\ &\quad + \|w_\delta - w\|_{L^2(\mathbb{T})} \\ &\leq \left\|\overline{P}_h w_\delta - w_\delta\right\|_{L^2(\mathbb{T})} + 2\|w_\delta - w\|_{L^2(\mathbb{T})} \\ &< \left\|\overline{P}_h w_\delta - w_\delta\right\|_{L^2(\mathbb{T})} + 2\delta.\end{aligned}$$

By the first part of the proof of (iii) we have that $\lim_{h\to 0}\left\|\overline{P}_h w_\delta - w_\delta\right\|_{L^2(\mathbb{T})} = 0$. Hence, if we first take the limit superior for $h \to 0$ in the inequality above and then the limit for $\delta \to 0$, the required result follows.

(iv) Let (h_j) be a sequence of positive real numbers which converges to zero and let (h_{j_k}) be a subsequence. By Morrey's inequality [44, Theorem 11.34], (2.52), and part (i) of this theorem, there exists some constants $\tilde{C}_1, \tilde{C}_2 > 0$ such that

$$\|\overline{P}_{h_{j_k}} w\|_{L^\infty(\mathbb{T})} \leq \tilde{C}_1 \|\overline{P}_{h_{j_k}} w\|_{H^1(\mathbb{T})} \leq \tilde{C}_2 \|w\|_{H^1(\mathbb{T})},$$

thus the sequence $\left(\overline{P}_{h_{j_k}} w\right)$ is uniformly bounded.
Let $x, y \in \mathbb{T}$ and assume $[x, y] \subset \mathbb{T}$ with appropriate periodicity taken into account. By Hölder's inequality and part (i) of this theorem we have

$$
\begin{aligned}
\left|\overline{P}_{h_{j_k}} w(x) - \overline{P}_{h_{j_k}} w(y)\right| &= \left|\int_{[x,y]} \overline{P}_{h_{j_k}} w'(x)\, dx\right| \\
&\le \int_{[x,y]} \left|\overline{P}_{h_{j_k}} w'(x)\right| dx \le \int_{\mathbb{T}} \left|\overline{P}_{h_{j_k}} w'(x)\right| dx \\
&\le \|1\|_{L^2(\mathbb{T})} \left\|\overline{P}_{h_{j_k}} w'\right\|_{L^2(\mathbb{T})} = \left\|\overline{P}_{h_{j_k}} w'\right\|_{L^2(\mathbb{T})} \\
&\le \left\|w'\right\|_{L^2(\mathbb{T})}.
\end{aligned}
$$

Thus the sequence $\left(\overline{P}_{h_{j_k}} w\right)$ is equicontinuous. By the Arzelà–Ascoli theorem, it follows that there exists a $\overline{w} \in C(\mathbb{T})$ and a subsubsequence $\left(\overline{P}_{h_{j_{k_l}}} w\right)$ such that $\overline{P}_{h_{j_{k_l}}} w \to \overline{w}$ uniformly as $l \to \infty$. Hence, since $\mathbb{T}$ has finite measure, we also have $\overline{P}_{h_{j_{k_l}}} w \to \overline{w}$ in $L^2(\mathbb{T})$ and thus, by part (iii), $\overline{w} = w$. Since every subsequence of $\left(\overline{P}_{h_j} w\right)$ has a further subsubsequence which converges uniformly to w, it follows that $\left(\overline{P}_{h_j} w\right)$ itself also converges to w uniformly. From this we conclude the desired result.

□

2.5.8 Proof of Theorem 2.7

We shall prove Theorem 2.7 as an application of Theorem 2.2.

Proof (of Theorem 2.7) Let $T > 0$. In the notation of Theorem 2.2, we set $M_h = \overline{L}^2_h$, $M = L^2(\mathbb{T})$, $i_h = I_h$, $p_h = P_h$, $\Phi_h = \varphi^h_{AC}$, and $\Phi = \varphi_{AC}$. Since $\overline{L}^2_h$ is a Hilbert space, it is closed and thus $\overline{\mathcal{D}\left(\varphi^h_{AC}\right)} = \overline{L}^2_h$. As derived at the end of Sect. 2.5.2, we also have $\overline{\mathcal{D}\left(\varphi_{AC}\right)} = L^2(\mathbb{T})$. By our discussion in Sect. 2.5.1 we know that φ^h_{AC} and φ_{AC} satisfy condition ($-\alpha$-AGS). In Sect. 2.5.7 we established that condition (2.7) is satisfied. It remains to show that conditions (2.9), (2.10), and (2.11) are satisfied.

First we shall prove that (2.9) holds. Let $w \in L^2(\mathbb{T})$. Without loss of generality we may assume that $\varphi_{AC}(w) < +\infty$ and thus $w \in H^1(\mathbb{T})$. By (2.29) and Theorem 2.10 (i i), we see that

$$
\lim_{h\to 0} \varphi^h_D(P_h w) = \lim_{h\to 0} \varphi_D(I_h P_h w) = \lim_{h\to 0} \varphi_D(\overline{P}_h w) = \varphi_D(w).
$$

By Theorem 2.10 (iv) we have that $P_h w \to w$ uniformly in $\mathbb{T}$, thus

$$
\lim_{h\to 0} \varphi_W(\overline{P}_h w) = \varphi_W(w).
$$

We claim that

$$
\lim_{h\to 0} \left|\varphi^h_W(P_h w) - \varphi_W(\overline{P}_h w)\right| = 0. \tag{2.54}
$$

Combining this with the limit above gives

$$\lim_{h\to 0} \varphi_W^h(P_h w) = \varphi_W(w),$$

which together with the convergence of $\varphi_D^h(P_h w)$ above establishes (2.9). To prove the claim in (2.54), we recall that $\varphi_W^h(P_h w) = \varphi_W^h(\pi_h \overline{P}_h w) = \sum_{k=0}^{\frac{1}{h}-1} hW\left(\overline{P}_h w(kh)\right)$ and thus

$$\begin{aligned}
\left|\varphi_W^h(P_h w) - \varphi_W(\overline{P}_h w)\right| &\leq \sum_{k=0}^{\frac{1}{h}-1} \left| hW\left(\overline{P}_h w(kh)\right) - \int_{kh}^{(k+1)h} W\left(\overline{P}_h w(x)\right) dx \right| \\
&= \sum_{k=0}^{\frac{1}{h}-1} \left| \int_{kh}^{(k+1)h} \left[W\left(\overline{P}_h w(kh)\right) - W\left(\overline{P}_h w(x)\right)\right] dx \right| \\
&\leq \sum_{k=0}^{\frac{1}{h}-1} \int_{kh}^{(k+1)h} \left|W\left(\overline{P}_h w(kh)\right) - W\left(\overline{P}_h w(x)\right)\right| dx.
\end{aligned} \tag{2.55}$$

For the integrand on the right-hand side we estimate

$$\begin{aligned}
\left|W\left(\overline{P}_h w(kh)\right) - W\left(\overline{P}_h w(x)\right)\right| \leq &\left|W\left(\overline{P}_h w(kh)\right) - W\left(w(kh)\right)\right| \\
&+ \left|W\left(w(kh)\right) - W\left(w(x)\right)\right| \\
&+ \left|W\left(w(x)\right) - W\left(\overline{P}_h w(x)\right)\right|,
\end{aligned}$$

where we remember that in the integral $x \in [kh, (k+1)h]$ and thus $|x - kh| \leq h$. Let $\eta > 0$. Since $w \in H^1(\mathbb{T})$, we know by the Sobolev embedding that $w \in C(\mathbb{T})$ and thus, since W is continuous, $W \circ w \in C(\mathbb{T})$. Since $\mathbb{T}$ is compact this means that $W \circ w$ is uniformly continuous. Hence there exists a $h_1 > 0$ such that, if $0 < h < h_1$, then $|W(w(kh)) - W(w(x))| < \eta/3$. Moreover, since $\overline{P}_h w \to w$ uniformly and since W is continuous, we have that $W(\overline{P}_h w) \to W(w)$ uniformly. Hence, there exists a $h_2 > 0$ such that, if $0 < h < h_2$, then $|W(w(kh)) - W(w(x))| < \eta/3$ and $\left|W\left(\overline{P}_h w(kh)\right) - W\left(\overline{P}_h w(x)\right)\right| < \eta/3$. Defining $h^* := \min(h_1, h_2)$ we find that, if $0 < h < h^*$, then $\left|W\left(\overline{P}_h w(kh)\right) - W\left(\overline{P}_h w(x)\right)\right| < \eta$. Substituting this into (2.55), we find, for $h \in (0, h^*)$,

$$\left|\varphi_W^h(P_h w) - \varphi_W(\overline{P}_h w)\right| < \sum_{k=0}^{\frac{1}{h}-1} \int_{kh}^{(k+1)h} \eta \, dx = \frac{1}{h} \cdot h \cdot \eta = \eta,$$

which proves the claim in (2.54) and thereby concludes the proof of (2.9).

Next we turn our attention to establishing the existence of a $\delta > 0$ and a nonnegative function $\Psi : L^2(\mathbb{T}) \to \mathbb{R} \cup \{+\infty\}$, such that $\Psi(u(\cdot)) \in L^1(0, T)$ and (2.10) is satisfied for all $h \in (0, \delta) \cap \mathfrak{H}$. In fact, we will show that for our choice of Ψ, (2.10) is satisfied for all $h \in \mathfrak{H}$, so that any choice of $\delta > 0$ is valid. Hence, let $h \in \mathfrak{H}$.

By Proposition 2.1 we know that for all $t > 0$, $u(t) \in \mathcal{D}(\varphi_{AC}) \subset H^1(\mathbb{T})$. Therefore, we define $\Psi := +\infty$ on $L^2(\mathbb{T}) \setminus H^1(\mathbb{T})$. Now let $w \in H^1(\mathbb{T})$. By (2.29) and Theorem 2.10 (i), we have

$$\varphi_D^h(P_h w) = \varphi_D(\overline{P}_h w) \leq \frac{C^2}{2} \|w'\|^2_{L^2(\mathbb{T})}.$$

Recalling that $W(x) = \alpha(x^2 - 1)^2/4$, we observe that

$$\varphi_W(\overline{P}_h w) \leq \frac{\alpha}{4} \left(\int_{\mathbb{T}} |\overline{P}_h w|^4 dx + 1 \right).$$

By Theorem 2.10 (i), the one-dimensional Sobolev (Morrey) inequality, and (2.52), there exists a constant $c_0 > 0$ such that

$$\begin{aligned} \|\overline{P}_h w\|_{L^\infty(\mathbb{T})} &\leq c_0 \left(\left\|(\overline{P}_h w)'\right\|_{L^2(\mathbb{T})} + \|\overline{P}_h w\|_{L^2(\mathbb{T})} \right) \leq c_0 \left(C \left\|w'\right\|_{L^2(\mathbb{T})} + \|w\|_{L^2(\mathbb{T})} \right) \\ &\leq c_1 \|w\|_{H^1(\mathbb{T})}, \end{aligned}$$

with $c_1 := c_0 C$.

By the interpolation inequality for L^p spaces [1, Theorem 2.11] (a special case of the Gagliardo–Nirenberg interpolation inequality), (2.52), and the inequality above, we have

$$\|\overline{P}_h w\|_{L^4(\mathbb{T})} \leq \|\overline{P}_h w\|^{1/2}_{L^2(\mathbb{T})} \|\overline{P}_h w\|^{1/2}_{L^\infty(\mathbb{T})} \leq C_0 \|w\|^{1/2}_{L^2(\mathbb{T})} \|w\|^{1/2}_{H^1(\mathbb{T})}, \tag{2.56}$$

with $C_0 := c_1^{1/2}$.

Thus

$$\varphi_W(\overline{P}_h w) \leq C_1 \|w\|^2_{L^2(\mathbb{T})} \|w\|^2_{H^1(\mathbb{T})} + \frac{\alpha}{4},$$

where $C_1 := \alpha C_0^4/4$. We note that C_1 is independent of h and w.

Using the explicit form of W, Lemma 2.4 (or, equivalently, (2.31) with $m = 2$), (2.32) with $m = 4$, and (2.56), we find

$$\begin{aligned} \varphi_W^h(P_h w) - \varphi_W(\overline{P}_h w) &= \varphi_W^h(P_h w) - \varphi_W(I_h P_h w) \\ &= \frac{\alpha}{4} \left(\|P_h w\|^4_{L_h^4} - \|I_h P_h w\|^4_{L^4(\mathbb{T})} - 2\|P_h w\|_h^2 + 2\|I_h P_h w\|^2_{L^2(\mathbb{T})} \right) \\ &\leq \frac{\alpha}{2} \left(\|P_h w\|^4_{L_h^4} - \|I_h P_h w\|^4_{L^4(\mathbb{T})} \right) \\ &\leq \alpha \|I_h P_h w\|^4_{L^4(\mathbb{T})} = \alpha \|\overline{P}_h w\|^4_{L^4(\mathbb{T})} \\ &\leq C_2 \|w\|^2_{L^2(\mathbb{T})} \|w\|^2_{H^1(\mathbb{T})}, \end{aligned}$$

with $C_2 := \alpha C_0^4$. It thus follows that

$$\varphi_W^h(P_h w) \le \varphi_W(\overline{P}_h w) + C_2 \|w\|^2_{L^2(\mathbb{T})} \|w\|^2_{H^1(\mathbb{T})} \le C_3 \|w\|^2_{L^2(\mathbb{T})} \|w\|^2_{H^1(\mathbb{T})} + \frac{\alpha}{4},$$

where $C_3 := C_1 + C_2 = \frac{5}{4}\alpha C_0^4$.

We define, for all $w \in L^2(\mathbb{T})$,

$$\Psi(w) := \begin{cases} \frac{C^2}{2}\|w'\|^2_{L^2(\mathbb{T})} + C_3\|w\|^2_{L^2(\mathbb{T})}\|w\|^2_{H^1(\mathbb{T})} + \frac{\alpha}{4}, & \text{if } w \in H^1(\mathbb{T}), \\ +\infty, & \text{otherwise}. \end{cases}$$

From the inequalities above, it follows that for all $w \in L^2(\mathbb{T})$ and all $h \in \mathfrak{H}$ the inequality in (2.10) is satisfied. Moreover, since $\alpha > 0$, Ψ is nonnegative.

As we observed earlier, by Proposition 2.1 we have that for all $t > 0$, $u(t) \in H^1(\mathbb{T})$, thus the functions $t \mapsto \|u(t)\|_{L^2(\mathbb{T})}$ and $t \mapsto \|\nabla u(t)\|_{L^2(\mathbb{T})}$ are well-defined from $(0, \infty)$ to $[0, \infty)$. By the same proposition we know that $t \mapsto \|u(t)\|_{L^2(\mathbb{T})}$ is continuous on $[0, \infty)$ (and hence measurable). It is also bounded on $[0, T]$ since, by the generalised contraction principle in (2.3) and the geodesic $(-\alpha)$-convexity of φ_{AC} with α, we have

$$\|u(t) - 0\|_{L^2(\mathbb{T})} \le e^{\alpha t} \|u_0 - 0\|_{L^2(\mathbb{T})} \le e^{\alpha T} \|u_0 - 0\|_{L^2(\mathbb{T})}. \tag{2.57}$$

By Lemma 2.2 we have that $t \mapsto \varphi_{AC}(u(t))$ is a non-increasing function on $[0, \infty)$ and thus $\|\nabla u(t)\|_{L^2(\mathbb{T})}$ is also bounded on $[0, T]$. Moreover, by monotonicity, $t \mapsto \varphi_{AC}(u(t))$ is measurable on $[0, \infty)$ and, since $u \in C([0, T], L^2(\mathbb{T}))$ by Proposition 2.1, if we view u as real-valued function on $\mathbb{T} \times [0, T]$, it is also measurable.[16] Thus so is $W \circ u$ and hence, by Fubini's theorem [28, Sect. 1.4, Theorem 1], $t \mapsto \int_{\mathbb{T}} W \circ (u(t))\, dx$ is measurable on $[0, T]$. Hence,

$$t \mapsto \frac{1}{2}\|\nabla u(t)\|^2_{L^2(\mathbb{T})} = \varphi_{AC}(u(t)) - \int_{\mathbb{T}} W \circ (u(t))\, dx$$

is measurable on $[0, T]$ as well.

---

[16] To prove this, partition $[0, T]$ into intervals $[a_0, a_1), [a_1, a_2), \ldots, [a_{m-1}, a_m]$ with $m \in \mathbb{N}$ and $a_{i+1} - a_i = T/m$. Set

$$u_m(t) := u(a_j) \quad \text{for} \quad a_j \le t < a_{j+1}, \quad j = 0, \ldots, m-1.$$

Viewed as function on $\mathbb{T} \times [0, T]$, u_m is measurable and it is in $L^2(\mathbb{T} \times (0, T))$. Since $u \in C([0, T], L^2(\mathbb{T}))$ and $[0, T]$ is compact, the function u is uniformly continuous on $[0, T]$ with values in $L^2(\mathbb{T})$. Because, for all $t \in [0, T]$, there is an a_j such that $\|u(t) - u_m(t)\|_{L^2(\mathbb{T})} = \|u(t) - u(a_j)\|_{L^2(\mathbb{T})}$, uniform continuity of u implies convergence in $C([0, T], L^2(\mathbb{T}))$ of the sequence (u_m) to u. In particular, (u_m) is a Cauchy sequence in $L^2(\mathbb{T} \times (0, T))$. As $L^2(\mathbb{T} \times (0, T))$ is complete by the Fisher–Riesz theorem [10, Theorem 4.8], [64, Sect. I.9, Proposition 2], we have $u \in L^2(\mathbb{R} \times (0, T))$ and thus u is measurable.

It follows that $t \mapsto \Psi(u(t))$ is well-defined on $[0, T]$ and bounded on the same interval, so that

$$\int_0^T e^{-2\alpha t} \Psi(u(t))\, dt < \infty.$$

It remains to prove that there exists a function $\varepsilon : (0, 1) \to (0, \infty)$ such that $\varepsilon(h) \to 0$ as $h \to 0$ and (2.11) is satisfied, i.e., for all $t \in [0, T]$,

$$\varphi_{AC}\left(I_h U^h(t)\right) \leq \varphi_{AC}^h\left(U^h(t)\right) + \varepsilon(h). \tag{2.58}$$

Let $t \in [0, T]$. By (2.29) we have $\varphi_D\left(I_h U^h(t)\right) = \varphi_D^h\left(U^h(t)\right)$, and (2.31) with $m = 4$ shows that $\left\| I_h U^h(t) \right\|_{L^4(\mathbb{T})}^4 \leq \left\| U^h(t) \right\|_{L_h^4}^4$. Combined with the explicit form of W, which we recalled earlier in this proof, this implies that

$$\begin{aligned}
&\varphi_{AC}\left(I_h U^h(t)\right) - \varphi_{AC}^h\left(U^h(t)\right) \\
=&\frac{\alpha}{4}\left(2\left\|U^h(t)\right\|_h^2 - 2\left\|I^h U^h(t)\right\|_{L^2(\mathbb{T})}^2 - \left\|U^h(t)\right\|_{L_h^4}^4 + \left\|I^h U^h(t)\right\|_{L^4(\mathbb{T})}^4\right) \\
\leq&\frac{\alpha}{2}\left(\left\|U^h(t)\right\|_h^2 - \left\|I^h U^h(t)\right\|_{L^2(\mathbb{T})}^2\right) \\
=&\frac{\alpha h^2}{12}\left\|\nabla_h U^h(t)\right\|_{h,h}^2.
\end{aligned}$$

For the last inequality we invoked (2.40). To estimate $\left\|\nabla_h U^h(t)\right\|_{h,h}^2$, we recall equation (2.42) in the form (2.44), with $\mu > \alpha$ arbitrary. We apply ∇_h to both sides of (2.44) and take the inner product $\langle \cdot, \cdot \rangle_h$ with $-\Delta_h U^h$. By (2.27), we obtain

$$\begin{aligned}
\left\langle \nabla_h U^h, \nabla_h \Gamma \dot{U}^h \right\rangle_{h,h} &= \left\langle -\Delta_h U^h, \Gamma \dot{U}^h \right\rangle_h \\
&= -\left\langle \Delta_h U^h, \Delta_h U^h \right\rangle_h - \mu \left\langle \Delta_h U^h, U^h \right\rangle_h + \left\langle \Delta_h U^h, F \circ U^h \right\rangle_h \\
&= -\left\|\Delta_h U^h\right\|_h^2 + \mu \left\|\nabla_h U^h\right\|_{h,h}^2 - \left\langle \nabla_h U^h, \nabla_h\left(F \circ U^h\right)\right\rangle_{h,h}.
\end{aligned}$$

We note that, since $\left(F \circ U^h\right)_k = F(U_k^h)$,

$$\begin{aligned}
\left(\nabla_h\left(F \circ U^h\right)\right)_{k,k\pm1} &= \frac{1}{h}\left\{\left(F \circ U^h\right)_{k\pm1} - \left(F \circ U^h\right)_k\right\} \\
&= \left(\nabla_h U^h\right)_{k,k\pm1} \int_0^1 F'\left(\theta U_{k\pm1}^h + (1-\theta) U_k^h\right) d\theta,
\end{aligned}$$

so the monotonicity of F implies that

$$\left\langle \nabla_h U^h, \nabla_h\left(F \circ U^h\right)\right\rangle_{h,h} \geq 0.$$

Thus, we obtain

$$\left\langle \nabla_h U^h, \nabla_h \Gamma \dot{U}^h \right\rangle_{h,h} \leq - \left\| \Delta_h U^h \right\|_h^2 + \mu \left\| \nabla_h U^h \right\|_{h,h}^2 \leq \mu \left\| \nabla_h U^h \right\|_{h,h}^2 .$$

Since $\mu > \alpha$ can be taken arbitrary, we may replace μ by α in the right-hand side.

Since Γ commutes with ∇_h and Γ is symmetric, the left-hand side equals

$$\frac{1}{2}\frac{d}{dt}\left\langle \nabla_h U^h, \Gamma \nabla_h U^h \right\rangle_{h,h}$$

which equals

$$\frac{1}{6}\frac{d}{dt}\left\| \nabla_h U^h \right\|_{h,h}^2$$

by (2.37). We thus obtain

$$\frac{1}{6}\frac{d}{dt}\left\| \nabla_h U^h \right\|_{h,h}^2 \leq \alpha \left\| \nabla_h U^h \right\|_{h,h}^2 ,$$

which yields by the differential form of Grönwall's inequality:

$$\left\| \nabla_h U^h(t) \right\|_{h,h}^2 \leq e^{6\alpha t} \left\| \nabla_h u_0^h \right\|_{h,h}^2 .$$

Combining these observations, we conclude that

$$\varphi_{AC}\left(I_h U^h(t)\right) - \varphi_{AC}^h\left(U^h(t)\right) \leq \frac{\alpha h^2}{12} e^{6\alpha t} \left\| \nabla_h u_0^h \right\|_{h,h}^2 .$$

If we define

$$\varepsilon(h) := \frac{\alpha h^2}{12} e^{6\alpha T} \left\| \nabla_h u_0^h \right\|_{h,h}^2 ,$$

then (2.58) follows by the assumption that $h \left\| \nabla u_0^h \right\|_h \to 0$ as $h \to 0$.

The result now follows by Theorem 2.2 and the additional comments about the "finite T" case in Remark 2.6. □

2.6 Conclusions and Future Work

In this paper we proved discrete-to-continuum convergence for total variation flow on a discretised n-dimensional torus and for Allen–Cahn flow on a discretised one-dimensional torus. Two potential generalisations immediately suggest themselves: Allen–Cahn flow convergence on a discrete torus of higher dimension and, for both flows, convergence of the flows on other graphs.

Our proof of convergence for the Allen–Cahn flow is not trivially extended to higher dimensions, since it makes extensive use of the properties of the operator Γ, which depend on the use of piecewise-linear interpolation to associate discrete functions with continuum functions. If piecewise-n-linear interpolation (bilinear, trilinear, etc.) were to be used instead, a different operator Γ (and corresponding embedding and projection operators) would need to be constructed. In higher dimensions, the space H^1 also loses a lot of its regularity.

Other graphs, besides the discrete torus, that are being considered by Samuel Mercer and the second author, are random geometric graphs, obtained by sampling points from Euclidean space (or possibly from a differentiable manifold) according to some probability density in order to construct the vertex set and subsequently connecting nodes with (potentially weighted) edges depending on their pairwise distances. The identification of discrete functions with continuum functions will then require a method different from the piecewise-constant and piecewise-linear interpolation methods that we used in this paper, such as the method based on optimal transport techniques that was introduced in [33].

Other flows can also be considered. For example, in the field of machine learning there is much current interest in proving consistency of methods through continuum limits of discrete dynamics.

Another possible direction for future research is the relationship between the continuum limit and the singular limit, for those dynamics that have an interesting singular limit. For example, if the potential term in the Allen–Cahn equation of (2.24) is rescaled by a factor ε^{-2}, then in the sharp interface limit $\varepsilon \downarrow 0$ solutions evolve according to flow by mean curvature [2, 14, 19, 20, 29, 30, 59, 60]. It cannot be expected that the continuum limit and singular limit commute, since in the limit $\varepsilon \downarrow 0$ the graph-based Allen–Cahn dynamics decouples per vertex and simply evolves into the well of W that is nearest the initial condition at the node [62]. A simultaneous limit, taking both $\varepsilon \downarrow 0$ and $h \downarrow 0$ (or, in a more general setting, the number of nodes $|V| \to \infty$), could potentially lead to nontrivial dynamics if ε scales in the correct way with h. A related example of such a simultaneous limit is the convergence of graph Merriman–Bence–Osher dynamics to mean curvature flow that has been established recently in [43].

2.7 Appendix A: A Proof of (2.21)

This section culminates with the proof of (2.21) in Lemma 2.12. Even though (2.21) seems to be a standard result, the authors did not find a proof in the literature. Along the way we prove some other useful results.

Lemma 2.8 *Let $p \in [1, \infty)$. Assume $g, \tilde{g} \in L^p(\mathbb{T}^n)$ and let $(g^j)_{j\in\mathbb{N}}$ be a sequence in $L^p(\mathbb{T}^n)$. If $g^j \to g$ in $L^p(\mathbb{T}^n)$ and $g^j \rightharpoonup^* \tilde{g}$ in $L^\infty(\mathbb{T}^n)$ as $j \to \infty$, then $g = \tilde{g}$ a.e. on $\mathbb{T}^n$.*

Consequently, if $w \in L^p(\mathbb{T}^n)$ *and* $(w^j)_{j\in\mathbb{N}}$ *is a sequence in* $L^p(\mathbb{T}^n)$ *which is uniformly bounded in* $L^\infty(\mathbb{T}^n)$ *and such that* $w^j \to w$ *in* $L^p(\mathbb{T}^n)$*, then* $w^j \rightharpoonup^* w$ *in* $L^\infty(\mathbb{T}^n)$*.*

Proof Let $U \subset \mathbb{T}^n$ be such that $g - \tilde{g} \geq 0$ on U and $g - \tilde{g} < 0$ on $U^c := \mathbb{T}^n \setminus U$. We note that as preimages of the measurable sets $[0, \infty)$ and $(-\infty, 0)$, respectively, U and U^c are also measurable. Define $\chi_U = 1$ on U and $\chi_U = 0$ on U^c. Then $2\chi_U - 1 \in L^1(\mathbb{T}^n) \cap L^{p^*}(\mathbb{T}^n)$, where p^* is defined via $1/p + 1/p^* = 1$. We compute

$$\begin{aligned}\int_{\mathbb{T}^n} |g(x) - \tilde{g}(x)|\, dx &= \int_{\mathbb{T}^n} (2\chi_U(x) - 1)\,(g(x) - \tilde{g}(x))\, dx\\ &= \int_{\mathbb{T}^n} (2\chi_U(x) - 1)\,\left(g(x) - g^j(x)\right) dx\\ &\quad+ \int_{\mathbb{T}^n} (2\chi_U(x) - 1)\,\left(g^j(x) - \tilde{g}(x)\right) dx.\end{aligned}$$

The first term on the right-hand side converges to zero as $j \to \infty$, by Hölder's inequality and $g^j \to g$ in $L^p(\mathbb{T}^n)$. The second term converges to zero, since $g^j \rightharpoonup^* \tilde{g}$ in $L^\infty(\mathbb{T}^n)$. This concludes the proof of the first part.

For the proof of the second part, we note that by the Banach–Alaoglu theorem each bounded sequence in $L^\infty(\mathbb{T}^n)$ has a weakly*-convergent subsequence. In particular each subsequence of (w^j) has a further subsubsequence which converges weakly*. By the first part of this result, each such subsubsequence has the same limit, w. Hence (w^j) converges weakly* in $L^\infty(\mathbb{T}^n)$ to w. □

For $1 \leq p \leq \infty$ we denote by $L^p(\mathbb{T}; \mathbb{R}^n)$ the space of integrable functions $u : \mathbb{T}^n \to \mathbb{R}^n$ for which $\int_{\mathbb{T}^n} |u(x)|^p\, dx < \infty$ (if $p < \infty$), where $|\cdot|$ denotes the Euclidean norm in $\mathbb{R}^n$, or for which, for almost every $x \in \mathbb{T}^n$, $|g(x)|_\infty < \infty$ (if $p = \infty$).

Lemma 2.9 *Let* $g \in L^\infty(\mathbb{T}^n; \mathbb{R}^n)$ *be such that, for all* $i \in [n]$, $\frac{\partial g_i}{\partial x_i} \in L^\infty(\mathbb{T}^n)$ *as weak partial derivatives and, for almost every* $x \in \mathbb{T}^n$, $|g(x)|_\infty \leq 1$*. Then there exists a sequence* $(g^j)_{j\in\mathbb{N}}$ *in* $C^\infty(\mathbb{T}^n; \mathbb{R}^n)$ *such that, as* $j \to \infty$, $g^j \rightharpoonup^* g$ *in* $L^\infty(\mathbb{T}^n; \mathbb{R}^n)$ *and, for all* $i \in [n]$, $\frac{\partial g_i^j}{\partial x_i} \rightharpoonup^* \frac{\partial g_i}{\partial x_i}$*. Moreover, for all* $j \in \mathbb{N}$ *and for almost all* $x \in \mathbb{T}^n$, $|g^j(x)|_\infty \leq 1$*.*

Proof This proof is a variant of the proofs in [25, Proposition 3.3], [15, Lemma A.3].

Let $\eta \in C^\infty(\mathbb{T}^n)$ be the standard mollifier (see, for example, [27, Appendix C.4]). In particular, $\eta \geq 0$ and $\int_{\mathbb{T}^n} \eta(x)\, dx = 1$. Define, for all $j \in \mathbb{N}$, $\eta^j(x) := j^n \eta(jx)$ and $g^j := g * \eta^j$. We note that $\eta^j \geq 0$ and $\int_{\mathbb{T}^n} \eta^j(x)\, dx = 1$. Moreover, since the convolution preserves the periodicity, g^j is well-defined on $\mathbb{T}^n$ and by a standard result [27], $g^j \in C^\infty(\mathbb{T}^n)$.

Let $p \in [1, \infty)$. Since $g \in L^\infty(\mathbb{T}^n; \mathbb{R}^n)$ and $|\mathbb{T}^n| < \infty$, also $g \in L^p(\mathbb{T}^n; \mathbb{R}^n)$. Again by a standard result [27], $g^j \to g$ in $L^p(\mathbb{T}^n; \mathbb{R}^n)$. Moreover, since

$$\frac{\partial g_i^j}{\partial x_i} = \frac{\partial g_i}{\partial x_i} * \eta^j$$

and $\frac{\partial g_i}{\partial x_i} \in L^\infty(\mathbb{T}^n) \subset L^p(\mathbb{T}^n)$, by the same standard result as above, $\frac{\partial g_i^j}{\partial x_i} \to \frac{\partial g_i}{\partial x_i}$ in $L^p(\mathbb{T}^n)$.

By Lemma 2.8 it now suffices to prove uniform-in-j L^∞ bounds on g^j and $\frac{\partial g_i^j}{\partial x_i}$. For the former we use $|g(x)|_\infty \leq 1$ to compute

$$|g_i^j(x)| \leq j^n \int_{\mathbb{T}^n} \eta(jy)\, dy = 1.$$

For the latter,

$$\begin{aligned}\left|\frac{\partial g_i^j}{\partial x_i}(x)\right| &= \left|\int_{\mathbb{T}^n} \frac{\partial g_i}{\partial x_i}(y)\eta^j(x-y)\, dy\right| \leq \int_{\mathbb{T}^n} \eta^j(x-y)\, dy \left\|\frac{\partial g_i}{\partial x_i}\right\|_{L^\infty(\mathbb{T}^n)} \\ &= \left\|\frac{\partial g_i}{\partial x_i}\right\|_{L^\infty(\mathbb{T}^n)}.\end{aligned}$$

□

Using Lemma 2.9 we deduce (similar to [15, Corollary 3]) that the regularity conditions on the admissible vector fields g in the definition of anisotropic total variation can be relaxed:

$$\begin{aligned}\int_{\mathbb{T}^n} |Du|_{l^1} = \sup\Bigg\{\int_{\mathbb{T}^n} u(x)\operatorname{div} g(x)\, dx \;\Bigg|\; & g \in L^\infty(\mathbb{T}^n; \mathbb{R}^n), \\ & \forall i \in [n]\ \frac{\partial g_i}{\partial x_i} \in L^\infty(\mathbb{T}^n), \text{ and for a.e. } x \in \mathbb{T}^n\ |g(x)|_\infty \leq 1\Bigg\}.\end{aligned} \tag{2.59}$$

Moreover, since Lemma 2.9 tells us that each of the weak partial derivatives $\frac{\partial g_i}{\partial x_i}$ converges separately, the summation formula (2.20) also holds when the suprema in (2.20) are taken over all $g = (g_1, \ldots, g_n) \in L^\infty(\mathbb{T}^n; \mathbb{R}^n)$ that satisfy the conditions in (2.59). It follows that identity (2.59) yields

$$\int_{\mathbb{T}^n} |Du|_{\ell^1} = \sum_{i=1}^n \int_{\mathbb{T}^n} |D_{x_i} u|, \tag{2.60}$$

if we define

$$\begin{aligned}\int_U |D_{x_i} u| := \sup\Bigg\{\int_U u(x)\frac{\partial f}{\partial x_i}(x)dx \;\Bigg|\; & f \in L^\infty(U),\ \frac{\partial f}{\partial x_i} \in L^\infty(U), \text{ and} \\ & \text{for a.e. } x \in U\ |f(x)| \leq 1\Bigg\},\end{aligned} \tag{2.61}$$

for subsets $U \subset \mathbb{T}^n$ that are open (in the Euclidean topology on $\mathbb{T}^n$) and functions $u \in L^1(U)$. We note that we have chosen the notation $\int_{\overline{U}} |D_{x_i} u|$ (rather than $\int_U |D_{x_i} u|$) to emphasize that this is not the standard total variation, since its behavior at the boundary of U is different. For example, if U is an interval and proper subset of $\mathbb{T}$ and u is constantly equal to $c \in \mathbb{R}$ on U, then $\int_{\overline{U}} |D_{x_1} u| = 2c$, whereas the total variation of u on U is $\int_U |Du|_{\ell^1} = 0$. In general, if U has smooth boundary, we can interpret $\int_{\overline{U}} |D_{x_i} u|$ as the total variation on $\mathbb{T}^n$ "in the direction of x_i" of the function $\overline{u}$ that is defined by $\overline{u} := u$ on U and $\overline{u} := 0$ on $\mathbb{T}^n \setminus U$.

For an open subset $U \subset \mathbb{T}^n$, we say that the boundary of U *is parallel to the x_i-axis* if ∂U (i.e., the topological boundary of U as subset of $\mathbb{T}^n$) is a non-empty subset of the union of countably many $n-1$-dimensional hyperplanes in $\mathbb{R}^n$ (where we interpret $\mathbb{T}^n$ as a subset of $\mathbb{R}^n$ via identification with the hypercube Ω_h; see Sect. 2.3.1) whose normal vectors are perpendicular to the x_i-axis.

We say a function $u : U \to \mathbb{R}$ is *independent of x_j* if the distributional derivative $\frac{\partial u}{\partial x_j}$ equals zero.

Lemma 2.10 *Let $i \in [n]$ and $n \geq 2$.*

(i) If the boundary of U is parallel to the x_i-axis, then, for all $u \in L^1(\mathbb{T}^n)$,

$$\int_{\mathbb{T}^n} |D_{x_i} u| = \int_{\overline{U}} |D_{x_i} u| + \int_{\overline{\mathbb{T}^n \setminus U}} |D_{x_i} u|. \tag{2.62}$$

(ii) Let $j \in [n]$, with $j \neq i$. If $u \in L^1(U)$ is independent of x_j, then

$$\int_{\overline{U}} |D_{x_i} u| = \sup\left\{ \int_U u(x) \frac{\partial f}{\partial x_i}(x)\, dx \;\middle|\; f \in L^\infty(U),\ \frac{\partial f}{\partial x_i} \in L^\infty(U), \right.$$
$$\left. \textit{for a.e. } x \in U\ |f(x)| \leq 1, \textit{ and } f \textit{ is independent of } x_j \right\}.$$

Proof (i) By definition, the inequality $\leq$ holds for every open set U. Assume that the boundary of U is parallel to the x_i-axis. For notational convenience, define $V := \mathbb{T}^n \setminus \overline{U}$. Let $f_U \in L^\infty(U)$ with $\frac{\partial f_U}{\partial x_i} \in L^\infty(U)$ and, for a.e. $x \in U$, $|f_U(x)| \leq 1$, and let $f_V \in L^\infty(V)$ with $\frac{\partial f_V}{\partial x_i} \in L^\infty(V)$ and, for a.e. $x \in V$, $|f_V(x)| \leq 1$. Let $f : \mathbb{T}^n \to \mathbb{R}$ be such that $f|_U = f_U$ and $f|_V = f_V$. Then $f \in L^\infty(\mathbb{T}^n)$ and, since the boundary of U is parallel to the x_i-axis, $\frac{\partial f_U}{\partial x_i} \in L^\infty(U)$ and $\frac{\partial f_V}{\partial x_i} \in L^\infty(V)$. Moreover, for a.e. $x \in \mathbb{T}^n$, $|f(x)| \leq 1$, Thus, if f_U and f_V are admissible functions in the suprema on the right-hand side of (2.62), then f is admissible in the supremum on the left-hand side. This yields the reverse inequality of the identity.[17]

(ii) The inequality $\geq$ follows immediately, since the admissible set of functions in the supremum on the right-hand side is a subset of the admissible set in (2.61).

[17] We note that this may not hold in general if the boundary of U is not parallel to the x_i-axis.

To prove the inequality $\leq$, we show that we can replace any admissible function f by another admissible function that is independent of x_j, without changing the value of $\int_U u(x) \frac{\partial f}{\partial x_i}(x)\, dx$.
Without loss of generality, we take $j = 1$ and $i > 1$. Let f be admissible in the supremum in (2.61). We define the x_1-average of f—which is independent of x_1—for $(x_1, \tilde{x}) \in U$ as

$$\overline{f}(x_1, \tilde{x}) := \frac{1}{|U_{\tilde{x}}|} \int_{U_{\tilde{x}}} f(y, \tilde{x})\, dy,$$

where $U_{\tilde{x}} := \left\{x_1 \in \mathbb{T} \mid (x_1, \tilde{x}) \in U\right\}$ and $|U_{\tilde{x}}|$ is the one-dimensional Lebesgue measure of $U_{\tilde{x}}$. We note that, for all $\tilde{x} \in \mathbb{T}^{n-1}$, $U_{\tilde{x}}$ is open in $\mathbb{T}$, since U is open in $\mathbb{T}^n$. In particular, if $(x_1, \tilde{x}) \in U$, then $U_{\tilde{x}} \neq \emptyset$ and thus $|U_{\tilde{x}}| > 0$. Hence $\overline{f}$ is well-defined. We define $X := \left\{\tilde{x} \in \mathbb{T}^{n-1} \mid U_{\tilde{x}} \neq \emptyset\right\}$.
By differentiating under the integral sign [39, Theorem 7.40 (or Sect. 11.12)],[18] we see that, for a.e. $(x_1, \tilde{x}) \in U$,

$$\frac{\partial \overline{f}}{\partial x_i}(x_1, \tilde{x}) = \overline{\frac{\partial f}{\partial x_i}}(x_1, \tilde{x}).$$

Hence $\overline{\frac{\partial f}{\partial x_i}} \in L^\infty(U)$. Moreover, we observe that the value of $\displaystyle\int_U u(x) \frac{\partial f}{\partial x_i}(x)\, dx$ remains unchanged if we replace f by $\overline{f}$:

$$\begin{aligned}
\int_U u(x) \frac{\partial \overline{f}}{\partial x_i}(x)\, dx &= \int_X \int_{U_{\tilde{x}}} u(x_1, \tilde{x}) \overline{\frac{\partial f}{\partial x_i}}(x_1, \tilde{x})\, dx_1\, d\tilde{x} \\
&= \int_X \int_{U_{\tilde{x}}} u(x_1, \tilde{x}) \frac{1}{|U_{\tilde{x}}|} \int_{U_{\tilde{x}}} \frac{\partial f(y, \tilde{x})}{\partial x_i}\, dy\, dx_1\, d\tilde{x} \\
&= \int_X \frac{1}{|U_{\tilde{x}}|} \int_{U_{\tilde{x}}} u(y, \tilde{x}) \frac{\partial f(y, \tilde{x})}{\partial x_i}\, dy \int_{U_{\tilde{x}}} dx_1\, d\tilde{x} \\
&= \int_X \int_{U_{\tilde{x}}} u(y, \tilde{x}) \frac{\partial f(y, \tilde{x})}{\partial x_i}\, dy\, d\tilde{x} \\
&= \int_U u(x) \frac{\partial f}{\partial x_i}(x)\, dx, \qquad (2.63)
\end{aligned}$$

where we have used that, for $x_1, y \in U_{\tilde{x}}$ and for $\tilde{x} \in X$, $u(x_1, \tilde{x}) = u(y, \tilde{x})$. Furthermore, for a.e. $x \in U$ we have $|\overline{f}(x)| \leq 1$ since, for a.e. $x \in U$, $|f(x)| \leq 1$. This completes the proof of (ii).

□

[18] For a more directly applicable result published in the Japanese language, we refer to [38, Theorem 19.4].

Remark 2.15 By iterating the result of Lemma 2.10 (ii), it follows that, if $u \in L^1(U)$ is independent of all elements of $\{x_j\}_{j \in J}$, for a $J \subset [n] \setminus \{i\}$, then

$$\int_{\overline{U}} |D_{x_i} u| = \sup\Bigg\{ \int_U u(x) \frac{\partial f}{\partial x_i}(x)\, dx \;\Bigg|\; f \in L^\infty(U),\ \frac{\partial f}{\partial x_i} \in L^\infty(U), \text{ for a.e. } x \in U\ |f(x)| \le 1, \text{ and } f \text{ is independent of } \{x_j\}_{j \in J} \Bigg\}.$$

In particular, extending the averaging argument from the proof of Lemma 2.10, if we average f over all elements of $\{x_j\}_{j \in J}$, the value of the integral remains unchanged as in (2.63). To be precise, without loss of generality let $\{x_j\}_{j \in J} = \{x_1, \dots, x_{|J|}\}$ and $i > |J|$, and define the $\{x_j\}_{j \in J}$-average of f as

$$\overline{f}(x_1, \dots, x_{|J|}, \tilde{x}) := \frac{1}{|U_{\tilde{x}}|} \int_{U_{\tilde{x}}} f(y_1, \dots, y_{|J|}, \tilde{x})\, dy_1 \dots dy_{|J|}, \tag{2.64}$$

for $(x_1, \dots, x_{|J|}, \tilde{x}) \in U$, $U_{\tilde{x}} := \left\{(x_1, \dots, x_{|J|}) \in \mathbb{T}^{|J|} \;\middle|\; (x_1, \dots, x_{|J|}, \tilde{x}) \in U\right\}$, and $|U_{\tilde{x}}|$ to be the $|J|$-dimensional Lebesgue measure of $U_{\tilde{x}}$, then by a similar argument as in (2.63) we obtain

$$\int_U u(x) \frac{\partial \overline{f}}{\partial x_i}(x)\, dx = \int_U u(x) \frac{\partial f}{\partial x_i}(x)\, dx. \tag{2.65}$$

We conclude this remark by noting that in the argument above, the fact that U is open is only used to guarantee that $|U_{\tilde{x}}| > 0$. The conclusion from (2.65) thus holds for any subset $U \subset \mathbb{T}^n$ on which f and its weak derivative $\frac{\partial f}{\partial x_i}$ are well-defined as elements of (equivalence classes in) $L^\infty(U)$ and for which, for all $\tilde{x} \in \mathbb{T}^{n-|J|}$ for which $U_{\tilde{x}} \neq \emptyset$, $|U_{\tilde{x}}| > 0$. In particular, if $U = Q_z^h$ for an $h \in \mathfrak{H}$ and a $z \in V_h$, then any non-empty $U_{\tilde{x}}$ will be the Cartesian product of sets $\{I_{z_i}^h\}_{i \in J}$ and thus have positive $|J|$-dimensional Lebesgue measure. This allows us to prove Corollary 2.2 below, which is used in turn in our proof of Lemma 2.3.

Corollary 2.2 *Let $h \in \mathfrak{H}$ and $i \in [n]$. If $u \in L_h^2$, then*

$$\int_{\mathbb{T}^n} |D_{x_i} i_h u| = \sup_{g \in H_{h,i}} \int_{\mathbb{T}^n} i_h u(x) \frac{\partial g}{\partial x_i}(x)\, dx,$$

where

$$H_{h,i} := \Big\{ f : \mathbb{T}^n \to \mathbb{R} \;\Big|\; f \in L^\infty(\mathbb{T}^n),\ \frac{\partial f}{\partial x_i} \in L^\infty(\mathbb{T}^n),\ \forall x \in \mathbb{T}\ |f(x)| \le 1, \text{ and } \forall z \in V_h\ \frac{\partial f}{\partial x_i} \text{ is constant a.e. on } \operatorname{int} Q_z^h \Big\}. \tag{2.66}$$

Proof Without loss of generality, we assume that $i = n$.

Since $H_{h,n}$ is a subset of the set of admissible functions in the supremum in the definition of $\int_{\mathbb{T}^n} |D_{x_n} i_h u|$ in (2.61) (with $i = n$), the inequality $\geq$ follows.

We prove the inequality $\leq$ by showing that for every admissible f in (2.61), there is an $\tilde{f} \in H_{h,n}$ such that the equality

$$\int_{\mathbb{T}^n} u(x) \frac{\partial \tilde{f}}{\partial x_n}(x)\, dx = \int_{\mathbb{T}^n} u(x) \frac{\partial f}{\partial x_n}(x)\, dx \tag{2.67}$$

holds and thus

$$\int_{\mathbb{T}^n} u(x) \frac{\partial \tilde{f}}{\partial x_n}(x)\, dx \leq \sup_{g \in H_{h,n}} \int_{\mathbb{T}^n} i_h u(x) \frac{\partial g}{\partial x_n}(x)\, dx,$$

from which the required inequality follows by first taking the supremum over all admissible f in the definition in (2.61).

To prove the equality (2.67), let f be admissible in (2.61). If $z \in V_h$ and $x \in Q_z^h$, then $i_h u(x) = u(z)$, thus in particular the function $i_h u$ is independent of all the variables in $\{x_j\}_{j \in [n-1]}$ on Q_z^h. For all $z \in V_h$ and for all $x \in Q_z^h$, define $\hat{f}(x) := \overline{f}_z(x)$, where $\overline{f}_z$ is the $\{x_j\}_{j \in [n-1]}$-average of f over Q_z^h, i.e., $\overline{f}_z$ is as $\overline{f}$ in (2.64) with $J = [n-1]$ and $U = Q_z^h$. This choice of U is allowed, by the argument at the end of Remark 2.15. Then, for all $z \in V_h$ (2.65) holds (with $i_h u$ instead of u and $\overline{f}_z$ instead of $\overline{f}$), i.e.,

$$\int_{Q_z^h} i_h u(x) \frac{\partial \hat{f}}{\partial x_n}(x)\, dx = \int_{Q_z^h} i_h u(x) \frac{\partial f}{\partial x_n}(x)\, dx.$$

Because $\hat{f}$ is independent of $\tilde{x} = (x_1, \ldots, x_{n-1})$ on each Q_z^h, we have that for a.e. $\tilde{x} \in \mathbb{T}^{n-1}$, $x_n \mapsto \hat{f}(\tilde{x}, x_n)$ is a Lipschitz-continuous function[19] on $\mathbb{T}$, and thus in particular pointwise evaluation of $\hat{f}(\cdot, \tilde{x})$ is well-defined for a.e. $\tilde{x} \in \mathbb{T}^{n-1}$. We recall the definitions of $I_{z_n}^h$ and $\tilde{Q}_{z,n}^h$ from (2.17) and (2.18), respectively. Then

[19] We recall that, if $h \in L^\infty(\mathbb{T})$ and $h' \in L^\infty(\mathbb{T})$, then h is Lipschitz-continuous (or, more accurately, has a representative in $L^\infty(\mathbb{T})$ that is Lipschitz-continuous) [44, Exercise 11.50], [27, Sect. 5.8, Theorem 4].

$$
\begin{aligned}
\int_{\mathbb{T}^n} i_h u(x) \frac{\partial f}{\partial x_n}(x)\,dx &= \sum_{z\in V_h} \int_{Q_z^h} i_h u(x) \frac{\partial f}{\partial x_n}(x)\,dx = \sum_{z\in V_h} \int_{Q_z^h} i_h u(x) \frac{\partial \hat{f}}{\partial x_n}(x)\,dx \\
&= \sum_{z\in V_h} u(z) \int_{\tilde{Q}_{z,n}^h} \int_{I_{z_n}^h} \frac{\partial \hat{f}}{\partial x_n}(\tilde{x}, x_n)\,dx_n\,d\tilde{x} \\
&= \sum_{z\in V_h} u(z) \int_{\tilde{Q}_{z,n}^h} \int_{I_{z_n}^h} \frac{\partial \hat{f}}{\partial x_n}(x)\,dx_n\,d\tilde{x} \\
&= \sum_{z\in V_h} u(z) \int_{\tilde{Q}_{z,n}^h} \left(\hat{f}(\tilde{x}, z_n + h/2) - \hat{f}(\tilde{x}, z_n - h/2)\right) d\tilde{x} \\
&= \sum_{z\in V_h} u(z) \int_{\tilde{Q}_{z,n}^h} \int_{I_{z_n}^h} \frac{\partial \tilde{f}}{\partial x_n}(\tilde{x}, x_n)\,dx_n\,d\tilde{x} \\
&= \sum_{z\in V_h} \int_{Q_z^h} u(x) \frac{\partial \tilde{f}}{\partial x_n}(x)\,dx \\
&= \int_{\mathbb{T}^n} u(x) \frac{\partial \tilde{f}}{\partial x_n}(x)\,dx.
\end{aligned}
$$

Here $\tilde{f} : \mathbb{T}^n \to \mathbb{R}$ is such that, for all $z \in V_h$, for all $x_n \in I_{z_n}^h$, and for all $\tilde{x} \in \tilde{Q}_{z,n}^h$, $\tilde{f}(\tilde{x}, x_n) := k_{\tilde{z}}(x_n)$, where $k_{\tilde{z}} : \mathbb{T} \to \mathbb{R}$ is the unique continuous piecewise-affine solution to

$$
k_{\tilde{z}}' = h^{-1}\left(\hat{f}(\tilde{x}, z_n + h/2) - \hat{f}(\tilde{x}, z_n - h/2)\right) \quad \text{on each int } I_{z_n}^h.
$$

We note that the function $k_{\tilde{z}}$ depends on $\tilde{z} := (z_1, \ldots, z_{n-1})$, but is independent of $\tilde{x} \in \tilde{Q}_{z,n}^h$, because $\hat{f}$ is independent of $\tilde{x}$ on each Q_z^h. Hence $\frac{\partial \tilde{f}}{\partial x_n}$ is constant on every Q_z^h, which immediately implies that $\frac{\partial \tilde{f}}{\partial x_n} \in L^\infty(\mathbb{T}^n)$.

Furthermore, on each Q_z^h, $\tilde{f}$ is independent of $\tilde{x}$ and $x_n \mapsto \tilde{f}(\tilde{x}, x_n)$ is affine with bounded derivative—after all, since $f \in L^\infty(\mathbb{T}^n)$ also $\hat{f} \in L^\infty(\mathbb{T}^n)$—and thus $\tilde{f} \in L^\infty(\mathbb{T}^n)$. Finally, since for a.e. $x \in \mathbb{T}^n$, $|f(x)| \leq 1$, we also have, for a.e. $x \in \mathbb{T}^n$, $|\tilde{f}(x)| \leq 1$. Thus, $\tilde{f} \in H_{h,n}$, which concludes the proof. □

It is well known[20] that in the one-dimensional case, for functions $u \in BV(\mathbb{T})$,

$$
\int_{\mathbb{T}} |Du|_{l^1} = \operatorname{ess Var} u := \inf_w \operatorname{Var} w, \quad \text{with}
$$

$$
\operatorname{Var} w := \sup_P \sum_{i=1}^{q} |w(x_i) - w(x_{i-1})| + |w(x_0) - w(x_q)|, \tag{2.68}
$$

[20] The results from [44, Theorem 7.9], [34, Sect. 1.30] should be adapted to the torus.

where the infimum in essVar is taken over all functions $w : \mathbb{T} \to \mathbb{R}$ that satisfy $w(x) = u(x)$ for Lebesgue almost every $x \in \mathbb{T}$, and the supremum in Var is taken over all partitions[21] P of $\mathbb{T}$.

Before we prove (2.21) in Lemma 2.12, we require the following result in the one-dimensional setting.

Lemma 2.11 *Let h be fixed, $q \in \mathbb{N}$, and let $\{I_i\}_{i=0}^q$ be a finite set of non-degenerate disjoint intervals in $[-h/2, 1-h/2)$ such that $\left|[-h/2, 1-h/2) \setminus \bigcup_{i=0}^q I_i\right| = 0$ and, for all $i \in \{0, \ldots, q-1\}$, for all $x \in I_i$, and for all $y \in I_{i+1}$, $x < y$. Assume $u \in BV(\mathbb{T})$ is constant (up to a null set) on each interval I_i (where we identify $\mathbb{T}$ with $[-h/2, 1-h/2)$), taking value c_i on interval I_i (up to a null set). Then*

$$\text{ess}\,\text{Var}\, u = \sum_{i=1}^q |c_i - c_{i-1}| + |c_0 - c_q|.$$

Proof Since each interval I_i has positive Lebesgue measure, for every function w that agrees almost everywhere with u and for all i, there are $x_i \in I_i$ such that $w(x_i) = c_i$. Hence

$$\text{Var}\, w \geq \sum_{i=1}^q |w(x_i) - w(x_{i-1})| + |w(x_0) - w(x_q)| = \sum_{i=1}^q |c_i - c_{i-1}| + |c_0 - c_q|$$

and thus

$$\text{ess}\,\text{Var}\, u \geq \sum_{i=1}^q |c_i - c_{i-1}| + |c_0 - c_q|.$$

The complement of the union of finitely many disjoint intervals in $[-h/2, 1-h/2)$ is itself a union of finitely many (possibly degenerate) disjoint intervals. Since $[-h/2, 1-h/2) \setminus \bigcup_{i=0}^q I_i$ has zero Lebesgue measure, it must consist of finitely many isolated points $\tilde{x}_j$. Let $\tilde{u}$ be equal to u on $\bigcup_{i=0}^q I_i$ and for each $\tilde{x}_j$ define $\tilde{u}(\tilde{x}_j) = c_i$, where I_i is the closest interval to $\tilde{x}_j$ (make an arbitrary choice between candidates in case of nonuniqueness). Then

$$\text{ess}\,\text{Var}\, u \leq \text{Var}\, \tilde{u} = \sum_{i=1}^q |c_i - c_{i-1}| + |c_0 - c_q|.$$

□

Now we are ready to prove (2.21).

---

[21] If h is fixed and we identify $\mathbb{T}$ with $[-h/2, 1-h/2)$ with periodic boundary conditions, then by a partition of $\mathbb{T}$ we mean a subset $P = \{x_0, \ldots x_q\} \subset [-h/2, 1-h/2)$, for some $q \in \mathbb{N}$, such that $x_0 < x_1 < \ldots < x_q$.

Lemma 2.12 *(Proof of (2.21)) Let $h \in \mathfrak{H}$. If $u \in L^2_h$, then*

$$\int_{\mathbb{T}^n} |Di_h u|_{l^1} = \varphi^h_{TV}(u)$$

and thus consequently $i_h u \in BV(\mathbb{T}^n)$.

Proof First we note that the second claim in the lemma follows directly from the first, since $\varphi^h_{TV}(u) < +\infty$ for $u \in L^2_h$.

For the proof of the first claim it is useful to introduce some additional notation. For all $i \in [n]$, we define $R^h_i(z) := \{\tilde{z} \in V_h : \forall j \in [n] \text{ with } j \neq i, \; \tilde{z}_j = z_j\}$. The set $R^h_i(z)$ contains all nodes in V_h that lie on the same "row" as z in the ith direction.

We decompose the graph total variation functional into functionals that only consider differences along the coordinate axes:

$$\varphi^h_{TV}(u) = \sum_{i=1}^n \varphi^h_{TV,i}(u), \qquad \text{with} \qquad \varphi^h_{TV,i}(u) := \frac{1}{2} \sum_{\substack{z;\overline{z}\sim z \\ \overline{z}\in R^h_i(z)}} h^{n-1} \left|u(z) - u(\overline{z})\right|,$$

where we use a variant of the notation from Sect. 2.4.1:

$$\sum_{\substack{z;\overline{z}\sim z \\ \overline{z}\in R^h_i(z)}} := \sum_{z\in V_h} \sum_{\overline{z}\in R^h_i(z):\overline{z}\sim z} .$$

By (2.60) it suffices to prove that, for all $i \in [n]$,

$$\int_{\mathbb{T}^n} |D_{x_i} i_h u| = \varphi^h_{TV,i}(u).$$

Let $i \in [n]$. By permutation of variables, we may assume that $i = n$. Given a $z \in V_h$, by $\tilde{x} = (x_1, \ldots, x_{n-1})$ we denote the coordinates on $\tilde{Q}^h_{z,n}$ (we recall the definition of $\tilde{Q}^h_{z,n}$ from (2.15)). We define V^0_h to be the set containing all nodes $z \in V_h$ with $z_n = 0$, i.e.,

$$V^0_h := \{z \in V_h \mid z_n = 0\} = \{0, h, 2h, \ldots, 1-h\}^{n-1} \times \{0\}.$$

The space $\mathbb{T}^n$ is represented as a disjoint union

$$\mathbb{T}^n = \bigcup_{z\in V^0_h} (\tilde{Q}^h_{z,n} \times \mathbb{T}).$$

For $z \in V^0_h$, we set

$$U^h_{z,n} := \operatorname{int} \tilde{Q}^h_{z,n} \times \mathbb{T}.$$

Since the boundary of $U^h_{z,n}$ is parallel to the x_n-axis, repeated application of Lemma 2.10 (i) leads to

$$\int_{\mathbb{T}^n} |D_{x_n} i_h u| = \sum_{z\in V_h^0} \int_{\overline{U^h_{z,n}}} |D_{x_n} i_h u|.$$

Because $i_h u$ is constant on each Q^h_z and thus independent of $\tilde{x}$ in each $\tilde{Q}^h_{z,n}$, it is independent of $\tilde{x}$ on each $U^h_{z,n}$. Temporarily fix $z \in V_h^0$. Then by Lemma 2.10 (ii) and Remark 2.15 we have that

$$\int_{\overline{U^h_{z,n}}} |D_{x_n} i_h u| = \sup\left\{ \int_{U^h_{z,n}} (i_h u)(x) f'(x_n)\, dx \;\middle|\; f \in L^\infty(\mathbb{T}),\ f' \in L^\infty(\mathbb{T}), \text{ and } \forall x \in \mathbb{T}\ |f(x)| \le 1 \right\}.$$

Moreover, writing $\overline{i_h u}^z(x_n) := (i_h u)(\tilde{x}, x_n)$ for $x = (\tilde{x}, x_n) \in U^h_{z,n}$, we get

$$\int_{U^h_{z,n}} (i_h u)(x) f'(x_n)\, dx = h^{n-1} \int_{\mathbb{T}} \overline{i_h u}^z(x_n) f'(x_n)\, dx_n,$$

and thus

$$\int_{\overline{U^h_{z,n}}} |D_{x_n} i_h u| = h^{n-1} \int_{\mathbb{T}} \left| D\overline{i_h u}^z \right|_{\ell^1}.$$

The problem is now reduced to a one-dimensional setting. Since $\overline{i_h u}^z$ is piecewise constant on $\mathbb{T}$, applying the one-dimensional result from Lemma 2.11 together with (2.68), we deduce that

$$\int_{\mathbb{T}} \left| D\overline{i_h u}^z \right|_{\ell^1} = \frac{1}{2} \sum_{\hat{z}\in R^h_n(z)} \sum_{\overline{z}\in R^h_n(z):\overline{z}\sim\hat{z}} \left| u(\tilde{z}, \hat{z}_n) - u(\tilde{z}, \overline{z}_n) \right|,$$

where $\hat{z} = (\tilde{z}, \hat{z}_n) \in V_h$ and $\overline{z} = (\tilde{z}, \overline{z}_n) \in V_h$. Now we unfix z and observe that

$$\begin{aligned}
\int_{\mathbb{T}^n} |D_{x_n} i_h u| &= \sum_{z\in V_h^0} \int_{\overline{U^h_{z,n}}} |D_{x_n} i_h u| = h^{n-1} \sum_{z\in V_h^0} \int_{\mathbb{T}} \left| D\overline{i_h u}^z \right|_{\ell^1} \\
&= \frac{1}{2} h^{n-1} \sum_{z\in V_h^0} \sum_{\hat{z}\in R^h_n(z)} \sum_{\overline{z}\in R^h_n(z):\overline{z}\sim\hat{z}} \left| u(\tilde{z}, \hat{z}_n) - u(\tilde{z}, \overline{z}_n) \right| \\
&= \frac{1}{2} h^{n-1} \sum_{\substack{z;z\sim\overline{z} \\ \overline{z}\in R^h_n(z)}} |u(z) - u(\overline{z})| = \varphi^h_{TV,n}(u).
\end{aligned}$$

This proves the lemma, so the proof of (2.18) is also complete now. □

2.8 Appendix B: Properties of Γ

In this section we prove some properties of the operator Γ, which is defined in (2.31), and its matrix exponential.

It will be useful to identify Γ with a circulant $1/h$ by $1/h$-matrix, which we will again denote by Γ. This matrix has entries, for all $i, j \in \left[\frac{1}{h}\right]$,

$$\Gamma_{ij} = \begin{cases} \frac{2}{3}, & \text{if } i = j, \\ \frac{1}{6}, & \text{if } i \equiv j - 1 \pmod{\frac{1}{h}} \text{ or } i \equiv j + 1 \pmod{\frac{1}{h}}, \\ 0, & \text{otherwise.} \end{cases}$$

The general form of the eigenvectors and eigenvalues of circulant matrices is known. In this case it can be checked via direct computation that the eigenvalues of Γ are, for $k \in \left[\frac{1}{h} - 1\right]_0$,

$$\gamma_k := \frac{1}{3}\left(2 + \cos(2\pi hk)\right).$$

The corresponding normalized eigenvectors of Γ (in $\mathbb{C}^{1/h}$) are

$$z^{(k)} := \sqrt{h} \begin{pmatrix} 1 \\ \omega^k \\ \omega^{2k} \\ \vdots \\ \omega^{(\frac{1}{h}-1)k} \end{pmatrix}, \quad \text{thus} \quad z_i^{(k)} = \sqrt{h}\omega^{(i-1)k}, \text{ for all } i \in \left[\frac{1}{h}\right],$$

where $\omega := e^{2\pi h\iota}$ is the $\frac{1}{h}$th root of unity and we recall that ι denotes the imaginary unit. We note that

$$\gamma_{\frac{1}{h}-k} = \gamma_k. \tag{2.69}$$

An alternative way to describe Γ is as the sum of three matrices,

$$\Gamma = \frac{2}{3}I + \frac{1}{6}A + \frac{1}{6}A^T, \tag{2.70}$$

where I is the $\frac{1}{h}$-by-$\frac{1}{h}$ identity matrix and A and A^T are the matrix representations of the operators τ_+ and τ_- from (2.33), respectively, i.e.,

$$A_{ij} := \delta_{i,j+1},$$

where we always interpret the indices modulo $\frac{1}{h}$, so $\delta_{i,\frac{1}{h}+1} = \delta_{i,1}$, etc. Remember that the Kronecker delta is defined by

$$\delta_{i,j} := \begin{cases} 1, \text{ if } i = j, \\ 0, \text{ otherwise.} \end{cases}$$

We note that, for all $n \in \mathbb{N}$,

$$A^n_{ij} = \delta_{i,j+n}, \tag{2.71}$$

as can be proved by induction on n: Taking as base case $n = 0$ or $n = 1$, we see it is true by definition of I or A, respectively. Assuming that it is true for $n = k > 1$, we find

$$A^{k+1}_{ij} = \sum_{l=1}^{\frac{1}{h}} A_{il} A^k_{lj} = \sum_{l=1}^{\frac{1}{h}} \delta_{i,l+1}\delta_{l,j+k} = \delta_{i,j+k+1}.$$

As a consequence of (2.71), we have that

$$A^{\frac{1}{h}} = A. \tag{2.72}$$

Furthermore $A^T = A^{-1}$, since

$$\left(AA^T\right)_{ij} = \sum_{l=1}^{\frac{1}{h}} A_{il} A_{jl} = \sum_{l=1}^{\frac{1}{h}} \delta_{i,l+1}\delta_{j,l+1} = \delta_{i,j}.$$

Thus we also have

$$(A^{-1})^{\frac{1}{h}} = (A^T)^{\frac{1}{h}} = (A^{\frac{1}{h}})^T = A^T = A^{-1}. \tag{2.73}$$

It is worth noting that

$$\left(A^{-1}\right)_{ij} = \left(A^T\right)_{ij} = A_{ji} = \delta_{j,i+1} = \delta_{i,j-1} \quad \text{and, for all } n \in \mathbb{N}, \ \left(A^{-n}\right)_{ij} = \delta_{i,j-n}.$$

These two characterizations of Γ, one through its spectrum and the other using the matrix A, lead to two expressions for the matrix exponential $e^{\Gamma x}$ in the following lemma.

Lemma 2.13 *Let $x \in \mathbb{R}$. For all $i, j \in \left[\frac{1}{h}\right]$ we have*

$$\left(e^{\Gamma x}\right)_{ij} = h e^{\frac{2}{3}x} \sum_{l=0}^{\frac{1}{h}-1} e^{\frac{1}{3}\cos(2\pi hl)x} \cos\left(2\pi hl(i-j)\right) \tag{2.74}$$

and

$$\left(e^{\Gamma x}\right)_{ij} = e^{\frac{2}{3}x} \sum_{l=0}^{\frac{1}{h}-1} B_l(x) B_{l+j-i}(x), \tag{2.75}$$

where

$$B_l(x) := \sum_{m=0}^{\infty} \frac{1}{(\frac{m}{h}+l)!} \left(\frac{x}{6}\right)^{\frac{m}{h}+l}$$

has infinite radius of convergence. The subscripts of B are interpreted modulo $\frac{1}{h}$.

Proof The proof of the first expression follows an ODE approach: Consider the equation

$$u'(x) = \Gamma u(x). \tag{2.76}$$

By standard ODE theory [37, Sect. III.4], this equation has $\frac{1}{h}$ linearly independent solutions $u^j : \mathbb{R} \to \mathbb{R}^{1/h}$, for $j \in \left[\frac{1}{h}\right]$. Our discussion above shows that Γ has $\frac{1}{h}$ linearly independent eigenvectors and thus, for all $j \in \left[\frac{1}{h}\right]$, $u^j(x) = e^{\gamma_{j-1}x} z^{(j-1)}$. We can now compute

$$e^{\Gamma x} = X(x)X(0)^{-1}, \tag{2.77}$$

where $X : \mathbb{R} \to \mathbb{R}^{1/h \times 1/h}$ can be any fundamental matrix solution of (2.76). We choose the matrix which has the solutions u^j as columns: for all $i, j \in \left[\frac{1}{h}\right]$,

$$X_{ij}(x) := u_i^j(x) = \sqrt{h} e^{\gamma_{j-1}x} \omega^{(i-1)(j-1)}.$$

Then

$$X_{ij}(0) = u_i^j(0) = z_i^{(j-1)} = \sqrt{h}\omega^{(i-1)(j-1)}.$$

We claim that the inverse $X(0)^{-1}$ is given by, for all $i, j \in \left[\frac{1}{h}\right]$,

$$Y_{ij} := \sqrt{h}\omega^{-(i-1)(j-1)}.$$

To prove this claim, we compute

$$\begin{aligned}(X(0)Y)_{ij} &= \sum_{l=1}^{\frac{1}{h}} X(0)_{il} Y_{lj} = \sqrt{h} \sum_{l=1}^{\frac{1}{h}} z_i^{(l-1)} \omega^{-(l-1)(j-1)} = h \sum_{l=1}^{\frac{1}{h}} \omega^{(i-1)(l-1)} \omega^{-(l-1)(j-1)} \\ &= h \sum_{l=1}^{\frac{1}{h}} \omega^{(l-1)(i-j)}.\end{aligned}$$

Taking $i = j$ shows that $(X(0)Y)_{ii} = 1$. If $i \neq j$, then we multiply by ω^{i-j}:

$$\omega^{i-j}(X(0)Y)_{ij} = h \sum_{l=1}^{\frac{1}{h}} \omega^{l(i-j)} = h \sum_{k=2}^{\frac{1}{h}+1} \omega^{(l-1)(i-j)} = h \sum_{l=1}^{\frac{1}{h}} \omega^{(l-1)(i-j)} = (X(0)Y)_{ij},$$

where the third equality follows from the fact that $\omega^{(1/h+1-1)(i-j)} = \left(\omega^{1/h}\right)^{i-j} = 1^{i-j} = 1 = \omega^{0(i-j)}$—recall that ω is the $\frac{1}{h}$th root of unity, and thus $\omega^l = 1$ if and only if $l \equiv 0 \pmod{\frac{1}{h}}$. Since i is not congruent to j modulo $\frac{1}{h}$ and thus $\omega^{i-j} \neq 1$, it follows that $(X(0)Y)_{ij} = 0$. Thus $Y = X(0)^{-1}$.

Using (2.77) we compute

$$\begin{aligned}
\left(e^{\Gamma x}\right)_{ij} &= h \sum_{l=1}^{\frac{1}{h}} e^{\gamma_{l-1} x} \omega^{(i-1)(l-1)} \omega^{-(l-1)(j-1)} = h \sum_{l=0}^{\frac{1}{h}-1} e^{\gamma_l x} \omega^{(i-j)l} \\
&= h + h \sum_{l=1}^{\frac{1}{h}-1} e^{\gamma_l x} \omega^{(i-j)l}.
\end{aligned}$$

By (2.69) it follows that, if $\frac{1}{h}$ is odd,

$$\begin{aligned}
\left(e^{\Gamma x}\right)_{ij} &= h + h \sum_{l=1}^{\frac{1}{2}\left(\frac{1}{h}-1\right)} e^{\gamma_l x} \left(\omega^{(i-j)l} + \omega^{(i-j)(1/h-l)}\right) \\
&= h + \frac{h}{2} \sum_{l=1}^{\frac{1}{h}-1} e^{\gamma_l x} \left(\omega^{(i-j)l} + \omega^{(i-j)(1/h-l)}\right),
\end{aligned}$$

where for the second equality we also used that $\omega^{(i-j)l} + \omega^{(i-j)(1/h-l)}$ remains unchanged when l is replaced by $\frac{1}{h} - l$. Since

$$\begin{aligned}
&\omega^{(i-j)l} + \omega^{(i-j)(1/h-l)} \\
&= \cos\big(2\pi h(i-j)l\big) + i_\iota \sin\big(2\pi h(i-j)l\big) + \cos\big(2\pi h(i-j)(1/h-l)\big) \\
&\qquad + i_\iota \sin\big(2\pi h(i-j)(1/h-l)\big) \\
&= \cos\big(2\pi h(i-j)l\big) + \cos\big(2\pi(i-j) - 2\pi(i-j)hl\big) \\
&\qquad + i_\iota \left[\sin\big(2\pi h(i-j)l\big) + \sin\big(2\pi(i-j) - 2\pi(i-j)hl\big)\right] \\
&= \cos\big(2\pi h(i-j)l\big) + \cos\big(-2\pi(i-j)hl\big) + i_\iota \left[\sin\big(2\pi h(i-j)l\big) + \sin\big(-2\pi(i-j)hl\big)\right] \\
&= 2\cos\big(2\pi h(i-j)l\big), \qquad (2.78)
\end{aligned}$$

the desired expression (2.74), for $\frac{1}{h}$ odd, now follows from the definition of γ_l. If, on the other hand, $\frac{1}{h}$ is even, we have

$$\begin{aligned}
\left(e^{\Gamma x}\right)_{ij} &= h + h \sum_{l=1}^{\frac{1}{2h}-1} e^{\gamma_l x} \left(\omega^{(i-j)l} + \omega^{(i-j)(1/h-l)}\right) + h e^{\gamma_{h/2} x} \omega^{(i-j)/(2h)} \\
&= h + \frac{h}{2} \sum_{\substack{l=1 \\ l \neq \frac{1}{2h}}}^{\frac{1}{h}-1} e^{\gamma_l x} \left(\omega^{(i-j)l} + \omega^{(i-j)(1/h-l)}\right) + h e^{\gamma_{h/2} x} \omega^{(i-j)/(2h)}.
\end{aligned}$$

For the middle terms on the right-hand side we use (2.78) again. For the last term, we note that $\omega^{1/(2h)} = -1$ and thus

$$\omega^{(i-j)/(2h)} = (-1)^{i-j} = 1 = \cos(\pi(i-j)) = \cos\left(\frac{2\pi h(i-j)}{2h}\right).$$

Combined with the definition of γ_l, this now also gives (2.74) when $\frac{1}{h}$ is even.

To prove (2.75), we use the description of Γ from (2.70). First note that, since

$$\sum_{m=0}^{\infty}\left|\frac{1}{(\frac{m}{h}+l)!}\left(\frac{x}{6}\right)^{\frac{m}{h}+l}\right| = \sum_{m=0}^{\infty}\frac{1}{(\frac{m}{h}+l)!}\left|\frac{x}{6}\right|^{\frac{m}{h}+l} \leq \sum_{m=0}^{\infty}\frac{1}{m!}|x|^m = e^x,$$

the series B_l has an infinite radius of convergence.

Because A and $A^T = A^{-1}$ commute, we have

$$e^{\Gamma x} = e^{(\frac{2}{3}I+\frac{1}{6}A+\frac{1}{6}A^T)x} = e^{\frac{2}{3}x}e^{\frac{1}{6}Ax}e^{\frac{1}{6}A^{-1}x}. \tag{2.79}$$

Next we employ the series definition of the matrix exponential in combination with (2.72):

$$\begin{aligned}
e^{\frac{1}{6}Ax} &= \sum_{m=0}^{\infty}\frac{1}{m!}A^m\left(\frac{x}{6}\right)^m \\
&= \left[\frac{1}{(0/h)!}\left(\frac{x}{6}\right)^{0/h} + \frac{1}{(1/h)!}\left(\frac{x}{6}\right)^{1/h} + \frac{1}{(2/h)!}\left(\frac{x}{6}\right)^{2/h} + \frac{1}{(3/h)!}\left(\frac{x}{6}\right)^{3/h} + \ldots\right] I \\
&\quad + \left[\frac{1}{(0/h+1)!}\left(\frac{x}{6}\right)^{0/h+1} + \frac{1}{(1/h+1)!}\left(\frac{x}{6}\right)^{1/h+1} + \frac{1}{(2/h+1)!}\left(\frac{x}{6}\right)^{2/h+1}\right. \\
&\qquad\qquad \left. + \frac{1}{(3/h+1)}\left(\frac{x}{6}\right)^{3/h+1} + \ldots\right] A \\
&\quad + \left[\frac{1}{(0/h+2)!}\left(\frac{x}{6}\right)^{0/h+2} + \frac{1}{(1/h+2)!}\left(\frac{x}{6}\right)^{1/h+2} + \frac{1}{(2/h+2)!}\left(\frac{x}{6}\right)^{2/h+2}\right. \\
&\qquad\qquad \left. + \frac{1}{(3/h+2)}\left(\frac{x}{6}\right)^{3/h+2} + \ldots\right] A^2 \\
&\quad + \ldots \\
&\quad + \left[\frac{1}{(0/h+1/h-1)!}\left(\frac{x}{6}\right)^{0/h+1/h-1} + \frac{1}{(1/h+1/h-1)!}\left(\frac{x}{6}\right)^{1/h+1/h-1}\right. \\
&\qquad + \frac{1}{(2/h+1/h-1)!}\left(\frac{x}{6}\right)^{2/h+1/h-1} + \frac{1}{(3/h+1/h-1)}\left(\frac{x}{6}\right)^{3/h+1/h-1} \\
&\qquad\qquad \left. + \ldots\right] A^{1/h-1} \\
&= \sum_{l=0}^{\frac{1}{h}-1} B_l(x)A^l.
\end{aligned}$$

Similarly, by (2.73),

$$e^{\frac{1}{6}A^{-1}x} = \sum_{l=0}^{\frac{1}{h}-1} B_l(x)A^{-l},$$

where we use the standard notation $A^{-l} := (A^{-1})^l$.

Substituting both these identities into (2.79) we find

$$e^{\Gamma x} = e^{\frac{2}{3}x} \sum_{l=0}^{\frac{1}{h}-1} \sum_{r=0}^{\frac{1}{h}-1} B_l(x)B_r(x)A^l A^{-r}.$$

Hence,

$$\left(e^{\Gamma x}\right)_{ij} = e^{\frac{2}{3}x} \sum_{l=0}^{\frac{1}{h}-1} \sum_{r=0}^{\frac{1}{h}-1} B_l(x)B_r(x) \sum_{k=1}^{\frac{1}{h}-1} \delta_{i,k+l}\delta_{k,j-r} = e^{\frac{2}{3}x} \sum_{l=0}^{\frac{1}{h}-1} \sum_{r=0}^{\frac{1}{h}-1} B_l(x)B_r(x)\delta_{i,j-r+l}$$
$$= e^{\frac{2}{3}x} \sum_{l=0}^{\frac{1}{h}-1} B_l(x)B_{l+j-i}(x).$$

This completes the proof of (2.75). □

Remark 2.16 In the proof above we have given the deductions by which the two expressions for $e^{\Gamma x}$ were derived. Alternatively, we can also directly check that each column of $e^{\Gamma x}$ as given in (2.74) and (2.75) satisfies (2.76) with initial condition $e^{\Gamma 0} = I$.

2.9 Appendix C: Poincaré–Wirtinger-Type Inequalities

In this section we prove the Poincaré–Wirtinger-type inequalities we used in the proof of Lemma 2.7. The first one is a classic result, and the second one a higher order extension which is proved using similar methods as the first.

Lemma 2.14 *(i) Let $u \in H^1((0,h))$ with $u(0) = u(h) = 0$. Then the Poincaré–Wirtinger inequality holds:*

$$\int_0^h |u(x)|^2\,dx \le (h/\pi)^2 \int_0^h |u'(x)|^2\,dx.$$

The constant $(h/\pi)^2$ is optimal.

(ii) Let $u \in H^2((0,h))$ with $u(0) = u(h) = 0$. Then

$$\int_0^h |u(x)|^2\,dx \le (h/\pi)^4 \int_0^h |u''(x)|^2\,dx.$$

The constant $(h/\pi)^4$ is optimal.

Proof (i) We note that the boundary conditions are well-defined, since $H^1((0,h))$ is compactly embedded in the Banach space $\{u \in C((0,h)) \mid u \text{ is bounded}\}$[22] by the Rellich–Kondrachov theorem [1, Theorem 6.3].
The Poincaré–Wirtinger inequality for symmetric elliptic operators is found in [27, Sect. 6.5, Theorem 2 and the following remark], from which it immediately follows that there exists a $C > 0$ such that, for all $u \in H^1((0,1))$ with $u(0) = u(h) = 0$, the inequality

$$\int_0^h |u(x)^2|\,dx \le C \int_0^h |u'(x)|^2\,dx$$

holds. It also tells us that the optimal constant is $C = \nu^{-1}$, where ν is obtained from its Rayleigh quotient formulation $\nu = \min\limits_{\substack{w\in H^1((0,h))\\ w\neq 0}} \frac{\int_0^h |w'(x)|^2\,dx}{\int_0^h |w(x)|^2\,dx}$ (where the minimum is achieved), or as the minimal[23] eigenvalue in the eigenvalue problem[24]

---

[22] Equipped with the supremum norm $\sup_{x\in(0,h)} |u(x)|$.

[23] We emphasize that $w = 0$ is not allowed as eigenfunction.

[24] Additionally, the set of unit eigenfunctions of the differential operator $T(u) := u''$ is a complete orthonormal basis of the Hilbert space $L^2((0,h))$. This can be shown as follows. We claim that the inverse T^{-1} of the densely defined unbounded differential operator T on $L^2((0,h))$ exists and is compact. If this claim is true, then by the Riesz–Schauder theory [64, Sect. X.5, Theorem 2] all nonzero elements of the spectrum of T^{-1} are eigenvalues of T^{-1} and thus their reciprocals are eigenvalues of T with the same eigenfunctions. Moreover, since T^{-1} is also self-adjoint, its eigenfunctions form a complete orthogonal basis for $L^2((0,h))$ [64, Sect. XI.6, Theorem 1, and Sect. XI.9, Theorem 1 and Corollary 2]. Hence, the same is true for the eigenfunctions of T.
To prove that T^{-1} exists and is compact, we observe that, for $f \in L^2((0,h))$, $T^{-1}f = u$, where u is the unique solution to $u'' = f$ on $(0,h)$ and $u(0) = u(h) = 0$. Existence follows from an explicit computation that shows that

$$(T^{-1}f)(x) = \frac{1}{h}\int_0^x \int_0^h \left[\int_0^w f(y)\,dy - \int_0^z f(y)\,dy\right] dz\,dw.$$

If $\|f\|_{L((0,h))} \le 1$ and $x, y \in (0,h)$ with $y > x$, then, by the Cauchy–Schwarz inequality,

$$\begin{aligned}\left|(T^{-1}f)(x)\right| &\le \frac{2}{h}\int_0^h \int_0^h \int_0^h |f(y)|\,dy\,dz\,dw = 2h\int_0^h |f(y)|\,dy\\ &\le 2h\|1\|_{L^2((0,h))}\|f\|_{L^2((0,h))} \le 2h^{3/2},\end{aligned}$$

$$\begin{cases} w'' + \tilde{\nu} w = 0, & \text{on } (0,h), \\ w(0) = w(h) = 0. \end{cases}$$

(Strict positivity of ν is given by [27, Sect. 6.5, Theorem 1].) This ordinary differential equation can be solved by standard ODE techniques [37, Theorem III.4.1]; its general solution (without imposing boundary conditions) is given by

$$w(x) = c_1 \cos(\sqrt{\tilde{\nu}}x) + c_2 \sin(\sqrt{\tilde{\nu}}x),$$

for constants $c_1, c_2 \in \mathbb{R}$. From $w(0) = 0$, it follows that $c_1 = 0$. Since $w(h) = 0$, we have $c_2 = 0$ or $\sin(\sqrt{\tilde{\nu}}h) = 0$. Because $w = 0$ is not allowed as eigenfunction, the option $c_2 = 0$ is excluded, hence ν is the smallest $\tilde{\nu} > 0$ such that $\sin(\sqrt{\tilde{\nu}}h) = 0$. This proves that $\nu = \pi^2/h^2$, as required.
We note that [23, Sect. 1.7(2)] provides an alternative proof for continuously differentiable functions u based on Fourier series.

(ii) Let $u \in \mathcal{A}_h := \{u \in H^2((0,h)) \mid u(0) = u(h) = 0\}$. We note that by the Rellich–Kondrachov theorem [1, Theorem 6.3] $H^2((0,h))$ is compactly embedded in the Banach space $\{u \in C^1((0,h)) \mid u, u' \text{ are bounded}\}$[25] and thus the boundary conditions in the definition of $\mathcal{A}_h$ are well-defined.
We define, for this proof,

$$\nu := \inf_{\substack{w\in\mathcal{A}_h \\ w\neq 0}} \frac{\int_0^h |w''(x)|^2\,dx}{\int_0^h |w(x)|^2\,dx} = \inf_{w\in\mathcal{A}_h} \left\{ \|w''\|^2_{L^2((0,h))} \;\middle|\; \|w\|^2_{L^2((0,h))} = 1 \right\}, \tag{2.22}$$

where the second expression follows by rescaling u by $\left(\int_0^h |w(x)|^2\,dx\right)^{-1}$. Since the quotient is nonnegative, $\nu \geq 0$ exists. If, as we will prove, $\nu > 0$, then it follows that

$$\int_0^h |u(x)|^2\,dx \leq \nu^{-1} \int_0^h |u''(x)|^2\,dx.$$

---

and

$$\begin{aligned} \left|(T^{-1}f)(x) - (T^{-1}f)(x)\right| &\leq \frac{1}{h}\int_{x_1}^{x_2}\int_0^h \left|\int_0^w f(y)\,dy - \int_0^z f(y)\,dy\right| dz\,dw \\ &\leq 2|x_2 - x_1| \int_0^h |f(y)|\,dz \\ &\leq 2h^{1/2}\|f\|_{L^2((0,h))}|x_2 - x_1|. \end{aligned}$$

Hence by the Arzelà–Ascoli theorem [1, Theorem 1.33], [44, Theorem 5.37], T^{-1} is a compact operator.

[25] Equipped with the supremum norm $\max(\sup_{x\in(0,h)}|u(x)|, \sup_{x\in(0,h)}|u'(x)|)$.

Moreover, a minimizer of the minimization problem in the definition of ν exists by the direct method of the calculus of variations, as the following argument shows. We use the second characterization of ν in (2.22). Consider a minimizing sequence $(w_m)_m$ in $\left\{u \in \mathcal{A}_h \mid \|u\|^2_{L^2((0,h))} = 1\right\}$. By definition $(\|w_m\|^2_{L^2((0,h))})_m$ and $(\|w''_m\|^2_{L^2((0,h))})_m$ are bounded. By part (i) of this lemma with $u = w'_m$, $(\|w'_m\|^2_{L^2((0,h))})_m$ is also bounded, hence $(w_m)_m$ is bounded in $H^2((0,h))$ and thus an $H^2((0,h))$-weakly converging subsequence exists with limit $w_* \in H^2((0,h))$. By the compact embedding of $H^2((0,h))$ into a space of continuous functions that was established above, $\|w_*\|^2_{L^2((0,h))} = 1$. Moreover, since $u \mapsto \|u''\|^2_{L^2((0,h))}$ is lower semicontinuous with respect to weak-$H^2((0,h))$ convergence [10, Proposition 3.5], w_* minimizes $w \mapsto \|w''\|^2_{L^2((0,h))}$ over the admissible set.

From the first expression for ν in (2.22), we see that the minimizer w_* has to satisfy

$$\frac{d}{d\varepsilon} \left. \frac{\int_0^h \left((w_* + \varepsilon f)''(x)\right)^2 dx}{\int_0^h (w_* + \varepsilon f)^2(x)\, dx} \right|_{\varepsilon=0} = 0,$$

for all $f \in C^\infty([0,h])$ with $f(0) = f(h) = 0$. Computing this derivative and using that $\nu = \frac{\|w''_*\|^2_{L^2((0,h))}}{\|w_*\|^2_{L^2((0,h))}}$, we find

$$\frac{2}{\|w_*\|^2_{L^2((0,h))}} \int_0^h \left[w''_*(x) f''(x) - \nu w_*(x) f(x)\right] dx = 0.$$

Integrating by parts, we see

$$\begin{aligned}\int_0^h w''_*(x) f''(x)\, dx &= -\int_0^h w'''_*(x) f'(x)\, dx + w''_* f' \Big|_0^h \\ &= \int_0^h w''''_*(x) f(x)\, dx - w'''_* f \Big|_0^h + w''_* f' \Big|_0^h .\end{aligned}$$

Since $f = 0$ on $\{0, h\}$, the term $w'''_* f\big|_0^h$ vanishes. Thus the weak form implies that

$$\int_0^h \left[w''''_*(x) - \nu(x) w_*(x)\right] f(x)\, dx + w''_* f' \Big|_0^h = 0,$$

for all $f \in C^\infty([0,h])$ with $f(0) = f(h) = 0$. Hence w_* solves

$$w'''' - \nu w = 0 \quad \text{on } (0,h).$$

Since f' can be chosen to have any two values at $\{0, h\}$, we get natural boundary conditions. Thus w_* solves

$$w''(0) = w''(h) = 0.$$

Since w_* satisfies $w_*|_{\{0,h\}} = 0$ and $w_* \neq 0$, we deduce that ν must be an eigenvalue of

$$\begin{cases} w'''' - \tilde{\nu} w = 0, & \text{on } (0, h), \\ w''(0) = w''(h) = 0, \\ w(0) = w(h) = 0. \end{cases} \tag{2.23}$$

We note that, if $\tilde{\nu} = 0$, then the ODE above implies that w is a polynomial of degree at most three and thus, due to the boundary conditions $w = 0$. Hence $\tilde{\nu} = 0$ is not an eigenvalue (since w is an eigenfunction and hence $w \neq 0$) and thus $\nu \neq 0$. Already we knew that $\nu \geq 0$, hence $\nu > 0$.
In fact, substituting the (admissible) eigenfunction $w(x) = \sin(\pi x/h)$ in the minimization problem in (2.22), we find that $\nu \geq (\pi/h)^4$.
Finally we prove that every eigenvalue $\tilde{\nu}$ in the eigenvalue problem in (2.23) is larger than or equal to $(\pi/h)^4$, which establishes that $\nu = (\pi/h)^4$, as required. Let $\tilde{\nu} > 0$ be an eigenvalue of (2.23) and let $\sigma > 0$ be the unique strictly positive real number such that $\tilde{\nu} = \sigma^4$. By standard ODE methods [37, Theorem III.4.1], the general solution to the ODE $w'''' - \tilde{\nu} w = 0$ is found to be

$$w(x) = c_1 e^{\sigma x} + c_2 e^{-\sigma x} + c_3 e^{\iota\sigma x} + c_4 e^{-\iota\sigma x},$$

for constants $c_1, c_2, c_3, c_4 \in \mathbb{R}$. We recall that ι is the imaginary unit. Since $w(0) = 0$ and $w''(0) = 0$ imply $c_1 + c_2 + c_3 + c_4 = 0$ and $\sigma^2(c_1 + c_2) - \sigma^2 (c_3 + c_4) = 0$, we deduce that $c_1 + c_2 = c_3 + c_4 = 0$. Therefore

$$w(x) = c_1(e^{\sigma x} - e^{-\sigma x}) + c_3(e^{\iota\sigma x} - e^{-\iota\sigma x}).$$

Thus

$$\begin{aligned} w(h) &= c_1(e^{\sigma h} - e^{-\sigma h}) + c_3(e^{\iota\sigma h} - e^{-\iota\sigma h}) = 0, \\ \sigma^{-2} w''(h) &= c_1(e^{\sigma h} - e^{-\sigma h}) - c_3(e^{\iota\sigma h} - e^{-\iota\sigma h}) = 0, \end{aligned}$$

which in matrix form reads

$$\begin{pmatrix} e^{\sigma h} - e^{-\sigma h} & e^{\iota\sigma h} - e^{-\iota\sigma h} \\ e^{\sigma h} - e^{-\sigma h} & -(e^{\iota\sigma h} - e^{-\iota\sigma h}) \end{pmatrix} \begin{pmatrix} c_1 \\ c_3 \end{pmatrix} = \begin{pmatrix} 0 \\ 0 \end{pmatrix}.$$

This equation has a nontrivial solution if and only if the determinant of the above matrix is zero: $-2(e^{\sigma h} - e^{-\sigma h})(e^{i\sigma h} - e^{-i\sigma h}) = 0$. Since $e^{\sigma h} \neq e^{-\sigma h}$, we obtain that $e^{i\sigma h} - e^{-i\sigma h} = 0$. In other words, $e^{2i\sigma h} = 1$. The minimum strictly positive σ for which this condition holds is π/h. Thus $\nu = (\pi/h)^4$. This concludes the proof.[26]

□

Acknowledgements The first author has received funding from the European Union's Horizon 2020 research and innovation programme under the Marie Skłodowska-Curie grant agreement No. 777826. The second author was partly supported by the Japan Society for the Promotion of Science through grants No. 19H00689 (Kiban A), No. 18H05323 (Kaitaku), and by Arithmer Inc. and Daikin Industries Ltd. through a collaborative grant. The work of the third author was done when he was a graduate student at the University of Tokyo and a Junior Research Associate at RIKEN. His work was partly supported by the Program for Leading Graduate Schools "Frontier of Mathematical Sciences and Physics", MEXT, Japan.

We thank the two anonymous referees whose feedback on an earlier manuscript of this work have improved the final version. We also thank Joshua Willems for his comments on a previous draft, as well as the audiences at talks given by the authors on the topic of this work whose questions and comments have refined the presentation of some of our results.

References

1. Adams, R.A., Fournier, J.J.F.: Sobolev Spaces. Pure and Applied Mathematics, vol. 140. Elsevier, Oxford, Amsterdam (2003)
2. Alfaro, M., Hilhorst, D., Matano, H.: The singular limit of the Allen-Cahn equation and the FitzHugh-Nagumo system. J. Differ. Equ. **245**(2), 505–565 (2008)
3. Allen, S.M., Cahn, J.W.: A microscopic theory for antiphase boundary motion and its application to antiphase domain coarsening. Acta Metall. **27**, 1085–1095 (1979)
4. Ambrosio, L., Brué, E., Semola, D.: Lectures on Optimal Transport, vol. 130 of UNITEXT, 1st edn. Springer, Cham (2021). La Matematica per il 3+2 (UNITEXTMAT)
5. Ambrosio, L., Gigli, N., Savaré, G.: Gradient Flows: in Metric Spaces and in the Space of Probability Measures, 2nd edn. Springer Science & Business Media (2008)
6. Andreu-Vaillo, F., Caselles, V., Mazón, J.M.: Parabolic quasilinear equations minimizing linear growth functionals. In: Progress in Mathematics, vol. 223. Birkhäuser Verlag, Basel (2004)
7. Bertozzi, A.L., Flenner, A.: Diffuse interface models on graphs for classification of high dimensional data. Multiscale Model. Simul. **10**(3), 1090–1118 (2012)
8. Braides, A.: Local Minimization, Variational Evolution and Γ-Convergence. Lecture Notes in Mathematics, vol. 2094. Springer International Publishing, Switzerland (2014)

[26] Additionally, we can conclude from our computations that the geometric multiplicity of the eigenvalue $\tilde{\nu} = (\pi/h)^4$ is equal to one, as any solution w to (2.23) with $\tilde{\nu} = (\pi/h)^4$ is of the form $C \sin \pi x$ for some constant $C \in \mathbb{R}$.

9. Brezis, H.: Opérateurs maximaux monotones et semi-groupes de contractions DANs les espaces de Hilbert, volume No. 5/ Notas de Matemática(50) of North-Holland Mathematics Studies. North-Holland Publishing Co./American Elsevier Publishing Co., Inc., Amsterdam-London/New York (1973)
10. Brezis, H.: Functional analysis. In: Sobolev Spaces and Partial Differential Equations. Universitext. Springer, New York (2011)
11. Budd, J., van Gennip, Y.: Graph Merriman-Bence-Osher as a semidiscrete implicit Euler scheme for graph Allen-Cahn flow. SIAM J. Math. Anal. **52**(5), 4101–4139 (2020)
12. Budd, J., van Gennip, Y.: Mass-conserving diffusion-based dynamics on graphs. Eur. J. Appl. Math. First View, 1–49 (2021)
13. Budd, J., van Gennip, Y., Latz, J.: Classification and image processing with a semi-discrete scheme for fidelity forced Allen-Cahn on graphs. GAMM Mitteilungen Spec. Issue: Sci. Mach. Learn.-Part I **44**, 1–43 (2021)
14. Chen, X.: Generation and propagation of interfaces in reaction-diffusion systems. Trans. Am. Math. Soc. **334**(2), 877–913 (1992)
15. Choksi, R., van Gennip, Y., Oberman, A.: Anisotropic total variation regularized L^1 approximation and denoising/deblurring of 2D bar codes. Inverse Probl. Imaging **5**(3), 591–617 (2011)
16. Chung, F.R.K.: Spectral Graph Theory, Volume 92 of CBMS Regional Conference Series in Mathematics. Published for the Conference Board of the Mathematical Sciences, Washington, DC, by the American Mathematical Society, Providence, Rhode Island (1997)
17. Cohen, I., Berkov, T., Gilboa, G.: Total-variation—fast gradient flow and relations to Koopman theory (2021). arXiv:2105.10044 [math.AP]
18. Crouzeix, M., Thomée, V.: The stability in l_p and w_p^1 of the l_2-projection onto finite element function spaces. Math. Comput. **48**, 521–532 (1987)
19. de Mottoni, P., Schatzman, M.: Development of interfaces in $\mathbb{R}^n$. Proc. R Soc. Edinburgh Sect. A **116**(3–4), 207–220 (1990)
20. de Mottoni, P., Schatzman, M.: Geometrical evolution of developed interfaces. Trans. Am. Math. Soc. **347**(5), 1533–1589 (1995)
21. Degiovanni, M., Marino, A., Tosques, M.: Evolution equations with lack of convexity. Nonlinear Anal. Theory Methods Appl. **9**(12), 1401–1443 (1985)
22. Dondl, P., Frenzel, T., Mielke, A.: A gradient system with a wiggly energy and relaxed EDP-convergence. ESAIM:COCV **25**, 68 (2019)
23. Dym, H., McKean, H.P.: Fourier Series and Integrals, 1st edn. Probability and Mathematical Statistics. Academic Press, New York and London (1972)
24. Ekeland, I., Temam, R.: Convex analysis and variational problems. North-Holland Publishing Co., Amsterdam (1976). Translated from the French, Studies in Mathematics and its Applications, vol. 1
25. Esedoglu, S., Osher, S.J.: Decomposition of images by the anisotropic Rudin-Osher-Fatemi model. Commnu. Pure Appl. Math. **57**(12), 1609–1626 (2004)
26. Esposito, A., Patacchini, Schlichting, F.S., Slepčev, A., Slepcev, D.: Nonlocal-interaction equation on graphs: gradient flow structure and continuum limit. Arch. Rational Mech. Anal. **240**, 699–760 (2021)
27. Evans, L.C.: Partial Differential Equations, vol. 19 of Graduate Studies in Mathematics, 2nd edn. American Mathematical Society, Providence, RI (2010)
28. Evans, L.C., Gariepy, R.F.: Measure Theory and Fine Properties of Functions. Studies in Advanced Mathematics, 1st edn. CRC Press LLC, Boca Raton, Florida (1992)
29. Evans, L.C., Soner, H.M., Souganidis, P.E.: Phase transitions and generalized motion by mean curvature. Commun. Pure Appl. Math. **45**(9), 1097–1123 (1992)
30. Fischer, J., Laux, T., Simon, T.M.: Convergence rates of the Allen-Cahn equation to mean curvature flow: a short proof based on relative entropies. SIAM J. Math. Anal. **52**(6), 6222–6233 (2020)
31. Fleißner, F.: Γ-convergence and relaxations for gradient flows in metric spaces: a minimizing movement approach. ESIAM:COCV **25**, 28 (2019)

32. Fujita, H., Saito, N., Suzuki, T.: Operator Theory and Numerical Methods. Studies in Mathematics and its Applications, vol. 30. Elsevier, Amsterdam (2001)
33. García Trillos, N., Slepčev, D.: Continuum limit of total variation on point clouds. Arch. Rat. Mech. Anal. **220**, 193–241 (2016)
34. Giusti, E.: Minimal Surfaces and Functions of Bounded Variation, vol. 80 of Monographs in Mathematics, 1st edn. Birkhäuser, Boston (1984)
35. Hafiene, Y., Fadili, J., Chesneau, C., Elmoataz, A.: Continuum limit of the nonlocal p-Laplacian evolution problem on random inhomogeneous graphs. ESAIM: Math. Model. Numer. Anal. **54**(2), 565–589 (2020)
36. Hafiene, Y., Fadili, J., Elmoataz, A.: Nonlocal p-Laplacian evolution problems on graphs. SIAM J. Numer. Anal. **56**(2), 1064–1090 (2018)
37. Hale, J.K.: Ordinary Differential Equations, 2nd edn. Dover Publications Inc., Mineola, New York (2009)
38. Itô, S.: Introduction to Lebesgue Integration, (Lebesgue sekibun nyu-mon). Shokabo (1965). in Japanese; reprint in 2017
39. Jones, D.S.: The Theory of Generalised Functions, 2nd edn. Cambridge University Press, Cambridge-New York (1982)
40. Keetch, B., van Gennip, Y.: A Max-Cut approximation using a graph based MBO scheme. Discrete Contin. Dyn. Syst. Ser. B **24**(11), 6091–6139 (2019)
41. Kōmura, Y.: Nonlinear semi-groups in Hilbert space. J. Math. Soc. Jpn. **19**, 493–507 (1967)
42. Michał, Ł, Moll, S., Mucha, P.B.: Total variation denoising in ℓ^1 anisotropy. SIAM J. Imaging Sci. **10**, 1619–1723 (2017)
43. Laux, T., Lelmi, J.: Large data limit of the MBO scheme for data clustering: convergence of the dynamics (2022). arXiv:2209.05837 [math.AP]
44. Leoni, G.: A First Course in Sobolev Spaces. Graduate Studies in Mathematics, vol. 105. American Mathematical Society, Providence, Rhode Island (2009)
45. Loeb, P.A.: Real Analysis. Birkhäuser/Springer, [Cham] (2016)
46. Mielke, A.: On evolutionary Γ-convergence for gradient systems. In: Muntean, A., Rademacher, J., Zagaris, A. (eds.) Macroscopic and Large Scale Phenomena: Coarse Graining. Mean Field Limits and Ergodicity, vol. 3 of Lecture Notes in Applied Mathematics and Mechanics, pp. 187–249. Springer, Cham (2016)
47. Mielke, A., Montefusco, A., Peletier, M.A.: Exploring families of energy-dissipation landscapes via tilting: three types of EDP convergence. Continuum Mech. Thermodyn. **33**, 611–637 (2021)
48. Mielke, A., Peletier, M.A., Stephan, A.: EDP-convergence for nonlinear fast-slow reaction systems with detailed balance. Nonlinearity **34**, 5762–5798 (2021)
49. Mielke, A., Rossi, R., Savaré, G.: Variational convergence of gradient flows and rate-independent evolutions in metric space. Milan J. Math. **80**, 381–410 (2012)
50. Mielke, A., Stephan, A.: Coarse-graining via EDP-convergence for linear fast-slow reaction systems. Math. Model. Methods Appl. Sci. **30**(9), 1765–1807 (2020)
51. Modica, L., Mortola, S.: Un esempio di Γ^--convergenza. Boll. Un. Mat. Ital. B **14**(1), 285–299 (1977)
52. Muratori, M., Savaré, G.: Gradient flows and evolution variational inequalities in metric spaces. I: structural properties. J. Funct. Anal. **278**(4), 108347 (2020)
53. Ortner, C.: Two variational techniques for the approximation of curves of maximal slope. Technical Report Technical Report NA-05/10, University of Oxford Numerical Analysis Group Research Reports, Oxford University Computing Laboratory, Numerical Analysis Group, Wolfson Building, Parks Road, Oxford, England OX1 3QD (2005)
54. Krishnan, P.: Topology, vol. 134 of UNITEXT, 1st edn. Springer, Singapore (2022). La Matematica per il 3+2 (UNITEXTMAT)
55. Roman, S.: Advanced Linear Algebra, vol. 135 of Graduate Texts in Mathematics, 3rd edn. Springer, New York, NY (2008)
56. Sandier, E., Serfaty, S.: Gamma-convergence of gradient flows with applications to Ginzburg-Landau. Commun. Pure Appl. Math. **57**, 1627–1672 (2004)

57. Serfaty, S.: Gamma-Convergence of Gradient Flows and Applications to Ginzburg–Landau Vortex Dynamics, vol. 2 of Lecture Notes of the Unione Matematica Italiana, pp. 267–292. Springer, Berlin (2006)
58. Serfaty, S.: Gamma-convergence of gradient flows on Hilbert and metric spaces and applications. Discrete Contin. Dyn. Syst. **31**, 1427–1451 (2011)
59. Soner, H.M.: Ginzburg-Landau equation and motion by mean curvature. I. Convergence. J. Geom. Anal. **7**(3), 437–475 (1997)
60. Soner, H.M.: Ginzburg-Landau equation and motion by mean curvature. II. Development of the initial interface. J. Geom. Anal. **7**(3), 477–491 (1997)
61. van Gennip, Y., Bertozzi, A.L.: Γ-convergence of graph Ginzburg-Landau functionals. Adv. Differ. Equ. **17**(11–12), 1115–1180 (2012)
62. van Gennip, Y., Guillen, N., Osting, B., Bertozzi, A.L.: Mean curvature, threshold dynamics, and phase field theory on finite graphs. Milan J. Math. **82**(1), 3–65 (2014)
63. von Luxburg, U.: A tutorial on spectral clustering. Stat. Comput. **17**(4), 395–416 (2007)
64. Yosida, K.: Functional analysis. In: Classics in Mathematics. Springer, Berlin (1995). Reprint of the sixth edition (1980)

Chapter 3
An Overview of Some Nonlinear Evolution Problems

Patrizia Pucci

Abstract This note is somehow a continuation of the previous survey [107] dedicated to the memory of Prof. Vasilii V. Zhikov. Here we present further results contained in [105] concerning the Kirchhoff wave initial boundary value problems in the so-called *degenerate* case, that is, when the involved Kirchhoff function M can be zero at zero, and involving the fractional Laplacian, nonlinear damping and source terms in $\Omega \times \mathbb{R}_0^+$, where $\Omega \subset \mathbb{R}^n$ is a bounded domain with Lipschitz boundary $\partial\Omega$. The first part of the survey briefly presents the main results of [105], concerning global existence, vacuum isolating, asymptotic behavior and blow-up of solutions. The second part is devoted to the the blow-up phenomenon of the initial boundary value problem for the periodic rotation-two-component Camassa–Holm (R2CH) system given in [79], as well as a precise blow-up criterion, two new blow-up statements and the exact blow-up rate for strong solutions of the R2CH system in question.

3.1 Introduction

The first part of this survey is dedicated to a degenerate Kirchhoff wave initial boundary value problem, with nonlinear damping and source terms, involving the fractional Laplacian, while the second part deals with the blow-up phenomenon of the initial boundary value problem for the periodic rotation-two-component Camassa–Holm (R2CH) system. This overview is somehow a continuation of the previous survey [107] dedicated to the memory of Prof. Vasilii V. Zhikov.

P. Pucci (✉)
Dipartimento di Matematica e Informatica—Università degli Studi di Perugia, Perugia, Italy
e-mail: patrizia.pucci@unipg.it

M. Jleli et al. (eds.), *Advances in Nonlinear Evolution Equations*, Infosys Science Foundation Series, https://doi.org/10.1007/978-981-97-9812-4_3

Let us first observe that degenerate evolutionary Kirchhoff problems are extremely interesting and delicate. Existence and qualitative properties of solutions already appear in famous well-known papers in the theory. We just mention only very few stone papers on this subject as [40, 100]. In [40, 100], as well as in the large literature in degenerate Kirchhoff problems, the transverse oscillations of a stretched string, with nonlocal flexural rigidity, depend continuously on the Sobolev deflection norm of the displacement u via M. Sometimes the Kirchhoff function M is assumed Lipschitz continuous, but not always monotone, even if the archetype proposed by *Kirchhoff* in 1883

$$M(\tau) = a + b\gamma\tau^{\gamma-1}, \quad a, b \geq 0, \quad a+b > 0, \ \gamma \geq 1, \ \tau \geq 0, \tag{3.1}$$

is clearly monotone. Let us refer to other two main contributions [87, 140], besides [40], in which M is not monotone and to the references therein.

Indeed, the original Kirchhoff equation is

$$\rho h u_{tt} - \left\{ p_0 + \frac{\mathcal{E}h}{2L} \int_0^L |u_x|^2 dx \right\} u_{xx} + \delta u_t + f(x, u) = 0 \tag{3.2}$$

for $t \geq 0$ and $0 < x < L$, where $u = u(t, x)$ is the lateral displacement at the time t and at the space coordinate x, $\mathcal{E}$ the Young modulus, ρ the mass density, h the cross section area, L the length of the string, p_0 the *initial axial tension*, δ the resistance modulus and f the external force. When $\delta = f = 0$ the Eq. (3.2) was introduced by *Kirchhoff* in [71]. Further details and physical phenomena described by Kirchhoff's classical theory can be found in [130]. The Kirchhoff equation was proposed as an extension of the classical D'Alembert wave equation for the vibration of elastic strings.

In the so-called *degenerate* Kirchhoff problems $M(0) = 0$, that is, the initial axial tension $p_0 = 0$ in (3.2), and $a = 0$ in (3.1). From a physical point of view the fact that $M(0) = 0$ means that the base tension of the string is zero and M measures the change of the tension on the string caused by the change of its length during the vibration. The presence of the nonlinear coefficient M is crucial to be considered when the changes in tension during the motion cannot be neglected.

In the case of linear string vibrations, the tension is constant that is $M(t) \equiv M(0)$. When the inertial effects of longitudinal modes can be neglected, the tension is spatially uniform along the string and can be directly computed from the elongation of the string according to the Hooke law. This approach gives the form of M proposed by *Kirchhoff* and derived properly by *Carrier*. Again the case $M(0) = 0$ means that the base tension of the string is zero, a very realistic model.

After the appearance of (3.2), several physicists considered such equations for their researches on nonlinear vibrations from both a theoretical and experimental point of view; see [20, 21, 97, 101].

In [20, 21] *Carrier* developed a more rigorous approach to model transverse vibrations via the coupled governing equation of planar vibration and recovered the nonlinear integro-partial differential equation, without quoting Kirchhoff. In [97]

Narasimha also obtained the equation, called nowadays *the Kirchhoff string equation* in the literature, using another approach.

In spite of very limited applications to stationary strings, the stationary Kirchhoff string equation, and modifications, have been widely used to analyze transverse vibration of axially moving strings. We mention [1, 29, 47, 48, 75, 78, 142] for problems involving the classical Laplace operator, [6, 31] for the p–Laplacian case. In [1, 47, 52] the authors use variational methods, as well as a concentration compactness argument. In [29, 75, 78] variational methods are still used, but the stationary Kirchhoff problems are set in the entire $\mathbb{R}^n$. In [6, 31] the *degenerate* case is covered, as well as in [8, 11, 142], but the involved Kirchhoff M is assumed nonnegative and nondecreasing as in (3.1). For evolution problems, we refer to [5, 8, 11] and to the references therein.

More recently, following [17] *Fiscella* and *Valdinoci* in [52] proposed a stationary Kirchhoff variational model, in bounded regular domains of $\mathbb{R}^n$, which takes into account the *nonlocal* aspect of the tension arising from nonlocal measurements of the *fractional length of the string*. In [52], however, only the *non-degenerate* case was covered. Since then, several papers have been devoted to stationary fractional Kirchhoff problems involving critical nonlinearities in the degenerate case. We refer, e.g., to [7, 41, 92, 134, 137] for degenerate problems in bounded regular domains of $\mathbb{R}^n$, to [74, 75, 89, 94, 136, 138] in all $\mathbb{R}^n$, and to [19, 49–51, 88, 104, 110, 117, 118, 120, 134, 136–138] for quasi-linear Kirchhoff problems involving the fractional p–Laplacian, as well as to the references therein. For further comments, we refer to the special issue *Progress in Nonlinear Kirchhoff Problems* and its Introduction [108] as well as the monograph [90].

Let us now turn back to Kirchhoff problems in the natural evolutionary case. In [111] the key result concerns global asymptotic stability for nonlinear damped Kirchhoff systems, governed by the fractional p–Laplacian operator, with homogeneous Dirichlet boundary conditions, under fairly natural assumptions on the external force $f = f(t, x, u)$ and the distributed damping $Q = Q(t, x, u, u_t)$. In [111] the Kirchhoff coefficient represents a nonlocal dissipative effect in the dynamical damping term, which models a frictional mechanism acting on the body, while in the Kirchhoff prototype (3.2) is a nonlocal coefficient of the elliptic part of the equation. In fact, in [22] the next initial value boundary problem for the damped wave equation, with a degenerate nonlocal weak damping, was considered

$$
\begin{cases}
u_{tt} - \Delta u + \|\nabla u\|_2^2 u_t + f(u) = 0 & \text{in } \Omega \times \mathbb{R}^+, \\
u = 0 & \text{on } \partial\Omega \times \mathbb{R}_0^+, \quad \|\nabla u\|_2^2 = \displaystyle\int_\Omega |\nabla u|^2 dx, \\
u(0, \cdot) = u_0, \;\; u_t(0, \cdot) = u_1 & \text{in } \Omega,
\end{cases}
\tag{3.3}
$$

where Ω is a bounded domain of $\mathbb{R}^n$, with smooth boundary $\partial\Omega$, $\mathbb{R}^+ = (0, \infty)$, $\mathbb{R}_0^+ = [0, \infty)$, the function $f = f(u)$ is a nonlinear source and the Kirchhoff function $M(\tau) = \tau$, $\tau \in \mathbb{R}_0^+$, corresponds to a nonlocal damping coefficient. Indeed, M is multiplied by u_t and not by the main elliptic operator, as in (3.2) and in the papers cited above.

This kind of nonlocal dissipative effect is introduced by *Lange* and *Perla Menzala* in [73] for the beam equation. The paper [73] deals with the model

$$u_{tt} + \Delta^2 u + M(\,\|\nabla u\|_2^2\,)u_t = 0, \tag{3.4}$$

where the function $M : \mathbb{R}_0^+ \to [1, \infty)$ was assumed to be of class C^1 and such that $M(\tau) \geq \tau + 1$ for all $\tau \in \mathbb{R}_0^+$. The nonlocal damping term in (3.4) has a dissipative effect, which implies the decay of solutions. Problem (3.4) is closely related to the nonlocal dissipative Schrödinger equation

$$i\,w_t + \Delta w + iM(\|\nabla(\operatorname{Im} w)\|_2^2)\big(\operatorname{Re} w\big) = 0, \tag{3.5}$$

where $x \in \mathbb{R}^n$, $t \in \mathbb{R}_0^+$, and $i = \sqrt{-1}$. Indeed, if w is a smooth solution of (3.5), then the imaginary part $u = \operatorname{Im} w$ of w solves (3.4). A problem involving this kind of nonlocal dissipative effect was treated in another context in [23].

Stability for damped wave problems has been widely studied, e.g., in [10, 12, 24, 58, 86, 96, 114, 127, 131, 141] for nonlinear damped equations, in [70] for a plate equation with nonlocal weak damping and in [66, 67, 76, 98, 99, 122] for semilinear damped wave equations.

In mechanical engineering, we often encounter structure composed of rigid and elastic components. The flexible parts are of course sensitive to disturbances and inserting an internal dissipation can lead to satisfactory results. Motivated by these considerations and inspired by the papers [9, 114, 115], in [111] we generalize the model proposed in [22] and analyze the fractional system

$$\begin{cases} u_{tt} + (-\Delta)_p^s u + \mu|u|^{p-2}u + \varrho(t)M([u]_{s,p}^p)|u_t|^{p-2}u_t & \\ \qquad +Q(t,x,u,u_t) + f(t,x,u) = 0 & \text{in } \mathbb{R}_0^+ \times \Omega, \\ u(t,x) = 0 & \text{on } \mathbb{R}_0^+ \times (\mathbb{R}^n \setminus \Omega), \end{cases} \tag{3.6}$$

where $s \in (0,1)$, $1 < p < n/s$, Ω is a bounded regular domain of $\mathbb{R}^n$, and the function $u = (u_1, \ldots, u_N) = u(t,x)$ stands for the vectorial displacement, with $N \geq 1$.

The vectorial fractional p–Laplacian operator $(-\Delta)_p^s$,

$$(-\Delta)_p^s u = \big((-\Delta)_p^s u_1, \ldots, (-\Delta)_p^s u_N\big),$$

is defined, up to a multiplicative constant depending only on n, s and p, for all $x \in \Omega$ by

$$(-\Delta)_p^s \varphi(x) = \int_{\mathbb{R}^n} \frac{|\varphi(x) - \varphi(y)|^{p-2}[\varphi(x) - \varphi(y)]}{|x-y|^{n+ps}} dy,$$

along any function $\varphi \in C_c^\infty(\Omega; \mathbb{R}^N)$. Here $|\cdot|$ denotes the Euclidean norm in $\mathbb{R}^N$.

In particular, $(-\Delta)_p^s$ is consistent with the fractional Laplacian $(-\Delta)^s = (-\Delta)_2^s$ as $p = 2$. Furthermore, in the scalar case $N = 1$ the definition of $(-\Delta)_p^s$ reduces, up to a constant, to the standard p–Laplacian as $s \to 1^-$ in the limit sense of Bourgain–Brezis–Mironescu described in [14, 15]; see also Theorem 3.1 of [16].

The solutions u of (3.6) are searched in suitable fractional Sobolev spaces, in which the Gagliardo p–seminorm of $u(t,\cdot)$, that is

$$[u(t,\cdot)]_{s,p} = \left(\iint_{\mathbb{R}^{2n}} |u(t,x) - u(t,y)|^p |x-y|^{-(n+ps)} dx dy\right)^{1/p}, \tag{3.7}$$

is well defined. In (3.7) and in what follows, we extend $u(t, \cdot)$ to be zero outside Ω and still denoted it by $u(t, \cdot)$.

The term $\mu|u|^{p-2}u$ in (3.6), where $\mu \geq 0$ plays the role of a *perturbation*, M is a dissipative Kirchhoff function and $\varrho \geq 0$ is assumed to be $L^1_{\text{loc}}(\mathbb{R}^+_0)$. In [111], the *Kirchhoff function* $M : \mathbb{R}^+_0 \to \mathbb{R}^+_0$ in (3.6), unless otherwise stated, *verifies the solely mild condition*

($\mathcal{M}$) *M is a continuous function in $\mathbb{R}^+_0$, with $M(\tau) \geq 0$ for $\tau \geq 0$.*

The *nonlinear external damping Q* represents the most common suppressions of the vibrations of an elastic structure of passive viscous type and absorbs vibration energy. However, it is worth noting that, in addition to the distributed damping Q, an *internal damping mechanism*, even if small, is always present in real materials, as long as the system vibrates. We refer to Chap. 4, *Dynamical Mechanical Properties*, of the book [56] by *Gent* for a complete physical justification on the importance of considering internal damping terms. The internal material damping of *Kelvin–Voigt* type is so represented by the term $\varrho(t)M([u]^p_{s,p})|u_t|^{p-2}u_t$ and strictly depends on the body structure. Usually, in standard literature, $M \equiv 1$, ϱ is a positive constant, while in [111] only the mild requests ($\mathcal{M}$), $\varrho \geq 0$ and $\varrho \in L^1_{\text{loc}}(\mathbb{R}^+_0)$ are assumed. Therefore, ϱ identically zero can be treated, thanks to the presence of Q and so a control from below and from above Q should be considered in order to avoid that the damping term Q is either too small (*underdamping*) or too large (*overdamping*). An interesting integral condition which prevents the damping term Q being either too small or too large as $t \to \infty$ was introduced in [112]; see also [113–116]. However, it is exactly thanks to the presence of the additional internal damping $\varrho(t)M([u]^p_{s,p})|u_t|^{p-2}u_t$, when non-trivial, that we can expect that global solutions of (3.6) exist under a suitable control on the external damping Q. Indeed, a strong action of dissipative terms could make easier the existence of global solutions, since they play the role of stabilizing terms and their smoothing effect makes more difficult the blow-up. In any case, the function ϱ makes the analysis more delicate though it helps in obtaining the stability of global solutions.

In [13, 121] global existence for semilinear wave systems is proved in the standard setting of the Sobolev spaces, without imposing any bound on the exponent q, which represents the growth at infinity of the source term f. This justifies the importance of considering for asymptotic stability also when the relation $q \leq p^*_s$ fails, where $p^*_s = pn/(n - ps)$ is the fractional critical Sobolev exponent associated with p. Indeed, in [111] even the case $q > p^*_s$ is broadly studied. We refer to [13, 121] for further details and to [11, 111] for a wide list of references and comments.

Even if (3.6) is somehow related to the problem studied in [9], since the perturbation, the damping term and the external force are structurally similar; the main difference occurs on the fact that in (3.6) the Kirchhoff function M is connected with the dynamical part of the system. Indeed, as in [22], in our model M does not multiply the fractional p–Laplacian operator, but a function of u_t, that is, in (3.6) the function M corresponds to a nonlocal damping coefficient. Furthermore, while in [22] the authors considered an equation in the special case $p = 2$, our system involves more general source and damping terms, and it is driven by the fractional p–Laplacian. For these reasons, we are not able to establish a local exponential stability as in [22], but in [111] we obtain local and global stability results in different regions of the potential valley, using completely different arguments than in [22].

Moreover, we also treat the degenerate case of (3.6), that is, the case when $M(0) = 0$. Actually, the main results on local and global asymptotic stability are proved in [111] without assuming the non-degeneracy of the problem (3.6). This fact represents a generalization of [9], where the non-degeneracy of the problem was assumed in the whole section concerning the local asymptotic stability. However, in [111] for some applications, where specified, we have to require non-degeneracy in order to overcome some technical difficulties, due to the Kirchhoff structure of (3.6).

Finally, regarding the structural hypothesis $(\mathcal{M})$ on M, in [111] we do not need to assume that M is linear and we do not impose even that M is nondecreasing, as requested in several other papers in the subject, such as [9], where M is taken of the standard form (3.1). Indeed, in problem (3.3) of [22] the function M is exactly the linear function $M(\tau) = \tau$ for $\tau \geq 0$, while we prove the main results under the sole condition $(\mathcal{M})$.

In [111] great attention is focused on the asymptotic behavior of the solutions also in the linear case of (3.6). In order to simplify the notation in the linear case, we restrict the interval of the time variable to $I = [1, \infty)$ instead of $\mathbb{R}_0^+$ and to the scalar case $N = 1$.

Following Sect. 5 of [114], where stability was proved in the simpler case of strongly damped systems, in Sect. 5 of [111] we consider an important special case of (3.6), that is the case when $p = 2$, $\mu \geq 0$, $\varrho \geq 0$ and in $L^1_{\text{loc}}(\mathbb{R}_0^+)$, M satisfying $(\mathcal{M})$, $f(t, x, u) = V(t, x)u$, where V is a bounded continuous function in $I \times \Omega$, and $Q(t, x, u, v) = q(t)t^\alpha v$, with $\alpha \in \mathbb{R}$ and q verifying

$$1/C \leq q(t) \leq C \qquad \text{in } I$$

for some $C > 0$. Thus, we study the asymptotic behavior as $t \to \infty$ of the solutions of

$$\begin{cases} u_{tt} + (-\Delta)^s u + \mu u + \varrho(t) M([u]_s^2) u_t & \\ \quad + q(t)t^\alpha u_t + V(t, x)u = 0 & \text{in } I \times \Omega, \\ u(t, x) = 0 & \text{on } I \times (\mathbb{R}^n \setminus \Omega), \end{cases} \tag{3.8}$$

where, for simplicity, $(-\Delta)^s = (-\Delta)_2^s$, $[u]_s = [u]_{s,2}$ and $N = 1$, that is in [111] the displacements u of (3.8) are solely scalar.

If $|\alpha| \leq 1$ and ϱ is sublinear for t sufficiently large, we get the stability of the solutions of (3.8). A more delicate argument is necessary when either $\alpha < -1$ or $\alpha > 1$. For this reason, we suppose that M is a positive constant function and that $V(t, x) = V(x) > -\mu$ a.e. in Ω when $|\alpha| > 1$. Here the proof techniques rely on an important result of independent interest on the properties of the eigenvalues and the eigenfunctions of the underlying perturbed problem of (3.8), that is of

$$\begin{cases} (-\Delta)^s u + a_0(x)u = \lambda u & \text{in } \Omega, \\ u = 0 & \text{in } \mathbb{R}^n \setminus \Omega, \end{cases} \tag{3.9}$$

where $a_0 = a_0(x)$ is a bounded nonnegative continuous function, with $a_0 > 0$ a.e. in Ω. Indeed, in the main application for (3.8) the function a_0 is exactly $V(x) + \mu$.

To this aim in the Appendix of [111], following the proof of Proposition 9 of [125] given for a related but different problem (see also Proposition 3.1 of the monograph [91]), we prove, besides other properties, that the first eigenvalue of (3.9) is positive and that the set of eigenfunctions is a basis of the natural solution space of (3.9). For further details, we refer to the original paper [111] as well as to the survey [107].

Let us comment in passing why the fractional approach is very powerful in the theory of solid mechanics. Indeed, the classical theory of elasticity, in the framework of the Cauchy continuum mechanics, is an adequate model to describe a large amount of phenomena within solid mechanics. Nevertheless, the results predicted by this approach are not in accordance with the experimental ones when the analyzed phenomena have a small scale. These aspects are relevant in case of dispersion of waves in one-dimensional solids [2], or composite mechanical behavior [4]. Preserving the framework of the continuum mechanics (i.e., avoiding the lattice theory), these shortcomings can be exceeded introducing the nonlocal elasticity theory; see, e.g., [4] and the references therein. Different from the local (classic) elasticity, where the point stress depends only of the strain at such a point, in the nonlocal elasticity the stress depends on a weighted average of the strain field around the point, as proposed by *Eringen* and *Edelen* in 1972. Under suitable natural assumptions the nonlocal elasticity can be also expressed in differential form and this approach has been further enhanced in different directions in recent papers, where it has been successfully applied in several fields of mechanics. In particular, it permits to analyze the structural behavior of nonlocal nanobeams, nonlocal Timoshenko nanobeams, cracked nanobeams, 2D nanoelements and functionally graded nonlocal beams; cf. [4]. Lately, the fractional calculus has been applied to describe materials and media with a nonlocal behavior. In this context a novel approach, based on the fractional Laplace operator, has been proposed in [3], where the static analysis of a nonlocal rod was considered. The fractional derivative approach has been utilized by several researchers to describe the dispersion law. In particular, in [4] the wave propagation in a fractional nonlocal one-dimensional elastic continuum has been considered where the nonlocal effects were modeled introducing long-range central body interactions applied to the centroids of the infinitesimal volume elements of the continuum.

In the first part of this survey we present some recent results obtained in [105] concerning degenerate wave Kirchhoff problems and under natural assumptions, we obtain the global existence, vacuum isolating, asymptotic behavior and blow-up of solutions combining the Galerkin method with potential wells theory. The significant feature and difficulty of the problem are that the coefficient of $(-\Delta)^s$ can vanish at zero, that is, that the problem is degenerate. More precisely, in Sect. 7.2 we summarize the main results of [105] for the degenerate fractional Kirchhoff wave problem

$$\begin{cases} u_{tt} + [u]_s^{2(\gamma-1)}(-\Delta)^s u + |u_t|^{a-2}u_t + u = |u|^{b-2}u, & \text{in } \Omega \times \mathbb{R}^+, \\ u(0,\cdot) = u_0, \quad u_t(0,\cdot) = u_1, & \text{in } \Omega, \\ u = 0, & \text{in } (\mathbb{R}^n \setminus \Omega) \times \mathbb{R}_0^+, \end{cases} \tag{3.10}$$

where $\Omega \subset \mathbb{R}^n$ is a bounded domain with Lipschitz boundary $\partial\Omega$. Clearly here $M(\tau) = \tau^{\gamma-1}$, $\tau \in \mathbb{R}_0^+$, that is $M(0) = 0$ and (4.12) is degenerate. Hence, when dealing with (4.12) we encounter serious difficulties by the degeneracy of (4.12) and the nonlocal nature of the fractional Laplacian $(-\Delta)^s$. To overcome the intrinsic complications induced by these new features, we need to produce more delicate estimates for (4.12) than the usual analysis available in the literature.

In (4.12) we assume that $\gamma \in [1, 2_s^*/2)$ and that

$$2 < a < 2\gamma < b < 2_s^* = \frac{2n}{n-2s},$$

and 2_s^* is the fractional critical Sobolev exponent associated with 2. A recent important paper on nonlocal porous-medium type equations involving the fractional p–Laplacian is [129], where *Vázquez* studied existence and uniqueness of a strong nonnegative solution to a related initial boundary value problem and presented enlightening comments and a series of intriguing open problems.

Turning back to Kirchhoff problems, in [119] we first investigate the space-fractional diffusion Kirchhoff problem

$$\begin{cases} u_t + M\left([u]_{s,p}^p\right)(-\Delta)_p^s u = f, & \text{in } \Omega \times \mathbb{R}^+, \\ u(0,\cdot) = u_0, & \text{in } \Omega \\ u = 0, & \text{in } (\mathbb{R}^n \setminus \Omega) \times \mathbb{R}_0^+, \end{cases} \tag{3.11}$$

where $[u]_{s,p}$ is the Gagliardo p–seminorm of $u(t,\cdot)$, introduced in (3.7). By applying the sub-differential approach, in the non-degenerate case, for problem (3.11), when the external force is of the type $f = f(t,x)$ instead of $f = f(u)$, we obtained in [119] the well-posedness of solutions, existence as well as the large-time behavior and extinction of solutions. However, [119] treats only the non-degenerate case, i.e., the Kirchhoff function is strictly positive in $\mathbb{R}_0^+$. In the canonical case (3.1) this means that $a > 0$.

In [55], we use the potential well theory to study the existence of global weak solutions and established the vacuum isolating and blow-up of strong solutions in the subcase of (3.11), in which $M \equiv 1$, $p = 2$ and $f = f(u) = |u|^{b-2}u$, $1 < b \le 2_s^*$. In [102] problem (3.11) is treated in the special degenerate case $M(\tau) = \tau^{\gamma-1}$, $\tau \in \mathbb{R}_0^+$, as in (4.12), with $\gamma \in [1, p_s^*/p)$, but with $f = f(u) = |u|^{b-2}u$ and $p < b < p_s^*$. Also in [102], existence of global solutions was obtained by the Galerkin method and potential well theory.

Indeed, the potential well theory is a great tool in order to prove global existence of solutions for nonlinear evolution equations and it was introduced by *Sattinger* [123] in 1968. Since then the existence of solutions for evolution equations has been derived via the potential well theory in several papers. We refer to [68, 81–84, 103, 139] and the references therein. However, the paper [103], due to *Payne* and *Sattinger*, can be regarded as the most important and typical work on the potential well. In [103] existence and finite time blow-up of solutions for the initial boundary value problem of semilinear hyperbolic equations and semilinear parabolic equations are proved by the potential well method. In [68], *Ikehata* and *Suzuki* treat stable and unstable sets for evolution equations of hyperbolic and parabolic type by the potential well method. In [81–84, 139] existence of global solutions for semilinear wave equations, parabolic equations and double dispersion equations was derived as an application of the potential well method.

In the previous paper [106], we study the Kirchhoff hyperbolic problem (4.12), but without the nonlinear damping, under some appropriate assumptions, and obtain global existence, vacuum isolating and blow-up of solutions by combining the Galerkin method with the potential well theory.

There is no doubt that in [105] we encounter serious difficulties when dealing with (4.12) because of the degeneracy of Kirchhoff function and the nonlocal feature of the fractional Laplacian. Of course, the interaction between the nonlinear damping and source terms also creates difficulties. The intrinsic complications, induced by these new features and the combined presence of these factors, cause and produce new interesting difficulties. These difficulties require the use of increasingly sophisticated nonlinear analysis tools, and they are overcome with new and more delicate estimates. In Sect. 7.2 we briefly present and comment the main results of [105], where the new difficulties encountered because of the degeneracy of (4.12) and of the nonlocal feature of the fractional Laplacian are overcome via accurate estimates. The conclusions also raise, and leave open, a number of other intriguing questions, which are briefly presented.

In the second part of this survey, we briefly present the main results obtained in [79] , such as the blow-up phenomenon of the initial boundary value problem for the periodic rotation-two-component Camassa–Holm (R2CH) system. The results in [79] include a precise blow-up criterion, when the involved real dimensionless parameter σ is positive. Two new blow-up statements and the exact blow-up rate for strong solutions of the system are also given in [79].

Indeed, in recent decades, shallow water wave equations have caught a great attention. The so-called shallow water wave means waves with length much greater than the depth of the water. Similar dynamics can also arise in nonlinear elasticity, specifically, in elastic rod. There are several equations modeling shallow water waves, the best known one being the Camassa–Holm (CH) equation

$$u_t - u_{txx} + 3uu_x = 2u_x u_{xx} + uu_{xxx},$$

where $u = u(t, x)$ describes the free surface of the water above a flat bottom. This equation is mathematically derived in [53] as a bi-Hamiltonian generalization of the Korteweg–de Vries (KdV) equation and physically proposed by *Camassa* and *Holm* in [18]. Since then several papers are devoted to the study of the CH equation such as [32–35, 37, 39, 44, 143], since it models both soliton interaction and wave breaking. Indeed, Camassa and Holm conjectured that an initial profile having a sufficiently negative slope steepens and verticality develops in finite time, and so wave breaking occurs. It is mathematically intriguing to know which shallow water wave PDE models include both the phenomena of soliton interaction and wave breaking. The Camassa–Holm equation reconciles these properties which have been known for different equations and has, with its several extensions, the potential to become the new master equation for shallow water wave theory. Recently, in [72] *Lakshmanan* points out that the CH equation can be used to model tsunami waves.

There are several multicomponent generalizations of CH equations. The most popular one is two-component Camassa–Holm (2CH) system

$$\begin{cases} llu_t - u_{txx} - Au_x + 3uu_x = 2u_x u_{xx} + uu_{xxx} - \rho\rho_x, & t > 0,\ x \in \mathbb{R}, \\ \rho_t + (u\rho)_x = 0, & t > 0,\ x \in \mathbb{R}, \end{cases} \tag{3.12}$$

where $A \geq 0$ characterizes a linear underlying shear flow. System (3.12), which first appears in [126], is completely integrable and has a bi-Hamiltonian structure, as shown in [45, 126]. However, (3.12) was mathematically introduced by *Chen, Liu* and *Zhang* in [26] by a reciprocal transformation and proposed by *Constantin* and *Ivanov* [38, 69] in the context of shallow water theory. Then, system (3.12) under Cauchy conditions has been extensively studied; see, e.g., [42, 43, 60–63, 132] and the references therein. In particular, in [43] it has been shown via Kato's semigroup theory that system (3.12) is locally well-posed when the initial data (u_0, ρ_0) are in $H^s(\mathbb{R}) \times H^{s-1}(\mathbb{R})$ and $s \geq 2$. In [61], *Gui* and *Liu* improve the local well-posedness result taking initial data (u_0, ρ_0) in Besov's spaces. Specifically, when $(u_0, \rho_0) \in H^s(\mathbb{R}) \times H^{s-1}(\mathbb{R})$ and $s > 3/2$. More appealingly, system (3.12) admits both global strong solutions [60, 62] and finite time blow-up solutions [43, 60, 61].

The interest in 2CH systems inspired the following two-component Dullin–Gottwald–Holm (2DGH) system [28, 64],

$$\begin{cases} u_t - u_{txx} - Au_x + 3uu_x \\ \quad = \sigma(2u_x u_{xx} + uu_{xxx}) - \mu u_{xxx} - \rho\rho_x, & t > 0,\ x \in \mathbb{R}, \\ \rho_t + (u\rho)_x = 0, & t > 0,\ x \in \mathbb{R}, \end{cases} \tag{3.13}$$

with $u(t, x) \to 0$ and $\rho(t, x) \to 1$ as $|x| \to \infty$. System (3.13) is derived in the shallow water regime, following the Ivanov modeling approach of [69]. Obviously, (3.13) is also a generalization of the CH equation. Mathematical properties of system (3.13) have been studied in many works; see, for example, [28, 64, 80, 146]. It has been shown that (3.13) can exhibit wave-breaking phenomenon for a class of initial data and the exact blow-up rate of such solutions is $-2/\sigma$. Meanwhile, in [28, 64] the authors present a sufficient condition for the existence of global solutions and study traveling wave solutions of (3.13). Moreover, in [80, 146] the authors discuss the blow-up phenomenon and global existence of strong solutions in the periodic case of system (3.13).

As we all know, the geophysical water waves are influenced by the gravity and the Coriolis force induced by the Earth rotation. *Fan, Guo* and *Liu* in [46] derive a modified two-component Camassa–Holm (2CH) system under the effect of the Coriolis force in the rotating fluid as follows:

$$\begin{cases} u_t - u_{txx} - Au_x + 3uu_x = \sigma(2u_x u_{xx} + uu_{xxx}) - \mu u_{xxx} & \\ \quad -(1-2\Omega A)\rho\rho_x + 2\Omega\rho(\rho u)_x, & t > 0,\ x \in \mathbb{R}, \\ \rho_t + (u\rho)_x = 0, & t > 0,\ x \in \mathbb{R}, \\ u(0,x) = u_0(x), & x \in \mathbb{R}, \\ \rho(0,x) = \rho_0(x), & x \in \mathbb{R}. \end{cases} \tag{3.14}$$

System (3.14) is a model for the equatorial water waves. The function $u = u(t, x)$ is the horizontal fluid velocity in the x-direction, $\rho = \rho(t, x)$ is in connection with the free surface elevation from equilibrium, the free parameter A describes a linear underlying shear flow, the real dimensionless constant σ provides the competition, or balance, in fluid convection between nonlinear steepening and amplification due to stretching, $\mu \in \mathbb{R}$ is a dimensionless parameter, and Ω characterizes the angular velocity of the Earth rotation. In the hydrodynamical derivation of system (3.14), it is required that $u(t, x) \to 0$ and $\rho(t, x) \to 1$ as $|x| \to \infty$. System (3.14) has received extensive attention and it has been recently studied on the real line $\mathbb{R}$, that is in the non-periodic case, in [27, 46, 144] and on the unit length circle $\mathbb{S} = \mathbb{R}/\mathbb{Z}$ in [93, 145]. It has been proved that (3.14) is locally well-posed in $H^s(\mathbb{R}(\mathbb{S})) \times H^{s-1}(\mathbb{R}(\mathbb{S}))$, $s > 3/2$, and the maximal existence time $T > 0$ of a classical solution (u, ρ) of (3.14), with $\sigma = 1$ and $\mu = 0$, is finite if and only if the slope of u becomes unbounded below in finite time (while the solution remains bounded). Moreover, solitary waves and the structure of solutions in the critical Besov space

$$B^{3/2}_{2,1} \times B^{1/2}_{2,1}$$

have been studied in [27] and in [144], respectively.

Now let us turn back to the second part of this survey, where we present the main results obtained in [79] concerning the periodic case of system (3.14), that is

$$\begin{cases} u_t - u_{txx} - Au_x + 3uu_x = \sigma(2u_x u_{xx} + uu_{xxx}) - \mu u_{xxx} & \\ \quad -(1-2\Omega A)\rho\rho_x + 2\Omega\rho(\rho u)_x, & t > 0,\ x \in \mathbb{R}, \\ \rho_t + (u\rho)_x = 0, & t > 0,\ x \in \mathbb{R}, \\ u(0,x) = u_0(x), & x \in \mathbb{R}, \\ \rho(0,x) = \rho_0(x), & x \in \mathbb{R}, \\ u(t,x+1) = u(t,x), & t \geq 0, x \in \mathbb{R}, \\ \rho(t,x+1) = \rho(t,x), & t \geq 0, x \in \mathbb{R}. \end{cases} \tag{3.15}$$

Since wave-breaking is one of the most intriguing long-standing problems in the water wave theory [133], we shall continue to study the wave-breaking phenomenon of (3.15). The aim of [79] is to derive sharp wave-breaking criterion for system (3.15) in the case $\sigma > 0$ (not just $\sigma = 1$ and $\mu = 0$), and to give two blow-up theorems and the corresponding blow-up rate result. A notable difference in the blow-up analysis for (3.15) stems from the cubic term $\Omega\rho(\rho u)_x$, which fails to be controlled by the conservation laws. Following the method in [28], we get in [79] a uniform upper bound for u_x, even if we do not know whether u_x is bounded from below.

3.2 Existence and Properties for (4.12)

In this section, we present the main results contained in [105] concerning the Kirchhoff wave initial boundary value problem (4.12) in the canonical degenerate case. The stationary function space associated with (4.12) is the Hilbert space W_0, which we introduce step by step. Following [124], let $Q = \mathbb{R}^n \backslash \mathcal{O}$, where $\mathcal{O} = \mathscr{C}(\Omega) \times \mathscr{C}(\Omega) \subset \mathbb{R}^{2n}$, $\mathscr{C}(\Omega) = \mathbb{R}^n \backslash \Omega$ and $\Omega \subset \mathbb{R}^n$ is a bounded domain with Lipschitz boundary $\partial\Omega$. Let W be the linear space of Lebesgue measurable functions u from $\mathbb{R}^n$ to $\mathbb{R}$ such that the restriction to Ω of any function u in W belongs to $L^2(\Omega)$ and

$$\iint_Q |u(x) - u(y)|^2 |x - y|^{-(n+2s)} dxdy < \infty.$$

The space W is equipped with the Hilbertian norm

$$\|u\|_W = \left(\|u\|^2_{L^2(\Omega)} + \iint_Q |u(x) - u(y)|^2 |x - y|^{-(n+2s)} dxdy \right)^{1/2}.$$

It is easy to prove that $W = (W, \langle \cdot, \cdot \rangle_W)$ is a Hilbert space and that

$$W_0 = \{u \in W \ : \ u = 0 \text{ a.e. in } \mathbb{R}^n \backslash \Omega\}$$

is a closed linear subspace of W.

We recall that the fractional Sobolev space $H^s(\Omega) = W^{s,2}(\Omega)$ is

$$H^s(\Omega) = \left\{ u \in L^2(\Omega) \ : \ \frac{|u(x) - u(y)|}{|x - y|^{s+n/2}} \in L^2(\Omega \times \Omega) \right\},$$

endowed with the Hilbertian norm

$$\|u\|_{H^s(\Omega)} = \left(\|u\|^2_{L^2(\Omega)} + \iint_{\Omega \times \Omega} \frac{|u(x) - u(y)|^2}{|x - y|^{n+2s}} dxdy \right)^{1/2}.$$

We briefly list the main properties on W and W_0, proved step by step in [124]. See also comments in [59, 109].

Lemma 3.1 *(i) For any $\nu \in [1, 2_s^*]$ there exists a positive constant $C_0 = C_0(n, \nu, s)$ such that for any $v \in W_0$*

$$\|v\|_{L^\nu(\Omega)}^2 \leq C_0 \iint_{\Omega\times\Omega} \frac{|v(x) - v(y)|^2}{|x - y|^{n+2s}} dxdy.$$

(ii) There exists a positive constant $\widetilde{C} = \widetilde{C}(n, s, \Omega)$ such that for any $v \in W_0$

$$\iint_Q \frac{|v(x) - v(y)|^2}{|x - y|^{n+2s}} dxdy \leq \|v\|_W^2 \leq \widetilde{C} \iint_Q \frac{|v(x) - v(y)|^2}{|x - y|^{n+2s}} dxdy.$$

(iii) For any $\nu \in [1, 2_s^*)$ and any bounded sequence $(v_k)_k$ in W_0 there exists a function $v \in L^\nu(\mathbb{R}^n)$, with $v = 0$ a.e. in $\mathbb{R}^n\setminus\Omega$, such that up to a subsequence, still denoted by $(v_k)_k$,

$$v_k \to v \quad \textit{strongly in } L^\nu(\Omega) \textit{ as } k \to \infty.$$

The main stationary space of [105] is W_0, which from now on is endowed with the equivalent norm

$$\|v\|_{W_0} = \left(\iint_Q \frac{|v(x) - v(y)|^2}{|x - y|^{n+2s}} dxdy\right)^{1/2} \quad \text{for all } v \in W_0,$$

see also [135] for further comments.

The rest of this section is devoted to the properties concerning the structural geometry of (4.12). To this aim, we begin with some useful notation. Define

$$A(u) = \frac{1}{2\gamma}\|u\|_{W_0}^{2\gamma} - \frac{1}{b}\|u\|_b^b, \quad B(u) = \|u\|_{W_0}^{2\gamma} - \|u\|_b^b, \tag{3.16}$$

and the potential well

$$V = \{u \in W_0 : B(u) > 0, \ A(u) < d\} \cup \{0\}, \quad \text{where } d = \inf_{\substack{u\in W_0\\ u\neq 0}} \sup_{\lambda\geq 0} A(\lambda u). \tag{3.17}$$

As in the proof of Theorem 3.1 in [102], in [105] we show that the potential depth d, defined in (3.17), is equivalent to the following definition:

$$d = \inf\{A(u) : u \in W_0 \text{ and } B(u) = 0\}.$$

We also define

$$\overline{V} = \{u \in W_0 : B(u) < 0, \ A(u) < d\}. \tag{3.18}$$

Furthermore, for all $\delta \in [0, 1]$, we put

$$A_\delta(u) = \frac{\delta}{2\gamma}\|u\|_{W_0}^{2\gamma} - \frac{1}{b}\|u\|_b^b,$$
$$d(\delta) = \frac{1-\delta}{2\gamma}\left(\frac{b}{2\gamma\, C_*^b}\,\delta\right)^{2\gamma/(b-2\gamma)}, \quad r(\delta) = \left(\frac{b}{2\gamma\, C_*^b}\,\delta\right)^{1/(b-2\gamma)}, \tag{3.19}$$

where C_* is the best embedding constant of the embedding W_0 into $L^b(\Omega)$, i.e.,

$$C_* = \sup_{\substack{u\in W_0\\ u\neq 0}} \frac{\|u\|_b}{\|u\|_{W_0}}.$$

In particular, (3.19) yields

$$A(u) = A_\delta(u) + \frac{1-\delta}{2\gamma}\|u\|_{W_0}^{2\gamma} \tag{3.20}$$

for all $\delta \in [0, 1]$. Thanks to (3.19) and (3.20), the first result gives the properties of $A_\delta(u)$ in terms of the level $A(u)$ with respect to the value $d(\delta)$.

Lemma 3.2 *Let $u \in W_0$ be fixed.*

- (i) *If $A(u) \le d(\delta)$, then $A_\delta(u) > 0$ if and only if $0 < \|u\|_{W_0} < r(\delta)$.*
- (ii) *If $A(u) \le d(\delta)$, then $A_\delta(u) < 0$ if and only if $\|u\|_{W_0} > r(\delta)$.*
- (iii) *If $A(u) \le d(\delta)$, then $A_\delta(u) = 0$ if and only if $\|u\|_{W_0} = r(\delta)$.*

The assumption (D.1) as well as the definition of the function $d : [0, 1] \to \mathbb{R}$ given in (3.19) give the qualitative properties of d, which will play an essential role in the application of the potential well method.

Lemma 3.3 *The function $d : [0, 1] \to \mathbb{R}$ given in* (3.19) *has the following properties.*

- (i) $d(0) = d(1) = 0$.
- (ii) *$\delta \mapsto d(\delta)$ is strictly increasing in $[0, \delta_0]$, strictly decreasing in $[\delta_0, 1]$ and takes the maximum value at $\delta_0 = 2\gamma/b$.*
- (iii) *For any $e \in [0, d(\delta_0)]$, the equation $d(\delta) = e$ has exactly two roots $\delta_1 \in [0, \delta_0]$ and $\delta_2 \in [\delta_0, 1]$.*
- (iv) *For all $\delta \in (0, 1)$*

$$d(\delta) = \inf\{A(u) : u \in W_0,\ \|u\|_{W_0} \neq 0,\ A_\delta(u) = 0\}.$$

- (v) $d = d(\delta_0) = \inf\{A(u) : u \in W_0,\ \|u\|_{W_0} \neq 0,\ A_{\delta_0}(u) = 0\}$.

The first and the last parts of the next result can be proved in the same way and are somehow a direct consequence of (3.19), while the central part follows from the assumptions and a delicate analysis.

Lemma 3.4 *Let $\delta \in (0, 1)$ and $u \in W_0$. If $A(u) \le d(\delta)$ and $A_\delta(u) > 0$, then*

$$0 \le \|u\|_{W_0}^{2\gamma} < \frac{d(\delta)}{a(\delta)}, \quad \text{where } a(\delta) = \frac{1-\delta}{2\gamma}.$$

In particular, if $A(u) \le d$ and $B(u) > 0$, then

$$0 \le \|u\|_{W_0}^{2\gamma} < \frac{2\gamma\, b}{b-2\gamma}\, d.$$

If $A(u) \le d(\delta)$ *and* $\|u\|_{W_0}^{2\gamma} > d(\delta)/a(\delta)$, *then* $A_\delta(u) < 0$. *Furthermore, if* $A(u) \le d$ *and*

$$\|u\|_{W_0}^{2\gamma} > \frac{2\gamma\, b}{b-2\gamma}\, d,$$

then $B(u) < 0$.

Let us define $F = \{u \in W_0 \,:\, A(u) < 0,\ A(u) < d\}$ and the family of potential wells in the stationary setting as follows. For all $\delta \in (0, 1)$, put

$$\begin{aligned}
V_\delta &= \{u \in W_0 \,:\, A_\delta(u) > 0,\ A(u) < d(\delta)\} \cup \{0\},\\
\overline{V}_\delta &= V_\delta \cup \partial V_\delta = \{u \in W_0 \,:\, A_\delta(u) \ge 0,\ A(u) \le d(\delta)\},\\
F_\delta &= \{u \in W_0 \,:\, A_\delta(u) < 0,\ A(u) < d(\delta)\},\\
\overline{F}_\delta &= F_\delta \cup \partial F_\delta = \{u \in W_0 \,:\, A_\delta(u) \le 0,\ A(u) \le d(\delta)\},\\
G_\delta &= \{u \in W_0 \,:\, \|u\|_{W_0} < r(\delta)\},\\
\overline{G}_\delta &= G_\delta \cup \partial G_\delta = \{u \in W_0 \,:\, \|u\|_{W_0} \le r(\delta)\},\\
G_\delta^c &= \{u \in W_0 \,:\, \|u\|_{W_0} > r(\delta)\},
\end{aligned}$$

where $d(\delta)$ and $r(\delta)$ are defined in (3.19).

If $A_\delta(u) > 0$, then $A(u) > 0$ by (3.20), and hence we have $F_{\delta_0} = F$. In view of Lemma 3.3–(v), we get $V_{\delta_0} = V$. Moreover, $A(u) \le \|u\|_{W_0}^{2\gamma}/2\gamma$ by (3.16). Consequently, if $u \in W_0$ and $0 < \|u\|_{W_0} < (1-\delta)^{1/2\gamma} r(\delta)$ for some $\delta \in (0, 1)$, then $A(u) < d(\delta)$ by (3.19). Therefore, $A_\delta(u) > 0$ by Lemma 3.2–(i). In conclusion,

$$G_{\overline{\delta}} \subset V_\delta, \quad \text{where } \overline{\delta} \text{ is such that} \quad r(\overline{\delta}) = (1-\delta)^{1/2\gamma} r(\delta).$$

In view of Lemma 3.2–(i) and –(ii) and Lemma 3.3 we obtain the following.

Lemma 3.5 *Let* $\delta \in (0, 1)$ *and let* V_δ, F_δ, G_δ G_c^δ *and* $\overline{\delta}$ *satisfy the definitions above. Then*

$$G_{\overline{\delta}} \subset V_\delta \subset G_\delta, \quad F_\delta \subset G_\delta^c.$$

If $\delta_0 = 2\gamma/b$ *is the number obtained in Lemma 3.3–(ii). Then*

$$G_{\overline{\delta_0}} \subset V \subset G_{\delta_0} \quad \text{and} \quad F \subset G_{\delta_0}^c,$$

where $G_{\overline{\delta_0}} = \big\{u \in W_0 \,:\, \|u\|_{W_0} < r(\overline{\delta_0})\big\}$, *and* $\overline{\delta_0}$ *is such that*

$$r(\overline{\delta_0}) = \left(\frac{b-2\gamma}{b}\right)^{1/2\gamma} r(\delta_0).$$

If $0 < \delta' < \delta'' < \delta_0$, *then* $V_{\delta'} \subset V_{\delta''}$. *While, if* $\delta_0 \le \delta' < \delta'' < 1$, *then* $F_{\delta''} \subset F_{\delta'}$.

As a consequence of Lemma 3.3-(ii) and -(iv) we derive the next result.

Lemma 3.6 *Let $0 < A(u) < d$ for some $u \in W_0$. Let δ_1 and δ_2, with $0 < \delta_1 < \delta_2$, be the two roots of the equation $d(\delta) = A(u)$. Then the sign of $\delta \mapsto A_\delta(u)$ does not change in (δ_1, δ_2).*

We are now in the position to state an existence result for global weak solutions of problem (4.12). We begin with some useful definitions.

Definition 3.1 A function $u = u(x,t)$ is said to be a *global* (weak) *solution of* problem (4.12), if

$$\begin{gathered} u \in L^\infty(0,\infty;W_0), \qquad u_t \in L^\infty(0,\infty;L^2(\Omega)), \\ u(\cdot,0) = u_0 \quad \text{in } W_0, \qquad u_t(\cdot,0) = u_1 \quad \text{in } L^2(\Omega) \quad \text{and} \\ (u_t(\cdot,t),\phi(\cdot,t)) + \int_0^t \langle u(\cdot,\tau),\phi(\cdot,\tau)\rangle_{W_0} d\tau + \int_0^t (|u_t(\cdot,\tau)|^{a-2}u_t(\cdot,\tau),\phi(\cdot,\tau))d\tau \\ + \int_0^t (u(\cdot,\tau),\phi(\cdot,\tau))d\tau = \int_0^t (|u(\cdot,\tau)|^{b-2}u(\cdot,\tau),\phi(\cdot,\tau))d\tau + (u_1,\phi(\cdot,0)), \end{gathered}$$

for any $\phi \in L^1(0,\infty;W_0)$ and for any $t \in \mathbb{R}_0^+$, where

$$\begin{aligned} \langle u(\cdot,t),\phi(\cdot,t)\rangle_{W_0} &= M(\|u(\cdot,t)\|_{W_0}^2) \iint_Q \frac{[u(x,t)-u(y,t)]\cdot[\phi(x)-\phi(y)]}{|x-y|^{n+2s}} dxdy \\ &= \|u(\cdot,t)\|_{W_0}^{2(\gamma-1)} \iint_Q \frac{[u(x,t)-u(y,t)]\cdot[\phi(x)-\phi(y)]}{|x-y|^{n+2s}} dxdy. \end{aligned}$$

If a (weak) global solution u belongs to $C(0,\infty;W_0)$, we say that u is a *strong global solution of* problem (4.12).

The energy function along a solution u of (4.12) is defined for all $t \in \mathbb{R}$ by

$$\begin{aligned} E(t) &= \frac{1}{2}\|u_t(\cdot,t)\|_2^2 + \frac{1}{2\gamma}\|u(\cdot,t)\|_{W_0}^{2\gamma} - \frac{1}{b}\|u(\cdot,t)\|_b^b + \frac{1}{2}\|u(\cdot,t)\|_2^2 \\ &= \frac{1}{2}\|u_t(\cdot,t)\|_2^2 + A(u(\cdot,t)) + \frac{1}{2}\|u(\cdot,t)\|_2^2. \end{aligned}$$

Theorem 3.1 *Let $u_0 \in W_0$ and $u_1 \in L^2(\Omega)$. Suppose that $0 < E(0) < d$, that δ_1 and δ_2, with $0 < \delta_1 < \delta_2$, are the two solutions of the equation $d(\delta) = E(0)$, that either $A_{\delta_2}(u_0) > 0$ or $\|u_0\|_{W_0} = 0$. Then problem (4.12) admits a global solution u, with $u \in L^\infty(0,\infty;W_0)$, $u_t \in L^\infty(0,\infty;L^2(\Omega))$ and $u(\cdot,t) \in V_\delta$ for all $\delta \in (\delta_1,\delta_2)$ and for all $t \in \mathbb{R}_0^+$.*

The essential part of the proof technique of Theorem 3.1 is the construction of a suitable Galerkin approximation sequence $(u_k)_k$, based on the existence of a sequence $(w_k)_k \subset C_0^\infty(\Omega)$ of eigenfunctions of the fractional Laplace operator $(-\Delta)^s$, which is an orthonormal basis of $L^2(\Omega)$ and an orthogonal basis of W_0 thanks to the results of [125]. A delicate analysis shows that, up to a subsequence, still denoted by $(u_k)_k$, the sequence $(u_k)_k$ weakly star converges in $L^\infty(0,\infty;W_0)$ and a.e. in $\Omega \times \mathbb{R}_0^+$ to a global solution u of (4.12). Moreover, involved estimates show that $u_k(\cdot,t) \in V_\delta$ for any $\delta \in (\delta_1,\delta_2)$, for any k and for any $t \in \mathbb{R}_0^+$, so that $u(\cdot,t) \in V_\delta$ for any $\delta \in (\delta_1,\delta_2)$ and any $t \in \mathbb{R}_0^+$, as required.

Finally, the powerful estimates and arguments of the proof of Theorem 3.1, as well as Lemma 3.3-(iii) and Lemma 3.4, allow us to prove also the next

Theorem 3.2 *Assume all the assumptions Theorem 3.1 and its notation.*

The global constructed solution u in Theorem 3.1 has the further property that $u(\cdot, t) \in \overline{V}_{\delta_1}$ *for all* $t \in \mathbb{R}_0^+$.

The conclusion of Theorem 3.1 continues to hold, when the assumption either $A_{\delta_2}(u_0) > 0$ *or* $\|u_0\|_{W_0} = 0$ *is replaced by the request that either* $B(u_0) > 0$ *or* $\|u_0\|_{W_0} = 0$.

If furthermore $E(0) = 0$ *and* $\|u_0\|_{W_0} = 0$, *problem (4.12) admits only the unique global solution* $u(x, t) \equiv 0$.

The proof arguments of Theorem 3.2 rely on a careful application of the definition of d in (3.19), the energy identity, Lemma 3.2–(ii) and –(iii), Lemma 3.3 and Lemma 3.6.

3.2.1 Vacuum Isolating of Solutions

Here we study the vacuum isolating behavior of solutions of problem (4.12), and assume that the initial data u_0 is fixed in W_0 and u_1 in $L^2(\Omega)$, without further mentioning.

Theorem 3.3 *Let* $0 < e < d$, δ_1 *and* δ_2, *with* $0 < \delta_1 < \delta_2$, *be the two roots of the equation* $d(\delta) = e$. *Then the following properties hold.*

(*i*) *All the solutions of (4.12), with* $E(0) = e$, *belong to* V_δ *for any* $\delta \in (\delta_1, \delta_2)$, *provided that either* $B(u_0) > 0$ *or* $\|u_0\|_{W_0} = 0$.
(*ii*) *All the solutions of (4.12), with* $E(0) = e$, *belong to* F_δ *for all* $\delta \in (\delta_1, \delta_2)$, *provided that* $B(u_0) < 0$.

In what follows, T denotes the maximal existence time of a solution u of (4.12).

Theorem 3.4 *Assume the hypotheses of Theorem 3.3 and use its notations.*

If the assumption $E(0) = e$ *is replaced by* $0 < E(0) \leq e$, *then the conclusion of Theorem 3.3 continues to hold.*

Any solution $u = u(x, t)$ *of (4.12), with initial energy* $E(0)$ *satisfying* $0 < E(0) \leq e$, *has the properties below.*

(*i*) $u(\cdot, t)$ *belongs to* $\overline{V}_{\delta_1}$ *for all* $t \in [0, T)$, *provided that* $B(u_0) > 0$;
(*ii*) $u(\cdot, t)$ *belongs to* $\overline{F}_{\delta_2}$ *for all* $t \in [0, T)$, *provided that* $B(u_0) < 0$.

Combining Theorem 3.4 and Lemma 3.2–(*i*)–(*iii*), we derive the next result.

Theorem 3.5 *Under the hypotheses and notation of Theorem 3.3, then for any solution u of (4.12), with initial energy* $E(0)$ *satisfying* $0 < E(0) \leq e$, *it results that*

(1) $u(\cdot, t)$ *lies inside the ball* $\overline{G}_{\delta_1}$ *for all* $t \in [0, T)$, *provided that* $u_0 \in G_{\delta_0}$;
(2) $u(\cdot, t)$ *lies outside the ball* G_{δ_2} *for all* $t \in [0, T)$, *provided that* $u_0 \in G^c_{\delta_0}$, *where the meaning of* $\delta_0 = 2\gamma/b$ *is clarified in Lemma 3.3–(ii).*

Theorem 3.5 shows that for the set of all solutions of (4.12), with $0 < E(0) \leq e$, there exists a *vacuum region* U_e, that is, a region which does not contain any solution of (4.12). The corresponding vacuum region is

$$U_e = \left\{u \in W_0 \ : \ r(\delta_1) < \|u\|_{W_0} < r(\delta_2)\right\},$$

where $r(\delta)$ is defined in (3.19).

The vacuum region U_e becomes bigger and bigger, as e decreases to 0^+. As limiting case, we obtain

$$U_1 = \{u \in W_0 \, : \, 0 < \|u\|_{W_0} < r(1)\}, \qquad r(1) = \left(\frac{b}{2\gamma \, C_*^b}\right)^{1/(b-2\gamma)}.$$

Indeed, we have the following:

Theorem 3.6 *Assume all the hypotheses of Theorem 3.3 and use its notation.*
Then all the nontrivial strong solutions of problem (4.12), with $E(0) = 0$, belong to

$$G^C_{r(1)} = \{u \in W_0 \, : \, \|u\|_{W_0} \geq r(1)\}, \qquad r(1) = \left(\frac{b}{2\gamma \, C_*^b}\right)^{1/(b-2\gamma)}$$

for all $t \in [0, T)$.
While all the nontrivial strong solutions u of problem (4.12), with $E(0) < 0$, satisfy $\|u(\cdot, t)\|_{W_0} > r(1)$ for all $t \in [0, T)$. Furthermore,

$$\|u(\cdot, t)\|_{W_0} \geq \left(\frac{b}{C_*^b}\sqrt{\frac{-2E(0)}{\gamma}}\right)^{1/(b-\gamma)}$$

for all $t \in [0, T)$.

The proof of Theorem 3.6 is obtained using the energy identity and the regularity of strong solutions of (4.12), that is, the fact that they are of class $C(0, T; W_0)$.

3.2.2 The Asymptotic Behavior of Global Solutions and the Blow-Up of the Solutions

Here, we study the asymptotic behavior of global (weak) solutions for problem (4.12) and to prove the decay estimate, in [105] we use the following useful lemma due to *Nakao* in [95].

Lemma 3.7 (Lemma 2 of [95]) *Let $\phi : \mathbb{R}_0^+ \to \mathbb{R}_0^+$ be a nonnegative non-increasing function, with the property that there exist an exponent $h > 0$ and a constant $C > 0$ such that*

$$\phi(t)^{1+h} \leq C\big[\phi(t) - \phi(t+1)\big] \quad \text{for all } t \in \mathbb{R}_0^+.$$

Then there exists a number $k > 0$ such that

$$\phi(t) \leq \left[\frac{h}{k}(t-1) + M^{-h}\right]^{-1/h} \qquad \text{for all } t \geq 1,$$

where $M = \max_{t \in [0,1]} \phi(t)$.

Theorem 3.7 *Let $u_0 \in V$, $u_1 \in L^2(\Omega)$. If $E(0) < d$, any global (weak) solution u of problem (4.12) has the following decay estimate:*

$$E(t) \le \left[\frac{a-2}{2k}(t-1) + M^{(2-a)/2}\right]^{2/(2-a)} \quad \textit{for all } t \ge 1,$$

where k is a positive constant and $M = \max_{t\in[0,1]} E(t)$.

To prove the blow-up of solutions of problem (4.12), we need some preliminary results.

Lemma 3.8 *The potential depth d, defined in* (3.17)*, is positive.*

The next result concerns properties of local solutions of (4.12) in their maximal interval of existence, which at some instant belong to the set $\overline{V}$, given in (3.18) and related to the potential well V, and have energy at that instant less than d—where both V and d are introduced in (3.17).

Lemma 3.9 *Let $u = u(x,t)$ be a local solution of (4.12) on its maximal existence interval $[0, T)$. If there exists $t_0 \in [0, T)$ such that $u(\cdot, t_0) \in \overline{V}$ and $E(t_0) < d$, then*

$$u(\cdot, t) \in \overline{V} \quad \textit{for all } t \in [t_0, T).$$

Moreover,

$$\|u(\cdot, t)\|_{W_0}^{2\gamma} > \frac{2\gamma\, b}{b - 2\gamma}\, d$$

for all $t \in [0, T)$.

Now we are ready to present the main blow-up result.

Theorem 3.8 *Let $u = u(x,t)$ be a local solution of (4.12) on $[0, T)$. If there exists $t_0 \in [0, T)$ such that $u(\cdot, t_0) \in \overline{V}$ and $E(t_0) < d$, then $T < \infty$, i.e., the solution u blows up in a finite time.*

The proof of Theorem 3.8 uses the definition (3.17) of the potential well V, the energy identity, the fact that energy E is non-increasing in $\mathbb{R}_0^+$, and a delicate convexity argument involving the convex function $x \mapsto \alpha^x/x$ in $\mathbb{R}^+$, with $\alpha > 0$.

3.2.3 Conclusions and Open Questions

Inspired by [128], it is natural to consider the following more general problem with decaying linear perturbation of the type $q^2(x)u$, that is

$$\begin{cases} u_{tt} + [u]_s^{2(\gamma-1)}(-\Delta)^s u + |u_t|^{a-2}u_t + q^2(x)u = |u|^{b-2}u & \text{in } \Omega \times \mathbb{R}^+, \\ u(\cdot, 0) = u_0, \quad u_t(\cdot, 0) = u_1, & \text{in } \Omega, \\ u = 0, & \text{in } (\mathbb{R}^n \setminus \Omega) \times \mathbb{R}_0^+, \end{cases} \tag{3.21}$$

where $2 < a < 2\gamma < b < 2_s^*$, as assumed for (4.12), while the function q is defined in $\mathbb{R}^n$, has limit zero as $|x| \to \infty$ and satisfies conditions (1.7)–(1.8) of [128] that is in our notation and for (3.21)

$$q(x) \geq \kappa(1 + |x|)^{-\eta} \quad \text{in } \Omega, \text{ for some contant } \kappa > 0, \text{ with}$$
$$0 \leq \eta \leq \min\left\{\frac{1}{b}, \frac{a-1}{2} \cdot \frac{b-2}{b-a}\right\}.$$

Using the proof arguments similar to those of [105], we get the existence and the vacuum isolating of the solutions for problem (3.21). However, the study of the asymptotic behavior of the global solutions of problem (3.21) seems to be fairly delicate. The main difficulty arises from the fact that the convexity in $\mathbb{R}^+$ of the function $\tau \mapsto \alpha^\tau/\tau$, with $\alpha > 0$, seems not to be useful for (3.21). Thus, new methods and strategies should be found to get the asymptotic behavior of the global solutions of (3.21). This is an interesting open question.

3.3 Wave-Breaking and Blow-Up Criteria for (3.15)

From here on, in presenting the main results of [79], we identify all spaces of periodic functions with function spaces over the unit circle $\mathbb{S} = \mathbb{R}/\mathbb{Z}$ and for simplicity, we drop $\mathbb{S}$ in notation. The norm in the Lebesgue space $L^p(\mathbb{S})$, for all $1 \leq p \leq \infty$, is written as $\|\cdot\|_{L^p}$, while $\|\cdot\|_{H^s}$, $s > 0$, stands for the norm in the classical Sobolev spaces $H^s(\mathbb{S})$ and $\|\cdot\|_{H^s \times H^s}$, $s > 0$, represents the norm in the classical Sobolev spaces $H^s(\mathbb{S}) \times H^s(\mathbb{S})$.

Let us reformulate system (3.15), introducing the function

$$G(x) = \frac{\cosh(x - [x] - 1/2)}{2\sinh(1/2)},$$

$x \in \mathbb{R}$. First note that $(1 - \partial_x^2)^{-1} f = G * f = \int_{\mathbb{S}} G(x - y) f(y) dy$ for all $f \in L^2$, that is, $*$ denotes the convolution operator. Using this identity, we can rewrite (3.15) as follows:

$$\begin{cases} u_t + (\sigma u - \mu)u_x = -\partial_x G * \Big((\mu - A)u + \dfrac{3-\sigma}{2}u^2 + \dfrac{\sigma}{2}u_x^2 \\ \qquad + \dfrac{1-2\Omega A}{2}\rho^2 - \Omega\rho^2 u\Big) + \Omega p * (\rho^2 u_x), & t > 0,\ x \in \mathbb{R}, \\ \rho_t + u\rho_x = -\rho u_x, & t > 0,\ x \in \mathbb{R}, \\ u(0,x) = u_0(x),\ \rho(0,x) = \rho_0(x), & x \in \mathbb{R}, \\ u(t, x+1) = u(t,x),\ \rho(t, x+1) = \rho(t,x), & t > 0,\ x \in \mathbb{R}. \end{cases} \tag{3.22}$$

For the sake of clarity, we recall some helpful lemmas which are used in the proofs of the main results of [79]. The first property is the local well-posedness result for the system (5.10). which is obtained as an application of the Kato theory.

Lemma 3.10 (Theorem 1 of [93] and Theorem 2.1 of [145]) *Given $z_0 = (u_0, \rho_0) \in H^s \times H^{s-1}$, $s > 3/2$, there exists a maximal $T = T(\|z_0\|_{H^s \times H^{s-1}}) > 0$ and a unique solution $z = (u, \rho)$ of the system (5.10) such that*

$$z = z(\cdot, z_0) \in C([0,T); H^s \times H^{s-1}) \cap C^1([0,T); H^{s-1} \times H^{s-2}).$$

Moreover, the solution z depends continuously on the initial data, i.e., the mapping

$$z_0 \to z(\cdot, z_0) : H^s \times H^{s-1} \to C([0,T); H^s \times H^{s-1}) \cap C^1([0,T); H^{s-1} \times H^{s-2})$$

is continuous.

The next result is a useful conservation law for strong solutions of (5.10).

Lemma 3.11 (Lemma 1 of [93]) *Let $z_0 = (u_0, \rho_0) \in H^s \times H^{s-1}$, $s > 3/2$, and let $T > 0$ be the maximal existence time of the corresponding solution $z = (u, \rho)$ of (5.10). Then*

$$\begin{aligned} E(t) &= \int_{\mathbb{S}} \big(u^2(t,x) + u_x^2(t,x) + (1 - 2\Omega A)(\rho - 1)^2(t,x)\big)dx \\ &= \int_{\mathbb{S}} \big(u_0^2(x) + u_{0,x}^2(x) + (1 - 2\Omega A)(\rho_0 - 1)^2(x)\big)dx = E(0) \end{aligned}$$

for all $t \in [0, T)$.

Adapting the Littlewood–Paley analysis of the transport equation and the Moser estimates, we can obtain the next blow-up criterion.

Lemma 3.12 (Theorem 2 of [93] and Theorem 2.2 of [145]) *Assume that $z = (u, \rho)$ is the solution of system (5.10), with initial data $z_0 = (u_0, \rho_0) \in H^s \times H^{s-1}$, $s > 3/2$, and let T be the maximal time of existence. Then*

$$T < \infty \quad \Rightarrow \quad \int_0^T \|u_x(\tau)\|_{L^\infty} d\tau = \infty.$$

Before establishing the precise blow-up mechanism of the strong solution, we claim that the solution of the system (5.10) can only have singularities which correspond to wave breaking. Let us state the following wave-breaking criterion.

Lemma 3.13 (Theorem 3 of [93] and Theorem 3.1 of [145]) *Assume that $1 - 2\Omega A > 0$. Let $z_0 = (u_0, \rho_0) \in H^s \times H^{s-1}$, with $s > 3/2$, and $T > 0$ be the maximal time of existence of the solution $z = (u, \rho)$ of system (5.10), with initial data $z_0 = (u_0, \rho_0)$. Then the corresponding solution $z = (u, \rho)$ blows up in finite time $T < \infty$ if and only if*

$$\lim_{t \to T^-} \left\{ \sup_{x \in \mathbb{S}} |u_x(t,x)| \right\} = \infty.$$

Now, let us consider the following initial value problem:

$$\begin{cases} \dfrac{\partial q}{\partial t} = u(t, q), & t \in [0, T), \\ q(0, x) = x, & x \in \mathbb{R}, \end{cases} \tag{3.23}$$

where $u \in C([0,T); H^s) \cap C^1([0,T); H^{s-1})$ is the first component of the solution $z = (u, \rho)$ of (5.10), with initial data $(u_0, \rho_0) \in H^s \times H^{s-1}$, $s > 3/2$, and $T > 0$ is the maximal time of existence of u.

Lemma 3.14 (Lemma 2.1 of [145]) *Let $u \in C([0,T); H^s) \cap C^1([0,T); H^{s-1})$, $s > 3/2$ as above. Then (5.26) has a unique solution $q \in C^1([0,T) \times \mathbb{R}; \mathbb{R})$. Moreover, the map $q(t, \cdot)$ is an increasing diffeomorphism of $\mathbb{R}$, with*

$$q_x(t,x) = \exp\left(\int_0^t u_x(s, q(s,x))ds\right) > 0 \quad \textit{for all } (t,x) \in [0,T) \times \mathbb{R}.$$

Consequently, for all $t \in [0,T)$ the L^∞-norm of any function $v(t,\cdot) \in L^\infty$ is preserved under the family of the diffeomorphisms $q(t,\cdot)$, i.e.,

$$\|v(t,\cdot)\|_{L^\infty} = \|v(t, q(t,\cdot))\|_{L^\infty}, \quad t \in [0,T).$$

Similarly,

$$\inf_{x\in\mathbb{S}} v(t,x) = \inf_{x\in\mathbb{R}} v(t, q(t,x)), \qquad \sup_{x\in\mathbb{S}} v(t,x) = \sup_{x\in\mathbb{R}} v(t, q(t,x)), \quad t \in [0,T).$$

The lemma below shows the time evolution of the slope in x at an inflection point for a function $v(t,\cdot) \in H^2$ with a C^1–dependence on the time parameter t.

Lemma 3.15 (Theorem 2.1 of [36]) *Let $T > 0$ and $v \in C^1([0,T); H^2)$. Then for every $t \in [0,T)$, there exists at least one point $\xi(t) \in \mathbb{R}$ with*

$$m(t) = \inf_{x\in\mathbb{R}}[v_x(t,x)] = v_x(t, \xi(t)),$$

and the function m is almost everywhere differentiable on $(0,T)$ with

$$\frac{d}{dt}m(t) = v_{xt}(t, \xi(t)) \quad \textit{a.e. on } (0,T).$$

The next sharp estimates play a crucial role in proving the precise blow-up rate.

Lemma 3.16 (Lemma 2.3 of [145], see also [143])
(i) $\max_{x\in[0,1]} f^2(x) \le \dfrac{e+1}{2(e-1)}\|f\|^2_{H^1}$ *for every $f \in H^1$ and the constant $\dfrac{e+1}{2(e-1)}$ is sharp.*
(ii) $\max_{x\in[0,1]} f^2(x) \le c\|f\|^2_{H^1}$ *for every $f \in H^3$, with the best possible constant c lying within the range $(1, 13/12]$. Actually, the best constant c is exactly $\dfrac{e+1}{2(e-1)}$.*

3.3.1 Precise Blow-Up Criteria of (3.15)

In [79] we improve the corresponding result in [93, 145], and obtain the precise blow-up criterion for strong solutions of (5.10), with $\sigma > 0$. Then we show that u_x is uniformly bounded from above on $\mathbb{S}$ even if we do not know whether u_x is bounded from below.

Lemma 3.17 *Suppose that* $1 - 2\Omega A > 0$ *and* $\sigma > 0$. *Let* $z = (u, \rho)$ *be the solution of (5.10), with initial data* $z_0 = (u_0, \rho_0) \in H^s \times H^{s-1}$, $s > 3/2$, *and* T *be the maximal time of existence. Then*

$$\sup_{x\in\mathbb{S}} u_x(t, x) \le \|u_{0,x}\|_{L^\infty} + \sqrt{\frac{(1 - 2\Omega A)\|\rho_0\|_{L^\infty}^2 + C^2}{\sigma}},$$

where C *is a positive constant depending on* $E(0)$ *and* $\|\rho_0\|_{L^\infty}$, *with*

$$C^2 = 2\left\{\kappa_1 + \kappa_2 E(0) + \frac{\Omega^2}{2}\|\rho_0\|_{L^\infty}^4 + \frac{2\Omega(1-\Omega A)}{4(1-2\Omega A)}\|\rho_0\|_{L^\infty} E(0) \right.$$
$$\left. + \frac{\Omega\cosh 1/2}{2\sinh 1/2}\sqrt{\frac{e+1}{2(e-1)}}\frac{1}{1-2\Omega A}(E(0))^{\frac{3}{2}}\right\}, \tag{3.24}$$

with

$$\kappa_1 = 1 - \Omega A + \frac{2(1 - 2\Omega A + \Omega^2)(\cosh 1/2)^2 + \Omega(\sinh 1 - 1)}{16(\sinh 1/2)^2},$$

$$\kappa_2 = 1 + \frac{(|3-\sigma|+1)(e+1)}{4(e-1)} + \frac{(A-\mu)^2(\sinh 1 - 1) + (4|3-\sigma| + 2|\sigma| + 4)\cosh 1/2}{16\sinh 1/2}$$
$$+ \frac{\Omega\cosh 1/2}{\sinh 1/2}\sqrt{\frac{e+1}{2(e-1)}}\frac{1}{\sqrt{1-2\Omega A}} + \frac{\Omega(2-3\Omega A)}{2(1-2\Omega A)}.$$

We are now able to give the first main result of [79], which provides a precise blow-up criterion for system (3.15), when $\sigma > 0$.

Theorem 3.9 *Assume that* $1 - 2\Omega A > 0$ *and* $\sigma > 0$. *Let* $z_0 = (u_0, \rho_0) \in H^s \times H^{s-1}$, *with* $s > 3/2$, *and let* $T > 0$ *be the maximal time of existence of the solution* $z = (u, \rho)$ *of* (3.15), *with initial data* z_0. *Then the corresponding solution* $z = (u, \rho)$ *blows up in finite time* $T < \infty$ *if and only if*

$$\lim_{t\to T}\left\{\inf_{x\in\mathbb{S}} u_x(t, x)\right\} = -\infty.$$

The proof of Theorem 3.9 is based on Lemmas 3.10, 3.12 and 3.17. For details we refer to Sect. 4 of [79].

3.3.2 Blow-Up Solutions of System (3.15)

We first present the blow-up phenomenon for system (5.10). Then, we prove the existence of strong solutions of system (5.10) which do not exist globally in time. The first sufficient conditions for blow-up solutions of system (3.15) are stated when $\sigma < 0$. More precisely, we have the following.

Theorem 3.10 *Assume that* $1 - 2\Omega A > 0$ *and* $\sigma < 0$. *Let* $z_0 = (u_0, \rho_0) \in H^s \times H^{s-1}$, *with* $s > 3/2$ *and* $T > 0$ *be the maximal time of existence of the solution* $z = (u, \rho)$ *of system* (3.15), *with initial data* z_0. *If there exists some* $x_0 \in \mathbb{S}$ *such that*

$$u_0'(x_0) > \frac{C}{\sqrt{-\sigma}},$$

where $C = C(E(0), \|\rho_0\|_{L^\infty}) > 0$ *is given in* (3.24) *of Lemma 3.17, then the corresponding solution* $z = (u, \rho)$ *of system* (3.15) *blows up in finite time in the following sense. There exists* T_1, *with*

$$0 < T_1 \leq \frac{-2}{\sigma u_0'(x_0) - \sqrt{C(-\sigma)^{\frac{3}{2}} u_0'(x_0)}},$$

such that

$$\lim_{t \to T_1} \sup_{x \in \mathbb{S}} u_x(t, x) = \infty.$$

The Sobolev imbedding theorem $H^s \hookrightarrow L^\infty$, with $s > 1/2$, implies at once that if $\rho = \rho(t, x)$ becomes unbounded in finite time, then the maximal time T of existence of the solution $z = (u, \rho)$ of system (5.10) is finite, so that u_x must be unbounded from below in finite time by Theorem 3.9. In order to study the fine structure of finite time singularities we shall assume that there is a number $L > 0$ such that $\|\rho(t, \cdot)\|_{L^\infty} \leq L$ for all $t \in [0, T)$.

Now, we state a sufficient condition for blow-up solutions of system (3.15), when $\sigma = 1$ and $\mu = 0$. More precisely, we give the following.

Theorem 3.11 *Assume that* $1 - 2\Omega A > 0$, $\sigma = 1$ *and* $\mu = 0$. *Let* $z_0 = (u_0, \rho_0)$ *in* $H^s \times H^{s-1}$, *with* $s > 3/2$, *and* $T > 0$ *be the maximal time of existence of the solution* $z = (u, \rho)$ *of system* (3.15), *with initial data* z_0. *Assume that there exists* $L > 0$ *such that* $\|\rho(t, \cdot)\|_{L^\infty} \leq L$ *for all* $t \in [0, T)$. *If*

$$\int_{\mathbb{S}} u_{0,x}^3 dx \leq -\sqrt{2E(0)N},$$

where

$$N = \frac{3\Omega}{1 - 2\Omega A} \cdot \frac{\cosh 1/2}{2 \sinh 1/2} \sqrt{\frac{e+1}{2(e-1)}} E(0)^{5/2} + \kappa_3 E(0)^2 + 3\Omega L^2 \sqrt{\frac{(e+1)}{2(e-1)}} E(0)^{3/2}$$
$$+ \left(\frac{3L^2(1 - 2\Omega A)}{2} + \frac{3}{2} + \frac{3\Omega^2}{8} \left(\frac{\cosh 1/2}{\sinh 1/2} \right)^2 + \frac{3\Omega(\sinh 1 - 1)}{16(\sinh 1/2)^2} \right) E(0),$$

with

$$\kappa_3 = \frac{3(e+1)}{2(e-1)} + \frac{3A^2(\sinh 1 - 1)}{16\sinh 1/2} + \frac{3\Omega}{\sqrt{1-2\Omega A}} \cdot \frac{\cosh 1/2}{\sinh 1/2}\sqrt{\frac{e+1}{2(e-1)}}$$
$$+ \frac{3(\Omega+1)}{2} + \frac{3\Omega(1-\Omega A)(L+1)}{2(1-2\Omega A)},$$

then the corresponding solution z *of system* (3.15) *blows up in finite time.*

The proof technique of Theorem is fairly delicate and the conclusion is derived from the use of also Lemma 3.13. For the actual proof we refer to Sect. 5 of [79].

The exact blow-up rate of the blow-up strong solutions of system (5.10), when $1 - 2\Omega A > 0$ and $\sigma < 0$ is given in the next statement, whose proof is established in Sect. 6 of [79]. This gives us more insight into the blow-up mechanism for the wave-breaking solution of system (5.10).

Theorem 3.12 *Assume that* $1 - 2\Omega A > 0$ *and* $\sigma < 0$. *Let* $z_0 = (u_0, \rho_0)$ *in* $H^s \times H^{s-1}$, *with* $s > 3/2$, *and* $T > 0$ *be the maximal time of existence of the solution* $z = (u, \rho)$ *of system* (3.15), *with initial data* z_0. *If* T *is finite in the following sense, if*

$$\lim_{t\to T} \sup_{x\in\mathbb{S}} u_x(t,x) = \infty,$$

then

$$\lim_{t\to T} [(T-t) \sup_{x\in\mathbb{S}} u_x(t,x)] = -\frac{2}{\sigma}.$$

For more up-to-dated references in the subject, we refer to the bibliographies of the recent papers [25, 54, 57, 65, 77, 85].

Acknowledgements The author is a member of the *Gruppo Nazionale per l'Analisi Matematica, la Probabilità e le loro Applicazioni* (GNAMPA) of the *Istituto Nazionale di Alta Matematica* (INdAM) and is partly supported by the INdAM—GNAMPA Project 2022 *Equazioni differenziali alle derivate parziali in fenomeni non lineari* (CUP_E55F22000270001).

The author was also partly supported by the *Fondo Ricerca di Base di Ateneo—Esercizio 2017–2019* of the University of Perugia, named *PDEs and Nonlinear Analysis.*

References

1. Alves, C.O., Corrêa, F.J.S.A., Figueiredo, G.M.: On a class of nonlocal elliptic problems with critical growth. Differ. Equ. Appl. **2**, 409–417 (2010)
2. Atanackovic, T.M., Stankovic, B.: Generalized wave equation in nonlocal elasticity. Acta Mech. **208**, 1–10 (2009)
3. Autuori, G., Cluni, F., Gusella, V., Pucci, P.: Mathematical models for nonlocal elastic composite materials. Adv. Nonlinear Anal. **6**, 355–382 (2017)
4. Autuori, G., Cluni, F., Gusella, V., Pucci, P.: Longitudinal waves in a nonlocal rod by fractional Laplacian. Mech. Adv. Mater. Struct. **27**, 599–604 (2020)
5. Autuori, G., Colasuonno, F., Pucci, P.: Lifespan estimates for solutions of polyharmonic Kirchhoff systems. Math. Models Methods Appl. Sci. **22**, 1150009, 36 pp. (2012)

6. Autuori, G., Colasuonno, F., Pucci, P.: On the existence of stationary solutions for higher–order p–Kirchhoff problems. Commun. Contemp. Math. **16**, 1450002, 43 pp. (2014)
7. Autuori, G., Fiscella, A., Pucci, P.: Stationary Kirchhoff problems involving a fractional operator and a critical nonlinearity. Nonlinear Anal. **125**, 699–714 (2015)
8. Autuori, G., Pucci, P.: Kirchhoff systems with dynamic boundary conditions. Nonlinear Anal. **73**, 1952–1965 (2010)
9. Autuori, G., Pucci, P.: Asymptotic stability for Kirchhoff systems in variable exponent Sobolev spaces. Complex Var. Elliptic Equ. **56**, 715–753 (2011)
10. Autuori, G., Pucci, P., Salvatori, M.C.: Asymptotic stability for anisotropic Kirchhoff systems. J. Math. Anal. Appl. **352**, 149–165 (2009)
11. Autuori, G., Pucci, P., Salvatori, M.C.: Global nonexistence for nonlinear Kirchhoff systems. Arch. Ration. Mech. Anal. **196**, 489–516 (2010)
12. Bellassoued, M.: Decay of solutions of the wave equation with arbitrary localized nonlinear damping. J. Differ. Equ. **211**, 303–332 (2005)
13. Bociu, L.: Local and global well-posedness of weak solutions for the wave equation with nonlinear boundary and interior sources of supercritical exponents and damping. Nonlinear Anal. **71**, e560–e575 (2009)
14. Bourgain, J., Brézis, H., Mironescu, P.: Another look at Sobolev spaces. In: Optimal Control and Partial Differential Equations, pp. 439–455. IOS, Amsterdam (2001)
15. Bourgain, J., Brézis, H., Mironescu, P.: Limiting embedding theorems for $W^{s,p}$ when $s \uparrow 1$ and applications. Dedicated to the memory of Thomas H. Wolff. J. Anal. Math. **87**, 77–101 (2002)
16. Brasco, L., Parini, E., Squassina, M.: Stability of variational eigenvalues for the fractional p-Laplacian. Discrete Contin. Dyn. Syst. **36**, 1813–1845 (2016)
17. Caffarelli, L., Roquejoffre, J.M., Savin, O.: Nonlocal minimal surfaces. Comm. Pure Appl. Math. **63**, 1111–1144 (2010)
18. Camassa, R., Holm, D.: An integrable shallow water equation with peaked solitons. Phys. Rev. Lett. **71**, 1661–1664 (1993)
19. Caponi, M., Pucci, P.: Existence theorems for entire solutions of stationary Kirchhoff fractional p-Laplacian equations. Ann. Mat. Pura Appl. **195**, 2099–2129 (2016)
20. Carrier, G.F.: On the nonlinear vibration problem of the elastic string. Quart. Appl. Math. **3**, 157–165 (1945)
21. Carrier, G.F.: A note on the vibrating string. Quart. Appl. Math. **7**, 97–101 (1949)
22. Cavalcanti, M.M., Domingos Cavalcanti, V.N., Jorge Silva, M.A., Webler, C.M.: Exponential stability for the wave equation with degenerate nonlocal weak damping. Israel J. Math. **219**, 189–213 (2017)
23. Cavalcanti, M.M., Domingos Cavalcanti, V.N., Ma, T.F.: Exponential decay of the viscoelastic Euler-Bernoulli equation with a nonlocal dissipation in general domains. Differ. Integr. Equ. **17**, 495–510 (2004)
24. Cavalcanti, M.M., Domingos Cavalcanti, V.N., Martinez, P.: Existence and decay rate estimates for the wave equation with nonlinear boundary damping and source term. J. Differ. Equ. **203**, 119–158 (2004)
25. Chen, A., Deng, T., Qiao, Z.: Stability of peakons and periodic peakons for a nonlinear quartic Camassa-Holm equation. Monatsh. Math. **198**, 251–288 (2022)
26. Chen, M., Liu, S.-Q., Zhang, Y.: A 2-component generalization of the Camassa-Holm equation and its solutions. Lett. Math. Phys. **75**, 1–15 (2006)
27. Chen, R.M., Fan, L., Gao, H., Liu, Y.: Breaking waves and solitary waves to the rotation-two-component Camassa-Holm system. SIAM J. Math. Anal. **49**, 3573–3602 (2017)
28. Chen, R.M., Liu, Y.: Wave breaking and global existence for a generalized two-component Camassa-Holm system. Int. Math. Res. Not. **6**, 1381–1416 (2011)
29. Chen, S.J., Li, L.: Multiple solutions for the nonhomogeneous Kirchhoff equation on $\mathbb{R}^N$. Nonlinear Anal. Real World Appl. **14**, 1477–1486 (2013)
30. Chueshov, I.: Long-time dynamics of Kirchhoff wave models with strong nonlinear damping. J. Differ. Equ. **252**, 1229–1262 (2012)

31. Colasuonno, F., Pucci, P.: Multiplicity of solutions for $p(x)$-polyharmonic elliptic Kirchhoff equations. Nonlinear Anal. **74**, 5962–5974 (2011)
32. Constantin, A.: On the Cauchy problem for the periodic Camassa-Holm equation. J. Differ. Equ. **141**, 218–235 (1997)
33. Constantin, A.: The Hamiltonian structure of the Camassa-Holm equation. Expo. Math. **15**, 53–85 (1997)
34. Constantin, A.: On the scattering problem for the Camassa-Holm equation. Proc. Roy. Soc. Lond. A **457**, 953–970 (2001)
35. Constantin, A., Escher, J.: Well-posedness, global existence and blow-up phenomena for a periodic quasi-linear hyperbolic equation. Comm. Pure Appl. Math. **51**, 475–504 (1998)
36. Constantin, A., Escher, J.: Wave breaking for nonlinear nonlocal shallow water equations. Acta Math. **181**, 229–243 (1998)
37. Constantin, A., Escher, J.: On the blow-up rate and the blow-up of breaking waves for a shallow water equation. Math. Z. **233**, 75–91 (2000)
38. Constantin, A., Ivanov, R.: On an integrable two-component Camassa-Holm shallow water system. Phys. Lett. A **372**, 7129–7132 (2008)
39. Constantin, A., Kolev, B.: Geodesic flow on the diffeomorphism group of the circle. Comment. Math. Helv. **78**, 787–804 (2003)
40. D'Ancona, P., Spagnolo, S.: Global solvability for the degenerate Kirchhoff equation with real analytic data. Invent. Math. **108**, 247–262 (1992)
41. D'Onofrio, L., Fiscella, A., Molica Bisci, G.: Perturbation methods for nonlocal Kirchhoff-type problems. Fract. Calc. Appl. Anal. **20**, 829–853 (2017)
42. Escher, J., Kohlmanna, M., Lenells, J.: The geometry of the two-component Camassa-Holm and Degasperis-Procesi equations. J. Geom. Phys. **61**, 436–452 (2011)
43. Escher, J., Lechtenfeld, O., Yin, Z.: Well-posedness and blow-up phenomena for the 2-component Camassa-Holm equation. Discrete Contin. Dyn. Syst. **19**, 493–513 (2007)
44. Escher, J., Yin, Z.: Initial boundary value problems of the Camassa-Holm equation. Commun. Partial Differ. Equ. **33**, 377–395 (2008)
45. Falqui, G.: On a Camassa-Holm type equation with two dependent variables. J. Phys. A: Math. Gen. **39**, 327–342 (2006)
46. Fan, L., Gao, H., Liu, Y.: On the rotation-two-component Camassa-Holm system modelling the equatorial water waves. Adv. Math. **291**, 59–89 (2016)
47. Figueiredo, G.M.: Existence of a positive solution for a Kirchhoff problem type with critical growth via truncation argument. J. Math. Anal. Appl. **401**, 706–713 (2013)
48. Figueiredo, G.M., Santos, J.R.: Multiplicity of solutions for a Kirchhoff equation with subcritical or critical growth. Differ. Integr. Equ. **25**, 853–868 (2012)
49. Fiscella, A., Pucci, P.: p-fractional Kirchhoff equations involving critical nonlinearities. Nonlinear Anal. Real World Appl. **35**, 350–378 (2017)
50. Fiscella, A., Pucci, P.: Kirchhoff Hardy fractional problems with lack of compactness. Adv. Nonlinear Stud. **17**, 429–456 (2017)
51. Fiscella, A., Pucci, P., Zhang, B.: p-fractional Hardy-Schrödinger-Kirchhoff systems with critical nonlinearities. Adv. Nonlinear Anal. **8**, 1111–1131 (2019)
52. Fiscella, A., Valdinoci, E.: A critical Kirchhoff type problem involving a nonlocal operator. Nonlinear Anal. **94**, 156–170 (2014)
53. Fokas, A., Fuchssteiner, B.: Symplectic structures, their Bäcklund transformation and hereditary symmetries. Physica D **4**, 47–66 (1981)
54. Fu, Y., Wang, H.: A note on the solution map for the periodic multi-dimensional Camassa-Holm-type system. Monatsh. Math. **197**, 435–461 (2022)
55. Fu, Y., Pucci, P.: On solutions of space–fractional diffusion equations by means of potential wells. Electron. J. Qual. Theory Differ. Equ., Special volume dedicated to Professor Tibor Krisztin on the occasion of his 60th birthday, Paper No. 70, p. 17 (2016)
56. Gent, A.N.: Engineering with Rubber: How to Design Rubber Components, 3rd ed., H. Gardner Publ. (2012)

57. Geyer, A., Martins, R.H., Natali, F., Pelinovsky, D.E.: Stability of smooth periodic travelling waves in the Camassa-Holm equation. Stud. Appl. Math. **148**, 27–61 (2022)
58. Gorain, G.C.: Boundary stabilization of nonlinear vibrations of a flexible structure in a bounded domain in $\mathbb{R}^n$. J. Math. Anal. Appl. **319**, 635–650 (2006)
59. Grisvard, P.: Elliptic Problems in Nonsmooth Domains. Monographs and Studies in Mathematics, vol. 24, Pitman, Boston, MA (1985)
60. Guan, C., Yin, Z.: Global existence and blow-up phenomena for an integrable two-component Camassa-Holm shallow water system. J. Differ. Equ. **248**, 2003–2014 (2010)
61. Gui, G., Liu, Y.: On the Cauchy problem for the two-component Camassa-Holm system. Math. Z. **268**, 45–66 (2011)
62. Gui, G., Liu, Y.: On the global existence and wave-breaking criteria for the two-component Camassa-Holm system. J. Funct. Anal. **258**, 4251–4278 (2010)
63. Guo, F., Peng, W.: Blow-up solutions for the generalized two-component Camassa-Holm system on the circle. Nonlinear Anal. **105**, 120–133 (2014)
64. Han, Y., Guo, F., Gao, H.: On solitary waves and wave-breaking phenomena for a generalized two-component integrable Dullin-Gottwald-Holm system. J. Nonlinear Sci. **23**, 617–656 (2013)
65. Hwang, G., Moon, B.: Stability of periodic peaked solitary waves for a cubic Camassa–Holm–type equation. J. Evol. Equ. **22**, 26. Paper No. 68 (2022)
66. Ikeda, M., Inui, T., Wakasugi, Y.: The Cauchy problem for the nonlinear damped wave equation with slowly decaying data. NoDEA Nonlinear Differ. Equ. Appl. **24**, 53. Art. 10 (2017)
67. Ikeda, M., Ogawa, T.: Lifespan of solutions to the damped wave equation with a critical nonlinearity. J. Differ. Equ. **261**, 1880–1903 (2016)
68. Ikehata, R., Suzuki, T.: Stable and unstable sets for evolution equations of parabolic and hyperbolic type. Hiroshima Math. J. **26**, 475–491 (1996)
69. Ivanov, R.: Two-component integrable systems modelling shallow water waves: the constant vorticity case. Wave Motion **46**, 389–396 (2009)
70. Jorge Silva, M.A., Narciso, V.: Long–time behavior for a plate equation with nonlocal weak damping. Differ. Integr. Equ. **27**, 931–948 (2014)
71. Kirchhoff, G.: Vorlesungen über Mechanik. Teubner, Leipzig (1897)
72. Lakshmanan, M.: Integrable nonlinear wave equations and possible connections to tsunami dynamics. In: Kundu, A. (ed.) Tsunami and Nonliner Waves, pp. 31–49. Springer, Berlin (2007)
73. Lange, H., Perla Menzala, G.: Rates of decay of a nonlocal beam equation. Differ. Integr. Equ. **10**, 1075–1092 (1997)
74. Liang, S., Pucci, P., Zhang, B.: Multiple solutions for critical Choquard-Kirchhoff type equations. Adv. Nonlinear Anal. **10**, 400–419 (2021)
75. Liang, S., Shi, S.: Soliton solutions to Kirchhoff type problems involving the critical growth in $\mathbb{R}^N$. Nonlinear Anal. **81**, 31–41 (2013)
76. Lin, J., Nishihara, K., Zhai, J.: L^2-estimates of solutions for damped wave equations with space-time dependent damping term. J. Differ. Equ. **248**, 403–422 (2010)
77. Liu, J.: Blow-up phenomena for the rotation-two-component Camassa-Holm system. Appl. Anal. **100**, 574–588 (2021)
78. Liu, J., Liao, J.-F., Tang, C.-L.: Positive solutions for Kirchhoff-type equations with critical exponent in $\mathbb{R}^N$. J. Math. Anal. Appl. **429**, 1153–1172 (2015)
79. Liu, J., Pucci, P., Zhang, Q.: Wave breaking analysis for the periodic rotation-two-component Camassa-Holm system. Nonlinear Anal. **187**, 214–228 (2019)
80. Liu, J., Zhang, D.: Blow-up phenomena and global existence for the periodic two-component Dullin-Gottwald-Holm system. Bound. Value Probl. **158**, 11 (2013)
81. Liu, Y.C.: On Potential wells and vacuum isolating of solutions for semilinear wave equations. J. Differ. Equ. **192**, 155–169 (2003)
82. Liu, Y.C., Xu, R.: Potential well method for Cauchy problem of generalized double dispersion equations. J. Math. Anal. Appl. **338**, 1169–1187 (2008)

83. Liu, Y.C., Zhao, J.S.: Nonlinear parabolic equations with critical initial conditions $J(u_0) = d$ or $I(u_0) = 0$. Nonlinear Anal. **58**, 873–883 (2004)
84. Liu, Y.C., Zhao, J.S.: On potential wells and applications to semilinear hyperbolic equations and parabolic equations. Nonlinear Anal. **64**, 2665–2687 (2006)
85. Luo, Z., Qiao, Z., Yin, Z.: Global existence and blow-up phenomena for a periodic modified Camassa-Holm equation (MOCH). Appl. Anal. **101**, 3432–3444 (2022)
86. Martinez, P.: Precise decay rate estimates for time-dependent dissipative systems. Israel J. Math. **119**, 291–324 (2000)
87. Milla Miranda, M., Lourêdo, A.T., Medeiros, L.A.: On nonlinear wave equations of Carrier type. J. Math. Anal. Appl. **432**, 565–582 (2015)
88. Mingqi, X., Pucci, P., Squassina, M., Zhang, B.: Nonlocal Schrödinger–Kirchhoff equations with external magnetic field. Discrete Contin. Dyn. Syst. Ser. A **37**, 1631–1649 (2017)
89. Miyagaki, O.H., Pucci, P.: Nonlocal Kirchhoff problems with Trudinger–Moser critical nonlinearity. NoDEA Nonlinear Differ. Equ. Appl. **26**, 26. Art. 27 (2019)
90. Molica Bisci, G., Pucci, P.: Nonlinear Problems with Lack of Compactness. De Gruyter Series in Nonlinear Analysis and Applications, vol. 36. De Gruyter, Berlin (2021)
91. Molica Bisci, G., Radulescu, V.D., Servadei, R.: Variational methods for nonlocal fractional problems. With a foreword by Jean Mawhin. Encyclopedia of Mathematics and its Applications, vol. 162. Cambridge University Press (2016)
92. Molica Bisci, G., Vilasi, L.: On a fractional degenerate Kirchhoff–type problem. Commun. Contemp. Math. **19**, 1550088, pp. 23 (2017)
93. Moon, B.: On the wave-breaking phenomena and global existence for the periodic rotation-two-component Camassa Holm system. J. Math. Anal. Appl. **451**, 84–101 (2017)
94. Mukherjee, T., Pucci, P., Xiang, M.: Combined effects of singular and exponential nonlinearities in fractional Kirchhoff problems. Discrete Contin. Dyn. Syst. **42**, 163–187 (2022)
95. Nakao, M.: Asymptotic stability of the bounded or almost periodic solutions of the wave equation with nonlinear dissipation term. J. Math. Anal. Appl. **58**, 336–343 (1977)
96. Nakao, M.: Decay of solutions of the wave equation with a local nonlinear dissipation. Math. Ann. **305**, 403–417 (1996)
97. Narashima, R.: Nonlinear vibration of an elastic string. J. Sound Vib. **8**, 134–146 (1968)
98. Nishihara, K., Wakasugi, Y.: Global existence of solutions for a weakly coupled system of semilinear damped wave equations. J. Differ. Equ. **259**, 4172–4201 (2015)
99. Ogawa, T., Takeda, H.: Large time behavior of solutions for a system of nonlinear damped wave equations. J. Differ. Equ. **251**, 3090–3113 (2011)
100. Ono, K.: Blowing up and global existence of solutions for some degenerate nonlinear wave equations with some dissipation. In: Proceedings of the Second World Congress of Nonlinear Analysts, Part 7 (Athens, 1996), Nonlinear Anal., vol. 30, pp. 4449–4457 (1997)
101. Oplinger, D.W.: Frequency response of a nonlinear stretched string. J. Acoust. Soc. Am. **32**, 1529–1538 (1960)
102. Pan, N., Zhang, B.L., Cao, J.: Degenerate Kirchhoff-type diffusion problems involving the fractional p-Laplacian. Nonlinear Anal. Real World Appl. **37**, 56–70 (2017)
103. Payne, L.E., Sattinger, D.H.: Saddle points and instability of nonlinear hyperbolic equations. Israel. J. Math. **22**, 273–303 (1975)
104. Piersanti, P., Pucci, P.: Entire solutions for critical fractional p-Kirchhoff equations. Publ. Mat. **62**, 3–36 (2018)
105. Pan, N., Pucci, P., Xu, R., Zhang, B.: Degenerate Kirchhoff-type wave problems involving the fractional Laplacian with nonlinear damping and source terms. J. Evol. Equ. **19**, 615–643 (2019)
106. Pan, N., Pucci, P., Zhang, B.: Degenerate Kirchhoff-type hyperbolic problems involving the fractional Laplacian. J. Evol. Equ. **18**, 385–409 (2018)
107. Pucci, P.: An overview on two degenerate Kirchhoff wave problems involving fractional Laplacian operators. Int. J. Evol. Equ., special volume dedicated to the memory of Professor Vasilii V. Zhikov, **10**, 411–436 (2018)

108. Pucci, P., Radulescu, V.: A brief introduction to the special issue progress in nonlinear Kirchhoff problems, as Guest editors. Nonlinear Anal. **186**, 1–5 (2019)
109. Pucci, P., Saldi, S.: Multiple solutions for an eigenvalue problem involving non–local elliptic p–Laplacian operators. In: Citti, G., Manfredini, M., Morbidelli, D., Polidoro, S., Uguzzoni, F. (eds.) Geometric Methods in PDE's, vol. 13, pp. 159–176. Springer INdAM Series (2015)
110. Pucci, P., Saldi, S.: Critical stationary Kirchhoff equations in $\mathbb{R}^N$ involving nonlocal operators. Rev. Mat. Iberoam. **32**, 1–22 (2016)
111. Pucci, P., Saldi, S.: Asymptotic stability for nonlinear damped Kirchhoff systems involving the fractional p-Laplacian operator. J. Differ. Equ. **263**, 2375–2418 (2017)
112. Pucci, P., Serrin, J.: Precise damping conditions for global asymptotic stability for nonlinear second order systems. Acta Math. **170**, 275–307 (1993)
113. Pucci, P., Serrin, J.: Continuation and limit behavior for damped quasi–variational systems. In: Ni, W.-M., Peletier, L.A., Vázquez, J.L. (eds.) Degenerate Diffusions, vol. 47, pp. 157–173. The IMA Volumes in Mathematics and its Applications. Springer, New York (1993)
114. Pucci, P., Serrin, J.: Asymptotic stability for nonautonomous dissipative wave systems. Comm. Pure Appl. Math. **49**, 177–216 (1996)
115. Pucci, P., Serrin, J.: Local asymptotic stability for dissipative wave systems. Israel J. Math. **104**, 29–50 (1998)
116. Pucci, P., Serrin, J.: Global nonexistence for abstract evolution equations with positive initial energy. J. Differ. Equ. **150**, 203–214 (1998)
117. Pucci, P., Xiang, M., Zhang, B.: Multiple solutions for nonhomogeneous Schrödinger-Kirchhoff type equations involving the fractional p-Laplacian in $\mathbb{R}^N$. Calc. Var. Part. Differ. Equ. **54**, 2785–2806 (2015)
118. Pucci, P., Xiang, M., Zhang, B.: Existence and multiplicity of entire solutions for fractional p-Kirchhoff equations. Adv. Nonlinear Anal. **5**, 27–55 (2016)
119. Pucci, P., Xiang, M., Zhang, B.: A diffusion problem of Kirchhoff type involving the nonlocal fractional p-Laplacian. Discrete Contin. Dyn. Syst. **37**, 4035–4051 (2017)
120. Pucci, P., Xiang, M., Zhang, B.: Existence results for Schrödinger-Choquard-Kirchhoff equations involving the fractional p-Laplacian. Adv. Calc. Var. **12**, 253–275 (2019)
121. Radu, P.: Weak solutions to the Cauchy problem of a semilinear wave equation with damping and source terms. Adv. Differ. Equ. **10**, 1261–1300 (2005)
122. Said-Houari, B.: Asymptotic behaviors of solutions for viscoelastic wave equation with space-time dependent damping term. J. Math. Anal. Appl. **387**, 1088–1105 (2012)
123. Sattinger, D.H.: On global solution of nonlinear hyperbolic equations. Arch. Rat. Mech. Anal. **30**, 147–172 (1968)
124. Servadei, R., Valdinoci, E.: Mountain pass solutions for non-local elliptic operators. J. Math. Anal. Appl. **389**, 887–898 (2012)
125. Servadei, R., Valdinoci, E.: Variational methods for non-local operators of elliptic type. Discrete Contin. Dyn. Syst. **33**, 2105–2137 (2013)
126. Shabat, A., Martínez Alonso, L.: On the prolongation of a hierarchy of hydrodynamic chains. In: Proceedings of the NATO Advanced Research Workshop, Cadiz, Spain 2002, pp. 263–280. NATO Science Series, Kluwer Academic Publishers, Dordrecht (2004)
127. Sobajima, M., Wakasugi, Y.: Diffusion phenomena for the wave equation with space-dependent damping in an exterior domain. J. Differ. Equ. **261**, 5690–5718 (2016)
128. Todorova, G.: Cauchy problem for a nonlinear wave equation with nonlinear damping source terms. Nonlinear Anal. **41**, 891–905 (2000)
129. Vázquez, J.L.: Existence of maximal solutions for some very singular nonlinear fractional diffusion equations in 1D. J. Evol. Equ. **16**, 723–758 (2016)
130. Villaggio, P.: Mathematical Models for Elastic Structures. Cambridge University Press (1997)
131. Wang, Z., Zhou, S.: Random attractor for stochastic non-autonomous damped wave equation with critical exponent. Discrete Contin. Dyn. Syst. **37**, 545–573 (2017)
132. Wei, L., Wang, Y., Zhang, H.: Breaking waves and persistence property for a two-component Camassa-Holm system. J. Math. Anal. Appl. **445**, 1084–1096 (2017)

133. Whitham, G.B.: Linear and Nonlinear Waves. Pure and Applied Mathematics, Wiley-Interscience, New York-London-Sydney (1974)
134. Xiang, M., Zhang, B.: Degenerate Kirchhoff problems involving the fracrional p-Laplacian without the (AR) condition. Complex Var. Elliptic Equ. **60**, 1–11 (2015)
135. Xiang, M.Q., Zhang, B.L., Ferrara, M.: Multiplicity results for the nonhomogeneous fractional p-Kirchhoff equations with concave-convex nonlinearities. Proc. Roy. Soc. A **471**, 14 (2015)
136. Xiang, M.Q., Zhang, B.L., Qiu, H.: Existence of solutions for a critical fractional Kirchhoff type problem in $\mathbb{R}^N$. Sci. China Math. **60**, 1647–1660 (2017)
137. Xiang, M., Zhang, B., Radulescu, V.D.: Multiplicity of solutions for a class of quasilinear Kirchhoff system involving the fractional p-Laplacian. Nonlinearity **29**, 3186–3205 (2016)
138. Xiang, M., Zhang, B., Zhang, X.: A nonhomogeneous fractional p-Kirchhoff type problem involving critical exponent in $\mathbb{R}^N$. Adv. Nonlinear Stud. **17**, 611–640 (2017)
139. Xu, R.Z., Liu, Y.C.: Global existence and nonexistence of solution for Cauchy problem of multidimensional double dispersion equations. Nonlinear Anal. **359**, 739–751 (2009)
140. Yamazaki, T.: Global solvability for the Kirchhoff equations in exterior domains of dimension three. J. Differ. Equ. **210**, 290–316 (2005)
141. Yang, M., Duan, J., Kloeden, P.: Asymptotic behavior of solutions for random wave equations with nonlinear damping and white noise. Nonlinear Anal. Real World Appl. **12**, 464–478 (2011)
142. Yang, Z.: Finite-dimensional attractors for the Kirchhoff models with critical exponents. J. Math. Phys. **53**, 15 (2012)
143. Yin, Z.: On the blow–up of solutions of the periodic Camassa–Holm equation. Dyn. Contin. Discrete Impuls. Syst. Ser. A Math. Anal. **12**, 375–381 (2005)
144. Zhang, L., Liu, B.: Well-posedness, blow-up criteria and Gevrey regularity for a rotation-two-component Camassa-Holm system. Discrete Contin. Dyn. Syst. **38**, 2655–2685 (2018)
145. Zhang, Y.: Wave breaking and global existence for the periodic rotation-Camassa-Holm system. Discrete Contin. Dyn. Syst. **37**, 2243–2257 (2017)
146. Zhu, M., Xu, J.: On the wave-breaking phenomena for the periodic two-component Dullin-Gottwald-Holm system. J. Math. Anal. Appl. **391**, 415–428 (2012)

Chapter 4
A Little Tribute to Zeev Nehari

Giovany Figueiredo

Abstract This article deals with a result due to Zeev Nehari, probably, one of the most impressive for the study of Differential Equations. Such a result, nowadays, known as Nehari's Method, has influenced studies all over the world since 1962. Up to our days, the number of published articles and mentored students is impressive related to this result. His personal life is also different from ordinary people. For example, he was born on February, 2, 1915, in Berlin, Germany, as Willy Weisbach. After finishing high school in Berlin, he moved to Palestine and changed his name to a Hebrew one. A little of all this will be seen in this article. Moreover, we will describe the generalizations and highlight the current results that use it.

4.1 Introduction

We begin these notes by describing Nehari's Method in an intuitive way. Let E be a reflexive Banach space and $\phi \in C^1(E)$. The Fréchet derivative of ϕ in $u \in E$, given by $\phi'(u)$, is an element of E' that applied for $v \in E$ will be denoted by $\phi'(u)v$. Now, if $u \in E$ is a critical point of ϕ with $u \neq 0$, then $\phi'(u) = 0$, which is equivalent to $\phi'(u)v = 0$, for all $v \in E$. Hence, necessarily, if $u \in E$ is a critical point of ϕ with $u \neq 0$, then u is in the set

$$\mathcal{N} := \{u \in E, u \neq 0 : \phi'(u)u = 0\}.$$

Such set is called **Nehari manifold**; sometimes, it is not indeed a manifold. The Nehari method consists of finding a minimizer of ϕ constrained to $\mathcal{N}$, i.e., to look for $u \in \mathcal{N}$ such that

$$\phi(u) = c := \inf_{v \in \mathcal{N}} \phi(v).$$

G. Figueiredo (✉)
Universidade de Brasília—UnB, Brasília, Brazil
e-mail: giovany@unb.br

M. Jleli et al. (eds.), *Advances in Nonlinear Evolution Equations*, Infosys Science Foundation Series, https://doi.org/10.1007/978-981-97-9812-4_4

After that, we must show that the minimum point of ϕ on the Nehari manifold $\mathcal{N}$ is a critical point of ϕ in whole space E. Note that this method gives a critical point $u \in E$ of ϕ with the following property:

$$\phi(u) \leq \phi(w),$$

for all nontrivial critical points $w \in E$ of ϕ. Such critical point u of ϕ is called **ground state**. When this critical point of ϕ is a solution of an elliptic problem, we say that the solution is a **ground state solution**.

Nehari introduced this method in [11, 12] considering a second order ODE in an interval (a, b) and showed the existence of nontrivial solution minimizing a functional ϕ of C^2 class on $\mathcal{N}$. After that, he used implicit function theorem in order to show that the minimizer u of ϕ on $\mathcal{N}$ satisfies the $\phi'(u) = 0$.

Nehari's result is impressive because instead of minimizing the functional throughout the space, Nehari minimized the functional in a smaller set consisting of all possible solutions. The set $\mathcal{N}$ is smaller than the entire space, the chances of finding the minimum increase.

The Nehari method allows to consider problems in which the functionals are not bounded from below in the whole space, but are bounded from below in the Nehari manifold. The literature on the Nehari manifold method is rather extensive. It would be impossible to cover all different aspects of this method or to provide a reasonably complete bibliography of it.

After this work by Zeev Nehari, another methods appeared as a consequence of his method, as so-called fibering method which has been introduced by Pohozaev [13]. The method still allows to show the existence of sign-changing solutions or infinitely many solutions for problem such that the associated functional ϕ is even.

Section 4.2 is devoted to a brief history of our honoree. We focus our attention on his academic career, his doctoral thesis, his supervisions and his legacy. In Sect. 4.3, in order to illustrate the pioneering work of Zeev Nehari, we will apply his method to the problem

$$-\Delta u = f(u) \text{ in } \Omega, \quad u = 0 \text{ on } \partial\Omega,$$

where Ω is a bounded domain in $\mathbb{R}^N$ under suitable hypotheses on the nonlinearity f. In the Nehari method, the geometry of a fiber application is studied. In the superlinear case this geometry coincides with the geometry of another, no less famous theorem, called Mountain Pass Theorem [1]. In Sect. 4.4 we will show this coincidence and show that in the sublinear case these two are different. To illustrate this difference, let us remember a very beautiful article due to Liu et al. [10] that shows the solution of a problem by the Nehari Method in which they cannot apply the Mountain Pass Theorem. In Sect. 4.5 we will show some generalizations of Nehari's method. We start with the most important of all which is the result of Szulkin and Weth. In [16] Szulkin and Weth rewrote Nehari's method in an abstract form. Besides, they put weaker hypotheses in the nonlinearity f. For example, for them it was enough that the nonlinearity was continuous. With this hypothesis the Nehari manifold was no longer a manifold and the implicit function theorem could no longer be used to show that the minimum point in the Nehari manifold was the critical point of the functional in the whole space. To get around this difficulty, they used the homeomorphism between the Nehari and the sphere. In [6] the authors improve Szulkin and Weth's results removing the homogeneity of the operator. In [8], once more, the authors improve Szulkin and Weth's results removing the regularity of the associated functional and the reflexivity of the space. In [7] the Nehari method was applied in cones. In such section, we show all the problems studied by these authors.

4.2 On Zeev Nehari

As may be realized in this section, Zeev Nehari was a teacher with all the grandeur that this word means. He was born on February 2, 1915, in Berlin, Germany, as Willy Weibach. After finishing high school in Berlin, he moved to Palestine and changed his name to a Hebrew one. He joined Kibbutz where he worked during several years. A kibbutz is a type of settlement which is unique to Israel, a collective community, traditionally agrarian. The first kibbutz was Deganya Aleph, founded in 1910. Today, there are over 270 kibbutzim in Israel. They have diversified greatly since their agricultural beginnings and now many are private. Regardless of their status, the kibbutz offers a unique insight into Israeli-society. Additionally, they are fascinating places to visit.

In 1935 he entered the Hebrew University in Jerusalem and began to study mathematics. He became an expert on elliptic problems, modular functions on special classes of univalent functions, and his Ph.D. thesis of 1941 as well as his first publications are devoted to these topics. In 1944 he became Instructor at the Hebrew University where he served until 1947. In 1947 he went to Harvard University and worked in a group of young mathematicians around Stefan Bergman. In 1948 Nehari went to Washington University in St. Louis as Associate Professor, where he rose to Full Professor in 1951 and taught until 1954. In 1954 he moved to Carnegie-Mellon University and stayed until his untimely death on September 1, 1978.

Between 1941, the year of his doctoral thesis defense and 1978, the year of his death, Zeev Nehari published eighty-seven articles. In addition, Zeev Nehari has supervised twenty doctoral theses.

These numbers are exceptional, even today when we have computers or text editors at our disposal.

4.3 A Description of the Nehari Method

As stated in the introduction, Zeev Nehari introduced this method through two articles [11, 12] by considering a second order ODE in an interval (a, b). But as a matter of personal preference, we will illustrate this method with the following problem:

$$(P) \qquad \begin{cases} -\Delta u = f(u) \text{ in } \Omega, \\ u = 0 \text{ on } \partial\Omega, \end{cases}$$

where $\Omega \subset \mathbb{R}^N$ is a smooth bounded domain. Before stating the main result of this section, we need some hypotheses on the nonlinearity f.

The function f is a C^1 class function and there are constants $C, r \in \mathbb{R}$ with $2 < r < 2^* := \frac{2N}{N-2}$, $N \geq 3$, such that

$$(f_1) \qquad |f(t)| \leq C(1 + |t|^{r-1}).$$

In addition, we consider at the origin and at infinity the following hypotheses on the nonlinearity f:

$$(f_2) \qquad \lim_{|t|\to 0} \frac{|f(t)|}{|t|} = 0$$

and

$$(f_3) \qquad \lim_{t\to+\infty} \frac{F(t)}{t^2} = +\infty,$$

where $F(t) = \displaystyle\int_0^t f(s)ds$.

Note that, by $(f_1) - (f_2)$, given $\varepsilon > 0$, there exists a constant $C_\varepsilon > 0$, such that

$$|f(t)| \leq \varepsilon|t| + C_\varepsilon|t|^{r-1}. \tag{4.1}$$

In addition, we suppose that the function

$$(f_4) \qquad t \mapsto \frac{f(t)}{t} \text{ is increasing for } t > 0.$$

For simplicity, in this paper we consider $f(t) = 0$ for $t \leq 0$. The main result of this section is given by

Theorem 4.1 *Assume that (f_1)–(f_4) hold. Then problem (P) has a ground state solution.*

4.3.1 The Variational Framework

We say that $u \in H_0^1(\Omega)$ is a weak solution of problem (P) if

$$\int_\Omega \nabla u \nabla \phi \, dx - \int_\Omega f(u)\phi \, dx = 0,$$

for all $\phi \in H_0^1(\Omega)$. Let us find solutions of (P) by looking for critical points of the C^1 class functional ϕ given by

$$\phi(u) = \frac{1}{2}\|u\|^2 - \int_\Omega F(u)\, dx,$$

where $\|\cdot\|$ is the usual norm in $H_0^1(\Omega)$. Note that

$$\phi'(u)\phi = \int_\Omega \nabla u \nabla \phi\, dx - \int_\Omega f(u)\phi\, dx,$$

for all $\phi \in H_0^1(\Omega)$. Hence, critical points of ϕ are weak solutions of (P). Now we define $\mathcal{N}$ as the Nehari manifold associated with ϕ by

$$\mathcal{N} = \{u \in H_0^1(\Omega)\backslash\{0\} : \phi'(u)u = 0\} = J^{-1}\{0\}\backslash\{0\},$$

where $J(u) = \phi'(u)u = \|u\|^2 - \int_\Omega f(u)u dx.$

4.3.2 Preliminary Results

The next result, that we will prove below, is a key point in Nehari's method. Essentially, it says that every $u \in H_0^1(\Omega)$, $u \neq 0$ can be projected on the Nehari manifold uniquely. As a consequence of this result, we have that the Nehari manifold is non-empty.

Lemma 4.1 *Assume that (f_1)–(f_4) are true, then for $u \in H_0^1(\Omega)$ with $u \neq 0$, there is a unique $t_0 = t_0(u) > 0$ such that $t_0 u \in \mathcal{N}$ and $\phi(t_0 u) = \max_{t \geq 0} \phi(tu)$.*

Proof Consider $u \in H_0^1(\Omega)$ with $u \neq 0$ and $h(t) = \phi(tu)$. Using (4.1) and the Sobolev continuous embedding, there are positive constants $C_1, C_2 > 0$ such that

$$h(t) = \phi(tu) \geq (1 - \varepsilon C_1)\frac{t^2}{2}\|u\|^2 - \frac{t^r}{r} C_\varepsilon C_2 \|u\|^r.$$

Since $2 < r$, there is a $t_1 > 0$ sufficiently small such that $h(t) > 0$, for all $0 < t < t_1$.

Now, from (f_3) we derive that given a constant $K > 0$, there is $M > 0$ such that, for all $t \geq M$, we obtain

$$F(t) \geq K t^2.$$

Hence,

$$h(t) \leq \frac{t^2}{2}\|u\|^2 - K t^2 \int_\Omega |u|^2\, dx.$$

Choosing a constant K, sufficiently large, we have

$$h(t) < 0, \quad \text{for all} \ \ t \geq M.$$

So, there is $t_0 > 0$ such that

$$h(t_0) = \max_{t \geq 0} h(t) = \max_{t \geq 0} \phi(tu).$$

Therefore, $h'(t_0) = 0$, i.e.,

$$t_0^2 \|u\|^2 = \int_\Omega f(t_0 u) t_0 u dx$$

implying $t_0 u \in \mathcal{N}$.

In what follows, we show that t_0 is unique. For this, we suppose that there is $s > 0$ such that $su \in \mathcal{N}$. Then,

$$\|u\|^2 = \int_\Omega \frac{f(t_0 u)}{t_0 u} u^2 dx$$

and

$$\|u\|^2 = \int_\Omega \frac{f(su)}{su} u^2 dx.$$

In consequence,

$$\int_\Omega \left(\frac{f(t_0 u)}{t_0 u} - \frac{f(su)}{su} \right) u^2 dx = 0.$$

The hypothesis (f_4) implies $t_0 = s$. □

The h function introduced in Lemma 4.1 is known in the literature as **fiber**.

The next lemma states that sequences in the Nehari manifold cannot converge to zero and that there is the real number c such that $c =: \inf_{\mathcal{N}} \phi > 0$.

Lemma 4.2 *For all $u \in \mathcal{N}$, there is a positive constant C independent of u such that*

$$0 < C \leq \|u\|$$

and

$$\phi(u) \geq 0.$$

Proof Note that by (4.1) and from Sobolev continuous embedding, for all $u \in \mathcal{N}$, we get

$$\|u\|^2 = \int_\Omega f(u)u\,dx \leq \varepsilon \int_\Omega |u|^2\,dx + C_\varepsilon \int_\Omega |u|^r\,dx \leq \varepsilon C_1\|u\|^2 + C_\varepsilon C_2\|u\|^r.$$

Hence, for ε sufficiently small

$$0 < \left[\frac{(1-\varepsilon C_1)}{C_\varepsilon C_2}\right]^{1/(r-2)} \leq \|u\|.$$

Now note that by (f_4), we obtain

$$f'(t)t - f(t) > 0, \tag{4.2}$$

for all $t > 0$. But the last inequality is equivalent to say that the function

$$\frac{1}{2}f(t)t - F(t) \text{ is increasing, for } t > 0. \tag{4.3}$$

Using (4.3) we conclude that

$$\phi(u) = \phi(u) - \frac{1}{2}\phi'(u)u \geq 0.$$

□

The next lemma states that all the minimizing sequences in the Nehari manifold are bounded.

Lemma 4.3 *If $(u_n) \subset \mathcal{N}$ is a minimizing sequence for ϕ, then (u_n) is bounded in $H_0^1(\Omega)$.*

Proof Suppose, by contradiction that, up to a subsequence, $\|u_n\| \to \infty$ and consider $v_n = \frac{u_n}{\|u_n\|} \rightharpoonup v_0$. If $v_0 = 0$, then for all $t > 0$, we obtain

$$\begin{aligned} c + o_n(1) = \phi(u_n) = &\ \phi(\|u_n\|v_n) \geq \phi(tv_n) \\ = &\ \frac{t^2}{2} - \int_\Omega F(tv_n)\,dx. \end{aligned}$$

Since

$$\int_\Omega F(tv_n)\,dx \to 0$$

we have

$$c \geq \frac{t^2}{2}, \quad \text{for all } t > 0,$$

which is a contradiction. Suppose now that $v_0 \neq 0$. In this case,

$$\frac{1}{\|u_n\|^2}\phi(u_n) = \frac{c}{\|u_n\|^2} = o_n(1).$$

Therefore,

$$\int_\Omega \frac{F(\|u_n\| v_n)}{\|u_n\|^2} = \frac{1}{2} + o_n(1),$$

which is a contradiction by (f_3). □

In the next lemma we will show that the infimum of ϕ restricted to Nehari manifold is reached. In this proof, the unique projection on the Nehari manifold and the weak lower semicontinuity of the functional ϕ are fundamental.

Lemma 4.4 *There exists $u \in \mathcal{N}$, such that*

$$\phi(u) = c.$$

Proof Let $(u_n) \subset \mathcal{N}$ be a minimizing sequence. Then, from Lemma 4.3, we derive that (u_n) is bounded in $H_0^1(\Omega)$ and, up to a subsequence,

$$u_n \rightharpoonup u_0 \text{ in } H_0^1(\Omega).$$

Note that $u_0 \neq 0$, otherwise, using (4.1) and the Sobolev compact embedding, we conclude

$$\|u_n\|^2 = \int_\Omega f(u_n)u_n \, dx = o_n(1),$$

which is a contradiction with Lemma 4.2. Let $t_0 > 0$ be the real constant such that $u = t_0 u_0 \in \mathcal{N}$. Since $\| \cdot \|$ is weak lower semicontinuous and $\int_\Omega F(u_n)dx = \int_\Omega F(u_0)dx + o_n(1)$, we get that ϕ is weak lower semicontinuous. Then,

$$c \leq \phi(u) = \phi(t_0 u_0) \leq \liminf_{n\to\infty} \phi(t_0 u_n) \leq \liminf_{n\to\infty} \phi(u_n) = c,$$

where it was used in the second inequality that if $u \in \mathcal{N}$, then $\phi(u) = \max_{t\geq 0} \phi(tu)$. □

Finally, we will show that this minimum point in the Nehari manifold for the functional ϕ is a critical point of ϕ in the whole space. We will prove this result in two different ways: The first one, using the Implicit Function Theorem, just like Nehari did. The second way is by using the Lagrange Multipliers Theorem, as it is commonly used in more recent works.

4.3.3 *Proof of Theorem 4.1 via Implicit Function Theorem*

We show that $\phi'(u)v = 0$, for all $v \in H_0^1(\Omega)$. Since $u \neq 0$, there exists $\varepsilon > 0$ such that $u + sv \neq 0$, for all $|s| < \varepsilon$. Then, by Lemma 4.1, there exists $t(s)$ such that $t(s)(u + sv) \in \mathcal{N}$. Now we define

$$\begin{aligned}\phi(s,t) = \phi'(t(u+sv))t(u+sv) &= \|t(u+sv)\|^2 \\ &- \int_\Omega f(t(u+sv))t(u+sv)\,dx.\end{aligned}$$

Note that $\phi \in C^1$, $\phi(0, 1) = 0$ and

$$\frac{\partial\phi}{\partial t}(0,1) = J'(u)u = 2\|u\|^2 - \int_\Omega \Big[f(u)u - f'(u)u^2\Big]\,dx,$$

where

$$J(u) = \|u\|^2 - \int_\Omega f(u)u\,dx.$$

Since $u \in \mathcal{N}$, we get

$$\frac{\partial\phi}{\partial t}(0,1) = J'(u)u = \int_\Omega \Big[f(u)u - f'(u)u^2\Big]\,dx. \tag{4.4}$$

Using (4.2) we obtain

$$\frac{\partial\phi}{\partial t}(0,1) < 0.$$

Using the Implicit Function Theorem, there exists $t(s) \in C^1$ such that $\phi(s, t(s)) = 0$ for all $|s| < \varepsilon$. So, $t(s) \neq 0$ and $t(s)(u + sv) \in \mathcal{N}$. Consider $\Upsilon(s) = \phi(t(s)(u + sv))$ with $|s| < \varepsilon$ and note that $\Upsilon \in C^1$ and $s = 0$ is a local minimum of Υ. Hence,

$$0 = \Upsilon'(0) = \phi'(u)v.$$

4.3.4 *Proof of Theorem 4.1 via Lagrange Multipliers Theorem*

By [17, Proposition 5.12], the derivative of ϕ restricted to $\mathcal{N}$ has the norm given by

$$\|\phi'(u)\|_* := \min_{\lambda\in\mathbf{R}} \|\phi'(u) - \lambda J'(u)\|_{H_0^1(\Omega)'}.$$

Since $u \in \mathcal{N}$ is a minimum point of ϕ restricted to $\mathcal{N}$, there exists $\lambda \in \mathbf{R}$ such that

$$\phi'(u) - \lambda J'(u) = 0. \tag{4.5}$$

Using u as a test function, we obtain

$$\phi'(u)u - \lambda J'(u)u = 0.$$

Hence, $\phi'(u)u = 0$ implies $\lambda J'(u)u = 0$. But by (4.4) and (4.2) we get $J'(u)u < 0$, which implies $\lambda = 0$. Therefore, by (4.5) we derive

$$\phi'(u) = 0.$$

4.4 Nehari Method Versus Mountain Pass Theorem

In this section we will make a comparison between the method introduced by Zeev Nehari and the famous Mountain Pass Theorem introduced by Ambrosetti and Rabinowitz [1] in 1973. In order to proceed, we need to define a Palais–Smale sequence for a functional ϕ of C^1 class.

Let X be a real Banach space. Let $C^1(X, \mathbb{R})$ be the set of functionals that are Fréchet differentiable with Fréchet derivatives continuous on X. For $\phi \in C^1(X, \mathbb{R})$ we say that ϕ satisfies the Palais–Smale condition if any sequence (u_n) in X, for which $\phi(u_n)$ is bounded and $\phi(u_n) \to 0$ as $n \to \infty$, has a convergent subsequence.

Theorem 4.2 *Let X be a real Banach space and $\phi \in C^1(X, \mathbb{R})$ satisfying the Palais–Smale condition. Suppose $\phi(0) = 0$ that there are $\alpha, \rho > 0$ such that*

$$(H_1) \qquad \phi(u) \geq \alpha > 0 \ \textit{for all} \ u \in X : \|u\| = \rho,$$

there is an $e \in X$ such that $\|e\| > \rho$ and

$$(H_2) \qquad I(e) < 0.$$

Then ϕ has a critical value $c \geq \alpha$. Moreover, c can be characterized as

$$0 < c = \inf_{\gamma \in \Gamma} \max_{[0,1]} I(\gamma(t)),$$

where

$$\Gamma = \{\gamma \in C([0, 1], X) \ : \gamma(0) = 0, \gamma(1) = e\}.$$

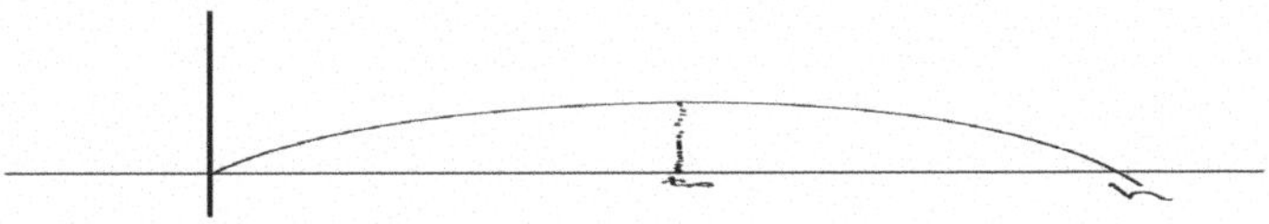

Fig. 4.1 Behavior of the fiber h

Conditions (H_1) and (H_2) are known in the literature as mountain pass geometry. It is observed that the geometry of the mountain pass coincides with the geometry of the fiber h, as seen in Fig. 4.1. Another important remark is that Nehari's method is from 1962 and the Ambrosetti and Rabinowitz's theorem is from 1973. However, the articles by Nehari [11, 12] were not quoted in the article [1] by Ambrosetti and Rabinowitz. This fact may seem strange these days, but it is perfectly understandable when we remember that the scientific dissemination of the 60s and 70s is quite different from the scientific dissemination of current times, with the emergence of the Internet and the possibility of online publications.

Although the absence of citation is perfectly understandable, the comparison of the mountain pass geometry (H_1) and (H_2) and the behavior of the fiber h in Fig. 4.1. Show that the theorems are related. This relationship was confirmed by Rabinowitz in [14], thirty years after Zeev Nehari's article and fourteen years after Zeev Nehari's death. In [14, Proposition 3.11] Rabinowitz shows that, in the superlinear case, the infimum of the functional ϕ in the Nehari manifold is equal to the level c of the Mountain Pass Theorem.

A very natural question is: was it the same theorem? The answer is: it wasn't the same theorem. The justifications are given in the paragraphs below.

Recently Bonheure, Moreira dos Santos, Parini, Tavares and Weth showed in [2] that considering a nonlinear second-order elliptic problems of the type

$$-\Delta u = f(u) \ in \ \Omega, \quad u = 0 \ on \ \partial\Omega,$$

where Ω is an open $C^{1,1}$ domain in $\mathbb{R}^N$, $N \geq 2$, under some general assumptions on the nonlinearity that include the case of a sublinear pure power $f(s) = |s|^{p-1}s$ with $0 < p < 1$ and of Allen–Cahn type $f(s) = \lambda(s - |s|^{p-1}s)$ with $p > 1$ and $\lambda > \lambda_2$ (the second Dirichlet eigenvalue of the Laplacian). They prove the existence of a least energy nodal (i.e., sign-changing) solution and of a nodal solution of mountain-pass type. Then they give explicit examples of domains where the associated levels do not coincide.

There is another article that also shows that the two theorems are not the same. Jia-Quan Liu, Ya-Qi Wang and Zhi-Qiang Wang in [10] study a quasi-linear Schrödinger equation whose prototype is given by

$$\begin{cases} -\Delta u - \Delta(u^2)u + V(x)u = |u|^{p-2}u \text{ in } \mathbb{R}^N, \\ \qquad 4 \leq p < 2 \cdot 2^* := \frac{4N}{N-2}, \quad u \in X, \end{cases} \tag{4.6}$$

where

$$X = \left\{ u \in H^1(\mathbb{R}^N) : \int_{\mathbb{R}^N} V(x)|u|^2\,dx < +\infty, \int_{\mathbb{R}^N} u^2|\nabla u|^2\,dx < +\infty \right\}.$$

They apply the Nehari method considering the functional

$$\phi(u) = \frac{1}{2}\int_{\mathbb{R}^N} (1+u^2)|\nabla u|^2\,dx + \frac{1}{2}\int_{\mathbb{R}^N} V(x)u^2\,dx - \frac{1}{p}\int_{\mathbb{R}^N} |u|^p\,dx,$$

defined in X. Note the condition $4 \le p < 2 \cdot 2^*$ characterizes the problem as superlinear. But the Mountain Pass Theorem cannot be applied. The reason is that X is not even a vector space and the functional ϕ is not Fréchet differentiable.

The problem was treated, for both the positive solution and the sign-changing solution, by a more unified approach that is the Nehari method. The authors used that ϕ is Gateaux differentiable. This was enough to show that Zeev Nehari's technique could be used here as well.

Jia-Quan Liu, Ya-Qi Wang and Zhi-Qiang Wang defined weak solution of the problem as critical point of the functional ϕ considering only the Gaetaux derivative of the functional as follows:

$$\begin{aligned}\langle \phi'(u), v\rangle = &\int_{\mathbb{R}^N} (1+u^2)\nabla u \nabla v\,dx + \int_{\mathbb{R}^N} u|\nabla u|^2 v\,dx \\ &+ \int_{\mathbb{R}^N} V(x)uv\,dx - \int_{\mathbb{R}^N} |u|^{p-2}uv\,dx\end{aligned}$$

for every $u \in X$ and for all $v \in C_0^\infty(\mathbb{R}^N)$.

An important point in the article by Jia-Quan Liu, Ya-Qi Wang, and Zhi-Qiang Wang [10]—which, in our view, has gone relatively unnoticed—concerns the proof that, even when X is not a vector space and ϕ is not Fréchet differentiable, the minimizer on the Nehari manifold is still a critical point of ϕ in the whole space. This is particularly noteworthy because, under such conditions, one cannot apply the implicit function theorem or the deformation lemma. The proof is remarkably ingenious and gained popularity after its reappearance in the work of Cingolani and Weth [5].

Lemma 4.5 *If $u \in X$ is such $\phi(u) = \inf_{\mathcal{N}} \phi$, then u is a critical point of ϕ.*

Proof Suppose, by contradiction, that u is not a critical point of ϕ. Then we find a function $\phi \in X$ such that

$$\phi'(u)\phi \le -1.$$

Then, there exist $\varepsilon, \delta > 0$ small enough such that

$$\phi'(t(u+\sigma\phi))\phi < 0, \quad \text{for all } t \in [1-\varepsilon, 1+\varepsilon] \text{ and for all } \sigma \in [-\delta, \delta]. \tag{4.7}$$

Moreover, since

$$\phi'((1-\varepsilon)u)u > 0 > \phi'((1+\varepsilon)u)u,$$

then there exists $\overline{\sigma} \in (0, \delta)$ such that

$$\phi'((1-\varepsilon)(u+\overline{\sigma}\phi)(u+\overline{\sigma}\phi) > 0 > \phi'((1+\varepsilon)(u+\overline{\sigma}\phi)(u+\overline{\sigma}\phi),$$

which implies that

$$\overline{t}(u+\overline{\sigma}\phi) \in \mathcal{N}, \quad \text{for some } \overline{t} \in (1-\varepsilon, 1+\varepsilon). \tag{4.8}$$

From (4.7) we have

$$\begin{aligned}\phi(\overline{t}(u+\overline{\sigma}\phi)) - \phi(u) &\leq \phi(\overline{t}(u+\overline{\sigma}\phi)) - \phi(\overline{t}u) \\ &= \int_0^{\overline{\sigma}} \phi'(\overline{t}(u+\sigma\phi))\phi d\sigma < 0,\end{aligned}$$

where we conclude

$$\phi(\overline{t}(u+\overline{\sigma}\phi)) < c = \inf_{u\in\mathcal{N}} \phi(u).$$

This, in turn, contradicts (4.8). □

4.5 Some Generalizations of Nehari's Method

In 2010, Szulkin and Weth [16] presented a unified approach to the method of Nehari manifold and illustrated it with a number of examples.

Theorem 4.3 [16] *Let ϕ be such that $\phi(0) = 0$ and $\phi = I_0 - I$ where I_0, I are C^1 functionals on X satisfying, for some $p > 1$:*

1. *$I'(u) = o(\|u\|^{p-1})$ as $u \to 0$.*
2. *$s \mapsto \frac{I'(su)}{s^{p-1}}$ is strictly increasing in $(0, \infty)$ for every $u \neq 0$.*
3. *$\frac{I(su)}{s^p} \to \infty$ uniformly for u on weakly compact subsets of $X\backslash\{0\}$ as $s \to \infty$.*
4. *I' is completely continuous. i.e., if $u_n \rightharpoonup u$ in X then $I'(u_n) \to I'(u)$ in X'.*
5. *I_0 is weakly lower semicontinuous, positively homogeneous of degree p, i.e., $I_0(su) = s^p I_0(u)$, and satisfies*

$$C_0\|u\|^p \leq I_0(u) \leq C_0^{-1}\|u\|^p$$

and

$$\left(I_0'(v) - I_0'(w)\right)(v-w) \geq C_1\left(\|v\|^{p-1} - \|w\|^{p-1}\right)(\|v\| - \|w\|) \tag{4.1}$$

for some constants $C_0, C_1 > 0$ and every $u, v \in X$.

Then the equation $\phi'(u) = 0$ has a ground state solution. Moreover, if ϕ is even then this equation has infinitely many pairs of solutions.

Some problems solved by Szulkin and Weth's abstract result:

$$\begin{cases} -\Delta u = f(u) & \text{in } \Omega, \\ u = 0 & \text{on } \partial\Omega, \end{cases} \tag{4.2}$$

where $I_0(u) = \frac{1}{2}\|u\|^2$ and $I(u) = \int_\Omega F(u)dx$,

$$\begin{cases} -\Delta_p u = f(u) & \text{in } \Omega, \\ u = 0 & \text{on } \partial\Omega, \end{cases} \tag{4.3}$$

where $I_0(u) = \frac{1}{p}\|u\|^p$ and $I(u) = \int_\Omega F(u)dx$.

$$\begin{cases} -\Delta u + V(x)u = f(u) & \text{in } \mathbb{R}^N, \\ u \in H^1(\mathbb{R}^N), \end{cases} \tag{4.4}$$

where $I_0(u) = \frac{1}{2}\int_{\mathbb{R}^N}\left(|\nabla u|^2 + V(x)u^2\right)dx$ and $I(u) = \int_{\mathbb{R}^N} F(u)dx$.

Moreover, they did not need to make customary assumptions which imply that $\phi \in C^2$ and $\phi''(u)uu < 0$ in $\mathcal{N}$. They removed these conditions using that the Nehari manifold is homeomorphic to Sphere. They also treated so-called indefinite problems in which 0 is a saddle point rather than a local minimum of ϕ using the generalized Nehari manifold $\mathcal{N}$ which has been introduced by Pankov.

In 2016, Figueiredo and Ramos [6] improved the Szulkin and Weth's abstract result for functionals of the form $\phi = I_0 - I$ in which I_0 is not necessarily homogeneous. More precisely they prove the following.

Theorem 4.4 [6] *Let X be a reflexive Banach space such that $\|\cdot\|$ is a C^1 functional on $X\backslash\{0\}$ and $\phi : X \to \mathbb{R}$ be a C^1 functional such that $\phi(0) = 0$. In addition, we assume that there exist $p, r > 1$ such that*

1. $\liminf_{u\to 0} \dfrac{\phi'(u)u}{\|u\|^r} > 0.$
2. *For every $u \in X$ we have $\phi(u) \geq C_0\|u\|^r - I(u)$ where $C_0 > 0$ and I is a weakly continuous functional on X.*

3. $\lim_{t\to\infty}\frac{\phi(tu)}{t^p} = -\infty$ *uniformly for u on weakly compact subsets of* $X\backslash\{0\}$.

4. *For every* $u \in X\backslash\{0\}$ *the maps* $t \mapsto \frac{\phi'(tu)u}{t^{p-1}}$ *and* $t \mapsto \phi(tu) - \frac{1}{p}\phi'(tu)tu$ *are decreasing and increasing on* $(0, \infty)$, *respectively.*

5. $u \mapsto \phi'(u)u$ *and* $u \mapsto \phi(u) - \frac{1}{p}\phi'(u)u$ *are weakly lower semicontinuous on* X.

Then $c := \inf_{\mathcal{N}} \phi$ *is positive and achieved by some* $u_0 \neq 0$, *i.e.,* ϕ *has a nontrivial ground state at a positive level. If, in addition,* ϕ *is even then we may choose* $u_0 \geq 0$.

They also illustrated it with a number of examples where it can be applied in order to show the existence of solutions to nonlinear boundary value problems. The principal difference between the two works was that Figueiredo and Ramos set the hypotheses on the functional ϕ and Szulkin and Weth in I_0 and I. Some problems solved by Figueiredo and Ramos's abstract result.

The first is a problem that involves the classical (p, q) operator

$$\begin{cases} -\Delta_p u - \Delta_q u = f(u) \ \text{ in } \Omega, \\ u = 0 \ \text{ on } \partial\Omega, \end{cases} \tag{4.5}$$

where $I_0(u) = \frac{1}{p}\|u\|^p + \frac{1}{q}\|u\|^q$ and $I(u) = \int_\Omega F(u)dx$.

The second problem involves the Kirchhoff operator

$$\begin{cases} -\left(a + b\int_\Omega |\nabla u|^2\right)\Delta u = f(u) \ \text{ in } \Omega, \\ u = 0 \ \text{ on } \partial\Omega, \end{cases} \tag{4.6}$$

where $I_0(u) = \frac{1}{2}a\|u\|^2 + \frac{1}{4}b\|u\|^4$ and $I(u) = \int_\Omega F(u)dx$.

Finally, the authors studied a problem that involves the anisotropic operator

$$\begin{cases} -\sum_{i=1}^{N} \partial_i\left(|\partial_i u|^{p_i-2}\partial_i u\right) = f(u) \ \text{ in } \Omega, \\ u \in \mathcal{D}_0^{1,\vec{p}}(\Omega) \end{cases} \tag{4.7}$$

where $I_0(u) = \sum_{i=1}^{N}\frac{1}{p_i}\int_\Omega |\partial_i u|^{p_i}$ and $I(u) = \int_\Omega F(u)dx$.

In 2018 Figueiredo and Pimenta [8] went back to Figueiredo and Ramos's abstract result and removed the differentiability of the functional ϕ and the reflexivity of the Banach space.

Theorem 4.5 [8] *Let E, F be Banach spaces, E compactly embedded into F and such that for all bounded sequence $(u_n) \subset E$ such that $u_n \to u$ in F, it holds that $u \in E$. Let $\phi, I_0 : E \to \mathbb{R}$, $I : F \to \mathbb{R}$ be functionals such that $\phi = I_0 - I|_E$, where I_0 is locally Lipschitz-continuous, $I \in C^1(E) \cap C^0(F)$, $I(0) = 0$ and for all $u \in E$, there exists the following limit:*

$$I_0'(su)u := \lim_{t \to 0} \frac{I_0(su + tu) - I_0(su)}{t}, \quad \forall s \in \mathbb{R}.$$

Suppose also the following conditions are satisfied:

(*i*) *I_0 is lower semicontinuous in the topology of F, i.e., if (u_n) is a bounded sequence in E, and $u \in E$ are such that, up to a subsequence, $u_n \to u$ in F, then $I_0(u) \leq \liminf_{n \to \infty} I_0(u_n)$.*

(*ii*) *There exist $\rho, \alpha_0 > 0$, such that $\phi(u) \geq \alpha_0 > \phi(0)$, for every $u \in E$ with $\|u\| = \rho$.*

(*iii*) *For all $u \in E$, $\phi(u) \geq \|u\| - I(u)$.*

(*iv*) *$t \mapsto \phi'(tu)u$ is strictly decreasing in $(0, +\infty)$.*

(*v*) *For each bounded sequence $(v_n) \subset E$ such that $v_n \to v \neq 0$ in F, it follows that*

$$\lim_{t \to \infty} \frac{\phi(tv_n)}{t} = -\infty, \quad \textit{uniformly for } n \in \mathbb{N}.$$

Then, the infimum of ϕ on the following set

$$\mathcal{N} := \{u \in E \backslash \{0\};\ I_0'(u)u = I'(u)u\}$$

is achieved.

In this paper, ϕ is locally Lipschitz and defined in BV, the space of functions of bounded variation, which is not reflexive. In order to show that the minimizer of ϕ on the Nehari manifold is a critical point of ϕ in whole space, the authors proved a version of Deformation lemma for locally Lipschitz functionals.

Theorem 4.6 [8] *Suppose all conditions of Theorem 4.5 hold, then if $u_0 \in \mathcal{N}$ is such that $\phi(u_0) = \min_{\mathcal{N}} \phi$, then u_0 is a critical point of ϕ in E.*

Proof Suppose by contradiction that $0 \notin \partial\phi(u_0)$, then $\beta(u_0) > 0$, where $\beta(u_0) = \inf\{\|z\|_*;\ z \in \partial\phi(u_0)\}$, since by [3][p. 105], such infimum is attained, where the dual norm and the subdifferential of ϕ_0 are taken with respect to E. Since $u \mapsto \beta(u)$ is lower semicontinuous (again by [3, p. 105] for instance), it follows that there exists $\kappa > 0$ such that

$$\beta(u) > \frac{\beta(u_0)}{2} > 0, \quad \forall u \in B_\kappa(u_0). \tag{4.8}$$

Let us denote $J = \left[1 - \frac{\kappa}{4}, 1 + \frac{\kappa}{4}\right] \subset \mathbb{R}$ and define $g : J \to E$ by

$$g(t) = tu_0.$$

Denoting $c := \phi(u_0) = \inf_{\mathcal{N}} \phi$, note that

$$\phi(g(t)) < c, \quad \forall t \neq 1.$$

Moreover, note that

$$\max\left\{\phi\left(g\left(1 - \frac{\kappa}{4}\right)\right), \phi\left(g\left(1 + \frac{\kappa}{4}\right)\right)\right\} = c_0 < c.$$

By using the version of the Deformation Lemma to locally Lipschitz functionals, without the Palais–Smale condition (see Lemma 2.3 of [4]), there exists $\varepsilon > 0$ such that

$$\varepsilon < \varepsilon_0 := \min\left\{\frac{c - c_0}{2}, \frac{\kappa\beta(u_0)}{16}\right\},$$

and a homeomorphism $\eta : E \to E$ such that

(*i*) $\eta(x) = x$ for all $x \notin \phi^{-1}([c - \varepsilon_0, c + \varepsilon_0]) \cap B_\kappa(u_0)$;

(*ii*) $\eta(\phi_{c+\varepsilon} \cap B_{\kappa/2}(u_0)) \subset \phi_{c-\varepsilon}$;

(*iii*) $\phi(\eta(x)) \le \phi(x)$, for all $x \in E$,

where, for $d \in \mathbb{R}$, $\phi_d = \{u \in E;\ \phi(u) \le d\}$.

Let us define now $h : J \to E$ by $h(t) = \eta(g(t))$ and two functions, $\Psi_0, \Psi_1 : J \to \mathbb{R}$ by

$$\Psi_0(t) = \phi'(tu_0)u_0$$

and

$$\Psi_1(t) = \frac{1}{t}\phi'(h(t))h(t).$$

Since for $t \in \left\{\left(1 - \frac{\kappa}{4}\right), \left(1 + \frac{\kappa}{4}\right)\right\}$, $\phi(g(t)) \le c_0 < c - \varepsilon_0$, then $h(t) = \eta(g(t)) = g(t) = tu_0$ for $t \in \left\{\left(1 - \frac{\kappa}{4}\right), \left(1 + \frac{\kappa}{4}\right)\right\}$. Hence

$$\Psi_0(t) = \Psi_1(t), \quad \forall t \in \left\{\left(1 - \frac{\kappa}{4}\right), \left(1 + \frac{\kappa}{4}\right)\right\}. \tag{4.9}$$

Let us briefly justify that the Brouwer Degree $d(\Psi_0, J, 0)$ is equal to -1. In fact, by the proof of Theorem 4.5, it follows that $\Psi_0(t) = \gamma'_{u_0}(t)$ is continuous, strictly decreasing, $\Psi_0(1-\kappa/4) > 0 > \Psi_0(1+\kappa/4)$, and $t=1$ is the unique point in J where Ψ_0 vanishes. Since there is no information about the smoothness of Ψ_0, in order to calculate $d(\Psi_0, J, 0)$, we should approximate Ψ_0 by some $C^1(J, \mathbb{R})$ function. By considering a strictly decreasing function $\overline{\Psi} \in C^1(J, \mathbb{R})$, such that $\overline{\Psi}(t-\kappa/4) = \Psi_0(t-\kappa/4)$, $\overline{\Psi}(t+\kappa/4) = \Psi_0(t+\kappa/4)$ and $\overline{\Psi}(1) = 0$, note that by the very definition of Brouwer Degree and by its invariance by homotopy, $d(\Psi_0, J, 0) = d(\overline{\Psi}, J, 0) = \mathrm{sgn}(\overline{\Psi}'(1)) = -1$.

Since $\Psi_0 = \Psi_1$ on ∂J and $d(\Psi_0, J, 0) = -1$, by degree theory, $d(\Psi_1, J, 0) = -1$. Then there exists $t \in J$ such that $h(t) \in \mathcal{N}$. This implies that

$$c \le \phi(h(t)) = \phi(\eta(g(t))).$$

But note that $\phi(g(t)) < c + \varepsilon$ and also $g(J) \subset B_{\kappa/2}(u_0)$. Then, by (ii)

$$\phi(\eta(g(t))) < c - \varepsilon,$$

which contradicts the last inequality. Then the result follows. □

Some problems solved by Figueiredo and Pimenta's abstract result: The first problem involves the 1−Laplacian operator. More precisely

$$\begin{cases} -\Delta_1 u = f(u) & \text{in } \Omega, \\ u = 0 & \text{on } \partial\Omega, \end{cases} \tag{4.10}$$

where $\Delta_1 u = \nabla\left(\frac{\nabla u}{|\nabla u|}\right)$ and can be seen as the limit case as $p \to 1$ of the p−Laplacian operator $\Delta_p u := \nabla\left(|\nabla u|^{p-2}\nabla u\right)$. The functional energy is given by $\phi : BV(\Omega) \to \mathbb{R}$, such that $\phi(u) = I_0(u) - I(u)$, where $I_0(u) = \int_\Omega |\nabla u| dx$, $\int_\Omega |\nabla u| dx$ is the total variation of the Radon measure ∇u and $BV(\Omega) = \{u \in L^1(\Omega) : \nabla u \in \mathcal{M}(\Omega)\}$.

The second problem involves the 1-biharmonic operator given by

$$\begin{cases} \Delta_1^2 u = f(u) \text{ in } \Omega, \\ u = \dfrac{\Delta u}{|\Delta u|} = 0 \text{ on } \partial\Omega, \end{cases} \tag{4.11}$$

where

$$\Delta_1^2 u = \Delta\left(\frac{\Delta u}{|\Delta u|}\right)$$

and can be seen as the limit case as $p \to 1$ of the p−biharmonic operator

$$\Delta_p^2 u := \Delta\left(|\Delta u|^{p-2}\Delta u\right).$$

The energy functional is given by $\phi : BL_0(\Omega) \to \mathbb{R}$, such that $\phi(u) = I_0(u) - I(u)$, where $I_0(u) = \int_\Omega |\Delta u| dx$, $\int_\Omega |\Delta u| dx$ is the total variation of the Radon measure Δu and $BL_0(\Omega) = \left\{u \in W_0^{1,1}(\Omega) : \Delta u \in \mathcal{M}(\Omega)\right\}$.

In 2020 Figueiredo and Pimenta go back to study problems involving 1−Laplacian operator and the Nehari method. In [9] they show existence of nodal solutions for two classes of problems, using two different arguments. One in $\mathbb{R}^N$ and another in bounded smooth domain.

The purpose of the first part of such work is to prove the existence of sign-changing solutions to a problem involving the 1−Laplacian operator, based on the analysis of the *Nehari set* associated with it, namely

$$-\text{div}\left(\frac{Du}{|Du|}\right) + \frac{u}{|u|} = f(u) \text{ in } \mathbb{R}^N. \tag{4.12}$$

They minimize the functional associated in the Nehari Nodal $\mathcal{M}$, which is a subset of the Nehari set such that the positive part and negative part of points of the elements in $\mathcal{N}$are nontrival and also are in the Nehari set $\mathcal{N}$.

In the second part of the paper, they show existence of nodal solution for this problem

$$\begin{cases} -\text{div}\left(\dfrac{Du}{|Du|}\right) = f(u) \text{ in } \Omega, \\ \qquad\qquad u = 0 \text{ on } \partial\Omega, \end{cases} \tag{4.13}$$

where f has a superlinear and subcritical growth at infinity. Their approach is based on an approximation of (4.13) by similar problems involving the p−Laplacian operator.

As has been repeated many times in this text, the main property that a functional must have in order for Nehari's method to be applied is

(H1) The map $t \mapsto \phi(tu)$, defined in $(0, \infty)$, has a unique critical point t_u, which is a global maximum point (in particular $\phi(t_u u) > 0$).

Property (H1) shows that the projection on the Nehari manifold is unique. This scenario typically arises in superlinear elliptic problems, as the prototype

$$-\Delta u = b(x)|u|^{r-2}u, \quad u \in H_0^1(\Omega), \tag{4.14}$$

where $\Omega \subset \mathbb{R}^N$ is a bounded domain, $r \in (2, 2^*)$, and b is a smooth coefficient such that $b(x) > 0$ for all $x \in \Omega$. This problem corresponds to a Lagrange equation for the functional given by

$$\phi(u) = \frac{1}{2}\int_\Omega |\nabla u|^2 - \frac{1}{r}\int_\Omega b(x)|u|^r, \quad u \in H_0^1(\Omega). \tag{4.15}$$

In [7] Figueiredo, Ramos Quorin and Silva studied situations in which the property (H1) is not true.

The main purpose of this article is to set up a framework for functionals that do not have a uniform geometry on X, in the sense that the map $t \mapsto \phi(tu)$ has not the same behavior for all $u \neq 0$. This is the case if (H1) does not hold in the whole of $X\backslash\{0\}$, but rather in some part of it. A simple example is given by (4.15) with b vanishing in some part of Ω or changing sign, in which case (H1) holds only for those u such that $\int_\Omega b(x)|u|^r dx > 0$. So it is reasonable to deal with functionals satisfying (H1) in an open cone of X, i.e., an open set $Y \subset X$ such that $tu \in Y$ for every $t > 0$ and $u \in Y$. Note that the condition $\int_\Omega b(x)|u|^r > 0$ clearly defines an open cone in X.

Furthermore, the authors consider other geometries than (H1). More concretely, we aim at functionals such that the map $t \mapsto \phi(tu)$ has at least a local minimum point whenever u lies in an open cone. This condition arises in several problems, as we shall see in the applications studied by the authors.

Overall they provide conditions ensuring that the infimum of ϕ over $\mathcal{N} \cap Y$ is achieved by a critical point of ϕ, which can then be considered as a *ground state relative to* Y, in the sense that it has the least energy among critical points in Y. Some further conditions on ϕ over $X\backslash Y$ shall imply that c is the ground state of ϕ.

4.6 Thank You so Much, Zeev Nehari

In this year of 2025, when Nehari's method is 63 years old, this author would like to say a big thank you to Zeev Nehari for his wonderful results, and for his career as supervisor of doctoral theses. We hope that these results and his life example continue to inspire a generation of young mathematicians.

References

1. Ambrosetti, A., Rabinowitz, P.: Dual variational methods in critical point theory and applications. J. Funct. Anal. **14**, 349–381 (1973)
2. Bonheure, D., dos Santos, E.M., Parini, E., Tavares, H., Weth, T.: Nodal solutions for sublinear-type problems with Dirichlet boundary conditions. Int. Math. Res. Not. **5**, 3760–3804 (2022)
3. Chang, K.: Variational methods for non-differentiable functionals and their applications to partial differential equations. J. Math. Anal. Appl. **80**, 102–129 (1981)
4. Dai, G.: Non-smooth version of fountain theorem and its application to a Dirichlet-type differential inclusion problem. Nonlinear Anal. **72**, 1454–1461 (2010)
5. Cingolani, S., Weth, T.: On the planar Schrödinger-Poisson system. Ann. I. Poincaré-AN **33**(1), 169–197 (2016)
6. Figueiredo, G.M., Ramos, H.Q.: Ground states of elliptic problems involving non homogeneous operators. Indiana Univ. Math. J. **65**, 779–795 (2016)
7. Figueiredo, G.M., Ramos, H.Q., Silva, K.: Ground states of elliptic problems over cones. Calc. Var. Part. Differ. Equ. **60**, 189 (2021)
8. Figueiredo, G.M., Pimenta, M.T.O.: Nehari method for locally Lischitz functionals with examples in problems in the space of bounded variation function. NoDEA **25**, 47 (2018)

9. Figueiredo, G.M., Pimenta, M.T.O.: Nodal solutions to quasilinear elliptic problems involving the 1-Laplacian operator via variational and approximation methods. Indiana J. Math **71**(2), 439–462 (2022)
10. Liu, Jia-quan, Wang, Ya.-qi, Wang, Zhi-Qiang.: Solutions for quasilinear equations via the Nehari method. Comm. in PDE **29**(5–6), 879–901 (2004)
11. Nehari, Z.: On a class of nonlinear second-order differential equations. Trans. Am. Math. Soc. **95**, 101–123 (1960)
12. Nehari, Z.: Characteristic values associated with a class of non-linear second-order dfferential equations. Acta Math. **105**, 141–175 (1961)
13. Pohozaev, S.I.: The fibration method for solving nonlinear boundary value problems. Trudy Mat. Inst. Steklov. **192**, 146–163 (1990). Translated
14. Rabinowitz, P.H.: On a class of nonlinear Schrödinger equations. Z. Angew Math. Phys. **43**, 270–291 (1992)
15. Rădulescu, V.D.: Mountain pass theorems for non-differentiable functions and applications. Proc. Japan. Acad. Ser. A **69**, 193–198 (1993)
16. Szulkin, A., Weth, T.: The method of Nehari manifold. In: Gao, D.Y., Montreanu, D. (eds.) Handbook of Nonconvex Analysis and Applications, pp. 597–632. International Press, Boston (2010)
17. Willem, M.: Minimax Theorems. Birkhäuser (1996)

Chapter 5
Fractional Constant Exponent or Variable-Order and Variable Exponent Problems

Jiabin Zuo

Abstract We discuss relatively complex and representative nonlinear fractional p-Laplace problems, in the setting of constant exponents and Kirchhoff terms. Two existence theorems of suitable solutions for a model problem with critical exponent are presented on the whole space $\mathbb{R}^n$. The Kirchhoff function of this problem may be zero at zero, and the lack of compactness in the critical case is the main feature and difficulty to overcome. Then, we deal with a critical fractional variable-order $p(\cdot)$-Kirchhoff type problem. We consider both the existence and asymptotic behavior of solution. The result is established combining appropriate continuous embeddings with the Palais–Smale property and a Brézis–Lieb type lemma. A dedicated framework setting is reviewed for a double nonlocal model problem, featured by two variable parameters. A particular attention is devoted to special continuous and compact embeddings about fractional Sobolev spaces with variable-order and variable exponent.

5.1 Research Background and Significance

Partial differential equation (PDE for short) is a very rich and ancient subject; its influence is extensive and profound. It can be used to describe various phenomena, such as acoustic, thermal, electrostatic, fluid dynamics, elastic mechanics, quantum mechanics and so on. Just as ordinary differential equations often simulate one-dimensional dynamical systems, PDEs often simulate multidimensional systems, meaning that these apparently distinct physical phenomena can be similarly formalized by appropriate PDEs.

The part of book is supported by the Guangdong Basic and Applied Basic Research Foundation (2022A1515110907).

J. Zuo (✉)
School of Mathematics and Information Science, Guangzhou University, Guangzhou, China
e-mail: zuojiabin88@163.com

M. Jleli et al. (eds.), *Advances in Nonlinear Evolution Equations*, Infosys Science Foundation Series, https://doi.org/10.1007/978-981-97-9812-4_5

PDEs can be divided into elliptic, parabolic and hyperbolic types. The Laplace equation, also known as Poisson equation, is the simplest elliptic partial differential equation, which is the most concerned one in our subject. Laplace operators are the basis of linear PDEs and linear theory in physics. However, as time goes by, at the end of the twentieth century, the study of nonlinear boundary value problems became extremely active, which attracted the attention of many scholars, and the p-Laplace equation became a typical representative of nonlinear elliptic equations.

On the one hand, we know that fractional derivatives with single variable go back to the origins of calculus, proposed in the nineteenth century by Liouville and Riemann. However, with the development of fraction-order linear and nonlinear equations with multiple variables, a prototype of fraction-order Laplace operator has appeared in recent decades. In the rich mathematical content, there are generally several ways to define it, including Fourier analysis, functional calculus, and singular integral. At the same time, there is a very interesting generalization in the theory of fractional calculus, which allows the fractional order to be variable; obviously this variable-order fractional calculus method is more obviously nonlocal, and has a wide range of applications in mechanics, control and signal processing. Therefore, by combining the above two aspects, a new kind of operator that is the fractional variable-order Laplace operator appears.

On the other hand, the variable-exponent problem in nonlinear elliptic equations reflects the physical phenomenon of "point-by-point anisotropy", and it was first produced in the model of electrorheology [39]. Therefore, in 1991 two distinguished mathematicians Kováčik and Rákosník gave some basic properties of variable exponent Lebesgue and Sobolev spaces in [40]. Subsequently, the space of variable exponents flourished, and Fan et al. [27] obtained some very important embedding theorems in variable exponent Lebesgue and Sobolev spaces; Fu and his team have done a lot of research on a class of $p(x)$-Laplace problems [17, 32–35, 70], and achieved fruitful results. More than ten years later, several mathematicians such as Rossi [36] extended integer order to the fractional order case, which laid a solid theoretical foundation for the study of fractional partial differential equations with variable exponent growth. When the $p(x)$-Laplace operator is replaced by the fractional $p(x)$-Laplace operator, it is still a natural and open question to see which results can be preserved. The change of this idea from integer to fractional order not only enables us to understand classical problems more deeply, but also provides a powerful support for solving complex problems in the real world.

Because many scholars have paid much attention to fractional elliptic partial differential equations and variational problems, most of the nonlinear functional analysis techniques, such as iterative methods, topological degree method, monotonicity methods and variational methods, and so on, have been adopted and adapted on these settings to improve performance in applications. Among them, the modern variational theory represented by minimax method [11], Morse theory [52], Lusternik–Schnirelmann theory [48, 49] is the most typical. It overcomes the classical variational theory of finding the extremum point of the functional and solves the existence problem of solutions to nonlinear partial differential equations in a larger scope. In the 1960s, Palais and Smale introduced the compact condition of the functional (PS condition for short), to make up for the lack of local compactness in work space. Their work opened the way for the modern critical point theory, followed by Ambrosetti and Rabinowitz's famous mountain pass lemma [3] in 1973. Twenty years later, the fountain theorem proposed by Bartsch [9], which is an extension of the symmetric mountain pass lemma, neatly solves the problem of infinite solutions of partial differential equations. Since then, many mathematicians have continuously enriched and improved the existing variational theory, producing a series of various forms of critical point theorems and variational methods.

5.2 Research Status and Progress

It is well known that the class of nonlocal problems is a hot issue with a very wide range of applications in mathematics, finance, physics, chemistry and other fields.

In the last few decades, there has been great interest in nonlocal equations such as the Schrodinger Poisson system [14, 16] Among them, a typical nonlocal equation that attracts much attention is the Kirchhoff type equation.

A physical model given by Kirchhoff [38] in observing the change in the length of a string:

$$\rho\frac{\partial^2 v}{\partial t^2}-\left(\frac{P_0}{h}+\frac{E}{2L}\int_0^L|\frac{\partial v}{\partial x}|^2dx\right)\frac{\partial^2 v}{\partial x^2}=0, \tag{5.1}$$

where v represents the displacement of the string, ρ is the mass density, P_0 is the initial tension, h is the area of the cross section, E is the Young modulus of the material, and L is the length of the string. For more details and physical phenomena described by the classical Kirchhoff theory, see [63].

Equation (5.1) is a generalization of the famous D'Alembert equation. Because it has interesting analytic structure and the existence of nonlocal Kirchhoff term $\frac{P_0}{h}+\frac{E}{2L}\int_0^L|\frac{\partial v}{\partial x}|^2dx$, this kind of equation is no longer the point identity, which makes our research more complex and challenging. In early times, Lions [43] gave the abstract functional analysis framework according to the model (5.1). Since then, the Kirchhoff function prototype ($M(t)=a+b\gamma t^{\gamma-1}$ for all $t\geq 0$, some $a,b\geq 0$, $a+b>0$, $\gamma\geq 1$) tends to be monotone. But because of the urgent need of practical problems, Kirchhoff functions are usually assumed to be Lipschitz-continuous, not always monotone. Here we list three important works in this regard [24, 51, 69]. Kirchhoff function M represents the change of tension caused by the change of string length in the process of vibration. When the change of tension in motion cannot be ignored, the nonlinear coefficient M needs to be considered. However, from a physical point of view, $M(0)=0$ means that the string is considered to be zero; hence for the initial tension in (5.1) we get $\frac{P_0}{h}=0$. In this case, it is called the degenerate Kirchhoff problem, namely $M(0)=0$ and for all $t>0$ we have $M(t)>0$.

After the appearance of the model (5.1), some physicists have considered such equations in the study of nonlinear vibrations both theoretically and experimentally [53, 54]. The state Kirchhoff type equation (without time t) and its revised form have been widely used to analyze the transverse vibration of strings, although it has very limited application on stationary strings. For more information about classical Laplace operator equation, we mention the works [8, 28, 42, 45]; for some papers concerning the p-Laplace operator we point out the works [12, 19, 23, 44].

In 2014, Fiscella and Valdinoci [31] provided the first example for constructing a steady state Kirchhoff function model containing the generalized differential integral operator and critical nonlinear terms when the length of the string was a fraction. We have

$$\begin{aligned} -\mathcal{M}\left(\iint_{\mathbb{R}^{2n}} |v(x)-v(y)|^2 K(x-y)dxdy\right)\mathcal{L}_K v \\ = \mu g(x,v) + |v|^{2_s^*-2}v,\ x\in\Omega, \\ v\Big|_{x\in\mathbb{R}^n\setminus\Omega} = 0, \end{aligned} \tag{5.2}$$

where for all $t \geq 0$ we have $\mathcal{M} = a + bt$, with $a > 0, b \geq 0$. Applying the mountain pass lemma, the authors prove that there is at least one nontrivial weak solution to the problem (5.2) and show the asymptotic behavior of this solution. Here they assume that the Kirchhoff function is non-degenerate. In the case when $a = 0$ and $b > 0$, then the Kirchhoff function is said to be degenerate. Readers can refer to [15, 56, 73] for non-degenerate Kirchhoff problems and [30, 58] for the latest progress in this field.

On the other hand, considering the addition of non-homogeneous perturbation term, Chen et al. [18] investigated the case of multiple solutions of a Dirichlet boundary value problem by using the mountain pass lemma and Ekeland's variational principle on a bound domain. In addition, taking the same approach, in the whole space, Xiang et al. [67] have given multiple solutions of a non-homogeneous fractional p-Kirchhoff type equation with concave–convex nonlinear terms. Later, Pucci et al. [57] established the Batsch–Wang compact embedding theorem in fractional Sobolev space. The existence and multiplicity of solutions of the nonhomogeneous fractional Schrödinger Kirchhoff equation with the p-Laplace operator are discussed.

Following the above arguments, in this chapter, we first focus on the nonhomogeneous fractional p-Kirchhoff type equation with critical nonlinear terms given by

$$\begin{aligned} \mathcal{M}\left(\iint_{\mathbb{R}^{2n}} |v(x)-v(y)|^p K(x-y)dxdy\right)\mathcal{L}_K^p \\ = \mu g(x)|v|^{q-2}v + |v|^{p_s^*-2}v + \mu f(x),\quad x\in\mathbb{R}^n, \end{aligned} \tag{5.3}$$

where $n > ps$, $p > 1$ and $s \in (0,1)$, $p_s^* = \frac{np}{n-ps}$ is the fractional critical Sobolev exponent to p, $\mathcal{M} : \mathbb{R}_0^+ \to \mathbb{R}_0^+$ is a continuous function (Kirchhoff type term), and the leading differential operator is denoted by $\mathcal{L}_K^p$, namely the nonlocal p-fractional type operator given by

$$\mathcal{L}_K^p\phi(x) = 2\lim_{\varepsilon\searrow 0}\int_{\mathbb{R}^n\setminus B_\varepsilon(x)} |\phi(x)-\phi(y)|^{p-2}(\phi(x)-\phi(y))K(x-y)dy,\ x\in\mathbb{R}^n. \tag{5.4}$$

Here, $\phi \in C_0^\infty(\mathbb{R}^n)$, $B_\varepsilon(x)$ means open ball in $\mathbb{R}^n$ of radius $\varepsilon > 0$ and center $x \in \mathbb{R}^n$. Also, by $K : \mathbb{R}^n\setminus\{0\} \to (0,\infty)$ we denote a suitable measurable function. On the right-hand side of the equation, for a positive parameter μ, we consider the competing effects of three terms, a parametric weighted term $\mu g(x)|v|^{q-2}v$ with $g(x) > 0$ on $\mathbb{R}^n$, a critical term $|v|^{p_s^*-2}v$ and a second parametric term $\mu f(x)$, where $f(x)$ may change sign on $\mathbb{R}^n$. We note the following assumptions:
(K_1) $\zeta K \in L^1(\mathbb{R}^n)$, where $\zeta(x) = \min\{|x|^p, 1\}$;
(K_2) for any $x \in \mathbb{R}^n\setminus\{0\}$, there exists $\kappa_0 > 0$, satisfying $K(x) \geq \kappa_0|x|^{-(n+ps)}$.

In the case when K is of standard type (namely $K(x) = |x|^{-(n+ps)}$), the problem (5.3) becomes of the form

$$\mathcal{M}\left(\iint_{\mathbb{R}^{2n}} \frac{|v(x)-v(y)|^p}{|x-y|^{n+ps}}dxdy\right)(-\Delta)_p^s v = \mu g(x)|v|^{q-2}v + |v|^{p_s^*-2}v + \mu f(x),$$

for $x \in \mathbb{R}^n$, where $(-\Delta)_p^s$ is a fractional p-Laplace operator defined by

$$(-\Delta)_p^s v(x) = 2\lim_{\varepsilon\to 0^+}\int_{\mathbb{R}^n\setminus B_\varepsilon(x)} \frac{|v(x)-v(y)|^{p-2}(v(x)-v(y))}{|x-y|^{n+ps}}dy,\ x\in\mathbb{R}^n. \tag{5.5}$$

Since the critical level value is assumed to be less than zero when the (PS) condition is verified, one can obtain the negative energy solution of problem (5.3) by using the Ekeland variational principle and fixing the parameter μ within a certain threshold range. The main features of this model are in considering certain critical nonlinear terms and in adopting a weighted approach to overcome the lack of embedding compactness in the whole space, which is not reflected, for example, in [67]. We note that the above considered Kirchhoff function $\mathcal{M}$ covers also the degenerate case.

Clearly, the natural setting to carry out the analysis of problem (5.3) is the fractional Sobolev space with constant value s. Hence, an interesting issue is to understand how a variable fractional order, namely $s(\cdot)$, impacts the solvability of the problem. In particular, the crucial topic is in confirming the well-known embedding theorems in the original Sobolev space. Consequently, the fractional variable-order Laplace problem is largely investigated from many points of view. The proposal of the variable-order fractional derivative can be traced back to 1993, when Samko and Ross [62] first proposed the fractional-order variable derivative in the nonlinear reaction diffusion process. Since the change in temperature can be better described by using the variable derivative of a nonlocal integral-differential operator, Lorenzo and

Hartley [47] apparently give several forms of it in an independent way. In addition, Kikuchi and Negoro [37] obtained the conditions under which pseudo-differential operators on variable-order fractional Sobolev spaces generate Feller semigroups with transition densities. Leopold [41] studied the embedding problem from variable-order differential function space to variable-order integral function space. It is worth noting that Ruiz-Medina et al. [60] also discuss isomorphism of their regenerated core Hilbert space as a random field class of fractional variable-order Sobolev space.

In 2019, Xiang and Zhang et al. studied an elliptic equation containing the fractional variable-order Laplace operator in [68] with the form:

$$\begin{aligned} &(-\Delta)^{s(\cdot)}u+\lambda V(x)u(x)=\alpha|u|^{p(x)-2}u+\beta|u|^{q(x)-2}u,\ x\in\Omega,\\ &u\Big|_{x\in\mathbb{R}^N\setminus\Omega}=0, \end{aligned}$$

where $(-\Delta)^{s(\cdot)}$ is the fractional variable-order Laplace operator, which is defined as

$$(-\Delta)^{s(\cdot)}u(x)=2P.V.\int_{\mathbb{R}^N}\frac{u(x)-u(y)}{|x-y|^{N+2s(x,y)}}dy,\ \forall u\in C_0^\infty(\mathbb{R}^N),x\in\mathbb{R}^N, \quad (5.6)$$

where $P.V.$ denotes the Cauchy principal value, $V(x)$ is a continuous potential function, and the arguments λ, α, β are all non-negative. Firstly, the authors establish the relevant continuous embedding theorems and compact embedding theorems in fractional variable-order Sobolev space. Then, by combining the classical mountain pass lemma and Ekeland's variational principle, two different non-trivial weak solutions of the above problems are given, and the asymptotic behavior of the two solutions is illustrated. Finally, when $\lambda=0$, the existence of infinitely many solutions to this problem is discussed by means of the symmetric mountain pass lemma. We note that the above problem contains concave–convex nonlinear terms, which puts forward a higher requirement for proving the boundedness of (PS) sequences. However, the usual methods to deal with such problems are Nehari manifold and fibering mapping. The basic idea of this method is that when the functional is not bounded below in the whole space, the critical point corresponding to the functional is found on a subset of the space. Reference [65] used this method to solve this kind of fractional variable-order problem with degenerate Kirchhoff functions, and they also obtained two different non-negative solutions and an infinite number of solutions to this kind of problem. In addition, the existence of infinitely many solutions to this class of Kirchhoff problems is proved again by Ref. [64] using four kinds of different critical point theorems.

Considering the problem mentioned above (see [64, 65, 68]), we point out that the exponent of the nonlinear term at the right end of the equation is variable growth, and such problems with non-standard growth conditions can be generically referred to as variable exponent problems. Now, the $p(x)$-Laplace equation is the most representative of nonlinear elliptic equations with variable exponent, which is obviously more complex than the p-Laplace equation and describes a variety of mathematical and physical phenomena. For example, Chen et al. [20] proposed a framework for solving the image restoration problem based on a $p(x)$-Laplace operator. Another major application relates to the modeling of electrorheological fluids; see [25, 39]. In addition, the literature on the problem of variable exponent is very rich. Here, we only list a few of the papers involving the existence and multiplicity of solutions to $p(x)$-Laplace equation, namely [2, 46, 50, 59], and the monograph [61].

It is natural for people to think about any difficulties that the fractional case of the $p(x)$-Laplace equation will bring us. What good results can be retained? These issues have been the focus of attention in this field in recent years. Indeed, once again, Rossi [36] and some Italian mathematical researchers first extended variable exponent spaces to the fractional case, and proved a compact embedding theorem from fractional variable exponent Sobolev space to variable exponent Lebesgue spaces. As an application, they proved the existence and uniqueness of a nonlocal solution to a fractional $p(x)$-Laplace equation. Subsequently, Rădulescu and Bahrouni [7] give some qualitative properties of fractional exponential Sobolev space, such as reflexivity, separability and density. The abstract results established by them can be applied to the variational analysis of nonlocal fractional problems with variable exponents. In addition, one can refer to the works [1, 5, 6], for more on the problem of fractional variable exponent.

An interesting fractional Kirchhoff type problem with variable exponent growth conditions to review, by considering the relevant literature on variable order and variable exponents, is given in the following form:

$$\begin{cases} M\left(\iint_{\mathbb{R}^{2N}} \frac{1}{p(x,y)} \frac{|v(x)-v(y)|^{p(x,y)}}{|x-y|^{N+p(x,y)s(x,y)}} dxdy\right)(-\Delta)^{s(\cdot)}_{p(\cdot)} v \\ \qquad = \mu |v|^{\overline{p}(x)-2} v + g(x,v),\ x \in \Omega, \\ v\Big|_{x\in\mathbb{R}^N\setminus\Omega} = 0, \end{cases} \tag{5.7}$$

where $\Omega \subset \mathbb{R}^N$ is a bounded and smooth with $N > p(x,y)s(x,y)$ for any $(x,y) \in \overline{\Omega} \times \overline{\Omega}$, where μ is a real parameter, g is a subcritical continuous function, M is a Kirchhoff type function, $s(\cdot) : \mathbb{R}^N \times \mathbb{R}^N \to (0,1)$ and $p(\cdot) : \mathbb{R}^N \times \mathbb{R}^N \to (1,\infty)$, set $\overline{p}(x) = p(x,x)$ for $x \in \overline{\Omega}$. Here, $(-\Delta)^{s(\cdot)}_{p(\cdot)}$ is a $p(\cdot)$-Laplace operator with fractional variable $s(\cdot)$-order, which is given by

$$\begin{aligned} &(-\Delta)^{s(\cdot)}_{p(\cdot)} \varphi(x) \\ &= P.V. \int_{\mathbb{R}^N} \frac{|\varphi(x)-\varphi(y)|^{p(x,y)-2}(\varphi(x)-\varphi(y))}{|x-y|^{N+p(x,y)s(x,y)}} dy, \quad x \in \mathbb{R}^N. \end{aligned} \tag{5.8}$$

To be precise, the existence and multiplicity of the solution to problem (5.7) are obtained by using the mountain pass lemma and the fountain theorem under appropriate conditions in [72].

Obviously, the main feature and difficulty of problem (5.7) are in the existence of a double nonlocal term containing two variable parameters. Hence, a careful understanding of the framework space settings, together with the special skills associated with the problem' formulation, makes this research more challenging. In fact, according to the present literature, both fractional constant exponent problems and integer variable exponent problems have continuous embedding theorems in critical cases. However, from the point of view of the fractional variable-order problem and the fractional variable exponent problem, the changes and growth of each term are subcritical, and the critical case is not often mentioned. So, it is interesting to provide some information and details about the impact of the critical terms on the geometry of the problem. Namely, we consider the fractional variable-order Kirchhoff equation including the critical variable exponent growth given as

$$\begin{aligned} &M\left(\iint_{\mathbb{R}^{2N}} \frac{1}{p(x,y)} \frac{|v(x)-v(y)|^{p(x,y)}}{|x-y|^{N+p(x,y)s(x,y)}} dxdy\right) (-\Delta)_{p(\cdot)}^{s(\cdot)} v \\ &\qquad\qquad = \mu g(x,v) + |v|^{r(x)-2} v,\ x \in \Omega, \\ &v\|_{x\in\mathbb{R}^N\setminus\Omega} = 0, \end{aligned} \tag{5.9}$$

where g is a continuous subcritical function, μ is a positive parameter, and the variable index $r(x)$ can be close to the critical variable index $p_s^*(x) = N\overline{p}(x)/(N - \overline{s}(x)\overline{p}(x))$. By adopting a technique step analysis method suitable for critical mountain pass level, it is proved that the problem (5.9) has at least one nontrivial weak solution, and the asymptotic behavior of this weak solution is commented.

It is also pointed out that the continuous embedding theorem from fractional variable-order Sobolev space with variable exponent to variable exponent Lebesgue space is no longer valid due to the presence of variable critical exponent $r(x)$ in problem (5.9). To overcome the situation successfully, we can show how to establish the continuous embedding theorem from fractional variable-order Sobolev space with variable exponent to variable exponent Lebesgue space in the critical case, and combine it with a new Brézis–Lieb lemma. This strategy leads to conclude the existence and asymptotic behavior of the solution to problem (5.9).

5.3 Analysis of Problem (5.3)

In this section, we will review the existence of solutions to certain nonhomogeneous fractional p-Kirchhoff problems involving critical nonlinearities and weight terms. Precisely, we cover the finding of the work [74]; hence we discuss problem (5.3). For readers' convenience, we recall and fix the assumptions on the Kirchhoff function as follows:

(M_1) There exists $\theta \in \left(1, \frac{n}{n-ps}\right)$, such that

$$\theta\widetilde{M}(t) = \theta\int_0^t M(s)ds \geq M(t)t \text{ for all } t \in \mathbb{R}_0^+;$$

(M_2) for every $\sigma > 0$ there exists $m = m(\sigma) > 0$ such that $\mathcal{M}(t) \geq m$ for all $t \geq \sigma$;
(M_3) there exists a_0 such that $\mathcal{M}(t) \geq a_0 t^{\theta-1}$ for all $t \in [0, 1]$.

Now, (M_1) and (M_2) can be seen as appropriate structural conditions for dealing with Kirchhoff weight on a subcritical $p(x)$-polyharmonic setting. We mention the work of Colasuonno and Pucci [22], where the above assumptions are introduced. Similarly, one can find some comments and applications of assumption (M_3) in several critical fractional works as [10, 29, 30]. Hence, (M_3) is required to overcome the lack of compactness at critical level.

Furthermore, concerning the parametric terms, the positive weight term g and the sign-changing term f, we impose the following regularities:
(Z_1) $g \in L^{q_1}(\mathbb{R}^n) \cap L^\infty_{loc}(\mathbb{R}^n)$, $f \in L^\xi(\mathbb{R}^n)$, with $q_1 = \frac{p_s^*}{p_s^*-q}$, $\xi = \frac{p_s^*}{p_s^*-1}$.
(Z_2) $1 < q < p < \theta p < p_s^*$.

According to the classical literature, in [74] the authors use the following notation. Let $L^q(\mathbb{R}^n, g)$ be the weighted Lebesgue space, endowed with the norm

$$\|v\|_{q,g}^q = \int_{\mathbb{R}^n} g(x)|v(x)|^q dx.$$

The Banach space $L^q(\mathbb{R}^n, g) = (L^q(\mathbb{R}^n, g), \|\cdot\|_{q,g})$ is uniformly convex according to Proposition A.6 of [4]. By $W_K^{s,p}(\mathbb{R}^n)$ we denote the closure of $C_0^\infty(\mathbb{R}^n)$; hence $W_K^{s,p}(\mathbb{R}^n)$ is a Banach space which can be endowed with the norm, defined as

$$\|\phi\|_{s,p} = \left(\iint_{\mathbb{R}^{2n}} |\phi(x) - \phi(y)|^p K(x-y) dxdy\right)^{\frac{1}{p}}$$

for all $\phi \in C_0^\infty(\mathbb{R}^n)$. Moreover, it follows from Lemma 2.1 of [15] that the embedding $W_K^{s,p}(\mathbb{R}^n) \hookrightarrow L^q(\mathbb{R}^n, g)$ is compact, so we have

$$\|v\|_{q,g} \leq C_g \|v\|_{s,p} \quad \text{for all } v \in W_K^{s,p}(\mathbb{R}^n), \tag{5.10}$$

and $C_g = S^{-\frac{1}{p}} \|g\|_{q_1}^{\frac{1}{q}} > 0$, where $S = S(n, p, s)$ is the best fractional critical Sobolev constant, given by

$$S = \inf_{v \in W_K^{s,p}(\mathbb{R}^n)\setminus\{0\}} \frac{\|v\|_{s,p}^p}{\|v\|_{L^{p_s^*}(\mathbb{R}^n)}^p}. \tag{5.11}$$

Obviously, (5.26) together with (K_1)–(K_2) leads to the well-known fractional Sobolev embedding inequality

$$\|v\|_{L^{p_s^*}(\mathbb{R}^n)} \leq S^{-\frac{1}{p}} \|v\|_{s,p} \quad \text{for all } v \in W_K^{s,p}(\mathbb{R}^n). \tag{5.12}$$

Similar to [66], we can also get that $(W_K^{s,p}(\mathbb{R}^n), \|v\|_{s,p})$ is a uniformly convex Banach space. Hence, it is a reflexive Banach space.

Since we are interested in studying the solutions of problem (5.3) in weak sense, then we provide the precise definition as follows. Namely, $u \in W_K^{s,p}(\mathbb{R}^n)$ is a weak solution to (5.3) provided that

$$\mathcal{M}(\|v(x)\|_{s,p}^p) \iint_{\mathbb{R}^{2n}} |v(x)-v(y)|^{p-2}(v(x)-v(y))(\phi(x)-\phi(y)K(x-y)dxdy$$
$$= \mu \int_{\mathbb{R}^n} g(x)|v(x)|^{q-2}v(x)\phi(x)dx + \int_{\mathbb{R}^n} |v(x)|^{p_s^*-2}v(x)\phi(x)dx$$
$$+ \mu \int_{\mathbb{R}^n} f(x)\phi(x)dx$$

for any $\phi \in X_0$. We shall break down the proof of existence of non-negative weak solutions into two steps. In fact, first we get (see also [74]) non-trivial weak solutions to problem (5.3), with negative energy in the framework space. Namely, we have

Theorem 5.1 *Assuming that conditions* (M_1)–(M_3) *as well as conditions* (Z_1)–(Z_2) *hold, then we can find two values* $\mu_0, \tau_0 > 0$ *such that for any* $\mu \in (0, \mu_0)$, *the problem* (5.3) *has at least one non-trivial solution with negative energy in* $W_K^{s,p}(\mathbb{R}^n)$ *when* $\|f\|_\xi \le \tau_0$.

Then, the non-negativity of such solutions can be obtained in the following result, under more restrictive hypothesis on the sign of f.

Theorem 5.2 *Assuming that conditions* (M_1)–(M_3) *as well as conditions* (Z_1)–(Z_2) *hold, then we can find two values* $\mu_0', \tau_0' > 0$ *such that for any* $\mu \in (0, \mu_0')$, *the problem* (5.3) *has at least one non-trivial non-negative solution with negative energy in* $W_K^{s,p}(\mathbb{R}^n)$, *provided that* $f \ge 0$ *a.e. in* $\mathbb{R}^n$ *and* $\|f\|_\xi \le \tau_0'$.

To apply the above theorems, the readers can consider the following problem (see [74]):

$$\left(a + b\left(\iint_{\mathbb{R}^{2n}} \frac{|v(x)-v(y)|^2}{|x-y|^{n+1}}dxdy\right)^{\theta-1}\right)(-\Delta)^{\frac{1}{2}}v(x)$$
$$= \mu g(x)v^{q-1} + v^{\frac{n+1}{n-1}} + \mu f(x) \quad \text{in } \mathbb{R}^n,$$

where $n, \theta > 1$, a, b are non-negative constants with $a+b>0$, g and f are given as follows:

$$g(x) = (1+|x|^2)^{-\frac{2_s^*-q}{2_s^*}}, \quad \text{for all } x \in \mathbb{R}^n,$$

and

$$f(x) = (1+|x|^2)^{-\frac{1}{\xi}}, \quad \text{for all } x \in \mathbb{R}^n.$$

Since $g \in L^{\frac{2_s^*}{2_s^*-q}}(\mathbb{R}^n) \cap L^\infty_{loc}(\mathbb{R}^n)$ and $f \in L^\xi(\mathbb{R}^n)$, the assumptions (Z_1) and (Z_2) are satisfied. It is clear that (M_1)–(M_3) hold, as for each $\sigma > 0$

$$\mathcal{M}(t) = a + bt^{\theta-1} \ge a + b\sigma^{\theta-1} =: m(\sigma) > 0 \quad \text{for all } t \ge \sigma,$$

and

$$\widetilde{\mathcal{M}}(t) = \int_0^t \mathcal{M}(s)ds \geq \frac{1}{\theta}\mathcal{M}(t)t \quad \text{for all } t \geq 0.$$

5.3.1 Proofs of Theorems 5.1 and 5.2

Let $\mathcal{I}_\mu : W_K^{s,p}(\mathbb{R}^n) \to \mathbb{R}$ be the energy functional associated with problem (5.3), defined by

$$\mathcal{I}_\mu(v) = \frac{1}{p}\widetilde{\mathcal{M}}\Big(\iint_{\mathbb{R}^{2n}} |v(x)-v(y)|^p K(x-y)dxdy\Big) - \frac{\mu}{q}\int_{\mathbb{R}^n} g(x)|v|^q dx$$
$$- \frac{1}{p_s^*}\int_{\mathbb{R}^n} |v|^{p_s^*}dx - \mu\int_{\mathbb{R}^n} f(x)vdx,$$

for all $v \in W_K^{s,p}(\mathbb{R}^n)$.

The conditions on K, g and f (namely, (K_1)–(K_2) and (Z_1)–(Z_2)) ensure that this $C^1(W_K^{s,p}(\mathbb{R}^n))$-functional is well defined, and we have

$$\langle \mathcal{I}'_\mu(v), u\rangle =$$
$$\mathcal{M}(\|v\|_{s,p}^p)\iint_{\mathbb{R}^{2n}} |v(x)-v(y)|^{p-2}(v(x)-v(y))(u(x)-u(y)K(x-y)dxdy$$
$$-\mu\int_{\mathbb{R}^n} g(x)|v|^{q-2}vudx - \int_{\mathbb{R}^n} |v|^{p_s^*-2}vudx - \mu\int_{\mathbb{R}^n} f(x)udx,$$

for all $v, u \in W_K^{s,p}(\mathbb{R}^n)$ (see [15], Lemma 4.2). Thus, the critical points of functional $\mathcal{I}_\mu$ are weak solutions of the problem (5.3).

Now, it is interesting to look at some special features of this functional. A first result is established in the following lemma.

Lemma 5.1 [74, Lemma 3.1] *Assuming that conditions (M_1)–(M_2) as well as (Z_1)–(Z_2) hold, then we can find $\rho, \alpha = \alpha(\rho), \tau_0, \mu_1 > 0$, such that $\mathcal{I}_\mu(v) \geq \alpha$ for any $\mu \in (0, \mu_0)$ with $\|v\|_{s,p} = \rho$ and $\|f\| \leq \tau_0$, for all $\mu \in (0, \mu_1)$.*

For convenience, we will use also the notation $\|\cdot\|_q$ for the norm of Lebesgue space $L^q(\mathbb{R}^n)$. Now, we recall the classical Palais–Smale condition at level c.

Definition 5.1 Let $\mathcal{I}_\mu \in C^1(X, \mathbb{R})$, we say that $\mathcal{I}_\mu$ satisfies the $(PS)_c$ condition at the level $c \in \mathbb{R}$, if any sequence $\{v_j\} \subset X$ such that

$$\mathcal{I}_\mu(v_j) \to c, \quad \mathcal{I}'_\mu(v_j) \to 0 \quad \text{as } j \to \infty \tag{5.13}$$

possesses a convergent subsequence in X.

This condition is involved in the following result.

Lemma 5.2 [74, Lemma 3.3] *Assuming that conditions* (M_1)–(M_2) *as well as* (Z_1)–(Z_2) *hold, then we can find* μ_2 *such that for any* $\mu \in (0, \mu_2)$, *the functional* $\mathcal{I}_\mu$ *fulfills* $(PS)_c$ *condition, for some* $c < 0$.

For a better understanding of the strategy, we design the proof of Theorem 5.1, setting $\mu_0 = \min\{\mu_1, \mu_2\}$. We note that the proof of Theorem 5.2 can be established in a similar fashion changing the functional $\mathcal{I}_\mu$ by $\mathcal{I}_\mu^+$ (which is defined in the same proof of Theorem 5.1). According to [74], we consider separately the situation when $\not\equiv 0$ and when $\equiv 0$, respectively.

Assuming $f \not\equiv 0$, it is possible to establish that one can find $\phi_2 \in C_0^\infty(\mathbb{R}^n)$, with $\|\phi_2\|_{s,p} = 1$ satisfying the inequality $\int_{\mathbb{R}^n} f(x)\phi_2(x)dx > 0$. We note that $C_0^\infty(\mathbb{R}^n)$ is dense in $L^{\xi'}(\mathbb{R}^n)$ and $|f|^{\xi-2}f \in L^{\xi'}(\mathbb{R}^n)$, so that we can find $j_0 > 0$ satisfying the inequality

$$\frac{1}{2}\|f\|^{\xi-1}_{L^\xi(\mathbb{R}^n)} \geq \|f_{j_0} - |f|^{\xi-2}f\|_{L^{\xi'}(\mathbb{R}^n)}.$$

Consequently, we get that

$$\begin{aligned}\int_{\mathbb{R}^n} f(x)f_{j_0}(x)dx &\geq -\|f_{j_0} - |f|^{\xi-2}f\|_{L^{\xi'}(\mathbb{R}^n)}\|f\|_{L^\xi(\mathbb{R}^n)} + \int_{\mathbb{R}^n}|f(x)|^\xi dx\\ &\geq \frac{1}{2}\int_{\mathbb{R}^n}|f(x)|^\xi dx > 0.\end{aligned}$$

Note that $f_{j_0} \in W_K^{s,p}(\mathbb{R}^n)$. For $\phi_2 = \frac{f_{j_0}}{\|f_{j_0}\|_{s,p}}$, we obtain

$$\int_{\mathbb{R}^n} f(x)\phi_2(x)dx > 0.$$

Recalling that $p, q, p_s^* > 1$, we can derive the non-positivity of $\mathcal{I}_\mu(t\phi_2)$ as follows:

$$\begin{aligned}\mathcal{I}_\mu(t\phi_2) &= \frac{1}{p}\widetilde{\mathcal{M}}(\|t\phi_1\|^p_{s,p}) - \frac{\mu t^q}{q}\int_{\mathbb{R}^n} g(x)|\phi_2|^q dx\\ &\quad - \frac{t^{p_s^*}}{p_s^*}\int_{\mathbb{R}^n}|\phi_2|^{p_s^*}dx - \mu t\int_{\mathbb{R}^n} f(x)\phi_2 dx\\ &\leq \frac{\max\limits_{0\leq\varrho\leq1}\mathcal{M}(\varrho)}{p}t^p - \frac{\mu t^q}{q}\int_{\mathbb{R}^n} g(x)|\phi_2|^q dx\\ &\quad - \frac{t^{p_s^*}}{p_s^*}\int_{\mathbb{R}^n}|\phi_2|^{p_s^*}dx - \mu t\int_{\mathbb{R}^n} f(x)\phi_2 dx\\ &< 0 \end{aligned} \tag{5.14}$$

for $t \in (0, 1)$ small enough.

On the other hand, assuming that $f \equiv 0$, condition (5.14) remains true as $\|\phi_2\|_{s,p} = 1$ provided that $t \in (0, 1)$ is sufficiently small (recall $1 < q < p < p_s^*$).

Now, for any open ball $B_{\mathbb{R}} \subset W_K^{s,p}(\mathbb{R}^n)$, we deduce that

$$0 > \inf_{v \in \widetilde{B}_{\mathbb{R}}} \mathcal{I}_\mu(v) = c_r > -\infty. \tag{5.15}$$

It follows that

$$c = \inf_{v \in \widetilde{B}_\rho} \mathcal{I}_\mu(v) < 0 \quad \text{and} \quad \inf_{v \in \partial B_\rho} \mathcal{I}_\mu(v) > 0, \tag{5.16}$$

where $\rho > 0$ is defined in Lemma 5.1. For $\varepsilon_j \downarrow 0$ satisfying the condition

$$0 < \varepsilon_j < \inf_{v \in \partial B_\rho} \mathcal{I}_\mu(v) - \inf_{v \in B_\rho} \mathcal{I}_\mu(v), \tag{5.17}$$

we can find $\{v_j\}_j \subset \widetilde{B}_\rho$ satisfying

$$c \leq \mathcal{I}_\mu(v_j) \leq c + \varepsilon_j \tag{5.18}$$

and

$$\mathcal{I}_\mu(v_j) \leq \mathcal{I}_\mu(v) + \varepsilon_j \|v_j - v\|_{s,p} \quad \text{for all } v \in \widetilde{B}_\rho,\, v \neq v_j \tag{5.19}$$

(recall Ekeland's variational principle in [26]). Combining the bound conditions (5.17)–(5.19), one can deduce that

$$\mathcal{I}_\mu(v_j) \leq c + \varepsilon_j \leq \inf_{v \in B_\rho} \mathcal{I}_\mu(v) + \varepsilon_j < \inf_{v \in \partial B_\rho} \mathcal{I}_\mu(v), \tag{5.20}$$

and hence $v_j \in B_\rho$. Now, let $T : \widetilde{B}_\rho \to \mathbb{R}$ be the functional constructed over the inequality (5.19) and given as

$$T(v) = \mathcal{I}_\mu(v) + \varepsilon_j \|v_j - v\|_{s,p} \quad \text{for all } v \in \widetilde{B}_\rho.$$

It follows easily

$$T(v_j) < T(v) \text{ for all } v \in \widetilde{B}_\rho, \text{ with } v \neq v_j$$

$$\Rightarrow \quad v_j \text{ is a strict local minimum of } T.$$

We note that

$$\frac{T(v_j + tu) - T(v_j)}{t} \geq 0 \quad \text{for small } t > 0, \text{ all } u \in B_1,$$

$$\Rightarrow \quad \frac{\mathcal{I}_\mu(v_j + tu) - \mathcal{I}_\mu(v_j)}{t} + \varepsilon_j \|u\|_{s,p} \geq 0.$$

$$\Rightarrow \quad \langle \mathcal{I}_\mu^{'}(v_j), u \rangle + \varepsilon_j \|u\|_{s,p} \geq 0,\ \forall u \in B_1 \text{ (letting } t \to 0^+).$$

Similarly, one can obtain the inequality

$$-\langle \mathcal{I}'_{\mu}(v_j), u\rangle + \varepsilon_j \|u\|_{s,p} \geq 0, \ \forall u \in B_1.$$

We conclude that $\|\mathcal{I}'_{\mu}(v_j)\|_{(W_K^{s,p}(\mathbb{R}^n))'} \leq \varepsilon_j$.

To conclude the proof of Theorem 5.1, we just observe that we can find $\{v_j\}_j \subset B_\rho$ satisfying $\mathcal{I}_\mu(v_j) \to c < 0$ and $\mathcal{I}'_\mu(v_j) \to 0$ in $(W_K^{s,p}(\mathbb{R}^n))'$ as $j \to \infty$. Now, $\{v_j\}_j$ admits a convergent subsequence in $W_K^{s,p}(\mathbb{R}^n)$ (by Lemma 5.2), hence $v_j \to v_\infty$ in $W_K^{s,p}(\mathbb{R}^n)$. This means that v_∞ solves problem (5.3), furthermore $\mathcal{I}_\mu(v_\infty) < 0$.

If we like to conclude the existence of the non-trivial non-negative solution, namely we aim to add details about the sign of solutions, we use, like in [74] the functional $\mathcal{I}_\mu^+ : W_K^{s,p}(\mathbb{R}^n) \to \mathbb{R}$ given as

$$\mathcal{I}_\mu^+(v) = \frac{1}{p}\widetilde{\mathcal{M}}\Big(\|v\|_{s,p}^p\Big) - \frac{\mu}{q}\int_{\mathbb{R}^n} g(x)|v^+|^q dx - \frac{1}{p_s^*}\int_{\mathbb{R}^n} |v^+|^{p_s^*} dx - \mu\int_{\mathbb{R}^n} f(x)v dx,$$

for all $v \in W_K^{s,p}(\mathbb{R}^n)$, where $v^+ = \max\{v, 0\}$. The main issue of the proof is in establishing that $u^+ \in W_K^{s,p}(\mathbb{R}^n)$, where $u^+ = \max\{u, 0\} = \frac{|u|+u}{2}$, as $u \in W_K^{s,p}(\mathbb{R}^n)$. By calculation, we get

$$\iint_{\mathbb{R}^{2n}} |u^+(x) - u^+(y)|^p K(x-y) dx dy \leq \iint_{\mathbb{R}^{2n}} |u(x) - u(y)|^p K(x-y) dx dy,$$

hence the thesis follows. The same way, one can conclude that $u^- \in W_K^{s,p}(\mathbb{R}^n)$ ($u^- = \max\{-u, 0\}$).

Thus the C^1-functional $\mathcal{I}_\mu^+$ is well defined such that

$$\begin{aligned}
&\langle (\mathcal{I}_\mu^+(v))', \phi\rangle = \\
&\mathcal{M}(\|v\|_{s,p}^p)\iint_{\mathbb{R}^{2n}} |v(x) - v(y)|^{p-2}(v(x) - v(y))(\phi(x) - \phi(y)K(x-y) dx dy \\
&- \mu\int_{\mathbb{R}^n} g(x)|v^+(x)|^{q-2}v^+(x)\phi(x) dx - \int_{\mathbb{R}^n} |v^+(x)|^{p_s^*-2}v^+(x)\phi(x) dx \\
&- \mu\int_{\mathbb{R}^n} f(x)\phi(x) dx.
\end{aligned} \tag{5.21}$$

For any (non-negative) critical point v of $\mathcal{I}_\mu^+$, we deduce by (5.21) that

$$\langle (\mathcal{I}_\mu^+(v))', -v^-\rangle = 0. \tag{5.22}$$

By (5.22), $\mu > 0$ and $f \geq 0$ a.e. in $\mathbb{R}^n$, and the inequality

$$\iint_{\mathbb{R}^{2n}} |v^-(x) - v^-(y)|^p K(x-y)dxdy$$
$$\leq \iint_{\mathbb{R}^{2n}} |v(x) - v(y)|^{p-2}(v(x) - v(y))(-v^-(x) + v^-(y)K(x-y)dxdy,$$

it follows that

$$\iint_{\mathbb{R}^{2n}} |v^-(x) - v^-(y)|^p K(x-y)dxdy \leq 0.$$

Hence, $v^- = 0$ a.e. in $\mathbb{R}^n$ that leads to $v \geq 0$ a.e. in $\mathbb{R}^n$. This concludes the proof of main theorems.

5.4 Analysis of Problem (5.9)

In this section, we will review the existence of solutions to a different class of critical fractional p-Kirchhoff problems, in comparison to the previous one. Precisely, we cover the finding of the work [74]; hence we discuss problem (5.9), in comparison to problem (5.3). Hence, we recall and fix the assumptions on the Kirchhoff function as follows (see also [74]):
$(M_1^{'})$: There exists $\theta \in [1, r^-/p^+)$ such that

$$tM(t) \leq \theta \widetilde{M}(t) \text{ for all } t \in \mathbb{R}_0^+,$$

where $\widetilde{M}(t) = \int_0^t M(\tau)d\tau$;
$(M_2^{'})$: $M(0) = 0$ and the rest of this assumption is same (M_2);
$(M_3^{'})$ is same as (M_3).

For readers' convenience, we also use the following notation, as given in [74]. Namely, we assume the following:

$$s^- = \min_{(x,y)\in\mathbb{R}^{2N}} s(x,y), \quad s^+ = \max_{(x,y)\in\mathbb{R}^{2N}} s(x,y),$$
$$p^- = \min_{(x,y)\in\mathbb{R}^{2N}} p(x,y), \quad p^+ = \max_{(x,y)\in\mathbb{R}^{2N}} p(x,y),$$
$$q^- = \min_{x\in\overline{\Omega}} q(x), \quad q^+ = \max_{x\in\overline{\Omega}} q(x),$$
$$r^- = \min_{x\in\overline{\Omega}} r(x), \quad r^+ = \max_{x\in\overline{\Omega}} r(x).$$

Furthermore, the exponent functions $s(\cdot) : \mathbb{R}^{2N} \to (0, 1)$ and $p(\cdot) : \mathbb{R}^{2N} \to (1, \infty)$ are continuous and possess the following regularities:

$(H_1): 0 < s^- \leq s^+ < 1 < p^- \leq p^+$;
(H_2) : $s(\cdot)$ and $p(\cdot)$ are symmetric, that is, $s(x, y) = s(y, x)$ and $p(x, y) = p(y, x)$ for all $(x, y) \in \mathbb{R}^{2N}$;
(H_3) : $p(\cdot)$ satisfies the following log-Hölder type continuity condition

$$\inf_{\varepsilon>0} \sup_{(x,y)\in\Omega\times\Omega,\ 0<|x-y|<1/2} \left| p(x, y) - p^-_{\Omega_{x,\varepsilon}\times\Omega_{y,\varepsilon}} \right| \log\frac{1}{|x-y|} < \infty,$$

where $\Omega_{z,\varepsilon} = B_\varepsilon(z) \bigcap \Omega$ for $z \in \Omega$ and $\varepsilon > 0$, and

$$p^-_{\Omega_{x,\varepsilon}\times\Omega_{y,\varepsilon}} = \inf_{(x',y')\in\Omega_{x,\varepsilon}\times\Omega_{y,\varepsilon}} p(x', y').$$

In addition, also the exponent function $r(\cdot) : \overline{\Omega} \to \mathbb{R}$ is continuous and satisfies certain properties as follows:
$(R_1): 1 < r^- \leq r^+$;
(R_2) : for all $x \in \overline{\Omega}$ we can find $\varepsilon = \varepsilon(x) > 0$ satisfying

$$\sup_{y\in\Omega_{x,\varepsilon}} r(y) \leq \frac{N \inf_{(y,z)\in\Omega_{x,\varepsilon}\times\Omega_{x,\varepsilon}} p(y, z)}{N - \inf_{(y,z)\in\Omega_{x,\varepsilon}\times\Omega_{x,\varepsilon}} s(y, z) \inf_{(y,z)\in\Omega_{x,\varepsilon}\times\Omega_{x,\varepsilon}} p(y, z)}$$

with $\Omega_{x,\varepsilon}$ as given in hypothesis (H_3).

The property (R_2) is built over the fractional critical Sobolev exponent $p_s^*(\cdot)$, hence $r(x) \leq p_s^*(x)$ for all $x \in \overline{\Omega}$. Since we look for critical problems, we are interested the case when $r(x) = p_s^*(x)$ for some $x \in \overline{\Omega}$.

Next, to solve the problem under investigation, we need to control the growth of the right-hand side of Eq. in (5.9). Hence, according to the classical literature, we assume the following conditions on g. Namely, $g : \overline{\Omega} \times \mathbb{R} \to \mathbb{R}$ is continuous and we have
(G_1) : there exist $\mathfrak{q}$, $q \in C(\overline{\Omega})$ satisfying

$$\theta p^+ < \mathfrak{q}^- \leq \mathfrak{q}(x) \leq \mathfrak{q}^+ \leq q^- \leq q(x) \leq q^+ < r^-, \quad \textit{for any} \quad x \in \overline{\Omega}$$

and for every $\varepsilon > 0$ we can find a positive constant $c(\varepsilon) > 0$ such that

$$|g(x, t)| \leq \varepsilon|t|^{\mathfrak{q}(x)-1} + c(\varepsilon)|t|^{q(x)-1}$$

is true for all $(x, t) \in \overline{\Omega} \times \mathbb{R}$;
(G_2) : there exists $\lambda \in (p^+\theta, r^-)$, where θ is given in (M_1'), satisfying

$$0 < \lambda G(x, t) \leq tg(x, t), \qquad \text{for all } x \in \overline{\Omega}, \quad t \neq 0,$$

where $G(x, t) = \int_0^t g(x, \tau)d\tau$.

According to the classical literature, in [27] the authors use the following notation. Let

$$C_+(\overline{\Omega}) = \left\{ h \in C(\overline{\Omega}) : 1 < h(x) \text{ for all } x \in \overline{\Omega} \right\},$$

hence the appropriate Lebesgue space is given as follows:

$$L^{h(\cdot)}(\Omega) = \left\{ v : \Omega \to \mathbb{R} \text{ is measurable with } \int_\Omega |v(x)|^{h(x)} dx < \infty \right\}.$$

A suitable Luxemburg norm for this space is

$$\|v\|_{h(\cdot)} = \inf \left\{ \gamma > 0 : \int_\Omega \left| \frac{v(x)}{\gamma} \right|^{h(x)} dx \leq 1 \right\}.$$

We remark (see [40]) that the space $(L^{h(\cdot)}(\Omega), \|\cdot\|_{h(\cdot)})$ is separable reflexive Banach space, provided that $1 < \inf_{x\in\overline{\Omega}} h(x) \leq \sup_{x\in\overline{\Omega}} h(x) < \infty$. Also we mention the modular function $\varrho_{h(\cdot)} : L^{h(\cdot)}(\Omega) \to \mathbb{R}$ given by

$$\varrho_{h(\cdot)}(v) = \int_\Omega |v(x)|^{h(x)} dx.$$

The modular is linked to the Luxemburg norm by the following proposition of [25].

Proposition 5.1 *For $v \in L^{h(\cdot)}(\Omega)$ and $\{v_j\}_j \subset L^{h(\cdot)}(\Omega)$, we have the following:*

$$\begin{aligned}
&(1)\ \|v\|_{h(\cdot)} < 1(resp. = 1, > 1) \Leftrightarrow \varrho_{h(\cdot)}(v) < 1(resp. = 1, > 1),\\
&(2)\ \|v\|_{h(\cdot)} < 1 \Rightarrow \|v\|_{h(\cdot)}^{h^+} \leq \varrho_{h(\cdot)}(v) \leq \|v\|_{h(\cdot)}^{h^-},\\
&(3)\ \|v\|_{h(\cdot)} > 1 \Rightarrow \|v\|_{h(\cdot)}^{h^-} \leq \varrho_{h(\cdot)}(v) \leq \|v\|_{h(\cdot)}^{h^+},\\
&(4)\ \lim_{j\to\infty} \|v_j\|_{h(\cdot)} = 0(\infty) \Leftrightarrow \lim_{j\to\infty} \varrho_{h(\cdot)}(v_j) = 0(\infty),\\
&(5)\ \lim_{j\to\infty} \|v_j - v\|_{h(\cdot)} = 0 \Leftrightarrow \lim_{j\to\infty} \varrho_{h(\cdot)}(v_j - v) = 0.
\end{aligned}$$

We recall a Hölder inequality, as follows.

Lemma 5.3 *For $u \in L^{h(\cdot)}(\Omega)$ and $v \in L^{h'(\cdot)}(\Omega)$, we get that*

$$\left| \int_\Omega uv dx \right| \leq \left(\frac{1}{h^-} + \frac{1}{(h')^-} \right) \|u\|_{h(\cdot)} \|v\|_{h'(\cdot)} \leq 2\|u\|_{h(\cdot)} \|v\|_{h'(\cdot)},$$

where $h' \in C_+(\overline{\Omega})$ is the conjugate exponent to h, namely

$$\frac{1}{h(x)} + \frac{1}{h'(x)} = 1, \qquad \textit{for all } x \in \overline{\Omega}.$$

Hence we introduce the fractional Sobolev space, variable in exponent and in order, defined by

$$W^{s(\cdot),p(\cdot)}(\Omega)=\left\{v\in L^{\overline{p}(\cdot)}(\Omega):\iint_{\Omega\times\Omega}\frac{|v(x)-v(y)|^{p(x,y)}}{|x-y|^{N+p(x,y)s(x,y)}}dxdy<\infty\right\}$$

endowed with the following norm $\|v\|_{s(\cdot),p(\cdot)}=\|v\|_{p(\cdot)}+[v]_{s(\cdot),p(\cdot)}$, where

$$[v]_{s(\cdot),p(\cdot)}=\inf\left\{\gamma>0:\iint_{\Omega\times\Omega}\frac{|v(x)-v(y)|^{p(x,y)}}{\gamma^{p(x,y)}|x-y|^{N+p(x,y)s(x,y)}}dxdy<1\right\}. \tag{5.23}$$

On this space, we construct the solution space (see again [74])

$$X=\left\{v:\mathbb{R}^N\to\mathbb{R}:\ v|_\Omega\in L^{\overline{p}(\cdot)}(\Omega),\iint_{Q}\frac{|v(x)-v(y)|^{p(x,y)}}{\gamma^{p(x,y)}|x-y|^{N+p(x,y)s(x,y)}}dxdy<\infty\right\},$$

where $Q=\mathbb{R}^{2N}\backslash(\Omega^c\times\Omega^c)$. Its norm follows by adapting (5.23) as follows:

$$\|v\|_X=\|v\|_{p(\cdot)}+[v]_X.$$

Recall that $\Omega\times\Omega\subset Q$ but $\Omega\times\Omega\neq Q$.

In the sequel we suppose always $N>p(x,y)s(x,y)$ for all $(x,y)\in\overline{\Omega}\times\overline{\Omega}$. Significant compact and critical embedding results (see [13, Theorem 3.6] and [71, Lemma 2.3], respectively) are given in the following form.

Lemma 5.4 *Assuming that conditions* (H_1)*–*(H_2) *hold, and setting*

$$1<h^-=\min_{x\in\overline{\Omega}}h(x)\le h(x)<p_s^*(x)=\frac{N\overline{p}(x)}{N-\overline{p}(x)\overline{s}(x)}$$

for all $x\in\overline{\Omega}$, $h\in C_+(\overline{\Omega})$, *then we can find* $C_h=C_h(N,s,p,h,\Omega)>0$ *satisfying*

$$\|v\|_{h(\cdot)}\le C_h\|v\|_{W^{s(\cdot),p(\cdot)}(\Omega)}$$

for all $v\in W^{s(\cdot),p(\cdot)}(\Omega)$. *Furthermore,* $W^{s(\cdot),p(\cdot)}(\Omega)\hookrightarrow L^{h(\cdot)}(\Omega)$ *is compact.*

Lemma 5.5 *Assuming that conditions* (H_1)*–*(H_3) *hold, as well as* (R_1)*–*(R_2), *then we can find* $C_r=C_r(N,s,p,r,\Omega)>0$ *satisfying*

$$\|v\|_{r(\cdot)}\le C_r\|v\|_{W^{s(\cdot),p(\cdot)}(\Omega)} \tag{5.24}$$

for all $v\in W^{s(\cdot),p(\cdot)}(\Omega)$. *Furthermore,* $W^{s(\cdot),p(\cdot)}(\Omega)\hookrightarrow L^{r(\cdot)}(\Omega)$ *is continuous.*

Since $r(\cdot)\le p_s^*(\cdot)$ for all $x\in\overline{\Omega}$ the precise proof of the embedding $W^{s(\cdot),p(\cdot)}(\Omega)\hookrightarrow L^{p_s^*(\cdot)}(\Omega)$ remains open. Additionally we can consider the space

$$X_0 = \Big\{v \in X : v = 0 \ a.e. \text{ in } \mathbb{R}^N \backslash \Omega\Big\},$$

which is a separable reflexive Banach space (refer to [6, Lemma 2.3]) for the norm

$$\|v\|_{X_0} = \iint_{\mathbb{R}^{2N}} \frac{|v(x) - v(y)|^{p(x,y)}}{\gamma^{p(x,y)}|x - y|^{N+p(x,y)s(x,y)}} dxdy < 1\Big\}.$$

Following standard arguments, one can conclude that $\|\cdot\|_X$ and $\|\cdot\|_{X_0}$ are equivalent norms on X_0. Consequently, Lemmas 7.1 and 5.5 remain true for X_0. Now, we consider the best fractional Sobolev constant S' defined by

$$S' = \inf_{v \in X_0 \backslash \{0\}} \frac{\|v\|_{X_0}}{\|v\|_{r(\cdot)}} > 0 \text{ (by Lemma F.5).} \tag{5.25}$$

With respect to the modular $\varrho_{p(\cdot)}^{s(\cdot)} : X_0 \to \mathbb{R}$ as

$$\varrho_{p(\cdot)}^{s(\cdot)}(v) = \iint_{\mathbb{R}^{2N}} \frac{|v(x) - v(y)|^{p(x,y)}}{|x - y|^{N+p(x,y)s(x,y)}} dxdy,$$

one can give the analogs of Proposition 5.1 in X_0. Referring to [55, Lemma 3.2], the authors of [74] established the following Brézis–Lieb type lemma.

Lemma 5.6 *Assuming that conditions* (H_1)–(H_2) *hold, and* $\{v_j\}_j, v \in X_0$ *satisfy* $v_j \rightharpoonup v$ *weakly in* X_0*, and* $v_j \to v$ *a.e. in* $\mathbb{R}^N$*, then we have*

$$\varrho_{p(\cdot)}^{s(\cdot)}(v_j) - \varrho_{p(\cdot)}^{s(\cdot)}(v_j - v) = \varrho_{p(\cdot)}^{s(\cdot)}(v) + o(1) \quad as \ j \to \infty.$$

Lemma 7.4 is used for establishing the compactness of energy functional related to (5.9). Since we are interested to study the solutions of problem (5.9) in weak sense, then we provide the precise definition as follows. Namely, $v \in X_0$ is a weak solution to (5.9) provided that

$$M\Big(\delta_{p(\cdot)}(v)\Big)\langle v, \phi\rangle_{X_0} = \mu \int_\Omega g(x, v)\phi dx + \int_\Omega |v|^{r(x)-2} v\phi dx, \tag{5.26}$$

for all $\phi \in X_0$, where

$$\delta_{p(\cdot)}(v) = \iint_{\mathbb{R}^{2N}} \frac{1}{p(x,y)} \frac{|v(x) - v(y)|^{p(x,y)}}{|x - y|^{N+p(x,y)s(x,y)}} dxdy$$

$$\langle v, \phi\rangle_{X_0} = \iint_{\mathbb{R}^{2N}} \frac{|v(x) - v(y)|^{p(x,y)-2}(v(x) - v(y))(\phi(x) - \phi(y))}{|x - y|^{N+p(x,y)s(x,y)}} dxdy.$$

Thus, the authors in [74] obtained the following result.

Theorem 5.3 *Assuming that conditions* (M_1')–(M_3'), *as well as* (H_1)–(H_2) *and* (G_1)–(G_2) *and* (R_1)–(R_2) *hold, then we can find* $\mu_0 > 0$ *such that for any* $\mu \geq \mu_0$, *problem* (5.9) *has at least one non-trivial weak solution* v_μ *in an appropriate separable reflexive Banach space* X_0. *Furthermore, if* M *is non-decreasing, we get* $\|v_\mu\|_{X_0} \to 0$ *as* $\mu \to \infty$.

5.4.1 Proof of Theorem 5.3

Let $\mathcal{I}_\mu : X_0 \to \mathbb{R}$ be the energy functional associated with problem (5.9), defined by

$$\mathcal{I}_\mu(v) = \widetilde{M}\Big(\delta_{p(\cdot)}(v)\Big) - \mu \int_\Omega G(x, v)dx - \int_\Omega \frac{1}{r(x)}|v|^{r(x)}dx.$$

The conditions on $\widetilde{M}$, g and $r(\cdot)$ ensure that this C^1-functional is well defined.

Now, it is interesting to look at some special features of this functional. Some results are established in the following lemmas.

Lemma 5.7 [71, Lemma 3.1] *Assuming that conditions* (M_1')–(M_2') *as well as* (G_1) *hold, then for every* $\mu > 0$ *we can find* ρ *and* $\alpha > 0$ *such that* $\mathcal{I}_\mu(v) \geq \alpha$ *for all* $v \in X_0$ *with* $\|v\|_{X_0} = \rho$.

Lemma 5.8 [71, Lemma 3.2] *Assuming that conditions* (M_1')–(M_2') *as well as* (G_2) *hold, then for every* $\mu > 0$ *we can find a non-negative function* $w \in X_0$ *satisfying* $\mathcal{I}_\mu(w) < 0$ *and* $\|w\|_{X_0} > \rho$, *with* ρ *given by Lemma 5.1.*

Since the strategy is dictated by the Palais–Smale condition and mountain pass geometry of the energy functional, then we revisit Definition 5.1 and related lemmas as follows. Fixed $\mu > 0$, we set

$$c_\mu = \inf_{\gamma \in \Gamma} \max_{t \in [0,1]} \mathcal{I}_\mu(\gamma(t)) > 0 \text{ (by Lemma 5.7)},$$

where

$$\Gamma = \{\gamma \in C^1([0, 1], X_0) : \gamma(0) = 0, \mathcal{I}_\mu(\gamma(1)) < 0\}.$$

Hence the functional $\mathcal{I}_\mu$ satisfies the Palais–Smale condition at level c_μ if any sequence $\{v_j\}_j \subset X_0$ such that

$$\mathcal{I}_\mu(v_j) \to c_\mu \text{ and } \mathcal{I}_\mu'(v_j) \to 0 \text{ in } X_0^* \text{ as } j \to \infty, \tag{5.27}$$

possesses a convergent subsequence in X_0.

The authors in [74] obtain the following asymptotic result for c_μ. We remark that we deal with a lack of compactness, produced by critical nonlinearity.

Lemma 5.9 *Assuming that conditions* $(M_1^{'})$ *as well as* $(G_1)-(G_2)$ *hold, then* $c_\mu \to 0$ *as* $\mu \to \infty$.

Thus we have the following result.

Lemma 5.10 [71, Lemma 4.2] *Assuming that conditions* $(M_1^{'})$–$(M_3^{'})$ *as well as* $(G_1)-(G_2)$ *hold, then we can find* $\mu_0 > 0$ *such that for any* $\mu \geq \mu_0$*, the functional* $\mathcal{I}_\mu$ *fulfills* $(PS)_{c_\mu}$*, for some* $c_\mu > 0$.

As a byproduct of the above lemmas, one can deduce that the energy functional $\mathcal{I}_\mu$ possesses the mountain pass geometry, for any $\mu \geq \mu_0$. Hence, we can find $v_\mu \in X_0 \not\equiv 0$, critical point to $\mathcal{I}_\mu$ at level c_μ, because of $\mathcal{I}_\mu(v_\mu) = c_\mu > 0 = \mathcal{I}_\mu(0)$. So, to conclude the proof of Theorem 5.3 we look at the asymptotic behavior of $\|v_\mu\|_{X_0}$ as $\mu \to \infty$. By the analogs of Proposition 5.1, stated for the modular $\varrho_{p(\cdot)}^{s(\cdot)}(\cdot)$ in X_0, it is sufficient to prove that

$$\varrho_{p(\cdot)}^{s(\cdot)}(v_\mu) \to 0 \text{ as } \mu \to \infty.$$

If this is not the case, namely we assume

$$\limsup_{\mu \to \infty} \varrho_{p(\cdot)}^{s(\cdot)}(v_\mu) = \upsilon > 0,$$

hence we can find a sequence $j \mapsto \mu_j$ going to infinity satisfying the convergence $\varrho_{p(\cdot)}^{s(\cdot)}(v_{\mu_j}) \to \upsilon$, by passing to the limit for $j \to \infty$. By easy calculations, the conditions $(M_1^{'})$ and (G_2) give us

$$c_{\mu_j} \geq \left(\frac{1}{\theta} - \frac{p^+}{\lambda}\right) M\left(\frac{\varrho_{p(\cdot)}^{s(\cdot)}(v_{\mu_j})}{p^+}\right) \frac{\varrho_{p(\cdot)}^{s(\cdot)}(v_{\mu_j})}{p^+}.$$

By using Lemma 7.3, together with $(M_2^{'})$ and $\lambda > p^+\theta$ (by (G_1)). We deduce that

$$0 \geq \left(\frac{1}{\theta} - \frac{p^+}{\lambda}\right) M\left(\frac{\upsilon}{p^+}\right) \frac{\upsilon}{p^+} > 0 \text{ as } j \to \infty \ (M \text{ non-decreasing}).$$

We reached a contradiction, which proves the theorem. Some complementary results were established in [21, 75, 76].

5.5 Open Problems

As shown in the above sections, the embedding results are key tools in the study of existence of solutions to different classes of elliptic PDEs. The literature pertaining to embedding result of fractional Sobolev spaces with constant exponent is very rich and has been extensively used by the researchers. A challenging situation in improving the finding about problems driven by a nonlocal operator of variable order and exponent can be depicted as follows.

Let $\Omega \subset \mathbb{R}^N$ be a bounded Lipschitz domain. Assume that s, p, q obey the following conditions:
(C_1): $s \in C(\bar{\Omega} \times \bar{\Omega}, (0, 1))$, $p \in C(\bar{\Omega} \times \bar{\Omega}, \mathbb{R})$ are symmetric such that

$$1 < p^- \leq p^+ =< \frac{N}{s^+},$$

$$\inf_{\delta>0} \sup_{(x,y)\in\Omega\times\Omega,\ 0<|x-y|<1/2} |p(x, y) - p^-_{\Omega_{x,\delta}\times\Omega_{y,\delta}}| \log \frac{1}{|x-y|} < \infty, \tag{5.28}$$

where $\Omega_{x,\delta} = B_\delta(x) \cap \Omega$ for some $x \in \Omega$, $\delta > 0$ and $p^-_{\Omega_{x,\delta}\times\Omega_{y,\delta}} = \inf_{(x',y')\in\Omega_{x,\delta}\times\Omega_{y,\delta}} p(x', y')$.

(C_2): Let $q \in C_+(\bar{\Omega})$, for any $\delta > 0$, we have

$$\sup_{y\in\Omega_{x,\delta}} q(y) \leq \frac{N \sup_{(y,z)\in\Omega_{x,\delta}\times\Omega_{x,\delta}} p(y, z)}{N - \inf_{(y,z)\in\Omega_{x,\delta}\times\Omega_{x,\delta}} s(y, z) \sup_{(y,z)\in\Omega_{x,\delta}\times\Omega_{x,\delta}} p(y, z)}, \quad \forall x \in \Omega. \tag{5.29}$$

Is problem

$$W^{s(\cdot,\cdot),p(\cdot,\cdot)}(\Omega) \hookrightarrow L^{q(\cdot)}(\Omega)$$

true?

Finally, we remark that in the above formulation, we involved the infimum of $s(\cdot, \cdot)$, but also in the case when we involve the supremum of $s(\cdot, \cdot)$, the problem also still remains open.

References

1. Ali, K.B., Hsini, M., Kefi, K., Chung, N.T.: On a nonlocal fractional $p(\cdot, \cdot)$-Laplacian problem with competing nonlinearities. Complex Anal. Oper. Theory **13**(3), 1377–1399 (2019)
2. Alves, C.O., Ferreira, M.C.: Existence of solutions for a class of $p(x)$-Laplacian equations involving a concave-convex nonlinearity with critical growth in $\mathbb{R}^N$. Topol. Methods Nonlinear Anal. **45**(2), 399–422 (2015)
3. Ambrosetti, A., Rabinowitz, P.H.: Dual variational methods in critical point theory and applications. J. Func. Anal. **14**, 349–831 (1973)
4. Autuori, G., Pucci, P.: Existence of entire solutions for a class of quasilinear elliptic equations. NoDEA Nonlinear Differ. Equ. Appl. **20**(3), 977–1009 (2013)

5. Azroul, E., Benkirane, A., Shimi, M.: Eigenvalue problems involving the fractional $p(x)$-Laplacian operator. Adv. Oper. Theory **4**(2), 539–555 (2019)
6. Azroul, E., Benkirane, A., Shimi, M., Srati, M.: On a class of fractional $p(x)$-Kirchhoff type problems. Appl. Anal. (2019)
7. Bahrouni, A., Rădulescu, V.: On a new fractional Sobolev space and applications to nonlocal variational problems with variable exponent. Discrete Contin. Dyn. Syst. Ser. S **11**(3), 379–389 (2018)
8. Baraket, S., Bisci, G.M.: Multiplicity results for elliptic Kirchhoff-type problems. Adv. Nonlinear Anal. **6**(1), 85–93 (2017)
9. Bartsch, T.: Infinitely many solutions of a symmetric Dirichlet problem. Nonlinear Anal. **20**(10), 1205–1216 (1993)
10. Binlin, Z., Fiscella, A., Liang, S.: Infinitely many solutions for critical degenerate Kirchhoff type equations involving the fractional p-Laplacian. Appl. Math. Optim. **80**(1), 63–80 (2019)
11. Birkhoff, G.: Dynamical systems with two degrees of freedom. Trans. Am. Math. Soc. **18**(2), 199–300 (1917)
12. Bisci, G.M., Pizzimenti, P.F.: Sequences of weak solutions for non-local elliptic problems with dirichlet boundary condition. Proc. Edinb. Math. Soc. (2), **57**(3), 779–809 (2014)
13. Biswas, R., Tiwari, S.: Variable order nonlocal choquard problem with variable exponents. Complex Var. Elliptic Equ. (2020)
14. Bonheure, D., Cingolani, S., Secchi, S.: Concentration phenomena for the Sdchrödinger-Poisson system in $\mathbb{R}^2$. Discrete Contin. Dyn. Syst. Ser. S **14**(5), 1631–1648 (2021)
15. Caponi, M., Pucci, P.: Existence theorems for entire solutions of stationary Kirchhoff fractional p-Laplacian equations. Ann. Mat. Pura Appl. **195**(6), 2099–2129 (2016)
16. Cerami, G., Vaira, G.: Positive solutions for some non-autonomous Schrödinger-Poisson systems. J. Differ. Equ. **248**(3), 521–543 (2010)
17. Chabrowski, J., Fu, Y.: Existence of solutions for $p(x)$-Laplacian problems on a bounded domain. J. Math. Anal. Appl. **306**(2), 604–618 (2005)
18. Chen, C., Huang, J., Liu, L.: Multiple solutions to the nonhomogeneous p-Kirchhoff elliptic equation with concave-convex nonlinearities. Appl. Math. Lett. **26**(7), 754–759 (2013)
19. Chen, C., Song, H., Xiu, Z.: Multiple solution for p-Kirchhoff equations in $\mathbb{R}^N$. Nonlinear Anal. **86**, 146–156 (2013)
20. Chen, Y., Levine, S., Rao, M.: Variable exponent, linear growth functionals in image restoration. SIAM J. Appl. Math. **66**(4), 1383–1406 (2006)
21. Choudhuri, D., Zuo, J.: Critical Kirchhoff p(.)&q(.)-fractional variable-order systems with variable exponent growth. Anal. Math. Phys. **12**(1), 29. (2022)
22. Colasuonno, F., Pucci, P.: Multiplicity of solutions for $p(x)$-polyharmonic elliptic Kirchhoff equations. Nonlinear Anal. **74**(17), 5962–5974 (2011)
23. Corrêa, F.J.S.A., Figueiredo, G.M.: On a p-Kirchhoff equation via Krasnoselskii's genus. Appl. Math. Lett. **22**(6), 819–822 (2009)
24. D'Ancona, P., Spagnolo, S.: Global solvability for the degenerate Kirchhoff equation with real analytic data. Invent. Math. **108**(2), 247–262 (1992)
25. Diening, L., Harjulehto, P., Höstö, P., Ružička, M.: Lebesgue and Sobolev Spaces with Variable Exponents. Springe, Heidelberg (2011)
26. Ekeland, I.: On the variational principle. J. Math. Anal. Appl. **47**, 324–353 (1974)
27. Fan, X., Shen, J., Zhao, D.: Sobolev embedding theorems for spaces $w^{k,p(x)}(\omega)$. J. Math. Anal. Appl. **262**(2), 749–760 (2001)
28. Figueiredo, G.M.: Existence of a positive solution for a Kirchhoff problem type with critical growth via truncation argument. J. Math. Anal. Appl. **401**(2), 706–713 (2013)
29. Fiscella, A., Pucci, P.: Kirchhoff-hardy fractional problems with lack of compactness. Appl. Math. Optim. **17**(3), 429–456 (2017)
30. Fiscella, A., Pucci, P.: p-fractional Kirchhoff equations involving critical nonlinearities. Nonlinear Anal. Real World Appl. **35**, 350–378 (2017)
31. Fiscella, A., Valdinoci, E.: A critical Kirchhoff type problem involving a nonlocal operator. Nonlinear Anal. **94**, 156–170 (2014)

32. Fu, Y.: Existence of solutions for $p(x)$-Laplacian problem on an unbounded. Topol. Methods Nonlinear Anal. **30**(2), 235–249 (2007)
33. Fu, Y., Yu, M.: The Dirichlet problem of higher order quasilinear elliptic equation. J. Math. Anal. Appl. **363**(2), 679–689 (2010)
34. Fu, Y., Zhang, X.: A multiplicity result for $p(x)$-Laplacian problem in $\mathbb{R}^N$. Nonlinear Anal. **70**(6), 2261–2269 (2009)
35. Fu, Y., Zhang, X.: Multiple solutions for a class of $p(x)$-Laplacian equations in $\mathbb{R}^N$ involving the critical exponent. Proc. R. Soc. Lond. Ser. A Math. Phys. Eng. Sci. **466**(2018), 1667–1686 (2010)
36. Kaufmann, U., Rossi, J., Vidal, R.: Fractional Sobolev spaces with variable exponents and fractional $p(x)$-Laplacians. Electron. J. Qual. Theory Differ. Equ. **76**, 1–10 (2017)
37. Kikuchi, K., Negoro, A.: On Markov process generated by pseudodifferential operator of variable order. Osaka J. Math. **34**(2), 319–335 (1997)
38. Kirchhoff, G.: Mechanik. Teubner, Leipzig (1883)
39. Kirchhoff, G.: Electrorheological Fluids: Modeling and Mathematical Theory. Springer, Berlin (2000)
40. Kováčik, O., Rákosník, J.: On spaces $l^{p(x)}$ and $w^{1,p(x)}$. Czechoslovak Math. J. **41**(116), 592–618 (1991)
41. Leopold, H.G.: Embedding of function spaces of variable order of differentiation in function spaces of variable order of integration. Czechoslov. Math. J. **49**(124), 633–644 (1999)
42. Liang, S., Shi, S.: Soliton solutions to Kirchhoff type problems involving the critical growth in $\mathbb{R}^N$. Nonlinear Anal. **81**, 31–41 (2013)
43. Lions, J.-L.: On Some Questions in Boundary Value Problems of Mathematical Physics, pp. 284–346. North-Holland, Amsterdam-New York (1978)
44. Liu, D.: On a p-Kirchhoff equation via fountain theorem and dual fountain theorem. Nonlinear Anal. **72**(1), 302–308 (2010)
45. Liu, J., Liao, J.-F., Tang, C.-L.: Positive solutions for Kirchhoff-type equations with critical exponent in $\mathbb{R}^N$. J. Math. Anal. Appl. **429**(2), 1153–1172 (2015)
46. Liu, J., Pucci, P., Wu, H., Zhang, Q.: Existence and blow-up rate of large solutions of $p(x)$-Laplacian equations with gradient terms. J. Math. Anal. Appl. **457**(1), 944–977 (2018)
47. Lorenzo, C.F., Hartley, T.T.: Variable order and distributed order fractional operators. Nonlinear Dyn. **29**(1–4), 57–98 (2002)
48. Lusternik, L., Schnirelmann, L.: Méthodes topologiques dans les problèmes variationnels. Actualites Sci. Industr., 188, Paris (1934)
49. Lusternik, L., Schnirelmann, L.: Sur le problème de trois géodésiques fermées sur les suefaces de genre 0. C. R. Math. Acad. Sci. Paris **189**, 269–271 (1929)
50. Mihăilescu, M., Rădulescu, V.: On a nonhomogeneous quasilinear eigenvalue problem in Sobolev spaces with variable exponent. Proc. Am. Math. Soc. **135**(9), 2929–2937 (2007)
51. Milla Miranda, M., Lourêdo, A.T., Medeiros, L.A.: On nonlinear wave equations of carrier type. J. Math. Anal. Appl. **432**(1), 565–582 (2015)
52. Morse, M.: The critical points of functions and the calculus of variations in the large. Bull. Am. Math. Soc. **35**(1), 38–54 (1929)
53. Narashima, R.: Non-linear vibration of an elastic string. J. Sound Vib. **8**(1), 134–146 (1968)
54. Oplinger, D.W.: Frequency response of a nonlinear stretched string. J. Acoust. Soc. Am. **32**, 1529–1538 (1960)
55. Perera, K., Squassina, M., Yang, Y.: Bifurcation and multiplicity results for critical fractional p-Laplacian problems. Math. Nachr. **289**(2–3), 332–342 (2016)
56. Pucci, P., Saldi, S.: Critical stationary Kirchhoff equations in $\mathbb{R}^N$ involving nonlocal operators. Rev. Mat. Iberoam. **32**(1), 1–22 (2016)
57. Pucci, P., Xiang, M., Zhang, B.: Multiple solutions for nonhomogeneous Schrödinger-Kirchhoff type equations involving the fractional p-laplacian in $\mathbb{R}^N$. Calc. Var. Part. Differ. Equ. **54**(3), 2785–2806 (2015)
58. Pucci, P., Xiang, M., Zhang, B.: Existence and multiplicity of entire solutions for fractional p-Kirchhoff equations. Adv. Nonlinear Anal. **5**(1), 27–55 (2016)

59. Pucci, P., Zhang, Q.: Existence of entire solutions for a class of variable exponent elliptic equations. J. Differ. Equ. **257**(5), 1529–1566 (2014)
60. Ruiz-Medina, M.D., Anh, V.V., Angulo, J.M.: Fractional generalized random fields of variable order. Stoch. Anal. Appl. **22**(3), 775–799 (2004)
61. Rădulescu, V., Repovš, D.: Partial Differential Equations with Variable Exponents: Variational Methods and Qualitative Analysis. CRC Press, Taylor & Francis Group, Boca Raton FL (2015)
62. Samko, S.G., Ross, B.: Integration and differentiation to a variable fractional order. Integral Transform. Spec. Funct. **1**(4), 277–300 (1993)
63. Villaggio, P.: Mathematical Models for Elastic Structures. Cambridge University Press (1997)
64. Wang, L., Zhang, B.L.: Infinitely many solutions for Kirchhoff-type variable-order fractional Laplacian problems involving variable exponents. Appl. Anal. (2019)
65. Xiang, M., Hu,D., Zhang, B., Wang, Y.: Multiplicity of solutions for variable-order fractional Kirchhoff equations with nonstandard growth. J. Math. Anal. Appl. (2020)
66. Xiang, M., Zhang, B., Ferrara, M.: Existence of solutions for Kirchhoff type problem involving the non-local fractional p-Laplacian. J. Math. Anal. Appl. **424**(2), 1021–1041 (2015)
67. Xiang, M., Zhang, B., Ferrara, M.: Multiplicity results for the non-homogeneous fractional p-Kirchhoff equations with concave-convex nonlinearities. Proc. R. Soc. A: Math. Phys. Eng. Sci. **471**(2177), 1–14 (2015)
68. Xiang, M., Zhang, B., Yang, D.: Multiplicity results for variable-order fractional Laplacian equations with variable growth. Nonlinear Anal. **178**, 190–204 (2019)
69. Yamazaki, T.: Global solvability for the Kirchhoff equations in exterior domains of dimension three. J. Differ. Equ. **210**(2), 290–316 (2005)
70. Zhang, X., Fu, Y.: Bifurcation results for a class of $p(x)$-Laplacian equation. Nonlinear Anal. **73**(11), 3641–3650 (2010)
71. Zuo, J., An, T., Fiscella, A.: A critical Kirchhoff-type problem driven by a $p(\cdot)$-fractional Laplace operator with variable $s(\cdot)$-order. Math. Methods Appl. Sci. **44**(1), 1071–1085 (2021)
72. Zuo, J., An, T., Fiscella, A., Liu, C.: Existence and multiplicity of solutions for a bi-non-local problem. Mathematics **10**(12), 12 (2022)
73. Zuo, J., An, T., Li, M.: Superlinear Kirchhoff-type problems of the fractional p-Laplacian without the (AR) condition. Bound. Value Probl. **2018**, 13 (2018)
74. Zuo, J., An, T., Ye, G., Qiao, Z.: Nonhomogeneous fractional p-Kirchhoff problems involving a critical nonlinearity. Electron. J. Qual. Theory Differ. Equ. **15**, 15 (2021)
75. Zuo, J., Choudhuri, D., Repovš, D.: On critical variable-order Kirchhoff type problems with variable singular exponent. J. Math. Anal. Appl. **514**(1), 18 (2022)
76. Zuo, J., Fiscella, A., Bahrouni, A.: Existence and multiplicity results for p(.)&q(.) fractional Choquard problems with variable order. Complex Var. Elliptic Equ. **67**(2), 500–516 (2022)

Chapter 6
Nonlinear Periodic Systems with Variable Exponents

Zhenhai Liu and Nikolaos S. Papageorgiou

Abstract We consider a second order periodic system driven by a nonhomogeneous differential operator with variable growth and a reaction that involves two competing nonlinearities. A parametric "concave (sublinear)" term and a "convex (superlinear)" term. Using variational tools (critical point theory) and critical groups, we show that the system has at least two distinct nontrivial solutions.

6.1 Introduction

In this paper we study the existence of nontrivial solutions for the following nonlinear anisotropic second order periodic system:

$$\left\{\begin{array}{l} -a(u'(t))' + |u(t)|^{p(t)-2}u(t) = \lambda|u(t)|^{\tau(t)-2}u(t) + f(t, u(t)) \\ \qquad\qquad\qquad\qquad\qquad\qquad\qquad\qquad \text{a.a. on } T = [0, b] \\ u(0) = u(b),\ u'(0) = u'(b),\ \lambda > 0. \end{array}\right. \qquad (p_\lambda)$$

In this system the map $a : \mathbb{R}^N \to \mathbb{R}^N$ involved in the differential operator is a homeomorphism which need not be homogeneous, has nonstandard growth and includes as a special case the $p(t)$-Laplacian (the anisotropic p-Laplacian). In the reaction (right-hand side of (p_λ)), we have the combined effects of two terms which are of different nature. One is the parametric term $u \to \lambda|u|^{\tau(t)-2}u$. The assumptions on the variable exponent $\tau(\cdot)$ (see the hypotheses in Sect. 2) make this term a "concave" $((p(t)-1)$-sublinear) contribution to the reaction of (p_λ). The other

Z. Liu
Guangxi Colleges and Universities Key Laboratory of Complex System Optimization and Big Data Processing, Yulin Normal University, Yulin, People's Republic of China

N. S. Papageorgiou (✉)
Department of Mathematics, National Technical University, Athens, Greece
e-mail: npapg@math.ntua.gr

M. Jleli et al. (eds.), *Advances in Nonlinear Evolution Equations*, Infosys Science Foundation Series, https://doi.org/10.1007/978-981-97-9812-4_6

term is the perturbation $f(t,x)$ which is a Carathéodory function (that is, for all $x \in \mathbb{R}^N$, $t \to f(t,x)$ is measurable and for a.a. $t \in T$, $x \to f(t,x)$ is continuous), and we assume that $f(t,\cdot)$ is $(p(t)-1)$-superlinear as $x \to \pm\infty$, making this perturbation a "convex" contribution to the reaction. The superlinearity of $f(t,\cdot)$ is not expressed using the Ambrosetti–Rabinowitz condition (AR-condition) which is common in the literature. Instead we use a weaker condition, allowing superlinear nonlinearities with "slower" growth as $x \to \pm\infty$ which fail to satisfy the AR-condition. So, problem (p_λ) is an anisotropic periodic "concave–convex" system. Periodic systems driven by the "homogeneous" ($p(\cdot)$ is a constant) p-Laplacian were examined by Gasiński–Papageorgiou [4], Tang-Wu [8], Teng-Wu [9] and for systems driven by the $p(t)$-Laplacian, we have the works of Chen-Tang [1] and Wang-Yuan [10]. The aforementioned works do not deal with concave–convex problems and do not prove multiplicity theorems.

Here using variational tools from the critical point theory and critical groups, we show that for all $\lambda > 0$ small, problem (p_λ) has at least two nontrivial solutions in $C^1(T,\mathbb{R}^N)$.

6.2 Mathematical Background-Hypotheses

We introduce the following set:

$$E_1 = \{p \in L^\infty(0,b) : 1 < \operatorname*{ess\,inf}_{(0,b)} p\}.$$

Given $p \in E_1$, we define

$$p_- = \operatorname*{ess\,inf}_{(0,b)} p, \quad p_+ = \operatorname*{ess\,sup}_{(0,b)} p.$$

Also by $L^0((0,b),\mathbb{R}^N) = L^0_N(0,b)$, we denote the space of all measurable functions $u : [0,b] \to \mathbb{R}^N$. As usual we identify two such functions which differ only on a Lebesgue-null set.

Given $p \in E_1$, the variable Lebesgue space $L^{p(t)}((0,b),\mathbb{R}^N)$ is defined by

$$L^{p(t)}((0,b),\mathbb{R}^N) = L^{p(t)}_N(0,b) = \{u \in L^0_N(0,b) : \int_0^b |u|^{p(t)} dt < \infty\}.$$

The function $\rho_p(u) = \int_0^b |u|^{p(t)} dt$ is called the "modular function". We equip $L^{p(t)}_N(0,b)$ with the so-called "Luxemburg norm" defined by

$$\|u\|_{p(t)} = \inf\{\lambda > 0 : \int_0^b \left(\frac{|u|}{\lambda}\right)^{p(t)} dt \le 1\}.$$

With this norm $L_N^{p(t)}(0,b)$ becomes a separable and reflexive (in fact uniformly convex) Banach space.

Recall that if $u \in L^1_{loc}((0,b),\mathbb{R}^N)$, then the "weak derivative" of u is defined by

$$\int_0^b (u(t),h'(t))_{\mathbb{R}^N}dt = -\int_0^b (u'(t),h(t))_{\mathbb{R}^N}dt$$

for all $h \in C_c^\infty((0,b),\mathbb{R}^N)$. Using this notion we can define the variable Sobolev space $W^{1,p(t)}((0,b),\mathbb{R}^N)$ as follows:

$$W^{1,p(t)}((0,b),\mathbb{R}^N) = W_N^{1,p(t)}(0,b) = \{u \in L_N^{p(t)}(0,b) : u' \in L_N^{p(t)}(0,b)\}.$$

We equip $W_N^{1,p(t)}(0,b)$ with the following norm:

$$\|u\|_{1,p(t)} = \|u\|_{p(t)} + \|u'\|_{p(t)}.$$

With this norm $W_N^{1,p(t)}(0,b)$ becomes a separable and reflexive (in fact uniformly convex) Banach space.

Let $C^\infty_{per}(T,\mathbb{R}^N) = \{u \in C^\infty(T,\mathbb{R}^N) : u(0) = u(b)\}$ and let

$$W^{1,p(t)}_{N,per}(0,b) = \overline{C^\infty_{per}(T,\mathbb{R}^N)}^{\|\cdot\|_{1,p(t)}}.$$

We have that

$$W^{1,p(t)}_{N,per}(0,b) \hookrightarrow C(T,\mathbb{R}^N) \text{ compactly}$$

and

$$C^\infty_{per}(T,\mathbb{R}^N) \text{ is dense in } W^{1,p(t)}_{N,per}(0,b).$$

From these facts it follows

$$W^{1,p(t)}_{N,per}(0,b) = \{u \in L_N^{p(t)}(0,b) : u' \in L_N^{p(t)}(0,b), u(0) = u(b)\}.$$

Moreover, for $u \in W^{1,p(t)}_{N,per}(0,b)$, we have

$$\int_0^b (u(t),h'(t))_{\mathbb{R}^N}dt = -\int_0^b (u'(t),h(t))_{\mathbb{R}^N}dt$$

for all $h \in W^{1,p(t)}_{N,per}(0,b)$, and u admits an absolutely continuous representative. Also, we have $u(t) = u(0) + \int_0^t u'(s)ds$ for all $t \in T$ (see Fan-Fan [3]).

Finally, we also have

$$W_0^{1,p(t)}((0,b),\mathbb{R}^N) = \overline{C_c^\infty(0,b),\mathbb{R}^N}^{\|\cdot\|_{1,p(t)}}.$$

Given $p \in E_1$, let $p' \in E_1$ be defined by

$$\frac{1}{p(t)} + \frac{1}{p'(t)} = 1 \quad \text{for all } t \in [0, b].$$

We know that $L_N^{p(t)}(0, b)^* = L_N^{p'(t)}(0, b)$. In what follows for notational simplicity, we write

$$W_{per}^{1,p(t)} = W_{N,per}^{1,p(t)}(0, b),$$
$$\|u\| = \|u\|_{1,p(t)} \quad \text{for all } u \in W_{per}^{1,p(t)}.$$

There is a close relation between the norm and the modular function.

Proposition 6.1 *We have the following:*

(a) $\|u\| < 1 (resp = 1, > 1) \Leftrightarrow \rho_p(u') + \rho_p(u) < 1 (resp = 1, > 1)$;

(b) *if* $\|u\| < 1$, *then* $\|u\|^{p_+} \leq \rho_p(u') + \rho_p(u) \leq \|u\|^{p_-}$;

(c) *if* $\|u\| > 1$, *then* $\|u\|^{p_-} \leq \rho_p(u') + \rho_p(u) \leq \|u\|^{p_+}$;

(d) $\|u\| \to 0\ (resp \to \infty) \Leftrightarrow \rho_p(u') + \rho_p(u) \to 0\ (resp \to \infty)$.

We consider exponents $p, q, \tau \in C_{per}(T, \mathbb{R}) = \{k \in C(T, \mathbb{R}) : k(0) = k(b)\}$ and $1 < \tau_- \leq \tau_+ < q_- \leq q_+ < p_-$. Our hypotheses on the map $a(\cdot)$ involved in the differential operator of (p_λ) are the following:

H_1 : $a(y) = a_0(|y|)y$ for all $y \in \mathbb{R}^N$ with $a_0 : \mathring{\mathbb{R}}_+ \to \mathring{\mathbb{R}}_+$ ($\mathring{\mathbb{R}}_+ = (0, \infty)$) strictly increasing, $a_0(s)s \to 0^+$ as $s \to 0^+$ and

(i) $c_0|y|^{p(t)} \leq (a(y), y)_{\mathbb{R}^N} \leq c_1[|y|^{p(t)} + |y|^{q(t)}]$ for all $t \in T$, all $y \in \mathbb{R}^N$, some $c_0, c_1 > 0$;

(ii) if $G_0(\theta) = \int_0^\theta a_0(s)sds$, then $\limsup_{\theta \to 0^+} \frac{G_0(\theta)}{\theta^{q(t)}} \leq \widehat{c}$ uniformly for all $t \in T$;

(iii) $(a(y), y)_{\mathbb{R}^N} \leq p_+ G(y)$ for all $y \in \mathbb{R}^N$.

Remark 6.1 Evidently $s \to G_0(s)$ is strictly increasing and strictly convex. We set

$$G(y) = G_0(|y|) \quad \text{for all } y \in \mathbb{R}^N.$$

We see that

$$\nabla G(y) = G_0'(|y|)\frac{y}{|y|} = a_0(|y|)y = a(y) \quad \text{for all } y \in \mathbb{R}^N \setminus \{0\}.$$

So, $G(\cdot)$ is the primitive of $a(\cdot)$, $G \in C^1(\mathbb{R}^N, \mathbb{R})$ and of course it is convex with $G(0) = 0$. So, it follows that $G(y) \le (a(y), y)_{\mathbb{R}^N}$ for all $y \in \mathbb{R}^N$ and

$$c_2|y|^{p(t)} \le G(y) \le c_3[|y|^{p(t)} + |y|^{q(t)}]$$

for all $y \in \mathbb{R}^N$, some $c_2, c_3 > 0$. These assumptions incorporate in our framework the $p(t)$-Laplacian which corresponds to the map

$$a(y) = |y|^{p(t)-2}y \quad \text{for all } y \in \mathbb{R}^N$$

and the $(p(t), q(t))$-Laplacian which corresponds to the map

$$a(y) = |y|^{p(t)-2}y + |y|^{q(t)-2}y.$$

Let $V : W^{1,p(t)}_{per} \to (W^{1,p(t)}_{per})^*$ be defined by

$$\langle V(u), h\rangle = \int_0^b (a(u'(t)), h'(t))_{\mathbb{R}^N} dt \quad \text{for all } u, h \in W^{1,p(t)}_{per}.$$

As in Gasiński–Papageorgiou [5], p. 279, for this operator we have the following properties.

Proposition 6.2 *The operator $V(\cdot)$ is bounded (that is, maps bounded sets to bounded sets), continuous, monotone (thus maximal monotone too) and of type $(S)_+$, that is, "$u_n \xrightarrow{w} u$ in $W^{1,p(t)}_{per}$ and $\limsup_{n\to\infty}\langle V(u_n), u_n - u\rangle \le 0$ imply that $u_n \to u$ in $W^{1,p(t)}_{per}$".*

The hypotheses on the reaction $f(t, x)$ are the following:
H_2 : $f(t, x) = \nabla F(t, x)$ with $F(\cdot, \cdot)$ a measurable function which is C^1 in the $x \in \mathbb{R}^N$ variable, $\nabla F(t, 0) = 0$ for a.a. $t \in T$ and

(i) $|f(t, x)| \le \hat{a}(t)[1 + |x|^{r(t)-1}]$ for a.a. $t \in T$, all $x \in \mathbb{R}^N$, with $\hat{a} \in L^\infty(T)$, $r \in C(T)$ and $p_+ < r_-$;

(ii) $\lim\limits_{|x|\to\infty} \frac{F(t,x)}{|x|^{p_+}} = +\infty$ uniformly for a.a. $t \in T$ and there exists $\mu \in C(T)$ such that $r_+ - p_- < \mu(t)$ for all $t \in T$, $\tau_+ < \mu_-$ and

$$0 < \beta_0 \leq \liminf_{|x|\to\infty} \frac{(f(t,x),x)_{\mathbb{R}^N} - p_+ F(t,x)}{|x|^{\mu(t)}} \quad \text{uniformly for a.a. } t \in T;$$

(iii) $\lim\limits_{x\to 0} \frac{F(t,x)}{|x|^{p(t)}} = 0$ uniformly for a.a. $t \in T$ and there exists $\delta_0 > 0$ such that $F(t,x) \geq 0$ for a.e. $t \in T$, all $|x| \leq \delta_0$.

Remark 6.2 Hypothesis $H_2(ii)$ implies that $f(t,\cdot)$ is $(p_+ - 1)$-superlinear as $|x| \to \infty$. However, the superlinearity of $f(t,\cdot)$ is not expressed in terms of the AR-condition (see Willem [11], p. 59). Instead we employ a less restrictive condition which fits in our framework superlinear nonlinearities with "slower" growth as $|x| \to \infty$, which fail to satisfy the AR-condition. For example, consider the function $F(t,x)$ defined by

$$F(t,x) = \frac{|x|^{p_+}}{p_+}[\ln|x| - \frac{1}{p_+}] + \frac{1}{p}|x|^{p(t)}$$

with $p \in C_{per}(T)$, $1 < p_-$. Then $f(t,x) = \nabla F(t,x)$ satisfies hypotheses H_2, but fails to satisfy the AR-condition.

As we mentioned in the introduction, we will also use critical groups in order to distinguish between solutions. So, let us recall the definition of critical groups at an isolated critical point of a C^1-functional.

Let X be a Banach space and $\varphi \in C^1(X,\mathbb{R})$. We set

$$K_\varphi = \{u \in X : \varphi'(u) = 0\} \quad \text{(the critical set of } \varphi).$$

If (Y_1, Y_2) is a topological pair with $Y_2 \subseteq Y_1 \subseteq X$ by $H_k(Y_1, Y_2)$ $(k \in \mathbb{N}_0)$, we denote the kth relative singular homology group for the pair (Y_1, Y_2) with integer coefficients. Consider $u \in K_\varphi$ isolated and let $c = \varphi(u)$ and $\varphi^c = \{u \in X : \varphi(u) \leq c\}$. The critical groups of φ at u are defined by

$$C_k(\varphi, u) = H_k(\varphi^c \cap U, \varphi^c \cap U \backslash \{u\}) \quad \text{for all } k \in \mathbb{N}_0,$$

with U being the neighborhood of u such that $K_\varphi \cap \varphi^c \cap U = \{u\}$. The excision property of singular homology theory implies that this definition of critical groups is independent of the isolating neighborhood U (see [6], Chap. 6).

Finally we say that φ satisfies the "C-condition" if it has the following property: "Every sequence $\{u_n\}_{n\in\mathbb{N}} \subseteq X$ such that $\{\varphi(u_n)\}_{n\in\mathbb{N}} \subseteq \mathbb{R}$ is bounded, $(1 + \|u_n\|_X)\varphi'(u_n) \to 0$ in X^* as $n \to \infty$, has a strongly convergent subsequence."

6.3 Multiplicity Theorem

For every $\lambda > 0$, we introduce the energy functional $\varphi_\lambda : W^{1,p(t)}_{per} \to \mathbb{R}$ for problem (p_λ) defined by

$$\varphi_\lambda(u) = \int_0^b G(u')dt + \int_0^b \frac{1}{p(t)}|u|^{p(t)}dt - \int_0^b \frac{\lambda}{\tau(t)}|u|^{\tau(t)}dt - \int_0^b F(t,u)dt \text{ for all } u \in W^{1,p(t)}_{per}.$$

Evidently $\varphi_\lambda \in C^1(W^{1,p(t)}_{per})$.

Proposition 6.3 *If hypotheses H_1, H_2 hold and $\lambda > 0$, then $\varphi_\lambda(\cdot)$ satisfies the C-condition.*

Proof We consider a sequence $\{u_n\}_{n\in\mathbb{N}} \subseteq W^{1,p(t)}_{per}$ such that

$$|\varphi_\lambda(u_n)| \le c_4 \quad \text{for some } c_4 > 0, \text{ all } n \in \mathbb{N}, \tag{6.1}$$

$$(1 + \|u_n\|)\varphi'_\lambda(u_n) \text{ in } (W^{1,p(t)}_{per})^* \text{ as } n \to \infty. \tag{6.2}$$

From (6.2) we have

$$\begin{aligned} &|\langle V(u_n), h\rangle + \int_0^b |u_n|^{p(t)-2}(u_n,h)_{\mathbb{R}^N}dt \\ &-\lambda\int_0^b |u_n|^{\tau(z)-2}(u_n,h)_{\mathbb{R}^N}dz \\ &-\int_0^b (\nabla F(t,u_n),h)_{\mathbb{R}^N}dt| \le \frac{\varepsilon_n\|h\|}{1+\|u_n\|} \\ &\text{for all } h \in W^{1,p(t)}_{per}, \text{ with } \varepsilon_n \to 0_+. \end{aligned} \tag{6.3}$$

In (6.3), we use the test function $h = u_n \in W^{1,p(t)}_{per}$ and obtain

$$\begin{aligned} &-\int_0^b (a(u_n'),u_n')_{\mathbb{R}^N}dt - \rho_p(u_n) + \lambda\rho_\tau(u_n) \\ &+\int_0^b (\nabla F(t,u_n),u_n)_{\mathbb{R}^N}dt \le \varepsilon_n \text{ for all } n \in \mathbb{N}. \end{aligned} \tag{6.4}$$

On the other hand, from (6.1) we have

$$\int_0^b G(u_n')dt + \frac{1}{p_+}\rho_p(u_n) - \frac{\lambda}{\tau_-}\rho_\tau(u_n) - \int_0^b F(t,u_n)d \le c_4,$$
$$\Rightarrow \int_0^b p_+ G(u_n')dt + \rho_p(u_n) - \frac{\lambda p_+}{\tau_-}\rho_\tau(u_n)$$
$$-\int_0^b p_+ F(t,u_n)dt \le p_+ c_4 \text{ for all } n \in \mathbb{N}. \tag{6.5}$$

We add (6.4) and (6.5) and obtain

$$\int_0^b [p_+ G(u_n') - (a(u_n'), u_n')_{\mathbb{R}^N}]dt$$
$$+ \int_0^b [(\nabla F(t,u_n), u_n)_{\mathbb{R}^N} - p_+ F(t,u_n)]dt$$
$$\le \lambda \left[\frac{p_+}{\tau_-} - 1\right] p_\tau(u_n) + c_5$$
$$\text{for some } c_5 > 0 \text{ , all } n \in \mathbb{N},$$

$$\Rightarrow \int_0^b [(\nabla F(t,u_n), u_n)_{\mathbb{R}^N} - p_+ F(t,u_n)]dt$$
$$\le \lambda \left[\frac{p_+}{\tau_-} - 1\right] p_\tau(u_n) + 1 \text{ for all } n \in \mathbb{N} \tag{6.6}$$
(see hypothesis $H_1(iii)$).

On account of hypotheses $H_2(i), (ii)$, we can find $\beta_1 \in (0, \beta_0)$ and $c_6 > 0$ such that

$$\beta_1 |x|^{\mu(t)} - c_6 \le (\nabla F(t,x), x)_{\mathbb{R}^N} - p_+ F(t,x) \tag{6.7}$$

for a.a. $t \in T$, all $x \in \mathbb{R}^N$. Using (6.7) in (6.6), we obtain

$$\beta_1 \rho_\mu(u_n) \le \lambda \left[\frac{p_+}{\tau_-} - 1\right] \rho_\tau(u_n) + c_7 \tag{6.8}$$

for some $c_7 > 0$, all $n \in \mathbb{N}$.

We may assume that $\|u_n\| \ge 1$ is big for all $n \in \mathbb{N}$ so that $\|u_n\|_{\tau(t)}, \|u_n\|_{\mu(t)} \ge 1$ for all $n \in \mathbb{N}$, since $W^{1,p(t)}_{per} \hookrightarrow L^{\tau(t)}_N(0,b), L^{\mu(t)}_N(0,b)$ continuously (in fact compactly). Using Proposition 6.1, we have

$$\begin{aligned}\beta_1\|u_n\|_{\mu(t)}^{\mu_-} &\leq \lambda\left[\frac{p_+}{\tau_-}-1\right]\|u_n\|_{\tau(t)}^{\tau_+}+c_7\\ &\leq c_8\left[\lambda\|u_n\|_{\mu(t)}^{\tau_+}+1\right] \qquad (6.9)\\ &\text{for some } c_8>0, \text{ all } n\in\mathbb{N},\end{aligned}$$

$$\text{(since } \tau_+<\mu_-, \text{ we have } L_N^{\mu(t)}(0,b)\hookrightarrow L_N^{\tau(t)}(0,b) \text{ continuously)}$$

$$\Rightarrow \{u_n\}_{n\in\mathbb{N}}\subseteq L_N^{\mu(t)}(0,b) \text{ is bounded (because } \tau_+<\mu_-)$$

$$\Rightarrow \{u_n\}_{n\in\mathbb{N}}\subseteq L_N^{\mu_-}(0,b) \text{ is bounded}$$

$$\text{(since } L_N^{p(t)}(0,b)\hookrightarrow L_N^{\mu_-}(0,b) \text{ continuously).}$$

From hypothesis $H_2(ii)$, we see that we can always assume that $\mu_- < r_+$. Then we can find $t\in(0,1)$ such that

$$\frac{1}{r_+}=\frac{1-t}{\mu_-}. \qquad (6.10)$$

Using the interpolation inequality (see Papageorgiou–Winkert [7], p. 116), we have

$$\begin{aligned}&\|u_n\|_{r_+}\leq\|u_n\|_{\mu_-}^{1-t}\|u_n\|_\infty^t,\\ &\Rightarrow \|u_n\|_{r_+}^{r_+}\leq c_9\|u_n\|^{tr_+} \quad \text{for some } c_9>0, \text{ all } n\in\mathbb{N} \qquad (6.11)\\ &\text{(see (6.3.9) and recall that } W_{per}^{1,p(t)}\hookrightarrow C_{per}(T,\mathbb{R}^N) \text{ compactly).}\end{aligned}$$

From (6.3) with $h=u_n\in W_{per}^{1,p(t)}$, we have

$$\begin{aligned}&\int_0^b (a(u_n'),u_n')_{\mathbb{R}^N}dt+\rho_p(u_n)\\ &=\lambda\rho_\tau(u_n)+\int_0^b(\nabla F(t,u_n),u_n)_{\mathbb{R}^N}dt+\varepsilon_n\\ &\leq\lambda\rho_\tau(u_n)+c_{10}[1+\rho_r(u_n)]\\ &\quad\text{for some } c_{10}>0, \text{ all } n\in\mathbb{N} \text{ (see hypothesis } H_2(i))\\ &\leq c_{11}[\lambda\|u_n\|^{\tau_+}+\|u_n\|_{r_+}^{r_+}+1]\\ &\quad\text{for some } c_{11}>0, \text{ all } n\in\mathbb{N}\\ &\leq c_{12}[\lambda\|u_n\|^{\tau_+}+\|u_n\|^{tr_+}+1]\\ &\quad\text{for some } c_{12}>0, \text{ all } n\in\mathbb{N} \text{ (see (6.3.11))},\\ &\Rightarrow \rho_p(u_n')+\rho_p(u_n)\leq c_{13}[\lambda\|u_n\|^{\tau_+}+\|u_n\|^{tr_+}+1]\\ &\text{for some } c_{13}>0, \text{ all } n\in\mathbb{N} \text{ (see hypothesis } H_1(i)),\\ &\Rightarrow \|u_n\|^{p_-}\leq c_{13}[\lambda\|u_n\|^{\tau_+}+\|u_n\|^{tr_+}+1]. \qquad (6.12)\end{aligned}$$

From (6.10) and hypothesis $H_2(iii)$, we have

$$tr_+ = r_+ - \mu_- < p_-.$$

Also recall that

$$\tau_+ < \mu_-, p_-.$$

Then from (6.12), we infer that

$$\{u_n\}_{n\in\mathbb{N}} \leq W^{1,p(t)}_{per} \text{ is bounded.}$$

So, we may assume that

$$u_n \xrightarrow{w} u \text{ in } W^{1,p(t)}_{per}, \quad u_n \to u \text{ in } C_{per}(T, \mathbb{R}^N). \tag{6.13}$$

In (6.3) we use the test function $h = u_n - u \in W^{1,p(t)}_{per}$, pass to the limit as $n \to \infty$ and use (6.13). We obtain

$$\begin{aligned}
&\lim_{n\to\infty} \langle V(u_n), u_n - u\rangle = 0, \\
&\Rightarrow u_n \to u \text{ in } W^{1,p(t)}_{per} \text{ (see Proposition 6.2)}, \\
&\Rightarrow \varphi_\lambda(\cdot) \text{ satisfies the } C\text{-condition.}
\end{aligned}$$

□

The next proposition together with the $(p_+ - 1)$-superlinearity of $F(t, \cdot)$ (see hypothesis $H_2(ii)$) implies that for all $\lambda > 0$ small the energy functional $\varphi_\lambda(\cdot)$ satisfies the mountain pass geometry.

Proposition 6.4 *If hypotheses H_1, H_2 hold, then there exists $\lambda^* > 0$ such that for all $\lambda \in (0, \lambda^*)$ and there exists $\rho_\lambda > 0$ for which we have*

$$\varphi_\lambda(u) \geq m_\lambda > 0 \quad \textit{for all } u \in W^{1,p(t)}_{per}, \ \|u\| = \rho_\lambda.$$

Proof On account of hypotheses $H_2(i)$, (ii), given $\varepsilon > 0$, we can find $c_{14} > 0$ such that

$$F(t, x) \leq \varepsilon |x|^{p(t)} + c_{14}|x|^{r(t)} \quad \text{for a.a. } t \in T, \ \text{all } x \in \mathbb{R}^N. \tag{6.14}$$

Consider $u \in W^{1,p(t)}_{per}$ with $\|u\| \leq 1$ small so that we have also

$$\|u\|_{\tau(t)} \leq 1, \ \|u\|_{r(t)} \leq 1$$

(recall that $W_{per}^{1,p(t)} \hookrightarrow L_N^{\tau(t)}(0,b)$ and $W_{per}^{1,p(t)} \hookrightarrow L_N^{r(t)}(0,b)$ compactly). Then, we have

$$\begin{aligned}\varphi_\lambda(u) &\geq c_2\rho_p(u') + \frac{1}{p_+}\rho_p(u) - \frac{\lambda}{\tau_-}\rho_\tau(u) - \varepsilon\rho_p(u) - c_{14}\rho_r(u) \quad \text{(see (6.3.14))}\\ &= c_2\rho_p(u') + [\frac{1}{p_+} - \varepsilon]\rho_p(u) - \frac{\lambda}{\tau_-}\rho_\tau(u) - c_{14}\rho_r(u).\end{aligned}$$

Choosing $\varepsilon \in (0, 1/p_+)$ and using Proposition 6.1, we obtain

$$\begin{aligned}\varphi_\lambda(u) &\geq c_{15}\|u\|^{p_+} - c_{16}(\lambda\|u\|^{\tau_-} + \|u\|^{r_-})\\ &\quad \text{for some } c_{15}, c_{16} > 0\\ &= [c_{15} - c_{16}(\lambda\|u\|^{\tau_- - p_+} + \|u\|^{r_- - p_+})]\|u\|^{p_+}. \end{aligned} \tag{6.15}$$

Consider the function

$$\gamma_\lambda(s) = \lambda s^{\tau_- - p_+} + s^{r_- - p_+} \quad \text{for all } s > 0.$$

Since $\tau_- < p_+ < r_-$, we see that

$$\gamma_\lambda(s) \to +\infty \quad \text{as } s \to 0^+ \quad \text{and as } s \to +\infty.$$

Therefore we can find $s_0 > 0$ such that

$$\begin{aligned}&\gamma_\lambda(s_0) = \inf\{\gamma_\lambda(s) : 0 < s\},\\ &\Rightarrow \gamma_\lambda'(s_0) = 0,\\ &\Rightarrow (r_- - p_+)s_0^{r_- - \tau_-} = \lambda(p_+ - \tau_-),\\ &\Rightarrow s_0 = s_0(\lambda) = \left[\frac{\lambda(p_+ - \tau_-)}{r_- - p_+}\right]^{\frac{1}{r_- - \tau_-}}.\end{aligned}$$

Then we have

$$\gamma_\lambda(s_0) = \lambda\left[\frac{r_- - p_+}{\lambda(p_+ - \tau_-)}\right]^{\frac{p_+ - \tau_-}{r_- - \tau_-}} + \left[\frac{\lambda(p_- - \tau_-)}{r_- - p_+}\right]^{\frac{r_- - p_+}{r_- - \tau_-}}.$$

Since $\tau_- < p_+ < r_-$, we see that

$$\gamma_\lambda(s_0) \to 0 \quad \text{as } \lambda \to 0_+.$$

Therefore, we can find $\lambda^* > 0$ such that for all $\lambda \in (0, \lambda^*)$, we have

$$\begin{aligned}
&\gamma_\lambda(s_0) < \frac{c_{15}}{c_{16}}, \\
\Rightarrow \quad &\varphi_\lambda(u) \geq m_\lambda > 0 \\
&\text{for all } u \in W^{1,p(t)}_{per},\ \|u\| = \rho_\lambda = s_0(\lambda).
\end{aligned}$$

□

As we already hinted earlier, one solution of (p_λ) will be obtained using the mountain pass theorem. The second nontrivial solution will be produced via a local minimization problem. The next proposition will guarantee the nontriviality of the second solution.

Proposition 6.5 *If hypotheses H_1, H_2 hold and $\lambda \in (0, \lambda^*)$, then*

$$\inf\{\varphi_\lambda(u) : u \in W^{1,p(t)}_{per},\ \|u\| \leq \rho_\lambda\} < 0.$$

Proof On account of hypothesis $H_1(iii)$, we can find $c_{17} > \hat{c}$ and $\delta \in (0, \delta_0]$ such that

$$G(y) \leq c_{17}|y|^{q(t)} \quad \text{for all } |y| \leq \delta. \tag{6.16}$$

Consider $u \in C^1_{per}(T, \mathbb{R}^N)$ and choose $\theta \in (0, 1)$ small such that

$$\theta\|u\|_{C^1_{per}} \leq \delta. \tag{6.17}$$

Then from hypothesis $H_2(iii)$ we have

$$F(t, \theta u(t)) \geq 0 \quad \text{for a.a. } t \in T. \tag{6.18}$$

From (6.16), (6.17) and (6.18) we have

$$\begin{aligned}
\varphi_\lambda(\theta u) &\leq c_{17}\theta^{q_-}\rho_q(u') + \frac{\theta^{q_-}}{q_-}\rho_p(u) - \frac{\lambda\theta^{\tau_+}}{\tau_+}\rho_\tau(u) \\
&\leq c_{18}\theta^{q_-} - \lambda c_{19}\theta^{\tau_+} \quad \text{for some } c_{18}, c_{19} > 0.
\end{aligned}$$

Since $\tau_+ < q_-$, choosing $\theta \in (0, 1)$ even smaller if necessary, we have

$$\varphi_\lambda(\theta u) < 0 \quad \text{and} \quad 0 < \|\theta u\| \leq \rho_\lambda. \tag{6.19}$$

From (6.19), it follows that

$$\inf\{\varphi_\lambda(u) : u \in W^{1,p(t)}_{per},\ \|u\| \leq \rho_\lambda\} < 0.$$

□

Now we are ready to state and prove the multiplicity theorem for problem (p_λ).

Theorem 6.1 *If hypotheses H_1, H_2 hold, then for all $\lambda > 0$ small problem (p_λ) has at least two nontrivial solutions*

$$u_\lambda, \hat{u}_\lambda \in C^1(T, \mathbb{R}^N), u_\lambda \neq \hat{u}_\lambda.$$

Proof Let $\bar{B}_\lambda = \{u \in W_{per}^{1,p(t)} : \|u\| \leq p_\lambda\}$. The reflexivity of $W_{per}^{1,p(t)}$ and the Eberlein–Smulian theorem imply that

$$\bar{B}_\lambda \text{ is sequentially weakly compact.} \tag{6.20}$$

Moreover, using the fact that $W_{per}^{1,p(t)} \hookrightarrow C_{per}(T, \mathbb{R}^N)$ compactly, we see that

$$\varphi_\lambda(\cdot) \text{ is sequentially weakly lower semicontinuous.} \tag{6.21}$$

Then from (6.20), (6.21) and the Weierstrass–Tonelli theorem, we can find $u_\lambda \in \bar{B}_\lambda$ such that

$$\varphi_\lambda(u_\lambda) = \inf\{\varphi_\lambda(u) : u \in \bar{B}_{p_\lambda}\} = \hat{m}_\lambda. \tag{6.22}$$

On account of Propositions 6.4 and 6.5, if $\lambda > 0$ is small, then

$$0 < \|u_\lambda\| < \rho_\lambda. \tag{6.23}$$

From (6.22) and (6.23), we infer that

$$\begin{aligned}
&\langle \varphi_\lambda'(u_\lambda), h\rangle = 0 \quad \text{for all } h \in W_{per}^{1,p(t)},\\
&\Rightarrow \langle V(u_\lambda), h\rangle + \int_0^b |u_\lambda|^{p(t)-2}(u_\lambda, h)_{\mathbb{R}^N} dt\\
&= \lambda \int_0^b |u_\lambda|^{\tau(t)-2}(u_\lambda, h)_{\mathbb{R}^N} dt + \int_0^b (f(t, u_\lambda), h)_{\mathbb{R}^N} dt \qquad (6.24)\\
&\text{for all } h \in W_{per}^{1,p(t)}.
\end{aligned}$$

From the representation theorem for $W^{-1,p'(t)}((0,b),\mathbb{R}^N) = (W_0^{1,p(t)}((0,b),\mathbb{R}^N))^*$ (see Diening-Harjulehto-Hästo-Ruzička [2], p. 378), we see that

$$a(u_\lambda')' \in W^{-1,p'(t)}((0,b),\mathbb{R}^N).$$

In what follows by $\langle \cdot, \cdot \rangle_0$ we denote the duality brackets for the pair $(W^{-1,p'(t)}((0,b),\mathbb{R}^N), (W_0^{1,p(t)}((0,b),\mathbb{R}^N))^*)$. In (6.24) we choose $h \in C_c^1((0,b),\mathbb{R}^N) \subseteq W_{per}^{1,p(t)}$ and after integration by parts, we obtain

$$\begin{aligned}&\langle -a(u'_\lambda)', h\rangle_0 + \int_0^b |u_\lambda|^{p(t)-2}(u_\lambda, h)_{\mathbb{R}^N} dt \\ &= \lambda \int_0^b |u_\lambda|^{\tau(t)-2}(u_\lambda, h)_{\mathbb{R}^N} dt + \int_0^b (f(t, u_\lambda), h)_{\mathbb{R}^N} dt.\end{aligned} \tag{6.25}$$

Recall that $C_c^1((0, b), \mathbb{R}^N)$ is dense in $W_0^{1,p(t)}((0, b), \mathbb{R}^N)$ (see Sect. 2). So, from (6.25) we infer that

$$\begin{cases} -a(u'_\lambda)' + |u_\lambda|^{p(t)-2}u_\lambda = \lambda|u_\lambda|^{\tau(t)-2}u_\lambda + f(t, u_\lambda) & \text{for a.a. } t \in T, \\ u_\lambda(0) = u_\lambda(b). \end{cases} \tag{6.26}$$

Note that $g_\lambda(t) = |u_\lambda|^{p(t)-2}u_\lambda - \lambda|u_\lambda|^{\tau(t)-2}u_\lambda - f(t, u_\lambda)$ belongs in $L_N^{r'(t)}(\Omega)$ (recall $\tau_+ < p_+ < r_-$). Integrating (6.26) we obtain

$$a(u'_\lambda(t)) = a(u'_\lambda(0)) + \int_0^t g_\lambda(s)ds \quad \text{for all } t \in T.$$

But $a(\cdot)$ is a homeomorphism. Hence we have

$$\begin{aligned}&u'_\lambda(t) = a^{-1}[a(u'_\lambda(0)) + \int_0^t g_\lambda(s)ds] \quad \text{for all } t \in T, \\ &\Rightarrow u'_\lambda \in C(T, \mathbb{R}^N).\end{aligned}$$

Therefore we have $u_\lambda \in C^1(T, \mathbb{R}^N)$. Given $h \in W_{per}^{1,p}$ and using integration by parts, we have

$$\begin{aligned}&a(u'_\lambda(b)) - a(u'_\lambda(0)) - \int_0^b ((a(u'_\lambda))', h)_{\mathbb{R}^N} dt \\ &= \langle V(u_\lambda), h\rangle \\ &= \int_0^b [\lambda|u_\lambda|^{\tau(t)-2}(u_\lambda, h)_{\mathbb{R}_N} - (f(t, u_\lambda), h)_{\mathbb{R}_N} - |u_\lambda|^{p(t)-2}(u_\lambda, h)_{\mathbb{R}^N}]dt\end{aligned}$$

(see (6.3.24))

$\Rightarrow a(u'_\lambda(b)) = a(u'_\lambda(0))$ (see (6.3.26)),

$\Rightarrow u'_\lambda(b) = u'_\lambda(0)$

(since $a(\cdot)$ is a homeomorphism),

$\Rightarrow u_\lambda \in C_{per}^1(T, \mathbb{R}^N)$ is a nontrivial solution of (p_λ).

From Proposition 6.4, we know that

$$\varphi_\lambda(0) = 0 < m_\lambda = \inf\{\varphi_\lambda(u) : \|u\| = p_\lambda\}. \tag{6.27}$$

In addition Proposition 6.3 says that

$$\varphi_\lambda(\cdot) \text{ satisfies the } C\text{-condition.} \tag{6.28}$$

If $u \in C^1_{per}(T, \mathbb{R}^N)$, $u \not\equiv 0$, then from hypothesis $H_2(ii)$ we have

$$\varphi_\lambda(\theta u) \to -\infty \text{ as } \theta \to +\infty. \tag{6.29}$$

From (6.27), (6.28) and (6.29), we see that we can apply the mountain pass theorem (see [6], p. 401) and find $\hat{u}_\lambda \in W^{1,p(t)}_{per}$ such that

$$\begin{aligned}
&\hat{u}_\lambda \in K_{\varphi_\lambda} \text{ and } \varphi_\lambda(0) = 0 < m_\lambda \leq \varphi_\lambda(\hat{u}_\lambda),\\
&\Rightarrow \hat{u}_\lambda \neq 0, \hat{u}_\lambda \in C^1_{per}(T, \mathbb{R}^N) \text{ and solves } (p_\lambda)\\
&\text{(as we did for } u_\lambda).
\end{aligned}$$

From Theorem 6.5.8, p. 527, of [6], we have

$$C_1(\varphi_\lambda, \hat{u}_\lambda) \neq 0. \tag{6.30}$$

On the other hand, we know that u_λ is a local minimizer of φ_λ. Therefore

$$C_k(\varphi_\lambda, u_\lambda) = \delta_{k,0}\mathbb{Z} \quad \text{for all } k \in \mathbb{N}_0. \tag{6.31}$$

Comparing (6.30) and (6.31), we conclude that $\hat{u}_\lambda \neq u_\lambda$. □

Acknowledgements The work was supported by NNSF of China Grant No. 12071413 and the European Union's Horizon 2020 Research and Innovation Programme under the Marie Sklodowska-Curie grant agreement No. 823731 CONMECH.

References

1. Cheng, P., Tang, X.: Periodic solutions for a differential inclusion problem involving the $p(t)$-Laplacian. Adv. Nonlin. Anal. **10**, 799–815 (2021)
2. Diening, L., Harjulehto, P., Hästo, P., Ruzicka, M.: Lebesgue and Sobolev Spaces with Variable Exponent. Lecture Notes in Math, vol. 2017, Springer, Heidelberg (2011)
3. Fan, X.L., Fan, X.: A Knobloch-type result of $p(t)$-Laplacian systems. J. Math. Anal. Appl. **282**, 453–464 (2003)
4. Gasiński, L., Papageorgiou, N.S.: Existence and multiplicity of second order periodic systems with the p-Laplacian and a nonsmooth potential. Monatshefte Für Math. **158**, 121–150 (2009)
5. Gasiński, L., Papageorgiou, N.S.: Exercises in Analysis, Part 2: Nonlinear Analysis, Springer Cham (2016)
6. Papageorgiou, N.S., Rădulescu, V.D., Repovš, D.D.: Nonlinear Analysis-Theory and Methods. Springer Nature, Swizerland AG (2019)

7. Papageorgiou, N.S., Winkert, P.: Applied Nonlinear Functional Analysis. De Gruyter, Berlin (2018)
8. Tang, C.L., Wu, X.: Periodic solutions for second order systems with non uniformly coercive potential. J. Math. Anal. Appl. **259**, 386–397 (2011)
9. Teng, K., Wu, X.: Existence and subharmonicity of solutions for nonlinear nonsmooth periodic systems with the p-Laplacian. Nonlinear Anal. **68**, 3742–3756 (2008)
10. Wang, X., Yuan, R.: Existence of periodic solutions for $p(t)$-Laplacian systems. Nonlinear Anal. **70**, 866–880 (2009)
11. Willem, M.: Minimax Theorems. Birkhäuser, Boston (1996)

Chapter 7
A Quasi-Linear Parabolic Inequality with Convection in Riemannian Manifolds

Mohamed Jleli, Bessem Samet, and Calogero Vetro

Abstract We consider an inhomogeneous quasilinear parabolic inequality involving a convection term, and posed in a complete noncompact Riemannian manifold. Under certain assumptions on the Ricci curvature of the manifold, we establish sufficient conditions for the nonexistence of weak solutions. Next, some special cases are discussed. Notice that no restriction on the sign of solutions is imposed.

7.1 Introduction

Let $\mathbb{M}$ be a complete noncompact Riemannian manifold of dimension N, equipped with a metric $g = (\cdot, \cdot)$. In this work, we consider the quasilinear parabolic inequality

$$u_t \geq \Delta(|u|^{m-1}u) + \big(b, \nabla(|u|^{q-1}u)\big) + V(x)|u|^p + W(x) \quad \text{in } (0, \infty) \times \mathbb{M}, \tag{7.1}$$

where $u = u(t, x)$, Δ is the Laplace–Beltrami operator, ∇ is the gradient operator, b is a constant vector field on $\mathbb{M}$, V is a measurable function, $V > 0$ almost everywhere, W is a nontrivial L^1_{loc}-function, $W \geq 0$ almost everywhere, $m, q \geq 1$ and $p > \max\{m, q\}$. Our aim is to obtain sufficient conditions for which inequality (7.1) admits no weak solution in the sense that will be specified later.

M. Jleli (✉) · B. Samet
Department of Mathematics, College of Science, King Saud University, P.O. Box 2455, Riyadh, Saudi Arabia
e-mail: jleli@ksu.edu.sa

B. Samet
e-mail: bsamet@ksu.edu.sa

C. Vetro
Department of Mathematics and Computer Science, University of Palermo, Palermo, Italy
e-mail: calogero.vetro@unipa.it

M. Jleli et al. (eds.), *Advances in Nonlinear Evolution Equations*, Infosys Science Foundation Series, https://doi.org/10.1007/978-981-97-9812-4_7

Let us recall some known results related to (7.1) in the Euclidean case $\mathbb{M} = \mathbb{R}^N$. In the special case $m = 1$, $b = 0$, $V \equiv 1$, $W \equiv 0$ and $u \geq 0$, (7.1) (with equality) reduces to the semilinear heat equation

$$u_t - \Delta u = u^p \ (u \geq 0) \quad \text{in } (0, \infty) \times \mathbb{R}^N. \tag{7.2}$$

In 1966 , Fujita [4] proved that Eq. (7.2) does not have any nontrivial global solution if $1 < p < p_F = 1 + \frac{2}{N}$, whereas if $p > p_F$, there exist both global (with small initial data) and non-global (with large initial data) solutions. The real number p_F is called the Fujita critical exponent. Later, it was shown that in the critical case $p = p_F$, the problem possesses no nontrivial global solution (see [9, 14]). In the special case $b = 0$, $V \equiv 1$, $W \equiv 0$ and $u \geq 0$, (7.1) (with equality) reduces to the Newtonian filtration equation with slow diffusion

$$u_t - \Delta u^m = u^p \ (u \geq 0) \quad \text{in } (0, \infty) \times \mathbb{R}^N, \tag{7.3}$$

where $p > m \geq 1$. Galaktionov [5] proved that equation (7.3) admits as Fujita-type critical exponent the real number $m + \frac{2}{N}$. In the special case $m = 1$, $b = 0$, $V \equiv 1$, (7.1) (with equality) reduces to the inhomogeneous parabolic equation

$$u_t - \Delta u = |u|^p + W(x) \quad \text{in } (0, \infty) \times \mathbb{R}^N. \tag{7.4}$$

This Eq. (7.4) has been investigated by Bandle, Levine and Zhang [2]. Namely, it was shown that, if $\int_{\mathbb{R}^N} W(x)\,dx > 0$, then the critical exponent of (7.4) jumps from p_F to the bigger exponent $\frac{N}{N-2}$ $(N \geq 3)$. An extension of this result to the case when the inhomogeneous term W depends on time and space can be found in [12]. When $V \equiv 1$, $W \equiv 0$ and $u \geq 0$, (7.1) (with equality) reduces to

$$u_t = \Delta u^m + b \cdot \nabla u^q + u^p \quad \text{in } (0, \infty) \times \mathbb{R}^N, \tag{7.5}$$

where $\cdot$ denotes the inner product in $\mathbb{R}^N$. Equation (7.5) has been investigated by Suzuki [22] when $m, p, q > 1$. Namely, let

$$p^*_{m,q} = \min\left\{m + \frac{2}{N}, m + \frac{q - m + 1}{N + 1}\right\},$$

then Suzuki proved the following results:

(i) If $q > m - 1$ and $\max\{m, q\} \leq p \leq p^*_{m,q}$; or $p \geq q \geq m + \frac{1}{N}$ and $p = p^*_{m,q}$, then all nonnegative solutions are non-global;
(ii) If $p > m + \frac{1}{N}$, there are nontrivial global solutions to (7.5).

We refer also to Kirane and Qafsaoui [13], where they investigated (among other problems) the parabolic equation

$$u_t = -(-\Delta)^{\frac{\alpha}{2}} u^m + b(x) \cdot \nabla u^q + V(t, x) u^p + W(t, x) \quad \text{in } (0, \infty) \times \mathbb{R}^N, \quad (u \geq 0) \tag{7.6}$$

where $0 < \alpha \leq 2$, $(-\Delta)^{\frac{\alpha}{2}}$ is the $\frac{\alpha}{2}$-fractional power of the Laplacian $-\Delta$ and $V(t, x)$, $W(t, x)$ are given nonnegative functions. Problem (7.6) has been investigated under the following assumptions:

(H1) $p > \max\{m, q\} > 1$;
(H2) $|V(R^2\tau, Ry)| \simeq R^{\nu}$, $\nu > -2$, for R large, $\tau \geq 0$ and y in a bounded domain;
(H3) The vector function $b(x) = (b_1(x), b_2(x), \ldots, b_N(x))$ satisfies

$$|b_i(R^{\frac{2}{\alpha}} y)| \simeq R^{\frac{2\delta_i}{\alpha}} |y|_i^{\delta}$$

for R large, y in a bounded domain, and $\delta_i \geq 2$;
(H4) $\int_{\mathbb{R}^N} W(t, x)\, dx\, dt \geq 0$, $u(0, x) \geq 0$.

Namely, the authors proved that under assumptions (H1)–(H4), if

$$N \leq \min\left\{\frac{\alpha(\nu + 2)}{2(p - 1)}, \frac{2p(1 - \alpha - \delta) + \alpha q(\nu + 2)}{2(p - q)}\right\},$$

where $\delta = \max\{\delta_i : i = 1, 2, \ldots, N\}$, then (7.6) admits no nontrivial global weak solution. Observe that the above result does not cover the case $b \equiv$ constant, since in this case, one has $\delta = 1$, which contradicts the assumption (H3).

Some interesting extensions of nonexistence results from the Euclidean case to Riemannian manifolds can be found in [7, 8, 15, 19–21, 23]. In those references, only nonnegative weak solutions have been considered, and the nonexistence results have been established by the construction of radial test functions. Moreover, only the gradient of such test functions is involved in the weak formulation, but not their Laplacian. In our situation, those techniques cannot be used due to the fact that no restriction on the sign of solutions is imposed. In this paper, our approach is based on a result due to Bianchi and Setti [3] about the construction of cut-off functions with controlled gradient and Laplacian, under certain assumptions on the Ricci curvature of the manifold. Such approach was used in [16] for hyperbolic inequalities, and [10] for elliptic inequalities involving generalized Baouendi–Grushin operator.

The rest of the paper is organized as follows. In Sect. 7.2, we recall some basic notions and results from Riemannian geometry. In Sect. 7.3, after defining weak solutions to (7.1), we state our main result and discuss some special cases. Section 7.4 is devoted to some auxiliary results. Finally, in Sect. 7.5, we prove our main result.

7.2 Preliminaries on Riemannian Manifolds

This section is devoted to some notions and results from Riemannian geometry. For more details, the reader is referred to [1, 6, 17]. Throughout this paper, the symbol C denotes always a generic positive constant. Its value could be changed from one line to another.

Let $\mathbb{M}$ be a Riemannian manifold of dimension N, equipped with a metric $g = (\cdot,\cdot)$ and a Riemannian distance $d(\cdot,\cdot)$. Let (U, ϕ) be a chart in $\mathbb{M}$, where $\phi : U \to \phi(U) \subset \mathbb{R}^N$ and

$$\phi(x) = (x^1(x), x^2(x), \dots, x^N(x)), \quad x \in U.$$

Let ∂_i, $i = 1, 2, \dots, N$, be the corresponding vector fields on U. The local components of the metric g are defined by

$$g_{ij} = (\partial_i, \partial_j), \quad i, j = 1, 2, \dots, N.$$

The inverse of the matrix (g_{ij}) is denoted by (g^{ij}). For a smooth function $u : \mathbb{M} \to \mathbb{R}$, the gradient of u, ∇u, is the vector field defined by

$$(\nabla u, X) = Xu$$

for each vector field X on $\mathbb{M}$. In local coordinates, we have

$$\nabla u = \sum_{i,j} g^{ij}(\partial_i u)\partial_j.$$

The divergence of a vector field X on $\mathbb{M}$ is defined by

$$\operatorname{div}(X)(x) = \operatorname{trace}\left(T_x\mathbb{M} \ni \xi \mapsto \widetilde{\nabla}_\xi X \in T_x\mathbb{M}\right), \quad x \in \mathbb{M},$$

where $T_x\mathbb{M}$ is the tangent vector space at $x \in \mathbb{M}$ and $\widetilde{\nabla}$ is the Levi-Civita connection associated with the metric g. In local coordinates, if $X = \sum_i X_i \partial_i$, the divergence of X can be expressed as

$$\operatorname{div}(X) = \frac{1}{\sqrt{\mathbf{g}}} \sum_j \partial_j \left(\sqrt{\mathbf{g}} X_j\right),$$

where $\mathbf{g} = \det(g_{ij})$. The Laplacian of u is the function

$$\Delta u = \operatorname{div}(\nabla u).$$

In local coordinates, we have

$$\Delta u = \frac{1}{\sqrt{\mathbf{g}}} \sum_{i,j} \partial_i \left(\sqrt{\mathbf{g}} g^{ij} \partial_j u \right).$$

We denote by Ric the Ricci tensor which is expressed in local coordinates as

$$R_{ij} = R_{ji} = \sum_{\ell} \partial_\ell \left(\Gamma_{ij}^{\ell} \right) - \sum_{\ell} \partial_j \left(\Gamma_{i\ell}^{\ell} \right) + \sum_{k,\ell} \left(\Gamma_{ij}^{k} \Gamma_{k\ell}^{\ell} - \Gamma_{i\ell}^{k} \Gamma_{kj}^{\ell} \right),$$

where Γ_{ij}^k are the Christoffel symbols. For a given function $f : \mathbb{M} \to \mathbb{R}$, the notation $\mathrm{Ric} \geq f(x)$ means that

$$\mathrm{Ric}(X, X) \geq f(x)|X|^2$$

for every vector field X on $\mathbb{M}$, where $|X| = \sqrt{(X, X)}$.

For $\delta > 0$, the geodesic ball of center $x_0 \in \mathbb{M}$ and radius δ is denoted by $B(x_0, \delta)$, that is,

$$B(x_0, \delta) = \{x \in \mathbb{M} : d(x_0, x) < \delta\}.$$

The volume of $B(x_0, \delta)$ is denoted by $|B(x_0, \delta)|$. Throughout this paper, for $x_0 \in \mathbb{M}$, we use the notation

$$r_{x_0}(x) = d(x_0, x), \quad x \in \mathbb{M}.$$

We recall below an important result for our study (see [3]).

Lemma 7.1 *Let $x_0 \in \mathbb{M}$ and suppose that for some $C_0 \geq 0$ and $\sigma \in [-2, 2]$ we have*

$$Ric \geq -C_0(N-1)\left(1 + r_{x_0}^2(x)\right)^{\frac{-\sigma}{2}}. \tag{7.7}$$

Then, there exists a family of functions $\{\psi_R\}_{R \geq 1} \subset C_c^\infty(\mathbb{M})$ (i.e., $\psi_R \in C^\infty(\mathbb{M})$ and $\mathrm{supp}(\psi_R) \subset\subset \mathbb{M}$*) satisfying the following:*

(i) $0 \leq \psi_R \leq 1$, $\psi_R|_{B(x_0,R)} \equiv 1$;
(ii) *there exists $\gamma_0 > 1$ (independent of R) such that*

$$\mathrm{supp}(\psi_R) \subset B(x_0, \gamma R), \quad \gamma > \gamma_0;$$

(iii) $|\nabla \psi_R| \leq C R^{-1}$;
(iv) $|\Delta \psi_R| \leq C R^{-\left(1+\frac{\sigma}{2}\right)}$.

Remark 7.1 Assume that condition (7.7) holds for some $\sigma > 2$. Then (7.7) holds for $\sigma = 2$. Hence, by Lemma 7.1, if (7.7) is satisfied for some $x_0 \in \mathbb{M}$, $C_0 \geq 0$ and $\sigma \geq -2$, then there exists a family of functions $\{\psi_R\}_{R \geq 1} \subset C_c^\infty(\mathbb{M})$ satisfying (i)–(iii) and

$$|\Delta \psi_R| \leq C R^{-\left(1+\frac{\min\{\sigma,2\}}{2}\right)}. \tag{7.8}$$

The following result can be found in [11].

Lemma 7.2 *Let* $x_0 \in \mathbb{M}$ *and suppose that there exist* $\alpha_{\mathbb{M}}, \beta_{\mathbb{M}} > 0$ *such that*

$$|B(x_0, R)| \leq \alpha_{\mathbb{M}} R^{\beta_{\mathbb{M}}}, \quad R > 0.$$

Let $\tau > 0$, $V = V(x) > 0$ *a.e. in* $\mathbb{M}$ *and* $V^{-\tau} \in L^1_{\text{loc}}(\mathbb{M})$. *Assume that*

$$V(x) \geq C_V r^{\kappa}_{x_0}(x), \quad a.e. \text{ in } \mathbb{M},$$

for some constants $C_V > 0$ *and* $\kappa \in \mathbb{R}$. *Then, the following holds as* $R \to \infty$:

$$\int_{B(x_0,R)} V^{-\tau}(x)\, dx = O\left(\ln R + R^{-\kappa\tau+\beta_{\mathbb{M}}}\right).$$

7.3 Main Result and Some Special Cases

We first define weak solutions to inequality (7.1). Hence, let $Q = (0, \infty) \times \mathbb{M}$ and introduce the test function space

$$\phi = \left\{\varphi \in C^2(Q) : \text{supp}(\varphi) \subset\subset Q,\ \varphi \geq 0\right\}.$$

Definition 7.1 Let $m, q \geq 1$, $p > \max\{m, q\}$, $V = V(x) > 0$ a.e. in $\mathbb{M}$, $W \in L^1_{\text{loc}}(\mathbb{M})$ and b be a constant vector field on $\mathbb{M}$. We say that $u \in L^p_{\text{loc}}(Q; V\,dx) \cap L^{\max\{m,q\}}_{\text{loc}}(Q)$ is a weak solution to inequality (7.1) if

$$\begin{aligned} &\int_Q V|u|^p \varphi\, dx\, dt + \int_Q W\varphi\, dx\, dt \\ &\leq -\int_Q u\varphi_t\, dx\, dt - \int_Q |u|^{m-1}u\Delta\varphi\, dx\, dt + \int_Q |u|^{q-1}u(\nabla\varphi, b)\, dx\, dt \end{aligned} \tag{7.9}$$

for every $\varphi = \varphi(t, x) \in \phi$.

We remark that the weak formulation (7.9) of (7.1) is obtained by standard integrations by parts.

For $m, q \geq 1$, $p > \max\{m, q\}$ and a vector field b, we introduce the finite set

$$\chi_b = \begin{cases} \left\{\frac{1}{p-1}, \frac{m}{p-m}\right\} & \text{if } |b| = 0, \\ \left\{\frac{1}{p-1}, \frac{m}{p-m}, \frac{q}{p-q}\right\} & \text{if } |b| > 0. \end{cases}$$

Let $x_0 \in \mathbb{M}$ be fixed. Our main result is stated in the following theorem.

Theorem 7.1 *Let $m, q \geq 1$, $p > \max\{m, q\}$, $W \in L^1_{\text{loc}}(\mathbb{M})$, $W(x) \geq 0$ a.e. in $\mathbb{M}$, $W \not\equiv 0$, b be a constant vector field on $\mathbb{M}$, $V(x) > 0$ a.e. in $\mathbb{M}$ and $V^{-\tau} \in L^1_{\text{loc}}(\mathbb{M})$ for all $\tau \in \chi_b$. Suppose that (7.7) holds for some $C_0 \geq 0$ and $\sigma > -2$. Assume also that there exists $\theta > 0$ such that for sufficiently large γ, we have*

$$\liminf_{R\to\infty} \left(R^{\frac{-\theta p}{p-1}} \int_{B(x_0,\gamma R)} V^{\frac{-1}{p-1}}(x)\, dx + R^{-(1+\frac{\widetilde{\sigma}}{2})\frac{p}{p-m}} \right.$$
$$\times \int_{B(x_0,\gamma R)\setminus B(x_0,R)} V^{\frac{-m}{p-m}}\, dx \tag{7.10}$$
$$\left. + |b|^{\frac{p}{p-1}} R^{\frac{-p}{p-q}} \int_{B(x_0,\gamma R)\setminus B(x_0,R)} V^{\frac{-q}{p-q}}\, dx \right) = 0,$$

where $\widetilde{\sigma} = \min\{\sigma, 2\}$. Then inequality (7.1) admits no weak solution.

Let us discuss some special cases of Theorem 7.1. We first assume that

$$V(x) > 0 \text{ a.e. in } \mathbb{M}, \quad V^{-\tau} \in L^1(\mathbb{M}),\ \tau \in \chi_b. \tag{7.11}$$

In this case, by the dominated convergence theorem, we obtain

$$\lim_{R\to\infty} \int_{B(x_0,\gamma R)\setminus B(x_0,R)} V^{\frac{-m}{p-m}}\, dx = 0,$$
$$\lim_{R\to\infty} \int_{B(x_0,\gamma R)\setminus B(x_0,R)} V^{\frac{-q}{p-q}}\, dx = 0,$$
$$\lim_{R\to\infty} \int_{B(x_0,\gamma R)} V^{\frac{-1}{p-1}}(x)\, dx = \int_{\mathbb{M}} V^{\frac{-1}{p-1}}(x)\, dx.$$

Hence, for all $\theta > 0$ and $\gamma > 1$, (7.10) holds. Thus, we deduce the following result.

Corollary 7.1 *Let $m, q \geq 1$, $p > \max\{m, q\}$, $W \in L^1_{\text{loc}}(\mathbb{M})$, $W(x) \geq 0$ a.e. in $\mathbb{M}$, $W \not\equiv 0$ and b be a constant vector field on $\mathbb{M}$. Suppose that (7.7) holds for some $C_0 \geq 0$ and $\sigma > -2$. Assume also that V satisfies (7.11). Then inequality (7.1) admits no weak solution.*

Next, we consider the case when (7.7) holds for some $C_0 \geq 0$ and $\sigma \geq 2$. In this case, by volume comparison theorems (see Proposition 2.11 and Theorem 2.14 in [18]), we have

$$|B(x_0, R)| \leq C R^{\nu}, \quad R > 0,$$

where

$$\nu = (N-1)\delta_\sigma + 1, \quad \delta_\sigma = \begin{cases} \frac{1+\sqrt{1+4C_0}}{2} & \text{if } \sigma = 2, \\ 1 & \text{if } \sigma > 2. \end{cases} \tag{7.12}$$

Assume that the potential function V satisfies

$$V^{-\tau} \in L^1_{\mathrm{loc}}(\mathbb{M}),\ \tau \in \chi_b, \quad V(x) \geq C r_{x_0}^{\kappa}(x) \text{ a.e. in } \mathbb{M}, \tag{7.13}$$

where $\kappa \in \mathbb{R}$. Then, by Lemma 7.2, the following hold as $R \to \infty$:

$$\int_{B(x_0,\gamma R)} V^{\frac{-1}{p-1}}(x)\,dx = O\left(\ln R + R^{\frac{-\kappa}{p-1}+\nu}\right),$$

$$\int_{B(x_0,\gamma R)\setminus B(x_0,R)} V^{\frac{-m}{p-m}}\,dx = O\left(\ln R + R^{\frac{-\kappa m}{p-m}+\nu}\right),$$

$$\int_{B(x_0,\gamma R)\setminus B(x_0,R)} V^{\frac{-q}{p-q}}\,dx = O\left(\ln R + R^{\frac{-\kappa q}{p-q}+\nu}\right).$$

Hence, for all $\theta > 0$, $\gamma > 1$ and sufficiently large R, we obtain (notice that $\widetilde{\sigma} = 2$)

$$\begin{aligned}
& R^{\frac{-\theta p}{p-1}} \int_{B(x_0,\gamma R)} V^{\frac{-1}{p-1}}(x)\,dx + R^{-\left(1+\frac{\widetilde{\sigma}}{2}\right)\frac{p}{p-m}} \int_{B(x_0,\gamma R)\setminus B(x_0,R)} V^{\frac{-m}{p-m}}\,dx \\
& + |b|^{\frac{p}{p-1}} R^{\frac{-p}{p-q}} \int_{B(x_0,\gamma R)\setminus B(x_0,R)} V^{\frac{-q}{p-q}}\,dx \\
& = R^{\frac{-\theta p}{p-1}} \int_{B(x_0,\gamma R)} V^{\frac{-1}{p-1}}(x)\,dx + R^{\frac{-2p}{p-m}} \int_{B(x_0,\gamma R)\setminus B(x_0,R)} V^{\frac{-m}{p-m}}\,dx \\
& + |b|^{\frac{p}{p-1}} R^{\frac{-p}{p-q}} \int_{B(x_0,\gamma R)\setminus B(x_0,R)} V^{\frac{-q}{p-q}}\,dx \\
& \leq C\Big[R^{\frac{-\theta p}{p-1}} \ln R + R^{\frac{-\theta p-\kappa}{p-1}+\nu} + R^{\frac{-2p}{p-m}} \ln R + R^{\frac{-2p-\kappa m}{p-m}+\nu} \\
& + |b|^{\frac{p}{p-1}} \left(R^{\frac{-p}{p-q}} \ln R + R^{\frac{-p-\kappa q}{p-q}+\nu} \right) \Big].
\end{aligned}$$

Then, for

$$\theta > \max\left\{0, \frac{(p-1)\nu - \kappa}{p}\right\},$$

we deduce that, if

$$\nu < \begin{cases} \dfrac{2p+\kappa m}{p-m} & \text{if } |b| = 0, \\[2mm] \min\left\{\dfrac{2p+\kappa m}{p-m}, \dfrac{p+\kappa q}{p-q}\right\} & \text{if } |b| > 0, \end{cases} \tag{7.14}$$

then (7.7) holds. Thus, by Theorem 7.1, we deduce the following result.

Corollary 7.2 *Let $m, q \geq 1$, $p > \max\{m, q\}$, $W \in L^1_{\mathrm{loc}}(\mathbb{M})$, $W(x) \geq 0$ a.e. in $\mathbb{M}$, $W \not\equiv 0$ and b be a constant vector field on $\mathbb{M}$. Suppose that V satisfies* (7.13), *and*

(7.7) *holds for some* $C_0 \geq 0$ *and* $\sigma \geq 2$. *If* (7.14) *is satisfied, where* ν *is given by* (7.12), *then inequality* (7.1) *admits no weak solution.*

Let us consider the special case of Corollary 7.2, when $\mathbb{M} = \mathbb{R}^N$ and $V \equiv 1$. In this case, $\nu = N$, $\kappa = 0$, and (7.7) holds for $\sigma = 2$. Then, (7.14) reduces to

$$\begin{cases} (N-2)p < mN & \text{if } |b| = 0, \\ (N-2)p < mN, \ (N-1)p < qN & \text{if } |b| > 0. \end{cases}$$

Hence, by Corollary 7.2, we deduce the following results.

Corollary 7.3 (The Euclidean case without convection) *Let* $\mathbb{M} = \mathbb{R}^N$, *where* $N \geq 1$. *Let* $p > m \geq 1$, $b = 0$, $V \equiv 1$ *and* $W \in L^1_{\text{loc}}(\mathbb{M})$, $W(x) \geq 0$ *a.e. in* $\mathbb{M}$, $W \not\equiv 0$. *We have the following:*

(I) If $N \in \{1, 2\}$, *then inequality* (7.1) *admits no weak solution.*
(II) If $N \geq 3$ *and*

$$p < \frac{mN}{N-2},$$

then inequality (7.1) *admits no weak solution.*

Remark 7.2 Taking $m = 1$ in Corollary 7.3, we retrieve [2, Theorem 2.1 (a)].

Corollary 7.4 (The Euclidean case with convection) *Let* $\mathbb{M} = \mathbb{R}^N$, *where* $N \geq 1$. *Let* $m, q \geq 1$, $p > \max\{m, q\}$, $b \neq 0$, $V \equiv 1$ *and* $W \in L^1_{\text{loc}}(\mathbb{M})$, $W(x) \geq 0$ *a.e. in* $\mathbb{M}$, $W \not\equiv 0$. *We have the following:*

(I) If $N = 1$, *then inequality* (7.1) *admits no weak solution.*
(II) If $N = 2$ *and* $p < 2q$, *then* (7.1) *admits no weak solution.*
(III) If $N \geq 3$ *and*

$$p < \min\left\{\frac{mN}{N-2}, \frac{qN}{N-1}\right\},$$

then inequality (7.1) *admits no weak solution.*

7.4 Auxiliary Results

In this section, we assume that $m, q \geq 1$, $p > \max\{m, q\}$, $V = V(x) > 0$ a.e. in $\mathbb{M}$, $W \in L^1_{\text{loc}}(\mathbb{M})$ and b is a constant vector field on $\mathbb{M}$. We now develop certain estimates useful to derive nonexistence of weak solutions to inequality (7.1).

7.4.1 A Priori Estimate

For $\varphi \in \phi$, let us introduce the integral terms

$$K_1(\varphi) = \int_{\mathrm{supp}(\varphi)} V^{\frac{-1}{p-1}} \varphi^{\frac{-1}{p-1}} |\varphi_t|^{\frac{p}{p-1}} \, dx \, dt, \tag{7.15}$$

$$K_2(\varphi) = \int_{\mathrm{supp}(\varphi)} V^{\frac{-m}{p-m}} \varphi^{\frac{-m}{p-m}} |\Delta\varphi|^{\frac{p}{p-m}} \, dx \, dt, \tag{7.16}$$

$$K_3(\varphi) = |b|^{\frac{p}{p-q}} \int_{\mathrm{supp}(\varphi)} V^{\frac{-q}{p-q}} \varphi^{\frac{-q}{p-q}} |\nabla\varphi|^{\frac{p}{p-q}} \, dx \, dt. \tag{7.17}$$

We have the following a priori estimate.

Lemma 7.3 *Assume that* $u \in L^p_{\mathrm{loc}}(Q; V\,dx) \cap L^{\max\{m,q\}}_{\mathrm{loc}}(Q)$ *is a weak solution to inequality* (7.1). *Then we have*

$$\int_Q W\varphi \, dx \, dt \le C \sum_{i=1}^{3} K_i(\varphi) \tag{7.18}$$

for every $\varphi \in \phi$, *provided that* $K_i(\varphi) < \infty$, $i = 1, 2, 3$.

Proof Let $u \in L^p_{\mathrm{loc}}(Q; V\,dx) \cap L^{\max\{m,q\}}_{\mathrm{loc}}(Q)$ be a weak solution to (7.1). By (7.9), there holds

$$\begin{aligned} &\int_Q V|u|^p \varphi \, dx \, dt + \int_Q W\varphi \, dx \, dt \\ &\le \int_Q |u||\varphi_t| \, dx \, dt + \int_Q |u|^m |\Delta\varphi| \, dx \, dt + |b| \int_Q |u|^q |\nabla\varphi| \, dx \, dt. \end{aligned} \tag{7.19}$$

On the other hand, using Young's inequality, we obtain

$$\begin{aligned} \int_Q |u||\varphi_t| \, dx \, dt &= \int_Q \left(V^{\frac{1}{p}} |u| \varphi^{\frac{1}{p}}\right) \left(V^{\frac{-1}{p}} \varphi^{\frac{-1}{p}} |\varphi_t|\right) dx \, dt \\ &\le \frac{1}{3} \int_Q V|u|^p \varphi \, dx \, dt + C K_1(\varphi). \end{aligned} \tag{7.20}$$

Similarly, we get

$$\int_Q |u|^m |\Delta\varphi| \, dx \, dt \le \frac{1}{3} \int_Q V|u|^p \varphi \, dx \, dt + C K_2(\varphi) \tag{7.21}$$

and

$$|b| \int_Q |u|^q |\nabla\varphi| \, dx \, dt \leq \frac{1}{3} \int_Q V|u|^p \varphi \, dx \, dt + C K_3(\varphi). \tag{7.22}$$

Combining the estimates (7.19), (7.20), (7.21) and (7.22), we obtain (7.18). □

7.4.2 Construction of a Family of Functions Belonging to ϕ

Let us consider a cut-off function $\zeta \in C^\infty([0, \infty))$ satisfying

$$\zeta \geq 0, \quad \text{supp}(\zeta) \subset\subset (0, 1). \tag{7.23}$$

For sufficiently large T and ℓ, let

$$\zeta_T(t) = \zeta^\ell\left(\frac{t}{T}\right), \qquad t \geq 0. \tag{7.24}$$

Now, assume that (7.7) holds for some $C_0 \geq 0$ and $\sigma > -2$. Then, by Lemma 7.1 and Remark 7.1, there exists a family of functions $\{\psi_R\}_{R\geq 1} \subset C_c^\infty(\mathbb{M})$ satisfying properties (i)–(iii) in Lemma 7.1, and property (7.8) too. Let us consider functions of the form

$$\varphi(t, x) = \zeta_T(t)\psi_R^\ell(x), \quad (t, x) \in Q. \tag{7.25}$$

Then, we obtain immediately the following result.

Lemma 7.4 *Assume that* (7.7) *holds for some* $C_0 \geq 0$ *and* $\sigma > -2$*. Then, for sufficiently large* T, R *and* ℓ*, the function* φ *defined by* (7.25) *belongs to* ϕ.

7.4.3 Estimates of $K_i(\varphi)$

Here we assume that (7.7) holds for some $C_0 \geq 0$ and $\sigma > -2$. For sufficiently large T, R and ℓ, let φ be the function defined by (7.25).

Lemma 7.5 *Let* $V^{\frac{-1}{p-1}} \in L^1_{\text{loc}}(\mathbb{M})$*. Then, there exists* $\gamma_0 > 1$ *such that for all* $\gamma > \gamma_0$ *we have*

$$K_1(\varphi) \leq C T^{1-\frac{p}{p-1}} \int_{B(x_0,\gamma R)} V^{\frac{-1}{p-1}} \, dx. \tag{7.26}$$

Proof By Lemma 7.1, Remark 7.1, (7.23), (7.24) and (7.25), there exists $\gamma_0 > 1$ such that for all $\gamma > \gamma_0$ we get

$$K_1(\varphi) = \left(\int_0^T \zeta_T^{\frac{-1}{p-1}} |\zeta_T'|^{\frac{p}{p-1}}\,dt\right)\left(\int_{B(x_0,\gamma R)} V^{\frac{-1}{p-1}} \psi_R^{\ell}\,dx\right). \tag{7.27}$$

On the other hand, by (7.23) and (7.24), we obtain the estimate

$$|\zeta_T'(t)| \le CT^{-1}\zeta^{\ell-1}\left(\frac{t}{T}\right), \quad 0 < t < T,$$

which yields

$$\zeta_T^{\frac{-1}{p-1}}(t)|\zeta_T'(t)|^{\frac{p}{p-1}} \le CT^{\frac{-p}{p-1}}\zeta^{\ell-\frac{p}{p-1}}\left(\frac{t}{T}\right), \quad 0 < t < T.$$

Integrating over $(0, T)$ this inequality, we deduce that

$$\begin{aligned}\int_0^T \zeta_T^{\frac{-1}{p-1}} |\zeta_T'|^{\frac{p}{p-1}}\,dt &\le CT^{\frac{-p}{p-1}} \int_0^T \zeta^{\ell-\frac{p}{p-1}}\left(\frac{t}{T}\right)\,dt\\ &= CT^{1-\frac{p}{p-1}} \int_0^1 \zeta^{\ell-\frac{p}{p-1}}(s)\,ds,\end{aligned}$$

that is,

$$\int_0^T \zeta_T^{\frac{-1}{p-1}} |\zeta_T'|^{\frac{p}{p-1}}\,dt \le CT^{1-\frac{p}{p-1}}. \tag{7.28}$$

Moreover, since $0 \le \psi_R \le 1$, we obtain

$$\int_{B(x_0,\gamma R)} V^{\frac{-1}{p-1}} \psi_R^{\ell}\,dx \le \int_{B(x_0,\gamma R)} V^{\frac{-1}{p-1}}\,dx. \tag{7.29}$$

Thus, (7.26) follows from (7.27), (7.28) and (7.29). □

Lemma 7.6 *Let $V^{\frac{-m}{p-m}} \in L^1_{\text{loc}}(\mathbb{M})$. Then, there exists $\gamma_0 > 1$ such that for all $\gamma > \gamma_0$ we have*

$$K_2(\varphi) \le CTR^{-\left(1+\frac{\tilde{\sigma}}{2}\right)\frac{p}{p-m}} \int_{B(x_0,\gamma R)\setminus B(x_0,R)} V^{\frac{-m}{p-m}}\,dx. \tag{7.30}$$

Proof By Lemma 7.1, Remark 7.1, (7.23), (7.24) and (7.25), there exists $\gamma_0 > 1$ such that for all $\gamma > \gamma_0$, we get

$$K_2(\varphi) = \left(\int_0^T \zeta_T\,dt\right)\left(\int_{B(x_0,\gamma R)\setminus B(x_0,R)} V^{\frac{-m}{p-m}} \psi_R^{\frac{-\ell m}{p-m}} |\Delta\psi_R^{\ell}|^{\frac{p}{p-m}}\,dx\right). \tag{7.31}$$

On the other hand, by (7.24), we obtain

$$\begin{aligned}\int_0^T \zeta_T\,dt &= \int_0^T \zeta^\ell\left(\frac{t}{T}\right)dt\\ &= T\int_0^1 \zeta^\ell(s)\,ds,\end{aligned}$$

that is,

$$\int_0^T \zeta_T\,dt = CT. \tag{7.32}$$

Moreover, by Lemma 7.1, Remark 7.1 and using the property

$$\Delta\psi_R^\ell = \ell(\ell-1)\xi_R^{\ell-2}|\nabla\xi_R|^2 + \ell\xi_R^{\ell-1}\Delta\xi_R,$$

we deduce that

$$|\Delta\psi_R^\ell(x)| \le CR^{-(1+\frac{\tilde{\sigma}}{2})}\xi_R^{\ell-2}(x), \quad x\in B(x_0,\gamma R)\backslash B(x_0,R),$$

which yields

$$\begin{aligned}V^{\frac{-m}{p-m}}\psi_R^{\frac{-\ell m}{p-m}}|\Delta\psi_R^\ell|^{\frac{p}{p-m}} &\le CR^{-(1+\frac{\tilde{\sigma}}{2})\frac{p}{p-m}}V^{\frac{-m}{p-m}}\xi_R^{\ell-\frac{2p}{p-m}}(x)\\ &\le CR^{-(1+\frac{\tilde{\sigma}}{2})\frac{p}{p-m}}V^{\frac{-m}{p-m}}, \quad x\in B(x_0,\gamma R)\backslash B(x_0,R).\end{aligned}$$

Integrating this inequality, we get

$$\begin{aligned}&\int_{B(x_0,\gamma R)\backslash B(x_0,R)} V^{\frac{-m}{p-m}}\psi_R^{\frac{-\ell m}{p-m}}|\Delta\psi_R^\ell|^{\frac{p}{p-m}}\,dx\\ &\le CR^{-(1+\frac{\tilde{\sigma}}{2})\frac{p}{p-m}}\int_{B(x_0,\gamma R)\backslash B(x_0,R)} V^{\frac{-m}{p-m}}\,dx.\end{aligned} \tag{7.33}$$

Hence, combining (7.31), (7.32) and (7.33), we obtain (7.30). □

Lemma 7.7 *Let* $V^{\frac{-q}{p-q}}\in L^1_{\mathrm{loc}}(\mathbb{M})$. *Then, there exists* $\gamma_0>1$ *such that for all* $\gamma>\gamma_0$ *we have*

$$K_3(\varphi) \le C|b|^{\frac{p}{p-1}}TR^{\frac{-p}{p-q}}\int_{B(x_0,\gamma R)\backslash B(x_0,R)} V^{\frac{-q}{p-q}}\,dx. \tag{7.34}$$

Proof By Lemma 7.1, Remark 7.1, (7.23), (7.24) and (7.25), there exists $\gamma_0 > 1$ such that for all $\gamma > \gamma_0$ we get

$$K_3(\varphi) = |b|^{\frac{p}{p-1}} \left(\int_0^T \zeta_T \, dt \right) \left(\int_{B(x_0,\gamma R)\setminus B(x_0,R)} V^{\frac{-q}{p-q}} \psi_R^{\frac{-\ell q}{p-q}} |\nabla \psi_R^\ell|^{\frac{p}{p-q}} \, dx \right). \tag{7.35}$$

On the other hand, one has

$$\nabla \psi_R^\ell(x) = \ell \psi_R^{\ell-1}(x) \nabla \psi_R(x), \quad x \in B(x_0, \gamma R)\setminus B(x_0, R),$$

which implies, by Lemma 7.1 and Remark 7.1, that

$$|\nabla \psi_R^\ell(x)| \le C R^{-1} \psi_R^{\ell-1}(x), \quad x \in B(x_0, \gamma R)\setminus B(x_0, R).$$

Hence, we deduce that

$$\begin{aligned} V^{\frac{-q}{p-q}} \psi_R^{\frac{-\ell q}{p-q}} |\nabla \psi_R^\ell|^{\frac{p}{p-q}} &\le C R^{\frac{-p}{p-q}} V^{\frac{-q}{p-q}} \psi_R^{\ell - \frac{p}{p-q}}(x) \\ &\le C R^{\frac{-p}{p-q}} V^{\frac{-q}{p-q}}, \quad x \in B(x_0, \gamma R)\setminus B(x_0, R). \end{aligned}$$

Integrating this inequality, we get

$$\begin{aligned} &\int_{B(x_0,\gamma R)\setminus B(x_0,R)} V^{\frac{-q}{p-q}} \psi_R^{\frac{-\ell q}{p-q}} |\nabla \psi_R^\ell|^{\frac{p}{p-q}} \, dx \\ &\le C R^{\frac{-p}{p-q}} \int_{B(x_0,\gamma R)\setminus B(x_0,R)} V^{\frac{-q}{p-q}} \, dx. \end{aligned} \tag{7.36}$$

Finally, (7.34) follows from (7.32), (7.35) and (7.36). □

7.5 Proof of Theorem 7.1

We are ready to combine the auxiliary results to establish the nonexistence of weak solutions to inequality (7.1). We make use of two main conditions. The first one is the inequality (7.7) imposed on the Ricci curvature of $\mathbb{M}$ (see Lemma 7.1), and the other is a convergence condition (see (7.10)). For convenience, we note that (7.7) was imposed in establishing Lemmas 7.4-7.7, and (7.10) is directly involved in the proof of Theorem 7.1 below.

Proof We argue by contradiction. So, we assume that

$$u \in L^p_{\text{loc}}(Q; V\, dx) \cap L^{\max\{m,q\}}_{\text{loc}}(Q)$$

is a weak solution to inequality (7.1). Then, by Lemmas 7.3 and 7.4, for sufficiently large T, R and ℓ, there holds

$$\int_Q W\varphi\, dx\, dt \le C\sum_{i=1}^{3} K_i(\varphi), \tag{7.37}$$

where φ is defined by (7.25). On the other hand, by (7.23), (7.24), (7.25), Lemma 7.1 and Remark 7.1, for sufficiently large γ, we obtain

$$\int_Q W\varphi\, dx\, dt = \left(\int_0^T \zeta_T(t)\, dt\right)\left(\int_{B(x_0,\gamma R)} W(x)\psi_R^\ell(x)\, dx\right). \tag{7.38}$$

Since $W \ge 0$ and $\psi_R \equiv 1$ in $B(x_0, R)$, we get

$$\begin{aligned}\int_{B(x_0,\gamma R)} W(x)\psi_R^\ell(x)\, dx &\ge \int_{B(x_0,R)} W(x)\psi_R^\ell(x)\, dx\\ &= \int_{B(x_0,R)} W(x)\, dx.\end{aligned} \tag{7.39}$$

Moreover, since $W \not\equiv 0$, there exists $R_0 > 0$ such that

$$\int_{B(x_0,R_0)} W(x)\, dx > 0. \tag{7.40}$$

Then, in view of (7.39) and (7.40), we get (for sufficiently large R)

$$\int_{B(x_0,\gamma R)} W(x)\psi_R^\ell(x)\, dx \ge \int_{B(x_0,R_0)} W(x)\, dx > 0. \tag{7.41}$$

Next, using (7.32), together with (7.38) and (7.41), we get

$$T\int_{B(x_0,R_0)} W(x)\, dx \le C\sum_{i=1}^{3} K_i(\varphi).$$

Hence, using Lemmas 7.5, 7.6 and 7.7, we obtain

$$\begin{aligned}&CT\int_{B(x_0,R_0)} W(x)\, dx\\ &\le T^{1-\frac{p}{p-1}}\int_{B(x_0,\gamma R)} V^{\frac{-1}{p-1}}\, dx + TR^{-\left(1+\frac{\sigma}{2}\right)\frac{p}{p-m}}\int_{B(x_0,\gamma R)\backslash B(x_0,R)} V^{\frac{-m}{p-m}}\, dx\\ &+ |b|^{\frac{p}{p-1}}TR^{\frac{-p}{p-q}}\int_{B(x_0,\gamma R)\backslash B(x_0,R)} V^{\frac{-q}{p-q}}\, dx,\end{aligned}$$

that is,

$$
\begin{aligned}
&C\int_{B(x_0,R_0)} W(x)\,dx \\
&\le T^{\frac{-p}{p-1}}\int_{B(x_0,\gamma R)} V^{\frac{-1}{p-1}}\,dx + R^{-\left(1+\frac{\widetilde{\sigma}}{2}\right)\frac{p}{p-m}}\int_{B(x_0,\gamma R)\setminus B(x_0,R)} V^{\frac{-m}{p-m}}\,dx \qquad (7.42)\\
&+|b|^{\frac{p}{p-1}}R^{\frac{-p}{p-q}}\int_{B(x_0,\gamma R)\setminus B(x_0,R)} V^{\frac{-q}{p-q}}\,dx.
\end{aligned}
$$

Next, let $\theta > 0$ be such that the limit inferior in (7.10) holds. Taking $T = R^\theta$ in (7.42), we get

$$
\begin{aligned}
&C\int_{B(x_0,R_0)} W(x)\,dx \\
&\le R^{\frac{-\theta p}{p-1}}\int_{B(x_0,\gamma R)} V^{\frac{-1}{p-1}}\,dx + R^{-\left(1+\frac{\widetilde{\sigma}}{2}\right)\frac{p}{p-m}}\int_{B(x_0,\gamma R)\setminus B(x_0,R)} V^{\frac{-m}{p-m}}\,dx \qquad (7.43)\\
&+|b|^{\frac{p}{p-1}}R^{\frac{-p}{p-q}}\int_{B(x_0,\gamma R)\setminus B(x_0,R)} V^{\frac{-q}{p-q}}\,dx.
\end{aligned}
$$

Then, passing to the infimum limit as $R \to \infty$ in (7.43) and using (7.10), we obtain

$$\int_{B(x_0,R_0)} W(x)\,dx \le 0,$$

which contradicts (7.40). Consequently, inequality (7.1) admits no weak solution. This completes the proof of Theorem 7.1. □

References

1. Alias, L.J., Mastrolia, P. Rigoli, M.: Maximum Principles and Geometric Applications. Springer (2016)
2. Bandle, C., Levine, H.A., Zhang, Q.S.: Critical exponents of Fujita type for inhomogeneous parabolic equations and systems. J. Math. Anal. Appl. **251**, 624–648 (2000)
3. Bianchi, D. Setti, A.: Laplacian cut-offs, porous and fast diffusion on manifolds and other applications. Calc. Var. Part. Diff. Eq. **57**, 4, 33 pp. (2018)
4. Fujita, H.: On the blowing up of solutions of the Cauchy problem for $u_t = \Delta u + u^{1+\alpha}$. J. Fac. Sci. Univ. Tokyo Sect. **I**(13), 109–124 (1966)
5. Galaktionov, V.A.: Blow-up for quasilinear heat equations with critical Fujita's exponents. Proc. R. Soc. Edinburgh Sect. A **124**, 517–525 (1994)
6. Grigor'yan, A.: Analytic and geometric background of recurrence and non-explosion of the Brownian motion on Riemannian manifolds. Bull. Amer. Math. Soc. **36**, 135–249 (1999)
7. Grigor'yan, A., Sun, Y.: On non-negative solutions of the inequality $\Delta u + u^\sigma \le 0$ on Riemannian manifolds. Comm. Pure Appl. Math. **67**, 1336–1352 (2014)
8. Gu, Q., Sun, Y., Xiao, J., Xu, F.: Global positive solution to a semi-linear parabolic equation with potential on Riemannian manifold. Calc. Var. Part. Diff. Eq. **59**, 170, 24 pp. (2020)

9. Hayakawa, K.: On nonexistence of global solutions of some semilinear parabolic equations. Proc. Japan Acad. **49**, 503–525 (1973)
10. Jleli, M., Ragusa, M.A., Samet, B.: Nonlinear Liouville-type theorems for generalized Baouendi-Grushin operator on Riemannian manifolds. Adv. Differ. Equ. **28**, 143–168 (2023)
11. Jleli, M., Samet, B.: Nonexistence for time-fractional wave inequalities on Riemannian manifolds. Discr. Cont. Dyn. Syst. Ser. S **16**, 1517–1536 (2023)
12. Jleli, M., Kawakami, K., Samet, B.: Critical behavior for a semilinear parabolic equation with forcing term depending on time and space. J. Math. Anal. Appl. **486**, 123931, 16 pp. (2020)
13. Kirane, M., Qafsaoui, M.: Global nonexistence for the Cauchy problem of some nonlinear reaction-diffusion systems. J. Math. Anal. Appl. **268**, 217–243 (2002)
14. Kobayashi, K., Siaro, T., Tanaka, H.: On the blowing up problem for semilinear heat equations. J. Math. Soc. Japan. **29**, 407–424 (1977)
15. Mastrolia, P., Monticelli, D.D., Punzo, F.: Nonexistence of solutions to parabolic differential inequalities with a potential on Riemannian manifolds. Math Ann. **367**, 929–963 (2017)
16. Monticelli, D.D., Punzo, F., Squassina, M.: Nonexistence for hyperbolic problems on Riemannian manifolds. Asympt. Anal. **120**, 87–101 (2020)
17. Petersen, P.: Riemannian Geometry. Graduate Texts in Mathematics. Springer, New York (2006)
18. Pigola, S., Rigoli, M., Setti, A.: Vanishing and Finiteness Results in Geometric Analysis. Birkhäuser Verlag, Berlin (2008)
19. Punzo, F.: Blow-up of solutions to semilinear parabolic equations on Riemannian manifolds with negative sectional curvature. J. Math. Anal. Appl. **387**, 815–827 (2012)
20. Sun, Y.: Uniqueness result for non-negative solutions of semi-linear inequalities on Riemannian manifolds. J. Math. Anal. Appl. **419**, 643–661 (2014)
21. Sun, Y.: On nonexistence of positive solutions to quasi-linear inequality on Riemannian manifolds. Proc. Am. Math. Soc. **143**, 2969–2984 (2015)
22. Suzuki, R.: Existence and nonexistence of global solutions to quasilinear parabolic equations with convection. Hokkaido Math. J. **27**, 147–196 (1998)
23. Zhang, Q.S.: Blow-up results for nonlinear parabolic equations on manifolds. Duke Math. J. **97**, 515–539 (1999)

The manufacturer's authorised representative in the EU is Springer Nature Customer Service Centre GmbH, Europaplatz 3, 69115 Heidelberg, Germany. If you have any concerns regarding our products, please contact ProductSafety@springernature.com

Printed and bound by CPI Group (UK) Ltd, Croydon, CR0 4YY
01/10/2026
02225950-0013